T0313918

ASTROPHYSICS

ASTROPHYSICS

Decoding the Cosmos

Second Edition

JUDITH A. IRWIN
Queen's University, Kingston, Canada

This second edition first published 2021

Edition History
John Wiley & Sons Ltd (1e, 2007)

Registered Offices
John Wiley & Sons, Inc., 111 River Street, Hoboken, NJ 07030, USA
John Wiley & Sons Ltd, The Atrium, Southern Gate, Chichester, West Sussex, PO19 8SQ, UK

Editorial Office
The Atrium, Southern Gate, Chichester, West Sussex, PO19 8SQ, UK

For details of our global editorial offices, customer services, and more information about Wiley products visit us at www.wiley.com.

Wiley also publishes its books in a variety of electronic formats and by print-on-demand. Some content that appears in standard print versions of this book may not be available in other formats.

Library of Congress Cataloging-in-Publication Data applied for
9781119623687 (hardback)

Cover Design: Wiley
Cover Image: © Courtesy of Caltech/MIT/LIGO Laboratory

Set in 9/13pt Ubuntu by SPi Global, Chennai, India

Printed and bound by CPI Group (UK) Ltd, Croydon, CR0 4YY

C9781119623687_300321

To my dear Richard – this one is for you

Contents

Preface to the 1st edition

Like many textbooks, this one originated from lectures delivered over a number of years to undergraduate students at my home institution – Queen's University in Kingston, Canada. These students had already taken a first year (or two) of physics and one introductory astronomy course. Thus, this book is aimed at an intermediate level and is meant to be a stepping stone to more sophisticated and focussed courses, such as stellar structure, physics of the interstellar medium, cosmology, or others. The text may also be of some help to beginning graduate students with little background in astronomy or those who would like to see how physics is applied, in a practical way, to astronomical objects.

The astronomy prerequisite is helpful, but perhaps not required for students at a more senior level, since I make few assumptions as to prior knowledge of astronomy. I *do* assume that students have some familiarity with celestial coordinate systems (e.g. Right Ascension and Declination or others), although it is not necessary to know the details of such systems to understand the material in this text. I also do not provide any explanation as to how astronomical distances are obtained. Distances are simply assumed to be known or not known, as the case may be. I provide some figures that are meant to help with 'astronomical geography', but a basic knowledge of astronomical scales would also be an asset, such as understanding that the Solar System is tiny in comparison to the Galaxy and rotates about the Galactic centre.

As for approach, I had several goals in mind while organizing this material. First of all, I did not want to make the book too 'object-oriented'. That is, I did not want to write a great deal of descriptive material about specific astronomical objects. For one thing, astronomy is such a fast-paced field that these descriptions could easily and quickly become out of date. And for another, in the age of the Internet, it is very easy for students to quickly download any number of descriptions of various astronomical objects at their leisure. What is more difficult is finding the thread of physics that links these objects, and it is this that I wanted to address.

Another goal was to keep the book practical, focussing on how we obtain information about our Universe from the signal that we actually detect. In the process, many equations are presented. While this might be a little intimidating to some students, the point should be made that the equations are our 'tools of the trade'. Without these tools, we would be quite helpless, but with them, we have access to the secrets that astronomical signals bring to us. With the increasing availability of computer algebra or other software, there is no longer any need to be encumbered by mathematics. Nevertheless, I have kept problems that require computer-based solutions to a minimum in this text and have tried to include problems over a range of difficulty.

A solutions manual to the problems is also available. I invite readers to visit the Wiley website. It is my sincere desire that this book will be a useful stepping-stone for students of astrophysics and, more importantly, that it may play a small part in illuminating this most remarkable and marvellous Universe that we live in.

Judith A. Irwin
Kingston, Ontario
October 2006

Preface to the 2nd edition

Much has happened in more than a decade, since the first edition of this textbook was printed. The sentiments from the 1st edition have not changed, but astronomy is fast-moving and it is sometimes difficult to keep pace. Fortunately, the physics is a constant, but the technology has moved ahead in leaps and bounds. The power of our telescopes and the detail, sensitivity and clarity with which we can probe astronomical sources has shown huge improvements. I hope that this text has adequately highlighted a few of them.

Along with technological improvements come discoveries and puzzles. Our knowledge of exoplanets has become a strong and fascinating subfield, to the point that atmospheres and magnetic fields of exoplanets are being probed. And mysterious γ-ray bursters and radio transients are opening up the world of 'temporal astronomy' as it relates to the high energy universe.

A change from the 1st edition is the introduction of Part I: *The Non-electromagnetic Signal.* This includes a discussion of 'particles', from rocks to neutrinos, in Chapter 1. Detections of these particles are now leading to important and sophisticated science. Also, Chapter 2 in this part discusses *gravitational radiation*. Predicted for decades, its long-awaited discovery in 2016 has opened up a new window that we can finally peak through.

The remainder of the text follows the format of the 1st edition for electromagnetic (EM) radiation, but with many updates and improved figures. We step through the parts, working 'backwards' along the signal path. Part II: *The EM Signal Observed*, Part III: *Matter and Radiation Essentials*, Part IV: *The EM Signal Perturbed*, Part V: *The EM Signal Emitted*, and Part VI: *The Signal Decoded.*

Although obtaining distances is still not a focus of the text, I do include some examples as to how distances can be found (e.g. Sects. 2.6.2, 9.1.1.1, and 9.1.2). Another change is that most of the appendices from the 1st edition have now been put *online* as supplementary material. Some new supplementary material is also included online (www.wiley.com/go/irwin/astrophysics2e). Whenever problems at the end of each chapter require information from the online material, the problem will start with '[*online*]'. Tables that are used often, however, are collected into a single appendix (Appendix T) at the end of the main text in this edition.

Finally, new to this edition are small sections at the end of the problems in each chapter that I've called, *Just for Fun*. These contain a few fanciful problems that are more open-ended. Hopefully, these will stir the imaginations of the students that engage with them.

Judith A. Irwin
Township of South Frontenac, Ontario
July 2020

Acknowledgments – 1st Edition

There are some important people that I feel honoured to thank for their help, patience, and critical assistance with this book. First of all, to the many people who generously allowed me to use their images and diagrams, I am very grateful. Astronomy is a visual science and the impact of these images cannot be overstated.

Thanks to the students, past and present, of Physics 315 for their questions and suggestions. I have more than once had to make corrections as a result of these queries and appreciate the keen and lively intelligence that these students have shown.

Thanks to Jeff Ross for his assistance with some of the 'nuts and bolts' of the references. And special thanks to Kris Marble and Aimee Burrows for their many contributions, working steadfastly through problems and offering scientific expertise.

With much gratitude, I wish to acknowledge those individuals who have read sections or chapters of this book and offered constructive criticism – Terry Bridges, Diego Casadei, Mark Chen, Roland Diehl, Martin Duncan, David Gray, Jim Peebles, Ernie Seaquist, David Turner, and Larry Widrow.

To my dear children, Alex and Irene, thank you for your understanding and good cheer when mom was working behind a closed door yet again, and also thanks to the encouragement of friends – Joanne, Wendy, and Carolyn.

My tenderest thanks to my husband, Richard Henriksen, who not only suggested the title to this book, but also read through and critiqued the *entire* manuscript. For patience, endurance, and gentle encouragement, I thank you. It would not have been accomplished without you.

Acknowledgments – 2nd Edition

I am deeply indebted to my generous friend and colleague, Theresa Wiegert, for her extraordinary support with proofreading and other scientific feedback. Many thanks to Dominic Rochfort for his careful reading of Chapter 2 and helpful comments about gravitational radiation. Peter Brown, a world expert on meteoritics, kindly reviewed that section for me. Thanks to Aaron Vincent for pointing me towards crucial information about neutrinos. Alex Wright sacrificed valuable time out of his busy schedule to read the cosmic ray and neutrino sections. I am lucky to be in a department where there is so much expertise on astroparticle physics.

Thank you to all the people who so kindly allowed me to use their amazing images. This text would be dull indeed, without such visual enhancement. And special thanks to Jayanne English, for providing her gorgeous picture of the cygnus star forming region in Chapter 12.

Last but not least to Richard for, again, reading the *whole thing*.

J.A.I

List of Symbols

Symbol	Meaning
a	radius, acceleration, radiation constant
$a(t)$	scale factor of the Universe
A	atomic weight, total extinction
A	area, albedo
$A_{\mathrm{j,i}}$	Einstein *A* coefficient between levels j and i
b	impact parameter, velocity parameter
B	magnetic flux density, magnetic field strength
B(T)	intensity of a black body (or specific intensity if subscripted with ν or λ)
c, c_s	speed of light, speed of sound, respectively
D_{p}	degree of polarization
e	charge of the electron, eccentricity
E, $E_{\lambda_1} - E_{\lambda_2}$	energy, selective extinction, respectively
E	electric field strength
$\mathcal{EM}$	emission measure
$f_{\mathrm{i,j}}$	oscillator strength between levels, i and j
f, **f**	correction factor, fraction, respectively
F, S, f	flux (or flux density, if subscripted with ν or λ)
F	force
f_{GW}, $\dot{f}_{GW}$	frequency and rate of change of frequency of a gravitational wave, respectively
g_n	statistical weight of level, n
g_{ff}, g_{bf}	Gaunt factor (free–free, bound–free, respectively)
G	universal gravitational constant
h	Planck's constant, gravitational strain
H	scale height
i	inclination
I	intensity (or specific intensity, if subscripted with ν or λ)
I	moment of intertia
j_ν	emission coefficient
J	mean intensity (or mean specific intensity, if subscripted with ν or λ)
$J(E)$	'specific intensity' of cosmic ray particles
k, k_{e}	Boltzmann constant, Coulomb's constant, respectively
$\bar{l}$	mean free path
L	luminosity (or spectral luminosity, if subscripted with ν or λ)
L_{n}	angular momentum of orbital, n
m, M	apparent magnitude, absolute magnitude, respectively
M, m	Mass
m	complex index of refraction
$\mathcal{M}$, $\mathcal{M}_c$	Mach number, chirp mass, respectively
n, N	number density and number of (object), respectively
n	index of refraction, principal quantum number

Symbol	Meaning
N	map noise
$\mathcal{N}$	number of moles, column density
p	momentum, electric dipole moment
P	power (or spectral power, if subscripted with ν or λ), probability
P_i	production rate of species, i
P	pressure
q	charge
Q	efficiency factor
$r, d, D, s, x, y, z,$ l, h, R	Position, separation, or distance
R	Rydberg constant
$\mathcal{R}$	collision rate
S_ν	source function
$t, t_{1/2}$	time, half-life, respectively
T	kinetic energy
T	temperature
$\mathcal{T}$	period
u	energy density (or spectral energy density, if subscripted with ν or λ)
U, V, B, etc.	apparent magnitudes
U	gravitational binding energy, partition function
$\mathcal{U}$	excitation parameter
v	velocity, speed
V	volume
W	work
X, Y, Z	mass fraction of hydrogen, helium, and heavier elements, respectively
z	redshift, zenith angle
Z	atomic number
α	synchrotron spectral index, fine structure constant
α_ν	absorption coefficient
α_r	recombination coefficient
β	speed relative to c
γ	Lorentz factor
γ_{coll}, γ_a	collision rate coefficient, adiabatic index, respectively
Γ	spectral index of cosmic ray power law distribution, damping constant
ε	permittivity, cosmological energy density
θ, ϕ	one-dimensional angle
κ_ν	mass absorption coefficient
λ, λ_i	wavelength, decay constant of species, i, respectively
μ	permeability, mean molecular weight, reduced mass, proper motion
ν	frequency, collision rate per unit volume
ρ	mass density
σ	cross-sectional area, Stefan–Boltzmann constant, Gaussian dispersion
τ_ν, τ	optical depth, timescale, respectively
Φ	line shape function
χ	ionization potential
ω	angular frequency
Ω	solid angle, cosmological mass, energy density

About the Companion Website

This book is accompanied by a companion website:

https://wiley.com/go/irwin/astrophysics2e

The website includes:

- Supplementary Material
- Solutions Manual

Introduction

Decode \dē–' kōd\ vb: to convert (a coded message) into ordinary language.

Knowledge of our Universe continues to grow exponentially in modern times. We see newly found planets around other stars, detections of powerful gamma ray bursts, galaxies in the process of formation in the infant Universe, evidence of a mysterious force that appears to be accelerating the expansion of the Universe, and now the discovery of gravitational radiation. From exotic black holes to the microwave background, the modern understanding of our larger cosmological home could barely be imagined just a generation ago. Headlines exclaim astonishing properties for astronomical objects – stars with densities equivalent to the mass of the sun compressed to the size of a city, energy sources of incredible power, luminosities as great as an entire galaxy from a single dying star, and distortions in the very fabric of space–time.

How could we possibly have reached these conclusions? How can we dare to describe objects so inconceivably distant that the only apparent influence they have on our lives is through our very astonishment at their existence? With the exception of the Moon, no human being has ever set foot on another astronomical object. Because of the vast distances involved, no space probe has ever reached a star other than the Sun, let alone returned with material evidence. Yet we continue to amass information about our Universe. How?

In contrast to our attempts to reach outwards into the expanse of space, the natural Universe itself has been continuously and quite effectively reaching down to us, communicating in its own language. Our challenge, in the absence of an ability to travel amongst the stars, is to find the best ways to detect and decipher such signals. These signals reach us via *matter, gravitational radiation,* and *electromagnetic radiation.*

Matter includes the high energy subatomic particles and nuclei which make up *cosmic rays* that continuously bombard the earth (Section 1.2) as well as an influx of meteoritic dust, occasional meteorites and those rare objects that are large enough to create impact craters. In this category, we also include the charge-less subatomic particle, the *neutrino.* Such incoming matter provides us with information on a variety of astronomical sources, including our Sun, our Solar System, supernovae in the Galaxy[1], and other more mysterious sources of the highest energy cosmic rays. Particles will be discussed more fully in Chapter 1.

Gravitational waves, weak perturbations in space predicted by Einstein's General Theory of Relativity, is a new and exciting field of astrophysics. The monumental discovery of gravitational waves, to be described more fully in Chapter 2, has opened up a window on our universe unlike any previously explored. As this field rapidly matures, we can expect a

[1] When 'Galaxy' is written with a capital G, it refers to our own Milky Way galaxy.

new understanding of a variety of phenomena, from regions around black holes to fundamental physics.

Electromagnetic (EM) radiation is emitted by all astronomical objects with the exception of the interior of black holes. When we say 'light' in this text, we mean any radiation in any waveband from the radio to the gamma-ray part of the spectrum. Electromagnetic radiation can be described as a wave and identified by its wavelength, λ or its frequency, ν. However, it can also be thought of as a massless particle called a *photon* which has an energy, E_{ph}. This energy can be expressed in terms of wavelength or frequency, $E_{ph} = h\,c/\lambda = h\nu$, where h is Planck's constant. The wavelengths, frequencies, and photon energies of various wavebands are given in Table I.1.

The wave–particle duality of light is a deep issue in physics and related via the concept of probability. Although it should be possible to understand a physical process involving light from both points of view, there are some problems that are more easily addressed by one approach rather than another. For example, it is sometimes more straightforward to consider waves when dealing with an interaction between light and an object that is small in comparison with the wavelength and to consider photons when dealing with an interaction between light and an object that is large in comparison to the photon's wavelength. In this text, we will apply whatever form is most useful for the task at hand. Some helpful expressions relating various properties of electromagnetic radiation are provided in Table I.2 and a diagram illustrating the wave nature of light is shown in Figure I.1.

If we now ask which of these information bearers provides us with most of our current knowledge of the universe, the answer is undoubtedly electromagnetic radiation. The world's astronomical volumes would be empty indeed were it not for an understanding of radiation and its interaction with matter. The radiation may come directly from the object of interest as when sunlight travels through a clear sky with little or no interaction en route, or it may be indirect, such as when we infer the presence of a black hole by the

Table I.1 The electromagnetic spectrum[a]

Waveband	Wavelength range (cm)	Frequency range (Hz)	Energy range (eV)
Radio	≥ 1	$\leq 3 \times 10^{10}$	$\leq 1.2 \times 10^{-4}$
(Microwave)	$(100 \rightarrow 0.1)$	$(3 \times 10^{8} \rightarrow 3 \times 10^{11})$	$(1.2 \times 10^{-6} \rightarrow 1.2 \times 10^{-3})$
Millimetre –Submillimetre[b]	$1 \rightarrow 0.01$	$3 \times 10^{10} \rightarrow 3 \times 10^{12}$	$1.2 \times 10^{-4} \rightarrow 1.2 \times 10^{-2}$
Infrared	$0.01 \rightarrow 10^{-4}$	$3 \times 10^{12} \rightarrow 3 \times 10^{14}$	$1.2 \times 10^{-2} \rightarrow 1.2$
Optical	$10^{-4} \rightarrow 3 \times 10^{-5}$	$3 \times 10^{14} \rightarrow 10^{15}$	$1.2 \rightarrow 4.1$
Ultraviolet	$3 \times 10^{-5} \rightarrow 10^{-6}$	$10^{15} \rightarrow 3 \times 10^{16}$	$4.1 \rightarrow 124$
X-ray	$10^{-6} \rightarrow 10^{-9}$	$3 \times 10^{16} \rightarrow 3 \times 10^{19}$	$124 \rightarrow 1.2 \times 10^{5}$
Gamma-ray	$\leq 10^{-9}$	$\geq 3 \times 10^{19}$	$\geq 1.2 \times 10^{5}$

[a]There is some variation as to where the 'boundaries' of the various wavebands lie.
[b]In astrophysics, the radio band is taken to include microwave frequencies and occasionally to include the millimetre – submillimetre bands as well.

Table I.2 Useful expressions relevant to light, matter, and fields.

Meaning	Equation
Wavelength and frequency relation	$c = \lambda\nu$
Lorentz factor	$\gamma = \dfrac{1}{\sqrt{1-\frac{v^2}{c^2}}}$
Energy of a photon	$E = h\nu = \dfrac{hc}{\lambda}$
Equivalent wavelength of a mass (de Broglie wavelength)	$\lambda = h/(mv)$
Momentum of a photon	$p = E/c$
Momentum of a particle	$p = \gamma m v$
Snell's law of refraction[a]	$n_1 \sin\theta_1 = n_2 \sin\theta_2$
Index of refraction[b]	$n = \dfrac{c}{v_l}$
Doppler shift[c]	$\dfrac{\Delta\lambda}{\lambda_0} = \dfrac{\lambda_{obs} - \lambda_0}{\lambda_0} = \left(\dfrac{1+\frac{v_r}{c}}{1-\frac{v_r}{c}}\right)^{1/2} - 1$ $\dfrac{\Delta\nu}{\nu_0} = \dfrac{\nu_{obs} - \nu_0}{\nu_0} = \left(\dfrac{1+\frac{v_r}{c}}{1-\frac{v_r}{c}}\right)^{1/2} - 1$ $\dfrac{\Delta\lambda}{\lambda_0} \approx \dfrac{v_r}{c},\ \dfrac{\Delta\nu}{\nu_0} \approx \dfrac{-v_r}{c},\ (v \ll c)$
Electric field vector	$\vec{E} = \vec{F}/q$
Electric field vector of a wave[d]	$\vec{E} = \vec{E}_0 \cos\left[2\pi\left(\dfrac{x}{\lambda} - \nu t\right) + \Delta\phi\right]$
Magnetic field vector of a wave[d]	$\vec{B} = \vec{B}_0 \cos\left[2\pi\left(\frac{x}{\lambda} - \nu t\right) + \Delta\phi\right]$
Poynting flux[e]	$\vec{S} = \left(\frac{c}{4\pi}\right)\vec{E} \times \vec{B}$
Time-averaged Poynting flux[e]	$\langle S \rangle = \frac{c}{8\pi} E_0 B_0$
Energy density of a magnetic field[f]	$u_B = \frac{B^2}{8\pi}$
Lorentz force[g]	$\vec{F} = q\left(\vec{E} + \dfrac{\vec{v}}{c} \times \vec{B}\right)$
Electric field magnitude in a parallel-plate capacitor[h]	$E = (4\pi N e)/A$
Electric dipole moment[i]	$\vec{p} = q\vec{r}$
Larmor's formula for power[j]	$P = \frac{dE}{dt} = \frac{2}{3}\frac{q^2}{c^3} \lvert\ddot{\vec{r}}\rvert^2$

(Continued)

Table I.2 *(Continued)*

Meaning	Equation
Heisenberg Uncertainty Principle[k]	$\Delta x \Delta p_x = \frac{h}{2\pi}, \quad \Delta E \Delta t = \frac{h}{2\pi}$
Universal law of gravitation[l]	$F_G = G\frac{M\,m}{r^2}$
Centripetal force[m]	$F_c = \frac{m\,v^2}{r}$

[a] If an incoming ray is travelling from medium 1 with index of refraction, n_1, into medium 2 with index of refraction, n_2, then θ_1 is the angle between the incoming ray and the normal to the surface dividing the two media and θ_2 is the angle between the outgoing ray and the normal to the surface.
[b] v_l is the speed of light in the medium, and c is the speed of light in a vacuum. Note that the index of refraction may also be expressed as a complex number whose real part is given by this equation and whose imaginary part corresponds to an absorbed component of light. See Appendix D.3 of the *online material* for an example.
[c] λ_0 is the wavelength of the light in the source's reference frame (the 'true' wavelength), λ_{obs} is the wavelength in the observer's reference frame (the measured wavelength), and v_r is the relative radial velocity between the source and the observer. v_r is taken to be positive if the source and observer are receding with respect to each other and negative if the source and observer are approaching each other.
[d] The wave is propagating in the x direction, and $\Delta\,\phi$ is an arbitrary phase shift. The magnetic field strength, H, is given by $B = H\mu$ where μ is the permeability of the substance through which the wave is travelling (unitless in the cgs system). For EM radiation in a vacuum (assumed here and throughout), this becomes $B = H$ since the permeability of free space takes the value, 1, in cgs units. Thus, B is often stated as the magnetic field strength, rather than the magnetic flux density and is commonly expressed in units of Gauss. In cgs units, E (dyn esu $^{-1}$) = B (Gauss).
[e] Energy flux carried by the wave in the direction of propagation. The cgs units are erg s^{-1} cm^{-2}. The time-averaged value is over one cycle (see Figure I.1).
[f] u_B has cgs units of erg cm^{-3} or dyn cm^{-2}.
[g] Force on a charge, q, with velocity, $\vec{v}$, by an electric field, $\vec{E}$ and magnetic field, $\vec{B}$.
[h] Here $N\,e$ is the charge on a plate, and A is its area. In SI units, this equation would be $E = \sigma/\varepsilon_0$, where σ is the charge per unit area on a plate, and ε_0 is the permittivity of free space (where we assume free space between the plates). In cgs units, $4\pi\varepsilon_0 = 1$.
[i] $\vec{r}$ is the separation between the two charges of the dipole, and q is the strength of one of the charges.
[j] Power emitted by a non-relativistic particle of charge, q, that is accelerating at a rate, $\ddot{\vec{r}}$.
[k] One cannot know the position and momentum (x, p) or the energy and time (E, t) of a particle or photon to arbitrary accuracy.
[l] Force between two masses, M and m, a distance, r, apart.
[m] Force on an object of mass, m, moving at speed, v, in a circular path of radius, r.

X-rays emitted from a surrounding accretion disk. Even when we send out exploratory astronomical probes, we still rely on man-made radiation to transfer the images and data back to earth.

This volume is thus largely (Parts II through VI) devoted to understanding radiative processes and how such an understanding informs us about our Universe and the astronomical objects that inhabit it. It is interesting that, in order to understand the largest and grandest objects in the Universe, we must very often appeal to microscopic physics, for it

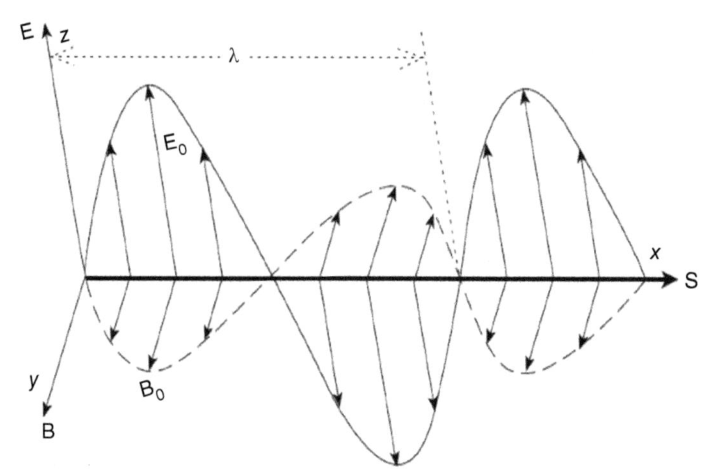

Figure I.1 Illustration of an electromagnetic wave showing the electric field and magnetic field perpendicular to each other and perpendicular to the direction of wave propagation which is in the *x* direction. The wavelength is denoted by λ.

is on such scales that the radiation is actually being generated and it is on such scales that matter interacts with it.

We should not proceed, however, without first making a brief excursion onto the road less travelled, that is, by considering incoming particles and gravitational waves (Part I). Our historical and necessarily heavy reliance on information from EM radiation can now be supported, confirmed, and expanded upon by appealing to these other messengers. Astronomy is not only multi-waveband, but also *multi-messenger* as well.

Throughout the text, there are references to *online material* which is meant to offer supplementary information for those who wish to take a deeper look at some of the concepts. In most cases, the supplementary material is not required to understand the concepts. However, some *Problems* at the ends of each chapter do require that *online material* be accessed. In those cases, (*Online*) is marked next to the problem number.

This text focuses on the 'how' of astronomy. How do we know the temperature of that asteroid? How do we find the speed of that star? How do we know the density of that interstellar cloud? How can we find the energy of that distant quasar? Most answers are hidden in the radiation that they emit and, earthbound, we have at least a few keys to unlock their secrets. We can truly think of the detected signal as a coded message. To understand the message requires careful decoding.

I.1 DIMENSIONS, UNITS AND EQUATIONS

The centimetre-gramme-second (cgs) system of units is widely used by astronomers internationally and is the system adopted in this text. A summary of the units is given in Table I.3

Table I.3 Selected cgs – SI conversions[a]

Dimension	cgs unit (abbrev.)	Factor	SI unit[b] (abbrev.)
Length	centimetre (cm)	10^{-2}	metre (m)
Mass	gramme (g)	10^{-3}	kilogramme (kg)
Time	second (s)	1	second (s)
Energy	erg (erg)	10^{-7}	joule (J)
Power	erg second^{-1} (erg s^{-1})	10^{-7}	watt (W)
Temperature	kelvin (K)	1	kelvin (K)
Force	dyne (dyn)	10^{-5}	newton (N)
Pressure	dyne centimetre^{-2} (dyn cm^{-2})	0.1	Newton metre^{-2} (N m^{-2})
	barye (ba)[c]	0.1	pascal (Pa)
Magnetic flux density (field)[d]	gauss (G)	10^{-4}	tesla (T)
Angle	radian (rad)	1	radian (rad)
Solid angle	steradian (sr)	1	steradian (sr)

[a] Value in cgs units times factor equals value in SI units.
[b] Système International d'Unités.
[c] This unit is rarely used in astronomy in favour of dyn cm^{-2}.
[d] See note *d* of Table I.2.

Table I.4 Examples of equivalent units.

Equation	Name	Units
$P = n k T$	Ideal Gas Law	dyn cm^{-2} $= \left(\frac{1}{\text{cm}^3}\right)\left(\frac{\text{erg}}{\text{K}}\right)(\text{K})$ $=$ erg cm^{-3}
$F = m\, a$	Newton's Second Law	dyn $=$ g cm s^{-2}
$W = F\, s$	Work Equation	erg $=$ dyn cm $=$ g cm^2 s^{-2}
$E_k = \frac{1}{2} m v^2$	Kinetic Energy Equation	erg $=$ g cm^2 s^{-2}
$E_{PG} = m\, g\, h$	Gravitational Potential Energy Equation	erg $=$ g cm^2 s^{-2}
$E_{Pe} = \frac{q_1 q_2}{r}$	Electrostatic Potential Energy Equation	erg $=$ esu^2 cm^{-1}

as well as corresponding conversions to Système International d'Unités (SI). If an equation is given without units, the cgs system is understood. The same symbols are generally used in both systems, and SI prefixes (e.g. mega, micro, see Table I.4 in the *online material*) are also equally applied to the cgs system (note that the base unit, cm, already has a prefix).

Almost all equations used in this text look identical in the two systems and one need only ensure that the constants and input parameters are all consistently used in the adopted system. There are, however, a few cases in which the equation itself changes between cgs and SI. An example is the Coulomb (electrostatic) force,

$$F = \frac{q_1 q_2}{r^2} \tag{I.1}$$

With the two charges, q_1 and q_2 expressed in *electrostatic units* (esu, see Table T.1), and the separation, r, in cm, the answer will be in dynes. Note that there is no constant of proportionality in this equation, unlike the SI equivalent (Problem I.1). Equations in the cgs system show the most difference with their SI equivalents when electric and magnetic quantities are used. For example, in the cgs system, the permittivity and permeability of free space, ε_0 and μ_0, respectively, are both unitless and equal to 1.

A very valuable tool for checking the answer to a problem, or to help understand an equation, is that of *dimensional analysis.* The dimensions of an equation (e.g. time, velocity, distance) must agree and therefore their units (s, cm s^{-1}, cm, respectively) must also agree. Two quantities can be added or subtracted only if they have the same units, and logarithms and exponentials are unitless. In this process, it is helpful to recall some *equivalent units* which are revealed by writing down some simple well-known equations in physics. A few examples are provided in Table I.4. The example of the Ideal Gas Law in this table also shows the process of dimensional analysis, which involves writing down the units to every term and then cancelling where possible. A more complex example of dimensional analysis is given in Example I.1.

Example I.1

For a gas in thermal equilibrium at some uniform temperature, T, and uniform density, n, the number density of particles with speeds[2] between v and $v + dv$ is given by the Maxwell–Boltzmann (or simply 'Maxwellian') velocity distribution,

$$n(v)dv = n\left(\frac{m}{2\pi kT}\right)^{3/2} \exp\left(-\frac{mv^2}{2kT}\right) 4\pi v^2 dv \tag{I.2}$$

where n(v) is the gas density per unit velocity interval, m is the mass of a gas particle (taken here to be the same for all particles), and v is the particle speed. A check of the units gives,

$$\frac{1}{\mathrm{cm^3 \frac{cm}{s}}}\frac{\mathrm{cm}}{\mathrm{s}} = \frac{1}{\mathrm{cm^3}}\left(\frac{\mathrm{g}}{\mathrm{\frac{erg}{K}K}}\right)^{3/2} \exp\left(-\frac{\mathrm{g}\left(\frac{cm}{s}\right)^2}{\mathrm{\frac{erg}{K}K}}\right)\left(\frac{\mathrm{cm}}{\mathrm{s}}\right)^2\frac{\mathrm{cm}}{\mathrm{s}} \tag{I.3}$$

Simplifying yields,

$$\frac{1}{\mathrm{cm^3}} = \frac{1}{\mathrm{s^3}}\left(\frac{\mathrm{g}}{\mathrm{erg}}\right)^{3/2} \exp\left(-\frac{\mathrm{g}\left(\frac{cm}{s}\right)^2}{\mathrm{erg}}\right) \tag{I.4}$$

Using the equivalent units for energy (Table I.4), the exponential is unitless, as required, and we find,

$$\frac{1}{\mathrm{cm^3}} = \frac{1}{\mathrm{cm^3}} \tag{I.5}$$

Figure 5.12 shows a plot of the function, Eq. (I.2). If this equation is integrated over all velocities, either on the left-hand side (LHS) or the right-hand side (RHS), the total density

[2] Speed and velocity are taken to be equivalent in this text unless otherwise indicated.

should result. Since an integration over all velocities is equivalent to a sum over the individual infinitesimal velocity intervals, the total density will also have the required units of $\frac{1}{\text{cm}^3}$. This dimensional analysis helps to clarify the fact that, because the total density, n, appears on the RHS of Eq. (I.2) and is a constant, an integration over all terms, except n, on the RHS must be unitless and equal 1. (In fact these remaining terms represent a *probability distribution function*, see Section 5.4.1 for a description.) Also, since the number density is just the number of particles, N, divided by a constant (the volume), we could have substituted N for n on the two sides of the equation. Similarly, since the density and temperature are constant, we could have multiplied Eq. (I.2) by kT to turn it into an equation for the particle pressure in a velocity interval $P(v)\,dv$. Thus, while dimensional analysis says nothing about the origin or fundamentals of an equation, it can go a long way in revealing the meaning of one and how it might be manipulated.

Astronomers also work in units that are specific to the discipline. One peculiarity is *velocity*, usually expressed in (km/s) which is neither cgs nor SI. The most common discipline-specific units result from a process of normalization. The value of some parameter is expressed in comparison to another known, or at least more familiar value. Some examples (see Table T.2) are the *astronomical unit* (AU) which is the distance between the Earth and the Sun. It is much easier to visualize the distance to Pluto as 40 AU than as 6×10^{14} cm. Expressing the masses of stars and galaxies in Solar masses ($M_\odot$) or an object's luminosity in Solar luminosities ($L_\odot$) is also very common. When such units are used, the parameter is often written as $M/M_\odot$, or $L/L_\odot$.

Some examples can be seen in Eqs. (10.16) through (10.19) and in many other equations in this book. An example showing how to convert an equation from cgs units to something more 'astronomically friendly' is given in Example I.2

Example I.2

A commonly used equation linking the luminosity, L, of a star (a spherical black body, see Section 6.1) with its radius, R, and its temperature, T_{eff}, is given by Eq. (6.13),

$$L = 4\pi R^2 \sigma T_{\text{eff}}^{\;4} \tag{I.6}$$

where $\sigma = 5.67 \times 10^{-5}$ erg s^{-1} cm^{-2} K^{-4} is called the Stefan–Boltzmann constant. This equation is in cgs units, so let us express it in units that allow us to make a quick comparison to our Sun. To do this, we will adopt the Solar luminosity, temperature, and radius to be, respectively, $L_\odot = 3.84 \times 10^{33}$ erg/s, $T_{\text{eff}\odot} = 5780$ K, and $R_\odot = 6.96 \times 10^{10}$ cm. Our conversion is then

$$\left[\frac{L}{L_\odot}\right] = \left[\frac{R}{R_\odot}\right]^2 \left[\frac{T_{\text{eff}}}{T_{\text{eff}\odot}}\right]^4 \left\{4\pi\sigma \frac{R_\odot^{\;2} T_{\text{eff}\odot}^{\;4}}{L_\odot}\right\} \tag{I.7}$$

$$= 1.0 \left[\frac{R}{R_\odot}\right]^2 \left[\frac{T_{\text{eff}}}{T_{\text{eff}\odot}}\right]^4 \tag{I.8}$$

Notice that in Eq. (I.7) we are simply dividing both sides by the Solar luminosity, and multiplying and dividing the RHS by the solar radius and temperature to their appropriate powers. We have put all constants to be evaluated in parentheses, the result being shown in Eq. (I.8). Notice the simplicity of this result which is unitless on both sides of the equals sign. Entering values for the Sun, of course, just returns the Solar luminosity.

Equation (I.8) easily shows how a star's luminosity scales with its radius and temperature. A star that is twice the size of the Sun will have a luminosity that is four times greater, all else being equal. Although we could have concluded this from Eq. (I.6) as well, the new normalized equation, being unencumbered by the various constants, shows the scaling in a straightforward way.

An important point is that one should be very careful of simplifying units without thinking about their meaning. A good example is the unit, erg s^{-1} Hz^{-1} which is a representation of a luminosity or power (see Section 3.1) per unit frequency (Hz) in some waveband. Since Hz can be represented as s^{-1}, the above could be written erg s^{-1} s^1 = erg which is simply an energy and does not really express the intended meaning of the term. Similar difficulties can arise when a term is expressed as 'per cm of waveband'. Units of frequency or wavelength should not be simplified with units of time or distance, respectively.

A final note regards *angles* which are seen repeatedly throughout this text, especially the *small angle formula*,

$$s = d\theta \tag{I.9}$$

Here d is the distance to a source, s is its linear diameter in the plane of the sky, and θ is the angle subtended by that diameter. In this equation, θ *must* be in *radians*! That being the case, s will simply take the same units as d. For example, a source that is at a distance of 5 pc and subtends 1 arcsec in the sky (4.848×10^{-6} rad) has a linear diameter of $s = 2.424 \times 10^{-5}$ pc. It is preferred, though, to represent the result in units that are more reasonable and it is also likely that fewer significant figures are required, depending on measurement uncertainties. Example results are to retain cgs units, $s = 7.5 \times 10^{13}$ cm, or to express the result in common astronomical units, $s = 5$ AU (see Table T.2 for conversions).

The two-dimensional analogy to Eq. (I.9) is

$$\sigma = d^2\Omega \tag{I.10}$$

where σ is the area of the source in the plane of the sky (cm^2), and Ω is the *solid angle* subtended by the source, in units of *steradians*.

For an ellipse on the sky of angular major axis and minor axis diameters of θ_1 and of θ_2, respectively, the relation between one-dimensional angles (in radians) and the two-dimensional solid angle, in steradians, is

$$\Omega = \frac{\pi}{4}\theta_1\theta_2 \tag{I.11}$$

Appendix B of the *online material* also presents these last three equations and provides further information about angles.

PROBLEMS

I.1. Calculate the repulsive force of an electron on another electron which is a distance 1 m away, in cgs units using Eq. (I.1) and also in SI units using the equation,

$$F = k_e \frac{q_1 q_2}{r^2} \tag{I.12}$$

Where the charge on an electron, in SI units, is 1.6×10^{-19} Coulombs (C) and the constant, $k_e = 8.988 \times 10^9$ N m^2 C^{-2}. Verify that the result is the same in the two systems.

I.2. [*Online*] Verify that the following equations have matching units on both sides of the equals sign: Eq. (D.2) (the expression that includes the mass of the electron, m_e), Eq. (11.8) in this text, and Eq. (F.19).

I.3. Verify that the equation for the *Ideal Gas Law* (Table I.4) is equivalent to $PV = \mathcal{N}\mathcal{R}T$, where $\mathcal{N}$ is the number of moles and $\mathcal{R}$ is the molar gas constant (see Table T.1).

I.4. Rewrite Eq. (10.14) with all quantities expressed in cgs units. What are the units of the new numerical constant in the equation?

I.5. Rewrite Eq. (6.16) so that both sides of the equation are unitless. Express the variables in units of the Sun's central pressure and temperature, $P_{c\odot} = 2 \times 10^{17}$ dyn cm^{-2}, and $T_{c\odot} = 1.5 \times 10^7$K, respectively.

JUST FOR FUN

I.6. Calculate your height in pc and in units of the Bohr radius (Table T.1). Calculate your jogging speed in units of the Earth's speed around the Sun, υ_E, and in units of a typical plate tectonic movement on the Earth's surface, $\upsilon_T \approx 50$ mm/year.

I.7. Look up the lyrics to the *Galaxy Song*, from Monty Python's *The Meaning of Life*. Convert all speeds to km/s and distances to pc and comment on how accurately the song portrays these quantities.

I.8. Look up the lyrics to *Seasons of Love* from the musical drama, *Rent*. Suggest how you might 'measure a year' in your life from something you do often.

PART I

The Non-electromagnetic Signal

Here, we consider signals that are *not* electromagnetic (EM) in nature. Historically, these have provided us with limited knowledge about our universe in comparison to EM radiation. This is still the case, but knowledge gleaned from such signals is increasing and it is useful to understand how information can be extracted from these non-EM messengers. The signals are also quite diverse, including *particles*, from the subatomic to larger meteoritic material, *sound waves* as detected from incoming meteors, and now gravitational waves (GWs) as well. In Chapter 1, we will look at a variety of *particulate matter*. Chapter 2 is devoted to *GWs* whose recent discovery has opened up a brand new window on our Universe.

Chapter 1

The Particles: Macroscopic to Subatomic

> It's not safe out here. It's wondrous, with treasures to satiate desires both subtle and gross, but it's not for the timid.
>
> –Star Trek, The Next Generation, Episode *Q Who?*

Before we look at signals that are coming to us in the form of electromagnetic (EM) radiation, let us consider the particles. The Earth is moving constantly through a stream of particles as it orbits the Sun, from the subatomic to the macroscopic. All together, these 'signals' have given us only a tiny amount of information compared to EM radiation, but as instrumentation and analysis improve, the particles are finding an important niche in the lexicon of astronomical information. The striking contrast in size between particulate matter is illustrated in Figure 1.1. Let us

Astrophysics: Decoding the Cosmos, Second Edition. Judith A. Irwin.

Companion website: www.wiley.com/go/irwin/astrophysics2e

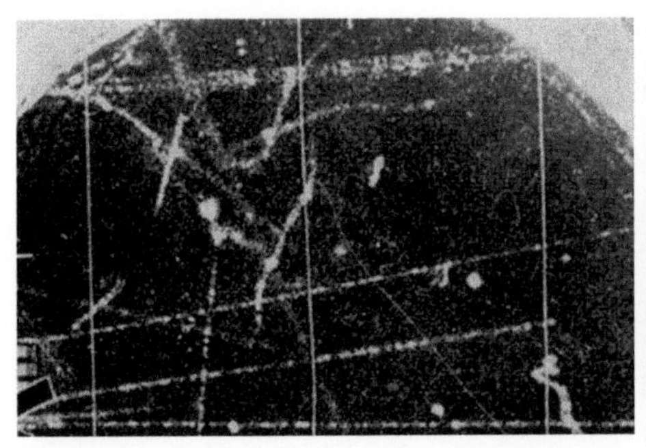

Figure 1.1 (Left) Early photograph of cosmic ray tracks in a cloud chamber. (Right) The 1.8 km diameter Lonar meteorite crater in India. Source: Cairns et al. [13], Judith A. Irwin.

start with the most tangible of these messengers – the meteoritic[1] particles.

1.1 METEORITICS

Of order 2×10^6 kg of meteoritic particles in the mass range, 10^{-5} to 10^2 mg, descend upon the Earth each year [43, 48]. The incoming particulate material covers an even wider mass range (10^{-12} to 1 g) with the largest contribution on a *daily* basis coming from particles with masses $\approx 10^{-2}$ mg. Thus, the total mass influx is likely an order of magnitude higher [47]. These values are uncertain because there is no single method of determining a mass influx over such a large range of masses. Small particles are swept-up interplanetary debris typically from asteroids or comets, called interplanetary dust particles (IDPs) or *micrometeorites.* Incoming meteors also produce smoke during ablation, called meteoric smoke particles (MSPs) which contribute to the mix. It is estimated that, if all of the dust in the Solar System between the Sun and Jupiter were rolled into a ball, the resulting sphere would be 25 km in diameter [47].

In the higher mass range of 0.01–10^9 kg, the *cumulative number* of particles, N, impacting the Earth per year (i.e. the total number that are *greater than* some energy, E) is shown in Figure 1.2. The corresponding diameter and mass are shown at the top [55] assuming a typical incoming velocity of 20.3 km s^{-1} and a density of 3 g cm^{-3}. The adopted density is typical of *chondrites*, which represent more than 90% of known meteorites. Chondrites are

[1] 'Meteoroid' refers to any rocky material before and during its passage through the Earth's atmosphere; a large meteoroid could also be called a small asteroid. 'Meteor' refers to the optical and related phenomena that are seen as the meteoroid travels through the atmosphere but the term is also commonly used to include the falling meteoroid as well. 'Meteorite' refers to rocky material that has reached the ground. A 'fireball' is a bright meteor, and a 'bolide' is a very bright meteor though these are often used interchangeably.

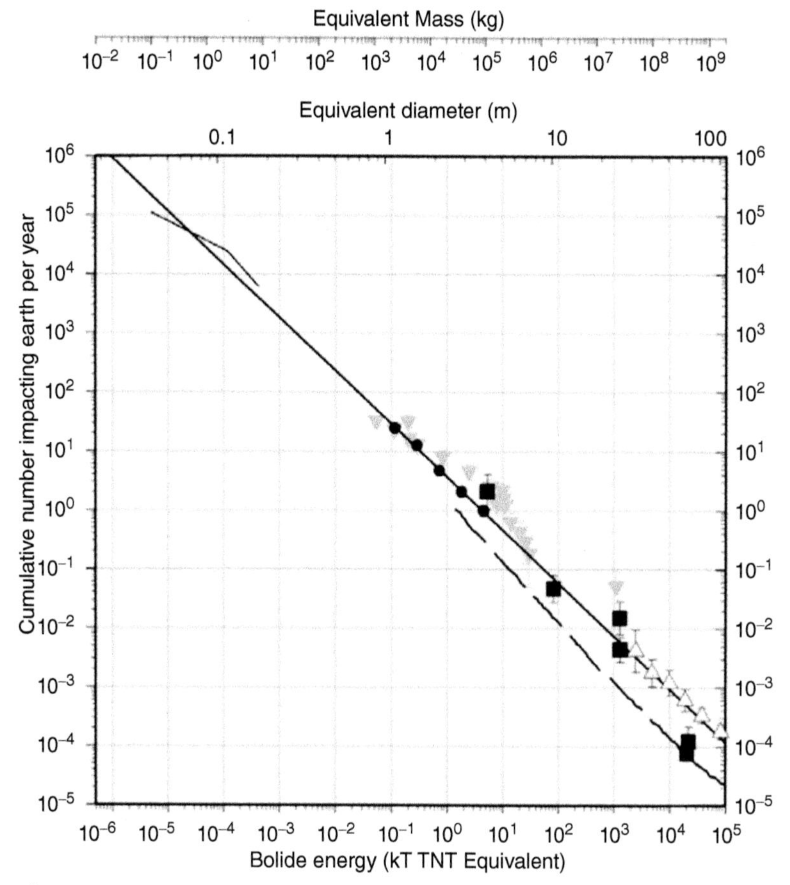

Figure 1.2 Cumulative number of atmospheric impacts per year, N, for bolides in the 0.01–100 m size range from a variety of ground-based and space-based observations (symbols). Mass and size are shown at the top of the figure, and bolide kinetic energy, E, is shown at the bottom in units of kT TNT equivalent (1kT TNT equivalent = 4.185×10^{12} J). For comparison, the atomic bomb that was dropped on Nagasaki on 9 August 1945, exploded with an energy of about 20 kT TNT equivalent. The long solid black line is a fit to satellite fireball data and has the form, $N = 3.7\,E^{-0.90}$ (for E in kT TNT), and the dashed line is from lunar cratering estimates. Source: Zolensky et al. [55]. © 2006, Zolensky, Michael E.

stony-type meteorites containing *chondrules* (Figure 1.3) which are spherical-appearing inclusions.

From Figure 1.2, for example, we can see that there are about 500 impacts per year from objects with masses greater than 100 kg (>40 cm diameter). There are approximately 10^{-3} per year (or one every 1000–10 000 years) for masses >10^8 kg (>40 m diameter). As

Figure 1.3 Picture of one piece of the 2007 Chergach fall in Mali, Africa. The meteorite has been sliced to expose the details of the rock. Notice the number of *inclusions* in the rock – those small features that appear distinct from the surrounding substance. Spherical inclusions are called *chondrules* and would have aggregated into the rocky material as molten droplets during formation. The exterior fusion crust, darkened from the trip through the Earth's atmosphere, is also apparent. Source: Reproduced with permission of Dieter Heinlein (https://www.meteorites.de/sale.htm). *Inset:* Two images of cone-shaped CR tracks in a different meteorite from [2]. These images are only $150 \times 80\,\mu$m (left) and $100 \times 55\,\mu$m (right) in size and the track of the CR is from right to left. Source: Alexandrov et al. [2], Reproduced with permission of Natalia Polukhina.

can be seen from the figure, however, this and similar plots suffer from small number statistics. Figure 1.2 shows a fit of $N \propto E^{-0.90}$, whereas [11] gives a power of −0.68 and an update by [25] reports ≈ −0.87, both over a narrower energy range.

A meteorite-producing fireball is typically decimetre to metre-sized. In the size range less than about 50–100 m, a bolide will typically detonate in the atmosphere and does not reach the ground as a single object [11]. At ≈ 50 m diameter or kinetic energy of $E \approx 10^4$ kiloTon TNT equivalent (1 kT TNT = 4.185×10^{12} J), the detonation can be severe, as occurred in 1908 near Tunguska, Siberia[2]. Larger objects can produce craters whose size depends on the kinetic energy of the bolide; these craters will be larger than the incoming rock itself. According to [35], the relationship between kinetic energy, E, and crater diameter, D, is

$$\left[\frac{E}{\text{erg}}\right] = 9.1 \times 10^{24} \left[\frac{\text{D}}{\text{km}}\right]^{2.59} \tag{1.1}$$

This calibration has been applied to craters of about 0.9–200 km diameter.

[2] On 30 June 1908, an explosion near the Tunguska River, Siberia, flattened 2000 square km of forest due to a meteor detonation at somewhere between 5 and 10 km altitude. No crater was formed.

The data from Figure 1.2 come from a variety of measurements of electromagnetic radiation. However, since in this chapter we are considering *non*-EM radiation, let us consider other ways in which we can glean information from these incoming messengers.

One is to study the constitutional make-up of meteorites (Figure 1.3), the remnants of the fall. Another is to examine meteors via sound waves, specifically *infrasound*. The final important source of information is that we are now physically returning rock and dust specimens from other locations in our Solar System. We will briefly examine these in the next three subsections.

1.1.1 Dating Meteorites

Once a meteorite is in hand, it can be probed and prodded to reveal a wealth of information. Meteorites have been examined by eye, crushed, heated, sliced, and irradiated. They have been bombarded with X-rays, examined via *magnetic resonance imaging* (MRI)[3], analysed using spectroscopy, and studied in a variety of other imaginative ways to extract information [27]. The results of such studies indicate that most meteorites originate from the asteroid belt, a region between 2 and 4 AU from the Sun between Mars and Jupiter. This is mainly determined from comparison of the optical properties of meteorites with those of various types of asteroids, as well as by studying trajectories of incoming meteors to determine orbital parameters. Material from the asteroid belt can be expelled because of thermal, gravitational, or collisional effects and/or because they drift into regions from which expulsion is easier [27]. A small number of meteorites are also known to have been ejected from Mars or the Moon after impacts on those bodies.

Asteroids, as with the other planets and the Sun, are believed to have been formed out of a *pre-solar nebula* – a collapsing fragment of a molecular cloud. Since meteorites are associated with asteroids, the oldest meteoritic material should date to the time at which that material solidified out of the nebula. The oldest known meteoritic matter is *calcium–aluminium-rich inclusions* (CAIs) which are submillimetre-to-centimetre in size and found within chondritic meteorites. CAIs have isotopic compositions that are similar to those expected from thermodynamic calculations for condensation from a hot pre-solar nebula. Thus, they likely condensed during nebular cooling, aggregating into rocky bodies [44]. Dating the oldest CAIs, therefore, effectively dates the formation of the Solar System.

The ages of CAIs can be found from *radioisotope dating*. An example is the use of different isotopes of lead. Three stable isotopes, ^{206}Pb, ^{207}Pb, and ^{208}Pb, are *radiogenic*, that is, they result from known radioactive decay chains of the parent elements: uranium (both

[3] MRI is a technique often used for the imaging of human tissue, but can also be used for inanimate objects such as rocks. A material is placed in a magnetic field that aligns atoms in the material with the field. Radio frequency EM pulses are then applied which produces misalignment; the time that it takes for atoms to re-align with the magnetic field when the pulse is off is related to the properties of the material.

^{235}U and ^{238}U) and thorium (^{232}Th). However, the stable isotope, ^{204}Pb, is not radiogenic. The half-lives[4] of the parent elements are very long ($\approx 10^9$–10^{10} year), making these elements good discriminants of long-lived phenomena. If a CAI, when formed, contained a mixture of elements typical of the pre-solar nebula, then over time, the ratios of the radiogenic lead isotopes compared to ^{204}Pb should increase. Measuring these ratios, then, gives the time since formation. The resulting measured ages of 4.56 billion years (e.g. [3]) confirm the primitive nature of CAIs. Other radioisotopes, such as ^{87}Rb and ^{147}Sm, have also been used with similar age results.

The age of CAIs is key to dating the Solar System. There is another way, however, of measuring the age of a meteorite. Meteorites that have landed on Earth (tens of thousands are known) are typically originally metre-sized meteoroids that have broken off of km-sized asteroidal material. How much time has elapsed since this breaking off has occurred? Key to answering this question is the *cosmic ray exposure* time. Cosmic rays (CRs) will be discussed in the next section (Section 1.2), but for our purposes here, they are a probe of how long a rock has been traversing interplanetary space. The longer the time, the greater will be the effects of CRs on it.

One way to measure the effects of CRs on meteoroids is to search for CR tracks. When CRs hit a solid object, they make tracks as they burrow into the material as can be seen in the *inset* to Figure 1.3 (similar tracks have even been observed in astronauts' helmets). Depending on energy, CRs can penetrate up to a few metres into a meteoroid [49] and some will go right through! With appropriate calibration, the number of tracks can be counted and related to the age (e.g. [41]).

An alternative measure of CR exposure ages relies on the fact that CRs can create *cosmogenic nuclides.* When a CR impacts a target meteoroid, its high energy can expel nucleons from nuclei in the rocky material, thus converting existing elements into new elements. The break-up of heavier nuclei into fragments by CRs, in general, is called *spallation.* The product nuclide could be stable or it could be radioactive in which case it will decay naturally with time. For any nuclide, i, that has been created in such a process, the rate of change of the number of such particles, N_i, per unit time, t, is

$$\frac{dN_i}{dt} = P_i - \lambda_i N_i \tag{1.2}$$

where P_i is the production rate of the nuclide, and the second term on the right hand side describes its radioactive decay rate. Here, λ_i (dimensions of inverse time) is the *decay constant* for that nuclide and is related to its half-life, $t_{1/2_i}$ via

$$\lambda_i = \frac{\ln(2)}{t_{1/2_i}} \tag{1.3}$$

[4] The half-life is the time it takes for a radioactive material to decay to half of its original abundance level.

If the production rate is constant (which assumes a constant CR flux over time)[5], then Eq. (1.2) can be integrated to find (Problem 1.3)

$$N_{i,rad} = \frac{P_i}{\lambda_i}(1 - e^{-\lambda_i t}) \tag{1.4}$$

The subscript, *rad*, refers to the relevant radioactive nuclide of type, *i*. At time $t = 0$, no nuclide has yet been formed, and as $t \rightarrow \infty$ (or effectively $t \gg \lambda^{-1}$), the number of nuclides reaches its maximum value of $N_i = P_i/\lambda_i$.

If the nuclide is stable, rather than radioactive, then λ_i is effectively zero ($t_{1/2_i} \rightarrow \infty$) and an integration of Eq. (1.2) just yields

$$N_{i,stab} = P_i t \tag{1.5}$$

The subscript, *stab*, means that the nuclide of type, *i*, is stable.

The production rate, P_i, depends on the CR intensity and spectrum, shielding (which depends on the depth of the nuclide from the surface of the meteoroid), the intensity and spectrum of secondary particles, and the production cross section as a function of energy for processes leading to the cosmogenic nuclide of interest [4]. For the most part, these are unknown quantities. However, most of the uncertainties cancel if the *ratio* of a radioactive and stable nuclide of the same element is formed. For example, the noble gas krypton has both stable and radioactive isotopes. Substituting ^{83}Kr for $N_{i,stab}$ and ^{81}Kr for $N_{i,rad}$ and forming the ratio of Eq. (1.5) to Eq. (1.4) gives

$$\frac{^{83}\text{Kr}}{^{81}\text{Kr}} = \lambda_{81} \frac{P_{83}}{P_{81}} \frac{t}{(1 - e^{-\lambda_{81} t})} \tag{1.6}$$

The ratio of production rates is generally known and tends to be of order, one, and the decay constant is also known. Therefore, if the isotopic ratio is measured, the cosmic ray exposure age, *t*, can be found.

A version of Eq. (1.6) can be written down for many different isotopic ratios, including argon, neon, and helium, among others, and has also been extended to ratios of different elements with appropriate calibration (e.g. ^{36}Ar/^{36}Cl) [19]. This means that the CR exposure age can be found using many different data points and therefore determined with greater confidence. Results suggest that meteorites can roam through interplanetary space over a wide range of times, from ≈ 0.1 to ≈ 100 Myr, but there are systematic differences between ages depending on the meteorite group being considered [32]. Similar meteorite types tend to cluster in age, and this suggests that a small number of asteroidal collisions have resulted in a large fraction of the known meteorites on Earth.

1.1.2 Infrasound

Our focus on the information that can be gleaned from *non*-EM radiation now leads us to *sound*. Sound waves (or *acoustic waves*) are longitudinal compressions and rarefactions

[5] If P_i varies in a known fashion, the solution can still be found but it is modified from Eq. (1.2) (see [41] for example).

of a medium, with a corresponding increase and decrease in pressure. Sound waves, therefore, require a medium to exist. *Infrasound* refers to sound waves whose frequencies are too low to be heard by the human ear (less than about 20 Hz)[6]. Such perturbations are typically very small. The human ear, for example, can detect tiny fractional changes in air pressure of order 10^{-6} [8]. Since

$$c_s = \lambda/\nu \tag{1.7}$$

where c_s is the speed of sound, λ is the wavelength, and ν is the frequency, and then, an infrasound frequency of 10 Hz corresponds to a wavelength of 34 m for a sound speed of 344 m s^{-1} in air. Let us consider the frequency range for infrasound.

The *lower limit* to the frequency of sound that can propagate through the atmosphere is called the *acoustic cut-off frequency*, ν_a (Hz) and for an isothermal case [9] is given by

$$\omega_a = 2\pi\nu_a = \frac{c_s}{2H} \tag{1.8}$$

where ω_a is the angular acoustic frequency, and H is the *scale height* of the atmosphere[7]. Equations (1.7) and (1.8) lead to a maximum wavelength for infrasound of

$$\lambda_a = 4\pi H \tag{1.9}$$

Thus, once the wavelength significantly exceeds the atmospheric scale height, the effects of gravity dominate and a perturbation will result in transverse *gravity waves* instead. That is, the atmosphere 'rings' at its very low natural frequency of ν_a and sound waves do not propagate. The scale height of the Earth's atmosphere is about 9 km so the longest infrasound wavelength is about 113 km, corresponding to a very low cut-off frequency of $\nu_a \approx 0.003$ Hz or period of $\tau_a = 1/\nu_a = 5.6$ minutes. In reality, the Earth's atmosphere is not isothermal and ν_a varies. However, the variation is less than about 30% up to an altitude of 100 km. From 100 to 300 km, ν_a decreases to $\approx$0.0015 Hz [9]. This means that very low frequency infrasound that propagates at extreme heights might not reach ground level.

What happens at higher frequencies? All sound waves lose some energy as they propagate because of atmospheric viscosity and thermal conduction. Essentially, sound waves heat the air when their organized gas motions become randomized. Therefore, sound waves weaken with distance. Since the *attenuation coefficient*[8] is proportional to ν^2, the energy loss is more acute at higher frequencies. For example, 99% of an audible 1000 Hz sound wave is absorbed after travelling 7 km at sea level, whereas for a 0.01 Hz sound wave, the distance is greater than the circumference of the Earth [8]! Although sound from bolides extends into the audible range, the higher frequencies are filtered out at large distances. Thus, in practical terms, the *effective* upper limit of sound waves that can

[6] Sound with frequencies higher than can be heard by the human ear is called *ultrasound.*
[7] A 'scale height' is a parameter that is defined for a distribution that is declining exponentially with height. For example, the pressure scale height, H, would be defined by $P \propto \exp(-z/H)$, where z is the height.
[8] Sound waves weaken exponentially with distance, x, i.e. the amplitude $\propto \exp(-\alpha_a x)$ where α_a is the frequency-dependent attenuation coefficient.

travel long distances and can therefore be 'heard' by infrasound-detecting instruments is about 20 Hz [9], just at the lower limit of the audible range for humans.

The International Monitoring System (IMS) was developed in the late 1990s in response to the Comprehensive Nuclear Test Ban Treaty (CTBT)[9]. The system's purpose was to detect explosive events from any location on the globe using a variety of methods, including seismic, hydro-acoustic, radionuclide, and *infrasound* measurements. The types of sources that produce infrasonic signals and are detectable by the IMS include avalanches, ocean waves from ocean storms, severe weather events such as tornadoes – and *bright meteors.* Consequently, the IMS has amassed an extensive data set on these extraterrestrial interlopers [26].

Bolide speeds, v, range between 11 km s^{-1} (the Earth's *escape velocity*)[10] to about 70 km s^{-1} as they enter the Earth's atmosphere [16]. Any object that exceeds the sound speed in air is *supersonic* and produces *shock waves* as it travels through the air. The ratio of the speed of an object to the speed of sound in a medium is called the *Mach* number, $\mathcal{M}$,

$$\mathcal{M} \equiv \frac{v}{c_s} \tag{1.10}$$

So, $\mathcal{M} > 1$ is supersonic. Figure 1.4 (left) shows the geometry, indicating that the cone angle of the shock front, ϕ, is given by

$$\sin \phi = \frac{c_s}{v} = \frac{1}{\mathcal{M}}. \tag{1.11}$$

The cone is formed by rotation about the x-axis, has its apex at the right, and shows how a shock would appear if it were illuminated in its entirety, like the shock of a supersonic jet with mist in the air, like the wake of a boat, or in current case of interest, like the shock of a bolide. Figure 1.4 (right) is a repeat of the left figure but illustrates the direction that the shock propagates, projecting as a ring in a plane on the far right where measurements might be made. In this case, the apex of the cone is at the left and the cone rotates about the x-axis with the apex angle θ, rather than ϕ.

Objects with very high Mach numbers (greater than about 5) are called *hypersonic* and, since bolide Mach numbers range from about 37 to over 200, even the slowest bolide would be hypersonic. This also means that the shock angle, ϕ, is highly acute, smaller than 1.5°. The angle shown in Figure 1.4 is much larger for the sake of clarity.

Hypersonic shock waves from a bolide imply that a measured signal will be 'pulse-like' rather than like a smooth wave and within the pulse are a range of infrasound frequencies. In practise, the *period*, τ, at the peak amplitude of the signal (the *dominant period*) is the most conveniently measured quantity.

Figure 1.5 shows some possibilities. As the bolide travels through the atmosphere hypersonically, it generates shock waves (a). From Eq. (1.11), the shock angle is very small, so the shock geometry can be considered 'line-like' and is approximated as a cylinder (inset) with a radius called the *blast radius*, R_0. The blast radius is larger than the rock itself.

[9] See www.ctbto.org.
[10] The velocity required to escape from the Earth's gravitational potential well.

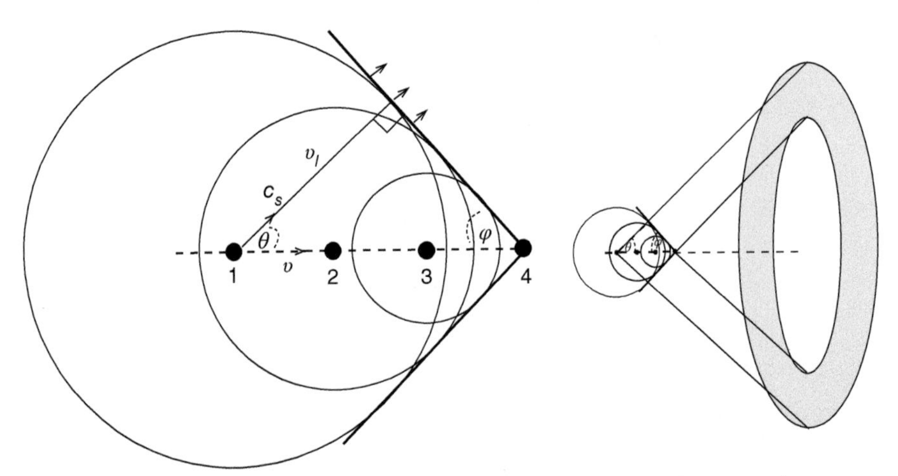

Figure 1.4 Illustration showing the conical shape of a shock wave when a particle's speed, v, is faster than the sound speed. (Left) As the particle moves along the x-axis (to the right) at times 1, then 2, 3, and 4, circles centred on those particles show how the signal propagates outwards and piles up at the shock front. Small arrows indicate the direction that the shock propagates. The shock forms a cone about the x-axis with an apex at time 4 where the cone angle is ϕ, and the cone opens to the left. The same geometry is applicable to a shock wave for sound (Section 1.1.2) and Cherenkov light (Section 1.3). In the former case, $\sin\phi = c_s/v$, where c_s is the speed of sound in the medium. In the latter case, $\sin\phi = v_l/v$, (or $\cos\theta = v_l/v$) where v_l is the speed of light in the medium. (Right) A miniaturized version of the left figure but showing the forward projection of the shock wave and the resulting ring in a plane where the shock could be measured at detectors. In this case, the apex of the cone is at time 1 with the cone opening up to the right. The relevant angle is now θ instead of ϕ.

Inside the cylinder is a strongly shocked and highly pressurized region that declines to the ambient atmospheric pressure at R_0. If the bolide explodes[11], a second shock wave can also result, this time being 'point-like' (b). In either case, an infrasound detector or array may detect the shock. Of course, if a detector is far enough away from the source, then a line-like shock will appear point-like since the meteor trail would be small in angular size.

Infrasound measurements provide source location, time of origin, and an estimate of the bolide energy [26]. Events with energies as low as 0.1 kT TNT equivalent (1 m diameter bolides, Figure 1.2) have currently been detected at multiple IMS stations. These stations are typically thousands of km from the event [1].

The bolide's energy is, in particular, a basic parameter that one would want to determine from infrasound measurements. The energy that is deposited in the cylinder is due

[11] The *explosion* is actually an exponential increase in the rate of ablation at the time of fragmentation; when this occurs, overlapping blast radii mimic a spherical blast shock.

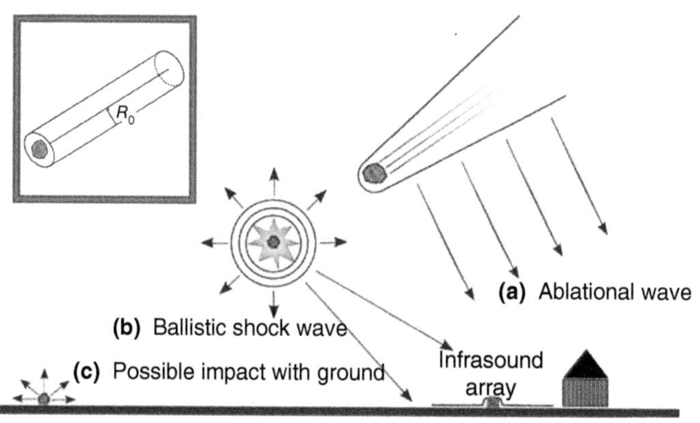

Figure 1.5 Illustration of two shock waves that are produced by bolides. (a) Shock waves are propagated perpendicular to the line-like trajectory of a hypersonic bolide. (b) If the bolide explodes, then a shock wave is generated radially from this point-like event. (c) The bolide might impact the ground. *Inset:* Approximation of the bolide trajectory as a cylinder showing the blast radius, R_0. Source: Adapted from Abbasi et al. [1].

to drag from collisions with atmospheric molecules and becomes visible as optical electromagnetic radiation. By requiring the deposited energy to be equal to the expansion work required to move the atmosphere out to the blast radius, this energy can be related to R_0. In addition, R_0 is related to the measured infrasound period; this period increases as the shock wave propagates, as the blast wave expands and the pulse weakens. The relation between period and R_0 is best determined experimentally. A standard benchmark is when the shock wave radius, R, has reached a distance such that $x \equiv R/R_0 = 10$. At this radius, the shock enters the 'weak shock domain' in which pressure perturbations become smaller than the ambient pressure. The dominant period at $x = 10$, from measurements, is

$$\tau_{x=10} = 2.81 \frac{R_0}{c_s}. \tag{1.12}$$

At $x \geq 10$ in the weak shock domain, the dominant period increases according to $\tau(x) \propto x^{1/4}$ [52] and the shock propagates weakly non-linearly[12]. Finally, there is a transition to the linearly propagating regime in which the period remains constant. From these relations, the measured period of the wave can be associated with the initial blast radius and the energy that is deposited during flight (Problem 1.5). The amplitude of the signal is also related to bolide energy. However, since the period shows little variation in the weak domain and is constant in the linear domain, the period is considered to be a better indicator of bolide energy than the more variable amplitude.

[12] 'Linear' waves have small perturbations and satisfy linear equations. Physically, this means that waves of different amplitude or frequency can be superimposed without interaction. This would not be the case for 'non-linear' waves which could interact energetically with each other.

In addition, since the deposited energy is related to the drag force, and the drag force depends on the physical diameter of the rock, d_m, and its speed, it should be possible to relate the initial blast radius to the meteor's size. For an isothermal atmosphere, this relation is

$$R_0 \sim \mathcal{M} d_m. \tag{1.13}$$

The assumptions in this equation are that the meteor has not fragmented, the rock is spherical, and there is no strong deceleration or strong ablation which would rapidly decrease the diameter. Fragmentation within the blast radius will actually *increase* R_0 and [52] suggest that Eq. (1.13) is not entirely reliable, although it could provide an estimate to within about a factor of 10.

If the meteor explodes, as illustrated in Figure 1.5, a large amount of energy is deposited at this point and the blast radius is largest. This occurred at a height of 30 km for the fireball that fell near Chelyabinsk, Russia, on 15 February 2013 [26]. The repercussions were significant, with the shock wave shattering windows in thousands of buildings and resulting in many injuries, mostly from broken glass and falling debris. Some 250 meteorites were recovered of which the largest fragment weighed 650 kg. This event, although rare, underscores the need for a good understanding of fireball energies and related shock waves. For Chelyabinsk, the incoming speed was 20 km s^{-1}, the diameter was $d_m \sim 19$ m, and blast radius was $R_0 = 9.5$ km. The meteor's best estimate kinetic energy was 500 kT TNT equivalent, from a variety of different measurements, including infrasound [11, 12, 26].

A challenge in linking an infrasound period with bolide energy is that of *calibrating* the signal. Given the different incoming speeds, wind speeds, angles of entry, noisiness of the detectors, and other variables, calibration has not been straightforward. Historically, ground level nuclear explosions have been used to relate periods to energies, for example those provided by the U.S. Air Force Technical Applications Center (AFTAC). Newer calibrations make use of electromagnetic radiation. For example, satellite measurements of optical light can be used to determine the deposited energy in the EM bandpass of the sensors, E_o, by assuming that the light can be represented by a *black body* (see Chapter 6) at 6000 K. The total kinetic energy, $E = (1/2)\, m v^2$, however, is higher than the energy that is radiated during entry. An estimate of the correction factor is [11]

$$E = 8.25 E_o^{0.885} \tag{1.14}$$

where the energies are in units of kT TNT equivalent.

Once infrasound periods of a sufficient number of identifiable satellite events can be measured, then the energies can be calibrated. In the linear regime in which the period should be constant, the calibration is

$$\log(E) = 3.84 \log(\overline{\tau}) - 2.21 \tag{1.15}$$

where the period is the dominant period (in seconds) averaged over infrasound stations, and the energy is in kT of TNT [26], that is, $E \propto \overline{\tau}^{3.84}$. This result is similar to that of AFTAC which found powers of 3.34 or 4.14, depending on energy. There is still a fairly significant scatter in the calibration, however. For example, even when the signal is in the linear regime in which the period should be constant, there will still be differences in the measured periods from station to station. As discussed earlier, stations that are farther from the event will measure longer periods simply because the higher frequencies (shorter periods) are more readily attenuated by the atmosphere (see [26]). It may also be necessary to ignore detectors that have noisier signals. As Eq. (1.15) also shows, small changes in period result in large changes in E; for example, if the period doubles, the implied energy is higher by a factor of 14. Clearly, a statistical approach is required so that average periods are measured for many bolides over a wide range of energies (e.g. [18]).

Infrasound is a unique approach to understanding cosmic messages and promises to reveal important information that cannot easily be found from electromagnetic radiation alone. It is a developing and dynamic field that encompasses a wide range of inquiry from meteoritics to meteorology. Infrasound has even recently been detected in the thin atmosphere of Mars [7]. For a review, see [51].

1.1.3 Gathering Dust

'Dust' generally refers to any small solid particle, icy, or rocky – and interplanetary space is a dusty place. Since dust exists in molecular clouds from which stars form, dust will inevitably be present in systems associated with stars, including our Solar System. Dust is on the surfaces of the rocky planets, is in the atmospheres of the gaseous planets, is 'shed' by asteroids along their orbits, and is present in planetary rings. Comets shed some of their dust as they orbit the Sun. The Earth passes through a number of different cometary debris trails regularly each year, resulting in predictable *meteor showers*. We will look at the absorptive and emissive properties of dust in Chapters 5 and 6, but what about *collecting* the dust for direct laboratory analysis?

Interplanetary dust has been collected in the Earth's stratosphere for many years (e.g. [20]), but the Earth's orbit restricts the region over which such particulate matter can be swept-up. Instead, we must reach out beyond our 'island' of Earth. The manned *Apollo* missions to the Moon in the late 1960s and early 1970s took the first pioneering steps to return samples from another body; about 380 kg of lunar rocks were recovered for study in Earth's laboratories. But unmanned missions can go farther.

For the first time, we have collected dust from a comet and returned the particles to the Earth for study (Figure 1.6). The *Stardust* mission scooped up dust from Comet Wild 2 in 2004 and landed its capsule containing these particles back on the Earth two years later. Wild 2 formed in the *Kuiper Belt*, a region of asteroids and comets beyond Neptune, which is thought to consist of early remnants from the formation of the Solar System. Thus, particles from Stardust were the first to be examined in the laboratory with a known origin in the outer Solar System. Surprisingly, the results showed that the icy outer regions were not as isolated as previously thought, suggesting that there was large-scale mixing of material in the early turbulent formation period [54].

Figure 1.6 The impact of a cometary dust grain onto aerogel (a sponge-like silicon-based solid that is 99.8% air) is shown here, captured when NASA's Stardust spacecraft flew through the dust and gas cloud surrounding Comet Wild 2 on 2 January 2004. Stardust returned to Earth on 15 January 2006. *Inset*: A tiny particle collected by the Stardust spacecraft. This particle is 2 μm across and is made of a silicate mineral called forsterite, known as peridot in its gem form. Source: Reproduced by permission of NASA/JPL-Caltech and the Stardust Outreach Team.

Stardust was only the beginning of the sample-returning missions. The Japanese *Hayabusa* (the 'falcon') landed on asteroid Itokawa and returned some particles to Earth in 2010; it has been followed by *Hayabusa 2*. NASA's OSIRIS-REx has visited asteroid Bennu, and Japan plans to return samples from Phobos, one of the moons of Mars. Recently, China's Change 5 satellite has also collected moon rocks. Many more missions that are not sample-returning, but involve flybys with imaging and analysis or landings, have taken place or are planned. These are knowledge-gathering enterprises. As we become more familiar with the conditions on these bodies, a search for samples may expand into mining operations or searches for water, as NASA's VIPER plans to do on the icy lunar south pole.

1.2 COSMIC RAYS

The existence of *cosmic rays* (CRs) has been known since 1912 when Victor Hess took an electrometer (a device for measuring the presence of charge) to a high altitude in a hot air balloon. Contrary to what was expected, the electrometer discharged more quickly at *higher* altitudes, leading Hess to conclude that the source of the discharge must be from above the atmosphere rather than the Earth itself. For this discovery, he shared the 1936 Nobel Prize for physics. Called cosmic *rays* because they were originally thought to be part of the electromagnetic spectrum, we know today that these are high energy (1 MeV to 10^{21} eV, [28]) *particles* coming from space. As noted in the Introduction, CRs, together with meteoritic material (see Section 1.1) and neutrinos (Section 1.3) represent the only known *particulate* matter reaching the Earth's surface. As such, they are important carriers of cosmic information beyond what is brought to us by light. The *relativistic electrons* in CRs also emit *synchrotron radiation*, an important emission mechanism which will be discussed in Chapter 10.

Figure 1.1 (Left) shows an early photograph of a cloud chamber containing CR tracks. One need only watch a real cloud chamber for a few seconds to see this subatomic world come to life. What is being viewed, however, are not the cosmic rays that impinge upon the Earth's atmosphere from above, but rather high energy particles that are created within the atmosphere from *cosmic ray showers*. These occur when a cosmic ray particle collides with an atmospheric nucleus, shattering it and producing secondary particles, some of which will undergo radioactive decay and some of which may collide with other atmospheric particles, creating a chain of subsequent events. Various subatomic particles are formed in the process, such as pions[13], muons[14], and neutrinos. Atmospheric cosmic rays are responsible for approximately one sixth of the naturally occurring radioactivity on Earth.

[13] A pion is a subnuclear particle. It is a type of *meson* which is a particle that consists of a *quark–antiquark pair. Quarks* (of which there are different types) are the basic 'building blocks' of nuclear matter; for example, protons and neutrons are each composed of three quarks, but the mixture of the types of quarks are different for the two particles.

[14] A muon is an elementary particle with a charge of -1e similar to the electron, but is 207 times heavier. Its mean lifetime is only a few microseconds.

To understand the origin of cosmic rays, it is therefore necessary either to make corrections for atmospheric interactions or to make measurements above the Earth's atmosphere. In the following sections, we assume that these corrections or measurements have been made and refer to such measurements as 'direct' (for a list of direct measurements, see [28]). An example of an 'indirect' measurement would be a measurement of EM radiation produced in some fashion by a CR.

1.2.1 Cosmic Ray Composition

What is the composition of cosmic rays? Approximately 99% of these particles are nucleons (protons and neutrons) as well as nuclei, and the remaining 1% are electrons and their positive counterparts, positrons. Of the electron/positron component, positrons account for only about 10%. Of the nucleonic component, 87% are hydrogen nuclei (protons), 12% are helium nuclei (also called *alpha particles*), and 1% are heavier nuclei [50]. CRs are charged particles and will therefore be affected by magnetic and electric fields.

Figure 1.7 shows the relative abundances of elements in *Galactic cosmic rays* (GCRs – those that originate from sources in the Milky Way outside of the Solar System) compared to the CR composition in the Solar System. 'Solar System abundances' are those that are typical of the Sun and also of other stars and clouds in the disc of our Galaxy (see also Section 5.3.4 and Figure 5.9). Compared to the Solar System, CRs are *underabundant* in hydrogen and helium and *overabundant* by many orders of magnitude in the light elements, lithium, beryllium, and boron (Li, Be, and B, respectively). They are also *overabundant* in elements with nuclear charges of 21–25 (scandium to magnesium) which is just below iron ($Z = 26$) and in elements with charges from 60 to 70 (neodymium to ytterbium) that are below lead ($Z = 82$). Thus, CR particles show important differences from what we normally see in the Solar System, differences that offer clues as to their origin and propagation.

Most compositional anomalies can be explained by models that take into account the interactions of CRs en route to Earth. The interstellar medium (ISM) acts like another 'atmosphere' through which CRs travel, and spallation (introduced in Section 1.1.1) creates cosmogenic nuclides in ISM material. In particular, the overabundance of Li, Be, and B can be understood from the break-up of carbon, nitrogen, and oxygen (C, N, and O, respectively) in the ISM. Similarly, the overabundance of the sub-iron and sub-lead species can be understood in terms of the interstellar fragmentation of heavier elements [34]. It is clear that, just as we saw for the Earth's atmosphere, in order to understand the original composition of these particles at the location where they are first accelerated (called *primary* CRs)[15], their interaction with the ISM must be understood. Many apparent anomalies disappear once the 'ISM atmosphere' has been taken into account.

Some differences from Solar abundances do remain, however. One is the relative underabundance of hydrogen. Supernova ejecta, for example, consist mostly of elements

[15] The term, 'primary cosmic ray', is also sometimes used to represent the CRs that impinge on the Earth's atmosphere. Here, we use 'primary' to indicate CRs where they are originally accelerated.

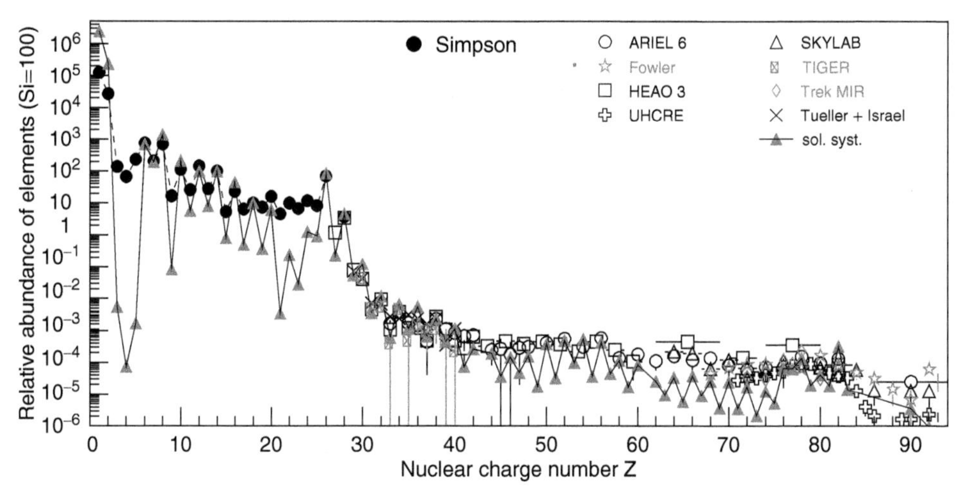

Figure 1.7 Abundances of the elements in Galactic cosmic rays (GCRs, various symbols) compared to Solar System abundances (solid line with triangles). Values have been normalized to silicon (Z = 14) which has been set to 100. Different symbols are listed in the legend at top right from direct measurements of GCRs. Notice the underabundance of hydrogen and helium (Z = 1 and 2, respectively), the overabundance of Li, Be, and B (Z = 3, 4, and 5, respectively), the overabundance of Sc, Ti, V, Cr, and Mn (Z = 21 to 25, respectively), and the overabundance of elements in the Z range between 60 and 70. Source: Courtesy of J. R. Hörandel [34]. © 2008, J. R. Hörandel.

heavier than hydrogen (Section 5.3.3). Also, acceleration mechanisms may be more effective for interstellar grains (which contain heavy elements) than for particles that are in the gas phase such as hydrogen [17]. The propagation of CRs through the ISM is currently an area of active research, but some models suggest that 80% of CRs result from interactions with the ISM, with only 20% being primary [28]. We will look at the origin of primary CRs in Section 1.2.3.

A remaining puzzle is why the number of electrons should constitute only about 1% of the number of nucleons. We will return to this question in the next section, once we look at the CR energy spectrum.

1.2.2 The Cosmic Ray Energy Spectrum

CRs move with relativistic speeds, and consequently, the kinetic energy of a particle, T, dominates its rest mass energy, E_0, over almost the entire spectrum. The rest mass energy is given by Einstein's famous mass–energy relation

$$E_0 = \mathrm{m}_0 c^2 \tag{1.16}$$

where m_0 is the mass of the particle, and c is the speed of light. Usually, the 'rest mass' of a particle is simply expressed as an energy in electron Volts (eV) rather than a mass in grammes. For example, the rest mass of an electron is 511 keV and the rest mass of a proton is 938 MeV. The total energy of a particle moving at speed, v, is the sum of its rest mass energy and kinetic energy,

$$E = E_0 + T = \gamma E_0 \tag{1.17}$$

where γ (Table I.2) is the *Lorentz factor*, defined as,

$$\gamma \equiv \frac{1}{\sqrt{1-\frac{v^2}{c^2}}} = \frac{1}{\sqrt{1-\beta^2}}. \tag{1.18}$$

Here, $\beta \equiv v/c$ is a unitless parameter that describes the speed in comparison with the speed of light.

If $v \ll c$, a binomial expansion (Eq. A.2 of the *online material*) can be used on Eq. (1.18) to find,

$$\gamma \approx 1 + \frac{1}{2}\frac{v^2}{c^2} \tag{1.19}$$

In this case, Eq. (1.17) reduces to,

$$E = \mathrm{m}_0 c^2 + \frac{1}{2}\mathrm{m}_0 v^2 \qquad (v \ll c) \tag{1.20}$$

which is a more familiar expression showing both the rest mass energy of the particle (first term) and its kinetic energy (second term). However, as we will see, the low velocity limit does not apply to CRs (Problem 1.9). For example, if we consider a 'relativistic speed' to be at least 0.3 c, then any particle with $\gamma > 1.05$ is relativistic. Yet, Lorentz factors as high as $\gamma \approx 10^{11}$ have now been measured for some high energy CRs, corresponding to $v = c$ to many decimal places.

The cosmic ray *energy spectrum* is shown in Figure 1.8. The particle kinetic energy (x-axis) spans an astonishing 13 orders of magnitude! Almost the entire energy range is much greater than the 938 MeV rest mass of a proton, so $E \approx T$ for most of the plot. Here, we see a beautiful power law spectrum, with several small but important changes in slope. Since this is a log–log plot, a straight line represents a *power law* distribution of energies which can be described by,

$$J(E) = KE^{-\Gamma} \tag{1.21}$$

where Γ is the *cosmic ray energy spectral index*, E is the energy of a particle, and K is a constant of proportionality. The quantity, $J(E)$, in units of number of particles $\mathrm{s^{-1}\ m^{-2}\ GeV^{-1}\ sr^{-1}}$, is a kind of *specific intensity*[16] for particles (compare this to a specific intensity for radiation which has units of $\mathrm{erg\ s^{-1}\ cm^{-2}\ Hz^{-1}\ sr^{-1}}$, as will be described in Section 3.3). Eq. (1.21) can be integrated to obtain the total number of CRs hitting an object per unit time; we defer such a calculation until Example 3.4 of Chapter 3, once intensities, fluxes, etc. have been introduced for radiation.

[16] Usually, 'number of particles' is left out of the units since the 'number of something' is not a unit!

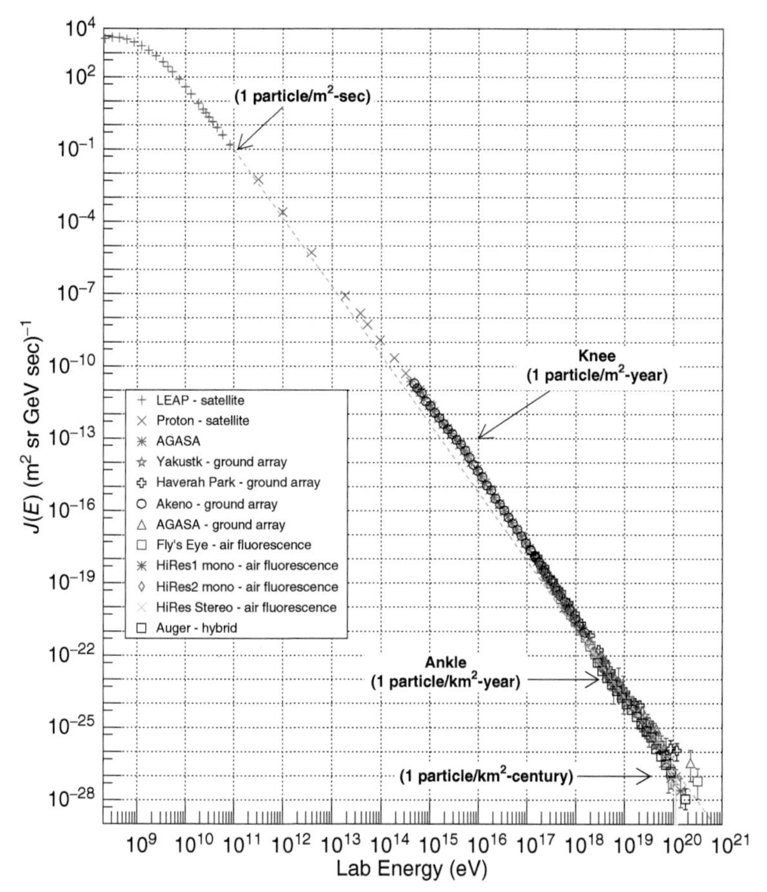

Figure 1.8 Cosmic ray energy spectrum for particles from 10^8 to 10^{21} eV. The ordinate represents $J(E)$ (Eq. (1.21) and the abscissa represents the kinetic energy per particle (Eq. (1.17)) for which $E \approx T$ over almost the entire energy range plotted. The dashed line shows a power law fit with a single spectral index of −3. Subtle changes in the spectral slope can be seen in two places: the 'knee' (at $E = 3 \times 10^{15}$ eV) and the 'ankle' ($E = 5 \times 10^{18}$ eV). Source: Courtesy of W. F. Hanlon [31]. © 2008, W. F. Hanlon.

A single value of Γ, as shown by the faint dashed curve in Figure 1.8, does not fit the data perfectly. There are changes in the slope, two of which are marked 'the knee' and 'the ankle' at energies (x-axis) of 3×10^{15} eV [28] and 5×10^{18} eV, respectively, the value of the ankle being less well-defined[17]. At the lowest energies, the spectrum does not follow a

[17] A 'second knee' near 10^{17} eV has also been identified though it is not obvious in Figure 1.8.

power law, but is curved in this log–log plot. Each of these changes in slope contain astrophysically interesting information.

For example, at less than about 10^{10} eV, CR energies are typical of particles ejected from our Sun in *flares* and *coronal mass ejections* [18]. These ejections are a high energy component of the continuous *Solar wind* that begins in the hot, tenuous Solar corona (Figure 8.6). Solar CRs are highly variable in time and so is the *interplanetary magnetic field* (IMF) [19] as it extends outwards from the Sun. Since CRs are charged particles, they will be affected by the strength and geometry of this IMF. Any additional CRs that originate from outside of the Solar System in this low energy regime (e.g. GCRs) will be strongly modulated by local effects. Incoming GCRs are affected by convection from the solar wind, drifts in direction due to the geometry of the IMF, diffusion when the IMF fluctuates, and adiabatic energy losses. They also show an 11-year cycle such that when the Sun is at Solar maximum, GCRs are at a minimum, and vice versa [45].

Above this energy, most CRs are accelerated in Galactic sources and follow a power law, with $\Gamma \approx 2.7$ a good fit to the data at energies lower than the knee and $\Gamma \approx 3.0$ above it. A value of K for the upper end of the spectrum is provided in Eq. (5.41). See Problem 5.13 for an estimate of K at energies less than this. At energies greater than the knee, extragalactic CRs may contribute, although it is not yet clear exactly at what energy the extragalactic particles become important [28]. At energies above the ankle, though, CRs are extragalactic.

The cosmic ray energy spectrum for *electrons* is also a power law but it shows some differences with respect to nucleons, especially at low energies because electrons are more strongly influenced by the Solar wind. It is likely that the electron spectrum is only free of this modulation at electron energies above 10 GeV [42]. After correcting for solar modulation, the electron spectrum can be fitted by the function [14],

$$J_e(E) = 412 E^{-3.44} \qquad (3\ \mathrm{GeV} < E < 2\ \mathrm{TeV}) \tag{1.22}$$

where E is the electron energy in GeV, and the units of $J_e(E)$ are the same as Eq. (1.21). The value of the constant of proportionality is less certain than that of the slope, due to difficulties in combining data from many different experiments using different instruments. Nevertheless, Eq. (1.22) indicates that the CR electron spectrum is significantly steeper than the $\Gamma \approx 2.7$ found for nucleons.

The difference in spectral index between electrons and protons can be accounted for by the greater susceptibility of electrons to interactions and energy losses in the ISM. A variety of energy loss mechanisms exists for both electrons and protons, but electrons lose energy more readily via *synchrotron radiation* (see Section 10.5) and *inverse Compton scattering* (Section) which are not important for nucleons. Since energy losses are *greater*

[18] A solar flare is an energetic release of particles associated with sunspots and related magnetic activity on the surface of the Sun. A coronal mass ejection is similar but much larger and more energetic; the high energies could result from reconnecting magnetic field lines.

[19] Some care is required when using this abbreviation. IMF, in astronomy, is also used for the completely unrelated 'initial mass function' which refers to the number of stars formed per unit mass interval.

for higher energy particles, the result is a *steepening* of the CR spectrum. In fact, the *initial* spectral indices, Γ_0, of *both* electrons and protons are believed to be the same and in the range, $2.1 \approx \Gamma_0 \approx 2.5$, depending on the specific acceleration mechanism [6].

We can now return to the puzzle of the observed low electron fraction of cosmic rays, i.e. a measured electron to proton number fraction of about 1%. This is highly underabundant because, in a normal ionized gas, there should be approximately equal numbers of protons and electrons. There are a number of subtleties, but according to [50], the following argument applies

The electron fraction is measured at some kinetic energy, T. Although a variety of acceleration mechanisms may be present, several of the better understood mechanisms in the intermediate energy range (e.g. shock acceleration, see next section) give a *distribution in momentum*, p, that is the *same* for electrons and protons except for a constant of proportionality, i.e. $N(p) \propto p^{\Gamma_0}$, that is, the *shape* of the momentum spectrum is the same for electrons and protons. The *total number* of particles above a fixed lower kinetic energy, T_0, is also expected to be the same for both electrons and protons, as in any ionized gas.

At a fixed lower energy, T_0, though, the electron momentum is much less than the proton momentum because the mass of an electron is much less than the mass of a proton. This means that the electron distribution is being sampled to lower momenta. Essentially, the electron momentum spectrum is shifted to lower energy in comparison with the proton spectrum. Therefore, at some given T at which a measurement is made, the number of electrons must be less than the number of protons, even though the total number of each particle is the same. This argument is outlined mathematically in Section 1.1 of the *online material* (see also [50] for further details).

1.2.3 The Origin of Primary Cosmic Rays

What kinds of astronomical sources can accelerate particles to such high energies? We have already seen some indications from the energy spectrum and abundances. At the lowest energies, as indicated above, the CR spectrum is dominated by Solar cosmic rays. Thus, we know that at least one star is responsible for a fraction of the CR flux!

At higher energies, CRs are believed to come from outside of the Solar System with most particles originating from sources within the Milky Way. Energetically, ordinary stars or isolated *neutron stars* (Section 5.3.3) in the Milky Way cannot account for the total flux of these higher energy CRs. Supernovae, however, provide sufficient energy. Only 10% of the kinetic energy of all supernovae in the Galaxy would be needed to explain the CRs in the Galaxy at energies up to about the knee of Figure 1.8. This connection has been strengthened by theoretical success at explaining the CR energies and spectral index via acceleration in shock waves created by supernovae. The detection of synchrotron radiation (see Section 10.5), which is emitted by the electron component of CRs in supernova remnants in the Milky Way, provides a further link, as does the detection of TeV γ-rays from supernova remnants. A supernova origin connects CRs with hot, massive stars since these are the only kinds of stars that produce supernovae (Section 5.3.3). It can also account for the underabundance of hydrogen as described above because the explosive ejecta will be *metal*-enriched (see Section 5.3.1). Therefore, supernovae are a source of mid-energy cosmic rays up to the knee (e.g. [10]) and probably an important contributor at higher energies as well.

An important more recent realization is that other Galactic sources are also needed to help explain the many different observations of CRs that are available, both direct and indirect. One clue, for example, is from the ratios of different isotopes. $^{22}Ne/^{20}Ne$, for instance, is about five times higher in GCRs than in the Solar System [28]. ^{22}Ne is known to be rich in the spectra of *Wolf–Rayet stars* (very massive, young hot stars, see Figure 7.10), hinting at a connection between CRs and shocks from the winds of massive stars that are rich in heavy metals [39]. Pulsars and pulsar-wind nebulae are also promising sources [28]. The reason for the change in slope at the knee is not entirely understood, and various proposals have been advanced. For example, there may be acceleration or reacceleration by a different combination of sources (a Galactic wind, pulsars, or gamma-ray bursts) or leakage of CRs from the Galaxy during propagation [23, 33]. There also appears to be a change in composition across the knee (higher fraction of heavier elements at higher energy), suggesting that the change in slope could be a result of a superposition of upper energy cut-offs for different elements from a given SN event. For example, particles with a higher charge (heavier elements) can achieve higher energies than particles with a lower charge [10].

It is challenging to interpret all of the measurements, and the study of CRs remains an active and dynamic area of research. Beyond corrections for the atmosphere, ISM interactions and subsequent energy losses during propagation, there is an additional problem in simply identifying the locations of the sources. Cosmic rays easily scatter from magnetic field lines in the galaxy (except at very high energy) and therefore propagate by diffusion, similar to the way a photon would take a random walk out of the Sun (Section 5.4.4). Therefore, CRs over most energies are *isotropic* (the same in all directions) as measured at the Earth.

The biggest mystery of all, though, is the ultra-high energy cosmic rays (UHECRs), i.e. those with energies above the ankle. Even when the most powerful of sources are considered, accelerating particles to $E \approx 10^{20}$ eV strains the limits of known acceleration mechanisms. The fact that particles with such high energies are observed at all is a topic of great current interest. These particles are likely extragalactic in origin and are extremely rare, with as few as 1 particle per square kilometre per century at the higher energies (Figure 1.8)!

Extragalactic CRs will meet *photons* from the cosmic microwave background (CMB, Section 5.1) while travelling through intergalactic space. Interactions of such high energy particles (mostly protons) with CMB photons can occur in a variety of ways, but most notably, pions can be formed with subsequent CR energy loss. The energy at which this occurs, about 6×10^{19} eV [1], is called the GZK (for Greisen–Zatsepin–Kuz'min) cut-off. This puts a limit on the region of space within which UHECRs could have originated, a distance that is approximately 50 Mpc (a few tens to a hundred, depending on energy). Possible candidates within this volume are galaxy clusters, *active galactic nuclei* (AGNs) or their jets, gamma-ray bursts, or *magnetars*[20].

On the positive side, at such high energies, these CRs do not scatter so strongly off of magnetic field lines, and therefore, their apparent angle in the sky should point back more closely to their place of origin. Evidence for anisotropies has already been reported, for example in directions towards the constellations of Ursa Major [1] and Canis Major [46] (the Big Bear and the Big Dog, respectively). Clearly, detecting more of these rare events would help. Pinning down the energy source for UHECRs is therefore only a matter of time.

1.3 NEUTRINOS

The elusive neutrino was originally postulated by Wolfgang Pauli in 1930 to explain the fact that a radioactive process called *beta decay*[21] resulted in electrons (beta particles) that had a range of energies; they were expected to have only a single energy. If another particle were involved in this decay, it could solve this energy problem and also conserve linear and angular momentum. The neutrino had to be neutral and of very low or zero mass (see Section 1.3.2.1). Neutrinos were first measured in 1956, and we now know that they come in 3 different 'flavours', the electron, muon, and tau neutrinos, designated ν_e, ν_μ, and ν_τ, respectively. Neutrino flavour depends on which charged particle, the electron, muon, or tau particle, that the neutrino interacts with[22]. Neutrinos along with these charged particles are called *leptons*. There are also *antiparticles* for each of the leptons. Electrons, muons, and taus are negatively charged (e^-, μ^-, and τ^-) but they have positively charged antiparticles (e^+, μ^+, and τ^+). *Antineutrinos* are designated $\overline{\nu_e}$, $\overline{\nu_\mu}$, and $\overline{\nu_\tau}$.

Neutrinos can be produced by a variety of processes but a common one is the radioactive beta decay discussed above. Such radioactivity and other neutrino interactions are governed by the *weak force*[23]. Any process that leaves an atomic nucleus in a state in which such a decay can occur is a source of neutrinos. There are therefore many potential origins for neutrinos, but in general, they are associated with *high energy phenomena*. A glance at Figure 5.5, which shows the main nuclear reactions inside our Sun, reveals several branches in which neutrinos are produced. Supernovae are also prolific sources of neutrinos. And in

[20] In brief, magnetars are pulsars with extremely high magnetic fields.

[21] There are two types of beta decay: β^- (with the emission of an electron, e^-, and an electron antineutrino, $\overline{\nu}_e$) and β^+ (with emission of a positron, e^+, and neutrino, ν_e). Simple examples are the decay of a free neutron, $n \rightarrow p^+ + e^- + \overline{\nu}_e$, or the conversion of a proton that is in a nucleus, $p^+ \rightarrow n + e^+ + \nu_e$.

[22] Technically, this would be through the 'charged current weak interaction'. However, all neutrinos can interact with all the charged leptons through neutral current interactions, i.e. exchange of a Z boson.

[23] During radioactive decay, an intermediate particle, called a W boson, is briefly formed. During neutrino scattering, an intermediate Z boson is involved. Both the W and Z bosons (there are different types) are 'carrier' particles and do not exist independently. They are also relatively massive elementary particles so the associated effective range of this process is small, hence the name, 'weak force'.

Section 1.2, we indicated that neutrinos could be made in cosmic ray showers when high energy cosmic rays collide with atmospheric nuclei.

Although both cosmic rays and neutrinos can result from high energy events, neutrinos need to be considered separately because they differ in two important ways. The first is their much lower mass (see Section 1.3.2.1), and secondly, neutrinos are neutral, uncharged particles. The lack of charge means that they will not be affected by magnetic or electric fields and rarely interact with normal matter. The probability of interaction depends on the effective cross-sectional area of a particle (see Section 5.4.3). For a neutrino of energy 1 GeV, for example, the cross section for scattering is only 10^{-38} cm^2 [21]. Compare this to the effective cross section for electron–proton scattering in a hot 10^4 K gas which is 10^{-15} cm^2 (Table 5.2). Neutrinos therefore pass through almost all matter (including human beings) as if the matter were not even there. Even the Earth is transparent to neutrinos up to about a TeV (10^{12} eV) in energy [53]. Only very dense material, such as in neutron stars, provides a strong barrier (Problem 5.9). More will be said in Section 7.4 about quantifying 'transparency' or 'opacity', but roughly, to say that a material is transparent to neutrinos means that most neutrinos will pass through without any interaction at all, but a minority of them can interact. This means that neutrinos are difficult, but not impossible, to detect. Given this challenge, a *neutrino observatory* must apply techniques that are markedly different from normal astronomical observatories. It needs to be placed, not on a high mountain top, but underground, and the target mass needs to be large.

Two examples are the Sudbury Neutrino Observatory (SNO) in Canada and IceCube at the South Pole (see Figure 1.9). SNO's tank, situated 2 km below the surface in an active nickel mine, is well shielded from naturally occurring above-ground radioactivity (mainly cosmic rays) that could be confused with the signal of interest. The tank contained a million kg of 'heavy water' (D_2O rather than H_2O) as the target material, where D refers to *deuterium* whose nucleus consists of one proton and one neutron. The choice of heavy water meant that reactions involving all flavours of neutrinos could be detected, not just electron neutrinos which had previously been measured in other experiments. SNO was designed to detect neutrinos from the Sun (see Section 1.3.2.1). IceCube, buried in Antarctica, uses the ice itself as the target material with a detector volume of one cubic km. It was designed to detect the highest energy neutrinos (higher than 10^{12} eV) that are believed to originate from outside of our Galaxy.

One technique, used at both observatories, is the detection of *Cherenkov radiation* (or *Cherenkov light*) named after its discoverer. The speed of light in a vacuum, c, is fixed and maximal. However, the speed of light in a medium, v_l, depends on the index of refraction, n, according to

$$v_l = \frac{c}{n} \tag{1.23}$$

(Table I.2). Ice, for example, has an index of refraction of about 1.3, so the speed of light in ice is 2.3×10^{10} cm s^{-1}, only 77% of the vacuum value. Essentially, all neutrinos travel at speeds close to the speed of light in a vacuum, so if a neutrino hits a charged particle in some medium, the charged particle could then move at a speed that is greater than the speed of light in that medium. In other words, a particle in the medium could move *super-luminally*. The charged particle can then radiate optical light (typically bluish) in a cone, i.e. an *electromagnetic shock wave* is formed, as depicted in Figure 1.4. The situation is entirely analogous to a sound shock wave discussed in Section 1.1.2, and just like sound, the cone of light points back in the direction in which the neutrino arrived, providing important directional information. A Cherenkov light cone is illustrated in the inset to the IceCube image on the right hand side of Figure 1.9. Because of the way that the detectors pick up the signal, the cone geometry is often described as in the right hand picture of Figure 1.4.

1.3.1 The Neutrino Spectrum

Figure 1.10 shows the Grand Unified Neutrino Spectrum (GUNS) which includes data from measurements and theoretical expectations [53]. Stepping through the labels provides a nice overview of the variety of sources from which neutrinos can originate:

1. CNB: Cosmic neutrino background. This contribution is a remnant of the hot thermal universe when it was about one second old. The CNB constitutes what is referred to as 'hot dark matter' although such a component is believed to be only a small fraction, if at all, of the unknown dark matter that is discussed in Section 5.2.
2. BBN: Big Bang nucleosynthesis involving (n) the decay of a neutron (see beta decay information in Footnote 21) or (^{3}H) the decay of an isotope of hydrogen into an isotope of helium via $^3\mathrm{H} \rightarrow\ ^3\mathrm{He} + \mathrm{e}^- + \overline{\nu}_\mathrm{e}$ (essentially the decay of a neutron within a nucleus). Such decays would occur within the first few minutes after the Big Bang when the light elements are formed.
3. Solar (thermal): Neutrinos that are formed within the Sun mostly from processes that involve photons. There are various possibilities but one example is when a high energy photon (designated γ) interacts with an electron: $\gamma + \mathrm{e}^- \rightarrow \mathrm{e}^- + \overline{\nu}_\mathrm{e} + \nu_\mathrm{e}$. This is a variant of *Compton scattering* that is discussed in Section 7.1.2.2.
4. Solar (nuclear): These neutrinos are a product of the main nuclear reactions in the core of the Sun that convert Hydrogen into Helium as shown in Figure 5.5 and will be discussed in Section 1.3.2.1.
5. Geoneutrinos: Neutrinos from naturally occurring radioactivity in the Earth, 99% of which are associated with the decay chains of ^{232}Th, ^{238}U, and ^{40}K.
6. Reactors: Neutrinos from nuclear power plants.
7. DSNB: Diffuse supernova neutrino background. This takes into account the roughly one supernova (SN) per century per galaxy in the visible universe. Each SN releases an enormous amount of energy in neutrinos equivalent to the gravitational binding energy[24] of a neutron star, i.e. $E_b = 3 \times 10^{53}$ erg. An average energy of 10 MeV (1.6×10^{-5} erg) for each neutrino (three flavours plus their corresponding

[24] Gravitational binding energy of a star or stellar remnant is $U = C_0 GM^2/R$ where M is the mass, R is the radius, and G is the universal gravitational constant. C_0 is a constant that depends on the mass distribution of the star. For a uniform density sphere, $C_0 = -3/5$.

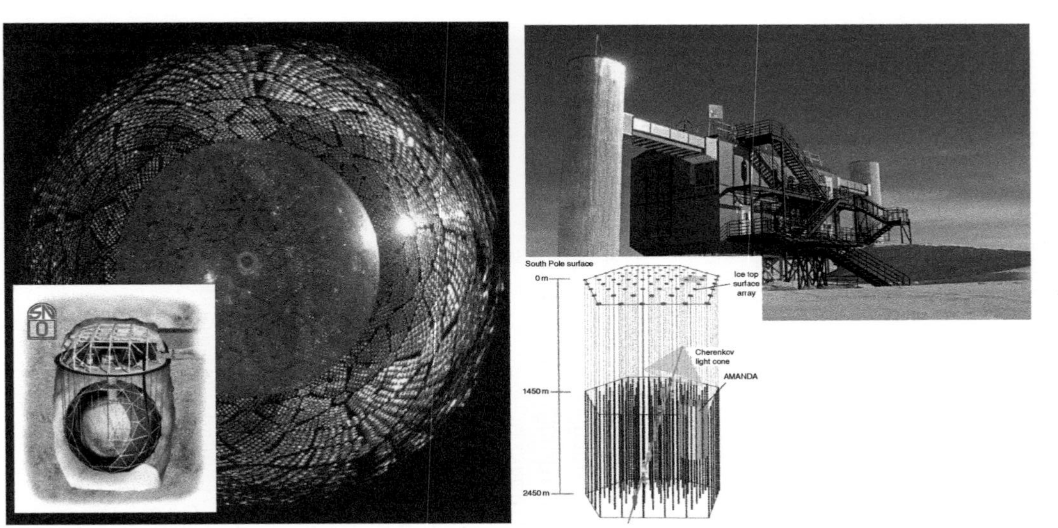

Figure 1.9 (Left) The tank of the Sudbury Neutrino Observatory (SNO) looks like a coiled snakeskin in this fish-eye view from the bottom of the tank before all photomultiplier tubes were in place. Photo credit: Ernest Orlando, Lawrence Berkeley National Laboratory. *Inset*: Schematic drawing of SNO showing the acrylic tank below ground [38]. Source: Image courtesy of A. McDonald. (Right) The IceCube Neutrino Observatory at NSF's Amundsen-Scott South Pole Station. Source: Mike Lucibella, *Antarctic Sun*. *Inset:* Diagram showing 86 cables drilled into the ice to a depth of 2.5 km. Optical modules (not visible) are connected to the cables [30]. The arrow-like cone is a depiction of Cherenkov light produced from a high velocity electron. Source: Image courtesy of F. Halzen.

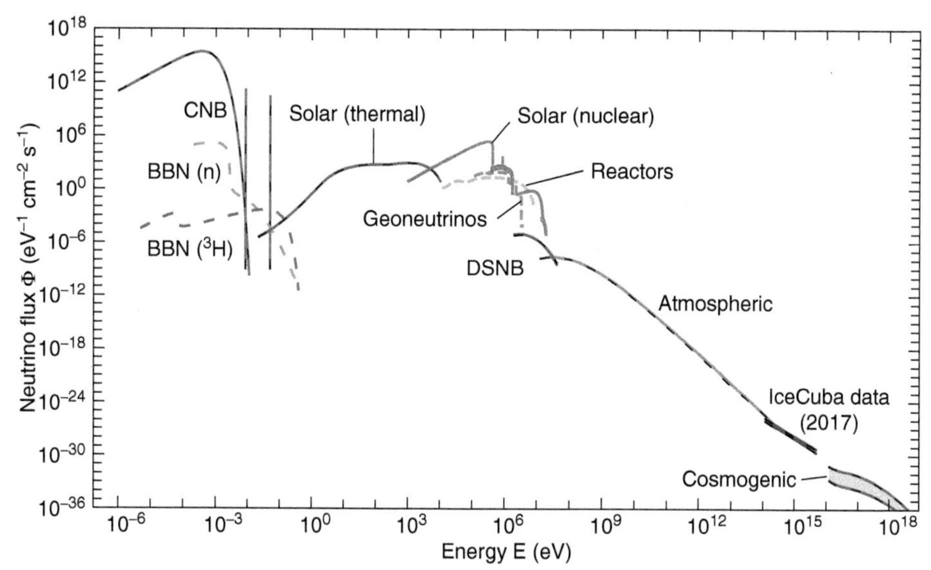

Figure 1.10 The Grand Unified Neutrino Spectrum (GUNS) at Earth plotted as a function of particle kinetic energy, T. Acronyms are explained in the text. Results include both measurements and expectations from theory. Solid lines refer to neutrinos, dashed or dotted lines refer to antineutrinos, and superimposed dashed-dotted lines are for both neutrinos and antineutrinos. Source: Courtesy of E. Vitagliano et al. from [53], © E. Vitagliano.

three antineutrinos) implies that there must be about 3×10^{57} neutrinos of each type emitted during any given SN.

8. Atmospheric: These are neutrinos that are produced in the Earth's atmosphere by interactions with cosmic rays. A contribution from neutrinos produced in the Sun's atmosphere by the same processes is included but the Sun's contribution is minor, up to $\approx 10^{12}$ eV. Linking CRs to their daughter neutrinos is a complex process that requires numerical modelling. A simple comparison of numbers of CRs and neutrinos at a given energy (*not* linking them physically) is given in Example 1.1.
9. Cosmogenic: As indicated in Section 1.1.1, cosmogenic implies that the particle is formed by impact with a CR. This is true for no. 8 as well, but here we refer to higher energy neutrinos produced by primary CRs in or near the source of origin or en route to Earth. Very high energy neutrinos are likely produced by very high energy CRs which we expect to be extragalactic in nature (see Section 1.2.3).

Example 1.1

Estimate the ratio of atmospheric neutrinos to incoming cosmic rays of the same energy.

From Figure 1.10, there is a peak in the curve for atmospheric neutrinos around 10^8 eV so we will adopt this energy for a comparison with CRs. At higher and lower energies, the number of atmospheric neutrinos declines rapidly (note the logarithmic scale) so this energy should be sufficient for a rough estimate. From the figure, we read that there are $\approx 10^{-8}$ neutrinos $eV^{-1}\,cm^{-2}\,s^{-1}$. Conversion to 'per GeV' gives us 10 neutrinos $GeV^{-1}\,cm^{-2}\,s^{-1}$ and to 'per square metre' gives 10^5 neutrinos $GeV^{-1}\,m^{-2}\,s^{-1}$. Atmospheric neutrinos are isotropic since they can pass through the 'transparent' Earth. We can therefore divide by 4π (see Appendix B of the *online material*) to convert to a value per unit solid angle ('per steradian') to find $\approx 8\times 10^3$ neutrinos $GeV^{-1}\,m^{-2}\,s^{-1}\,sr^{-1}$. We can now compare this value to CRs as given in the same units in Figure 1.8. At the same energy, there are $\approx 10^4$ $GeV^{-1}\,m^{-2}\,s^{-1}\,sr^{-1}$ incoming CRs. Thus, the ratio is close to one as a rough estimate.

1.3.2 Astrophysics with Neutrinos

What astrophysically important information can be found by measuring neutrinos on (or rather beneath) the Earth's surface? We have alluded to a number previously, from gaining knowledge on conditions shortly after the Big Bang, to understanding high energy phenomena such as active galactic nuclei (AGNs). The fact that neutrinos are unperturbed by intergalactic and interstellar magnetic fields also means that neutrinos should travel in straight lines, so directional information is potentially possible, provided atmospheric neutrinos have been ruled out. This has recently been claimed by IceCube which traced a high energy neutrino to an origin in a *blazar* [36, 37]. A blazar is an active galactic nucleus (AGN) that is believed to have a jet pointing almost directly towards the observer, making it very bright. Such connections are still in their infancy, and this booming field promises to reveal many more secrets of the high energy universe in the future. Historically, though, two areas of neutrino astrophysics stand out: a solution to the Solar Neutrino Problem and neutrinos from the nearby supernova, SN 1987A. We'll look at each of these next.

1.3.2.1 Solar Neutrinos and the Neutrino Mass

The Sun converts hydrogen into helium according to the proton–proton (PP) chain of nuclear reactions depicted in Figure 5.5. There are a number of branches to these chains with different probabilities, but in brief, the reactions can be summarized as

$$4\mathrm{H} \rightarrow {}^4\mathrm{He} + 2e^+ + 2\nu_e + 26.73\ \mathrm{MeV}. \tag{1.24}$$

The net 26.73 MeV energy is what powers the Sun. Here, we can see that two electron neutrinos are also produced in the process. At Solar nuclear energies, only electron neutrinos would be produced, not tau or muon neutrinos. The expected number of neutrinos and their energies can be predicted from a standard solar model which includes knowledge of the Sun's core temperature, temperature gradient, and elemental abundances. One can then compute the number of neutrinos that should be expected in Earth-based detectors. These values are usually expressed in *Solar Neutrino Units* (SNU), where 1 SNU = 1 interaction per 10^{36} target atoms per second.

However, there was a discrepancy between the measured number of neutrinos and the number expected from the solar model. The measured result, as originally measured in the Homestake Solar Neutrino Detector, was 2.6 ± 0.3 SNU, whereas the expected value was 7.8 ± 1.5 SNU [15]. This factor-of-three discrepancy became known as the *Solar Neutrino Problem.* The problem persisted when different types of neutrino detectors weighed in with their measurements. Could the Solar models be wrong, or were the neutrino measurements wrong?

This impasse persisted for decades until the heavy water experiments of SNO (Figure 1.9 Left). With D_2O, three different types of interaction could be measured, one that was sensitive to ν_e, one that was sensitive to all three flavours (ν_e, ν_μ, and ν_τ) but with lower cross sections for the second and third types, and one that was sensitive to all three flavours equally. The results were definitive. When all types of neutrino were counted, the results *agreed* with the theoretical expectation from the solar model. The Solar neutrino problem was no more.

But what happened to the fact that the Sun only made electron neutrinos? The new measurements *still* had the number of ν_e less than expected from models, but the deficit is made up by the other two flavours. Where did the other types come from? The best explanation is that the neutrinos were *changing flavour* in flight. The idea is that each neutrino flavour is actually a different superposition of the three neutrino states. If each neutrino has a different mass, then each mass propagates at a different rate through space. Subsequently, the superposition (i.e. the flavour) changes with distance from the source as the phase changes. A particle may originate as an electron neutrino, but as the propagation distance increases, it will cycle through the other types, return to an electron neutrino again, and the cycle continues as the propagation continues. This is called neutrino *oscillation,* and it is not a new idea, having originally been proposed by B. Pontecorvo in 1967 [38].

An important conclusion is that, for neutrino oscillations to exist, neutrinos must actually have mass. This was not originally obvious nor was it accounted for in the classic standard model of particle physics. Since neutrinos travel so close to the speed of light, prior to the recent experiments, it was thought that a neutrino could be a massless particle like a photon, travelling at exactly c in a vacuum. The now 'massive' neutrino has required a revision of the standard model of elementary particles [38] and has opened up a whole new area of physics that is being aggressively explored. For the discovery of neutrino oscillations, implying that neutrinos have mass, the 2015 Nobel Prize in physics was shared between Takaaki Kajita (for atmospheric neutrino oscillations) and SNO director Arthur B. McDonald (for Solar neutrino oscillations).

Then, what is the mass of a neutrino? The actual mass is not known, only that the sum of the masses is likely less than $\approx$0.3 eV [24]! Compare this to the rest mass of an electron at

511 keV – six orders of magnitude higher. Indeed, these particles are the lightest elementary particles known to exist, aside from the massless photon.

1.3.2.2 Neutrinos from SN 1987A

Supernova 1987A (SN 1987A) located in the nearby galaxy, the Large Magellanic Cloud (LMC, about 50 kpc distant) was the closest naked-eye supernova to us since Kepler's supernova in 1604. Its discovery and subsequent scientific scrutiny have revolutionized our understanding of supernova explosions. We now have more than 30 years of continued and regular monitoring of this event with modern equipment and sophisticated techniques. Never before has any supernova been studied in such exquisite detail, as illustrated by the optical images of Figure 1.11.

How are neutrinos involved? A supernova involves an almost instantaneous core implosion (less than a second) and subsequent explosion of the outer layers. When the core collapses, it becomes so dense that electrons combine with protons in nuclei, forming neutrons and neutrinos: $\mathrm{e}^- + \mathrm{p}^+ \rightarrow \mathrm{n} + \nu_\mathrm{e}$. The neutrons then constitute a neutron star, and the neutrinos escape. As indicated in Section 1.3.1 (point 7), these neutrinos carry away an enormous amount of energy, $\approx 10^{53}$ ergs.

SN 1987A marked the first and only time that neutrinos from a specific supernova were actually detected. Several hours before the light from the SN was seen, events were recorded at three neutrino observatories, namely Kamiokande II in Japan, Irvine-Michigan-Brookhaven in the U. S., and Baksan in Russia. The neutrinos were formed and escaped during the first few seconds during and after the collapse, whereas the visible brightening was associated with the later explosion of the outer layers. Only a few dozen neutrinos were detected, but they were convincingly associated with SN 1987A. Thus, *neutrino astronomy* began and allowed a direct link between models of core-collapse supernovae and the expected neutrino output to be probed.

Aside from confirmation of supernova models, one other indirect but important measurement was obtained about the neutrinos themselves, although only electron neutrinos could be measured at that time. If the neutrino had a mass, then one would expect that higher energy neutrinos would arrive at the Earth first and the lower energy neutrinos later. This would give rise to an energy-dependent dispersion in the arrival time of the neutrinos. With a few assumptions, an upper mass limit of 16 eV was found for ν_e [5]. This result was rather gross compared to the newer upper limits given in Section 1.3.2.1 but it remains the only neutrino measurement related to a specific supernova.

PROBLEMS

1.1. (a) The caption to Figure 1.2 provides an expression for the cumulative number of impacts on the Earth per year, N, as a function of energy. That is, $N = \int_E^\infty f(E)dE$. Find the function $f(E)$.

(b) How many impacts per year, N_{1-5}, are expected from meteoroids with diameters from 1 to 5 m?

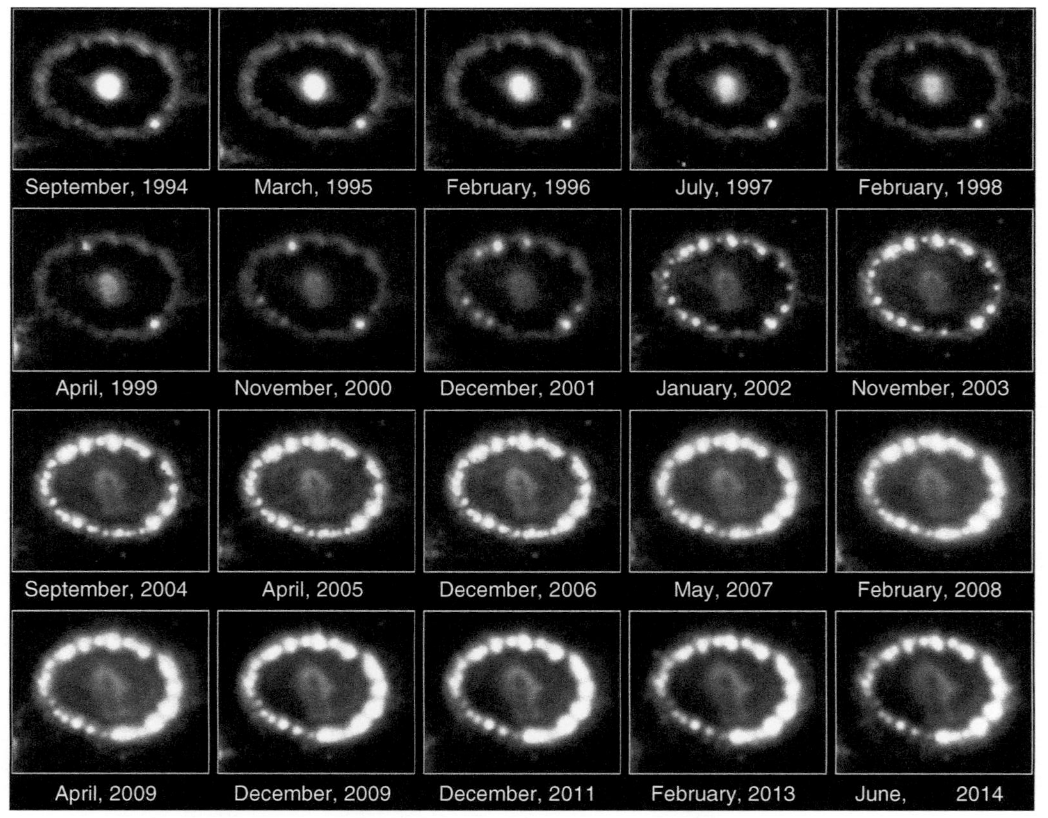

Figure 1.11 Optical Images of SN 1987A showing details of changes that have occurred over time. The main ring of material (1.3 light-years in diameter) was shed by the rotating progenitor star 20 000 years prior to the supernova explosion. Rapidly expanding ejecta from the SN itself can be seen at the centre. Shock waves from the SN have interacted with the ring, brightening it and making many hotspots visible [40]. Over later times, however, the ring has shown some fading and disruption, especially on the south-east (lower left) side. Source: Courtesy of Claes Fransson, from [22].

(c) How much energy is deposited from meteoroids in this range? Express your result in Joules.

1.2. Estimate the diameter of the meteoroid that formed the crater shown in Figure 1.1 (Right).

1.3. Work through the mathematical steps to show that Eq. (1.4) results from Eq. (1.2).

1.4. (a) Look up the half-life of ^{81}Kr and compute the decay constant (yr^{-1}).

(b) The Knyahinya stony meteorite was observed to fall in the year 1866, in what is today the Ukraine. For this meteorite, $^{83}\mathrm{Kr}/^{81}\mathrm{Kr} = 230.4$ and $^{83}P/^{81}P = 1.81$. Compute the cosmic ray exposure age (Myr) of Knyahinya.

1.5. When a gas expands, it does *work*, W (units of energy), against the surrounding medium

$$W = \int P dV \tag{1.25}$$

where P is the pressure of that medium and dV is a small change in volume.

(a) Find an expression for the blast radius, R_0, in terms of the meteor's energy per unit trail length, E_l, for the cylindrical geometry shown in the inset of Figure 1.5.

(b) Since a sound wave requires a medium within which to propagate, the maximum wavelength cannot be greater than the size of the medium. Find an expression for the maximum period, τ_m, of a sound wave in a shocked region of size, $x = 10$. Compare your result to Eq. (1.12) and comment.

(c) From your result of part (a) and Eq. (1.12), find a relation between E_l and $\tau_{x=10}$.

(d) Suppose we assume that the trail length, l, is $\approx R_0$ to order of magnitude. Repeat part (c) but for the total energy (work done) over the volume, E_W, rather than E_l. What is the power, β, such that $E_W \propto \tau_{x=10}{}^{\beta}$? Would you expect this power to be the same as implied by Eq. (1.15)? Why or why not?

1.6. Dominant infrasound periods from the shock wave of the Chelyabinsk fireball in Russia of 15 February2013, were measured at 15 stations with periods ranging from 17 to 82 s and an average of $\overline{\tau} = 37$ s [26].

(a) US government sensors measured an optical energy of $E_0 = 3.75 \times 10^{14}$ J. Determine the kinetic energy, E (in kT TNT).

(b) Find the expected average period from the calibration equation. Compare this to the measured average period and range.

1.7. For the Chelyabinsk fireball, refer to the text for its speed and best estimate kinetic energy and use an altitude of 30 km.

(a) From the definition of kinetic energy, find the mass of the fireball. Express your result in tonnes.

(b) Assuming that the rock is spherical with a typical density of 3 g cm^{-3}, find its diameter, d_m.

(c) The speed of sound in an ideal gas is

$$c_s = \sqrt{\frac{\gamma_a kT}{\mu m_H}} \tag{1.26}$$

where γ_a is the *adiabatic index* and equal to 1.4 for air, μ is the *mean molecular weight* defined in Section 5.4.2 and equal to 29 for dry air, and m_H is the mass of a hydrogen atom. Calculate c_s at the altitude of the fireball.

(d) What was the Mach number of the fireball?

(e) Calculate the blast radius and compare it to the given value.

(f) If (hypothetically) the fireball had not exploded in the air but instead hit the ground in a single piece, what size of crater might have been made?

1.8. Obtain information on the temperature of the Martian atmosphere and compute the following quantities for an incoming bolide on Mars:

(a) the escape velocity.

(b) the speed of sound ($\mu = 43, \gamma_a = 1.4$).

(c) the lowest Mach number of an incoming bolide.

(d) the maximum cone angle of the shock front. If the bolide had been on Venus instead (same μ, γ_a) would the maximum cone angle be larger or smaller?

1.9. Consider a typical CR particle at the *lower* energy limit and the *upper* energy limit of the measurements shown in Figure 1.8 and compute the following:

(a) i) the ratio of kinetic to total energy, *T/E*. Which energy dominates: the kinetic energy or the rest mass energy?
ii) At what kinetic energy is the rest mass energy only 10% of the total energy? Where does this fall in Figure 1.8?

(b) the Lorentz factor, γ.

(c) the ratio, v/c. Over what range of energies are CR particles relativistic?

(d) Estimate the ratio of the number of particles at low energy to the number of particles at high energy. Where do most CRs originate?

1.10. Estimate the minimum Lorentz factor and v/c of a) a neutrino from the CNB, and b) a cosmogenic neutrino.

1.11. Repeat Example 1.1 but for a cosmogenic neutrino. Assume isotropy.

1.12. (a) For Cherenkov radiation, find a simple expression for the angle, θ, shown in Figure 1.4, in terms of β and the index of refraction, n.

(b) If a charged particle is moving at $v \approx c$, find θ for (i) pure water, (ii) ice, and (iii) silicon.

JUST FOR FUN

1.13. Apparently micrometeorites can settle into the eavestroughs of houses over time (e.g. [29]). For particles in the range, 10^{-5} to 10^2 mg, estimate the mass of micrometeorites that settle into the eavestroughs of your house or apartment in one year. Assuming that these particles are chondrites, estimate the size of the particles at the lower and upper bounds of this mass range. If all the particles were low-mass particles, how many would you expect after one year? Repeat for the high mass particles. Do a little research to suggest how such particles could be distinguished from Earth-based particles.

1.14. Have you ever wondered whether a meteorite will land on a specific property in your lifetime? We'll adopt 5 m as the minimum size of a fireball in order for a fragment to hit the ground. Consider the ratio of areas of the property compared to the surface area of the Earth, and use the relation of Figure 1.2 (caption) to find the number of incoming meteors per unit time. We do not know how many pieces the incoming meteors will fragment into, but just assume that a single meteor will result in a single fragment on the property.

1.15. Human beings have evolved in an environment in which CRs are continually present. However, CRs do increase with altitude, so frequent airline flights increase the exposure rate. Visit the website of the Institut de Radioprotection et de Sûreté Nucléaire, https://www.sievert-system.org and compare the dosage received from a long flight over the pole to a short flight. How do your results compare with the maximum dosage guideline of the International Commission on Radiological Protection (ICRP) for flight personnel of 20 mSv (milli-Sieverts) per year?[25] How many long-haul flights could a pilot take per year before he or she reached this limit?

1.16. What is the height from which you would have to drop a 1 kg brick in order for its kinetic energy just before hitting the ground to be 10^{20} eV? It should now be clear why a relativistic CR particle can penetrate meteoroids – or astronauts' helmets (Section 1.1.1).

1.17. Estimate the number of Solar neutrinos that pass through your body every second.

[25] These guidelines may vary with country and type of worker. For example, European Union members use a guideline of 6 mSv/year for aircrew. The ICRP guidelines for pregnant radiation workers is 1 mSv/year (see https://www.cdc.gov/niosh/topics/aircrew/cosmicionizingradiation.html).

Chapter 2
Gravitational Radiation: A New Window

[The] cause of gravity is what I do not pretend to know · · ·That gravity should be innate inherent & essential to matter so [that] one body may act upon another at a distance through a vacuum [with]out the mediation of anything else · · · is to me so great an absurdity · · · Gravity must be caused by an agent acting constantly · · · but whether this agent be material or immaterial is a question I have left to [the] consideration of my readers.

– Isaac Newton, in letters to Richard Bentley [42]

2.1 CONCEPTS OF RELATIVITY

Isaac Newton admitted that he did not know the agent by which gravity acted between two bodies. In another famous quote, he said, 'I do not feign hypotheses'[1]. But he laid out the mathematical formulation of the effects of gravity, such as the universal law of gravitation (Table I.2) that has stood the test of time to this day. In most cases, Newtonian gravity is sufficient to describe phenomena on 'human scales'.

Albert Einstein went a step further, connecting matter to the 'vacuum' around it, and in so doing, found corrections to Newtonian gravity that are often small but can still be

[1] Hypotheses non-fingo.

Astrophysics: Decoding the Cosmos, Second Edition. Judith A. Irwin.

Companion website: www.wiley.com/go/irwin/astrophysics2e

important, from the Earth-based global positioning system (GPS), to supermassive black holes, to the broadscale structure of our Universe [14].

Einstein's *Special Theory of Relativity* is based on the postulate that the laws of physics are the same and the speed of light, c, is the same, regardless of the motion of the observer. The concept of *space–time* is a result of the intimate link between space and time through the constancy of c. His *General Theory of Relativity* (GR) later expanded these concepts to provide a geometrical description of space–time, its relation to gravitating masses, and its dynamic character, as we shall soon describe.

A geometrical description of space–time requires the use of coordinates. If two objects have the same spatial coordinates, x, y, z, then there is a spatial distance, 0, between them. However, if two objects are at these coordinates at different times, then they are 'separated' by that time, t, and will have no interaction with each other. This seems quite obvious, but the issue is much more acute for cosmology. Because of the finite speed of light and large distances in the Universe, we see the galaxies where they were in the past – not where they are now. And we see galaxies the way they appeared in the past – not the way they appear now. The more distant the galaxy, the farther in the past we see it because it has taken longer for the light to reach us. Thus, we are separated from these galaxies by *both* space and time. Our Universe is the ultimate time machine, allowing us to peer into the past, but not the future.

It is more useful, therefore, to describe a separation between *events in space–time*, rather than a separation between points in space or between instants in time. A mathematical measure of a distance between events in some geometry is called a *metric*. Given a distribution of matter and energy, certain symmetries, and large-scale curvature, a metric of the resulting space–time can be found. For example, a metric that describes a *homogeneous* (meaning the same at every location), *isotropic* (the same as viewed in any direction) and expanding universe is called the *Robertson–Walker metric*. This metric is important in the derivation of the Hubble Relation, as described in Appendix F of the *online material*. A metric that describes the curvature of space exterior to a spherically symmetric non-rotating mass, for example, is called the *Schwarzschild metric*. The Schwarzschild metric may be used in the development of the relevant equations for the *gravitational redshift* (Section 9.1.3) and *gravitational lenses* (Section 9.2.1)[2]. We do not present the various metrics, mathematically, in this text (see [9] for some examples). However, even without a mathematical characterization, it is possible to describe and visualize some of the concepts of relativity.

2.2 THE FABRIC OF SPACE–TIME

Central to GR is the concept of space–time as an 'entity'. The 'vacuum of space', even when devoid of all normal matter, radiation, magnetic or electric fields, still has properties and still has a residual vacuum energy. Indeed, particles can appear in and out of this vacuum for very brief periods of time. These are called *virtual particles,* and they cannot be

[2] If the mass were rotating, it would be the *Kerr metric* exterior to its boundary.

measured directly. Thus, space is not empty nothingness, but is an entity that can expand, curve, or distort. As the title to this section suggests, space–time is like a *fabric*, and a fabric is not rigid.

Considering only space for now, a helpful visualization is to represent three-dimensional space as a two-dimensional surface. This can be mentally challenging, but it does help in picturing how space can be interpreted geometrically. For example, if space is perfectly flat, like an unstretched sheet of paper, then we call it *Euclidean*. In Euclidean space, the interior angles of a triangle sum to exactly 180° and parallel lines remain parallel[3]. However, if space has positive curvature, that is, if space were convex like a section of the surface of a sphere, then interior angles sum to a value greater than 180° and lines that were initially parallel will eventually intersect. Space could also have negative curvature, like the interior of a saddle, where parallel lines will diverge and the interior angles of a triangle sum to a value less than 180°. One can then imagine any number of geometries for the space that we live in. Determining which one is correct, however, requires a knowledge of the *mass distribution* within it, because it is the distribution of mass that determines the curvature of space.

In turn, once a geometry is present, any mass or any light ray will move through that space, bending according to the local curvature of that space. The light still travels in 'straight lines' but through curved space. Figure 2.1 provides a picture. Here, a large central mass is strongly distorting the large-scale flat space around it, producing a gravitational *potential well*. Suppose two parallel light rays, such as the ones shown at left, are far from that mass. Then, they essentially travel through Euclidean space and they will remain parallel forever (unless they have a close encounter with another mass). If the sheet had some large-scale global curvature, then initially parallel beams would converge or diverge as described above.

This interplay between mass and the geometry of space can become complex, as one can imagine. Any mass travelling through flat or curved space will itself curve the space around it, but the degree of curvature depends on the magnitude of the mass and its kinetic energy. To avoid conceptual (and mathematical) 'feedback loops', it is helpful to consider a *test mass*. A test mass is a mass whose own distortions can be considered small in comparison to the broader global field through which it moves. A test mass need not be small; an entire galaxy could be considered a 'test mass' as it moves through the gravitational field of an even larger more massive structure.

On the largest cosmological scales, the best available evidence [31, 34, 38] suggests that space is flat, like the regions far from the central depression in Figure 2.1[4]. This does not mean, however, that space is static. A flat sheet can stretch outwards and still remain flat as it carries any embedded particles or photons along with it. Similarly, Euclidean space can expand and carry the galaxies with it, a property that is further described in Section 9.1.2.

[3] A rolled cylinder, however, would still be Euclidean because parallel lines around the circumference remain parallel.

[4] Time, however, can be thought of as curved as the Universe evolves.

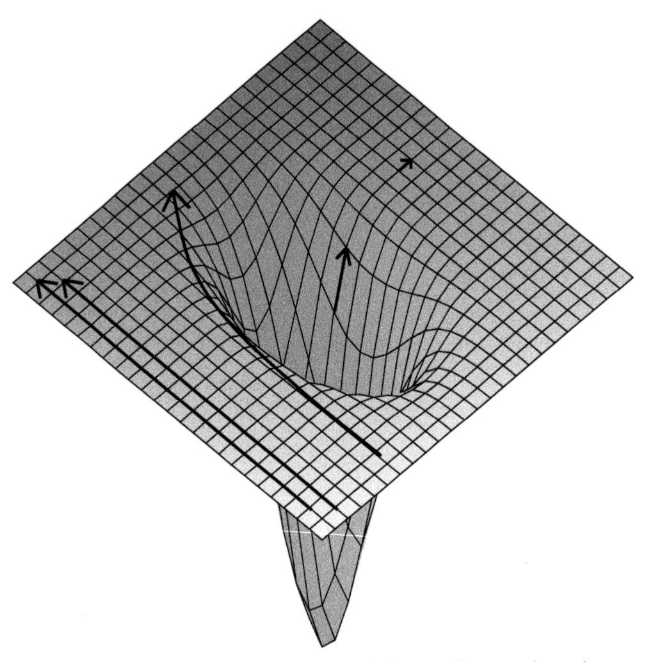

Figure 2.1 Conceptualization of three-dimensional space as a two-dimensional surface that can be curved by a gravitating mass. The two parallel lines at left represent parallel light beams that remain parallel in flat space. This is what our Universe is believed to be like on large scales. The curve represents light that travels near the mass and is deflected. This is what produces gravitational lensing (Section 9.2). The two short arrows represent the energy of a light ray that leaves the surface of the massive object in a radial direction, helping to illustrate the gravitational redshift (Section 9.1.3).

2.3 CURVED SPACE–TIME NEAR A MASS

Close to any mass, space is not flat but curved and the curvature is greater when the mass is greater. This is analogous to being close to our heavy ball at the centre of Figure 2.1. Any photon or any test mass travelling near the mass will experience a change in direction as it moves through this curvature. Since the effects close to the heavy mass are relatively strong, this is the regime in which GR has historically triumphed over its Newtonian counterpart. In the following paragraphs, we will review three important historical verifications of GR that led to it being the currently established theory of gravity. They all involve detection of light (electromagnetic radiation) which really belongs to Part II–VI of this text, but these steps are important en-route to our newly opened window on the GR universe.

The first involves light bending through curved space. When a photon travels from a background object near a mass that curves space as just described, it will appear to an observer at a different location in the sky than it otherwise would, as illustrated in

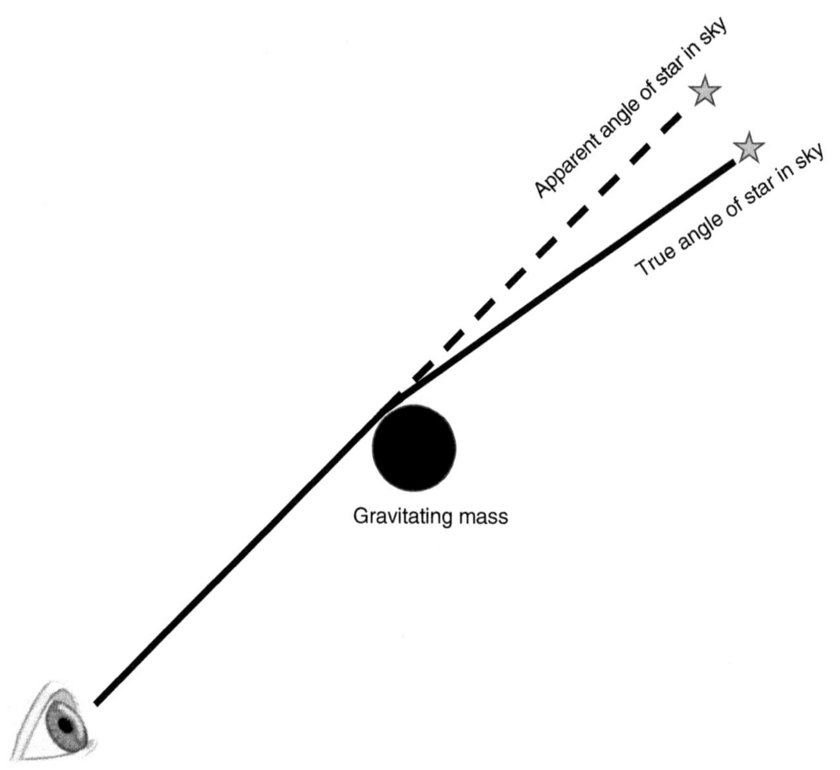

Figure 2.2 When a light ray travels by a mass, it is bent in the curved space–time around the mass (similar to the curve of Figure 2.1). This *gravitational refraction* makes the star appear to be at a different location in the sky to an observer on the Earth (sizes and angles exaggerated). The mass is called a *gravitational lens*.

Figure 2.2. This is called *gravitational bending* or *gravitational refraction*, and the mass is called a *gravitational lens*. We will discuss this quantitatively and in more detail in Section 9.2.1.

General relativity exploded into the world's consciousness in 1919 after Sir Arthur Eddington and others made observations of the positions of background stars during the total solar eclipse of 29 May. In that case, the gravitational lens was the Sun and the eclipse darkened the sky sufficiently that background stars in the vicinity could be observed and their shifted positions measured. Since the normal positions of the stars were known from night-time observations when there was no eclipse, the positional offsets could be measured. The results were in agreement with GR, but not with Newton. The eclipse expedition is a story in itself (e.g. [18]) but a New York Times headline on 10 November 1919 sums it up: 'Lights all askew in the Heavens; Men of Science More or Less Agog over the Results of Eclipse Observations'.

The second demonstration involves time. A profound consequence of GR is *time dilation*. A photon travelling near a mass is travelling a greater distance through curved space in comparison with a photon that is farther away (Figure 2.1). Since a local 'freely-falling' observer near the mass must measure the same speed of light, *c*, in comparison with an observer farther away, it follows that a unit of time near the mass is greater than a unit of time farther away. In a sense, time and space stretch together. If the observer near the mass sends out a signal at regular one second intervals according to his clock, a distant observer will measure the arriving signals as being greater than one second apart. Each observer's clocks are advancing 'normally' in their own frame of reference, but not when compared to each other. Similarly, the observer near the mass will age more slowly in comparison to the observer farther away. A consequence of time dilation[5] is that a photon that leaves a mass will experience a *gravitational redshift* (Section 9.1.3). The gravitational redshift was first measured on the Earth in 1959 [33].

The third important historical achievement involves the advance of the *perihelion* (position of closest approach to the Sun) of Mercury. Suppose that a test mass is in orbit about the central mass of Figure 2.1. We have known since the time of Kepler that planetary orbits are not perfectly circular, but rather ellipses with the attracting mass at one focus. In the absence of any other forces, an elliptical Newtonian orbit, which is described by an inverse square law force (Table I.2) would close. For example, the planet Mercury's *perihelion* would remain in the same position orbit after orbit[6], were it not for the gravitational influence of the other planets. The influence of other planets amounts to a perihelion advance of about 530 arcseconds per century[7]. However, once those planetary influences are taken into account, a discrepancy of 43 arcseconds per century still remained. This is a small, but measurable amount and had been known, but not understood, since the nineteenth century.

The discrepancy, however, can be beautifully accounted for by the general relativistic view of such an orbit. Mathematically, there is an extra term in the energy equation of the orbit, as described in Refs. [13] or [14]. Conceptually, the shape of the potential well shown in Figure 2.1 is simply not quite the same as would be produced by a $1/r^2$ Newtonian force, resulting in a steady regular advance of Mercury's perihelion. When Einstein realized that his theory could account for this discrepancy, he exclaimed in a letter to Paul Ehrenfest, 'Imagine my delight at realizing ... that the equations yield the correct perihelion motion of Mercury. I was beside myself with joy and excitement for days' [11].

In our modern day and age, relativity is well established and many tests have been performed that extend and expand upon those described above. Other stars in binary systems reveal an advance in *periastron* (closest approach to a star) similar to, but stronger than that seen for Mercury. Many gravitational lenses have been observed, showing patterns that provide information on the lens and the light source. And our GPS systems would not work without properly correcting for time dilation. A recent triumph is the imaging of the region around the supermassive black hole in the galaxy, M 87 – but

[5] Time dilation also occurs when an object moves, according to Special Relativity, but we restrict ourselves to the General Relativistic time dilation here.
[6] We ignore tidal effects.
[7] One might think that this number would depend on where the planets are in their orbits. However, since Mercury's perihelion advance is so slow compared to the orbital periods of the most influential planets, the influencing planetary masses can be represented as rings of mass around their orbits.

since this involves EM radiation, we return to it in Chapter 9. In the next sections, we will examine another profound consequence of GR – gravitational radiation.

2.4 GRAVITATIONAL WAVES

Now that we have a visualization of space–time as a two-dimensional flexible surface, it is not difficult to imagine waves in such a fabric. We will use the terms, 'gravitational waves' (GWs) and 'gravitational radiation' interchangeably although they have slightly different connotations since the former is a visualization of the latter[8]. It is important not to think of GWs as waves travelling through space the way that electromagnetic waves would travel through space. Rather, GWs are perturbations *of* space (but not time [36]). Space itself is stretching and squeezing. And this means that *every mass within it is also stretching and squeezing*, from the smallest particles, to human beings, to planets, stars and galaxies. It's just that the perturbations are usually too small to measure.

Just as accelerating charges generate electromagnetic radiation (e.g. the examples in Chapter 10), accelerating masses can generate gravitational radiation. Again, we see this close interplay between mass and space, with masses determining the curvature of space but also moving through it. Like electromagnetic (EM) waves, gravitational waves are transverse (i.e. perturbations are perpendicular to the direction of motion of the GW), and like EM waves, they have two polarizations although those polarizations are not orthogonal to each other (see Figure 2.3). GWs propagate at the speed of light as they perturb space, whereas EM waves travel at the speed of light *through* the vacuum of space.

Unlike EM waves that can be generated from an oscillating *dipole moment* (Table I.2), gravitational radiation is *quadrupolar* radiation, rather than dipolar[9]. Thus, rather than stretching and compressing in an oscillating fashion like a mass on a spring, space stretches and compresses in one direction, x, and then the other direction, y, in an x–y plane that is perpendicular to the direction of propagation, z. Figure 2.3 illustrates how a set of test masses, initially in a circle, would stretch and compress into an ellipse as a result of a GW, the top and middle rows showing the two different polarizations. The actual detected signal could be some combination of these waveforms, depending on the orientation of the source with respect to the observer. The GW behaves rather like a well-known child's paper toy as the bottom row of Figure 2.3 shows.

What kinds of objects produce GWs? Any accelerating non-spherically symmetric mass can do so, but as indicated above, the effects are usually too small to be measurable. Most favourable are binary star systems consisting of the collapsed remnants of stars, such as *neutron stars, pulsars*, or *black holes* (Section 5.3.3). These are objects within which substantial mass is contained in a small volume. Therefore, another mass can approach it very closely within its surrounding region of relatively strongly warped space (Figure 2.1).

[8] To make things a little more confusing, there are also 'gravity waves' in astrophysics, but these refer to material waves (cf. Section 1.1.2) and have nothing to do with the gravitational waves discussed here.

[9] In GR, a dipolar term relates to the total angular momentum which is a conserved quantity and therefore does not produce radiation.

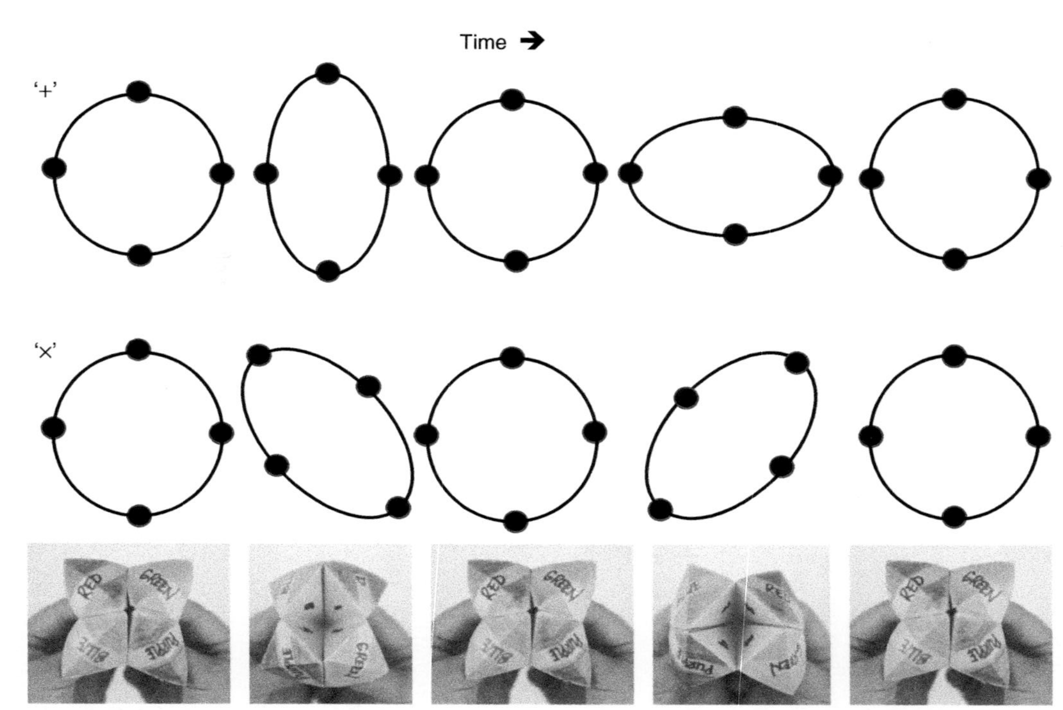

Figure 2.3 Dots represent masses that are being squeezed and stretched due to a passing GW. Time goes from left to right and one cycle of a wave is shown. The top row shows one polarization ('+'), and the centre row shows the other polarization ('x'). The third row shows a child's paper 'fortune teller' toy that follows the '+' polarization sequence. Source: From wikiHow (https://www.wikihow.com/Fold-a-Fortune-Teller) used under the Creative Commons Licence 3.0.

Collapsing, exploding, colliding, or rotating objects are also potential sources, provided they are not perfectly symmetric. So are violent processes in the early Universe. The bottom line is that, collectively, these objects and events are *common* in our Universe (e.g. [30]), and therefore, GWs should be as well. As we will see, the *energies* of GWs can be significant, but the *amplitudes* are typically very small.

In the next section, we will focus on binary orbits since, so far, these are the systems for which we have observational evidence for GWs.

2.5 GWs FROM BINARY ORBITS

In a binary system, each object orbits a common centre of mass, and this is true for any two masses, whether they be planets, stars, or stellar remnants such as white dwarfs[10], neutron stars, or black holes. Even the Sun shifts in its position around the centre of mass of our Solar System, depending on the positions of the various planets (Problem 2.2).

Figure 2.4 shows a visualization of a numerical simulation of a GW from a binary orbit. Since it is the continuously accelerating masses in the binary orbit that produce the continuous GW, the *frequency* of a GW results from the orbital motion of the source.

$$f_{GW} = 2f_{orbit} \tag{2.1}$$

The factor of 2 is because of the nature of quadrupolar radiation. A GW will go through two complete cycles (a single cycle is shown in Figure 2.3) for every orbit. For example, suppose an orbit has a period of one hour. Then, $f_{GW} = 2\,(1/\mathcal{T}) = 5.6 \times 10^{-4}$ Hz and $\lambda_{GW} = c/f_{GW} = 5.4 \times 10^{13}$ cm ≈ 3.6 AU! We are beginning to see some of the observational challenges of detecting GWs directly. Even though GWs travel at the speed of light, a GW that sweeps from one side of the Earth to the other would show a fractional wavelength change of only $d_{\oplus}/\lambda_{GW} = 2.4 \times 10^{-5}$. There would virtually be no variation in the amplitude of such a wave from one side of the Earth to the other. A wait of 15 minutes would be required before the wave went from maximum to minimum.

If the orbit is strongly elliptical, then there can also be a frequency that is equal to the lower orbital frequency (plus *harmonics*) because the GW emission will be strongest when the two objects are closest together at periastron[11]. This orbital frequency, if present and detectable, would produce a modulation of the quadrupolar signal and possibly even represent the strongest signal if the orbital eccentricity is high. As we will note in Section 2.6, however, such modulation is not generally observed, arguing for circular orbits during the time period when binaries are about to merge and GWs are directly observed.

Gravitational radiation, as the word 'radiation' implies, involves a loss of energy from the source. Usually, such energy loss is expressed as luminosity, L, or power, P, (erg s^{-1}) (see Section 3.1), but since we do not want to confuse it with EM luminosity, the

[10] A white dwarf is a remnant of a low mass star like our Sun. It is typically $\approx 1 M_\odot$ but with a radius $\approx R_\oplus$.

[11] I will continue to use *periastron* for the point of closest approach even though the two components could be stellar remnants rather than stars.

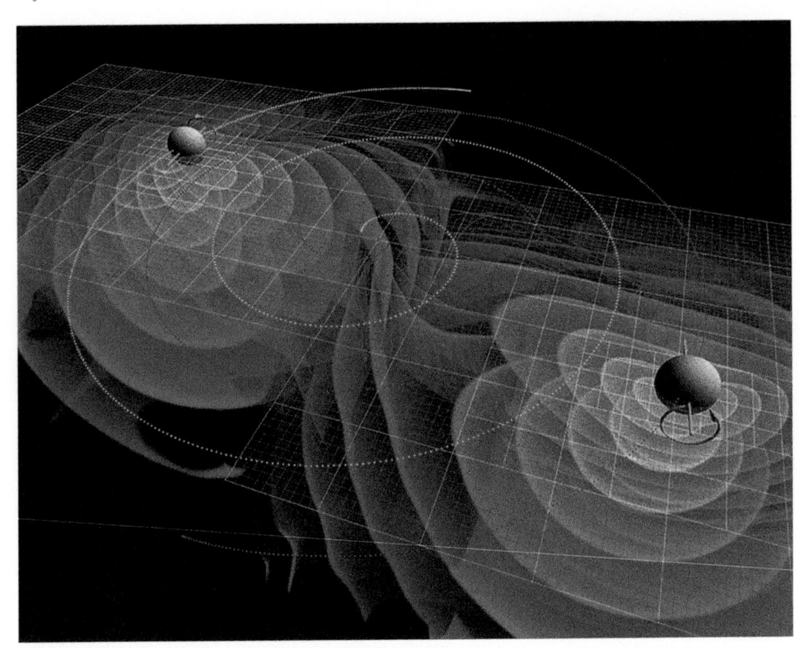

Figure 2.4 Numerical simulations of the gravitational waves emitted by the inspiral and merger of two black holes. The coloured contours around each black hole represent the amplitude of the gravitational radiation; the cyan and magenta curves represent the orbits of the black holes, and the green arrows represent their spins. Credit: NASA/Ames Research Centre/C. Henze. Source: https://physics.aps.org/articles/v9/17.

GW 'luminosity' will be represented here as $\dot{E}$, where the 'dot' represents a change of a quantity with time. The equation describing the energy loss has been known for many decades [28], long before any direct or indirect evidence for GWs existed. For a binary system containing 'point' masses (that is, small compared to the orbital size) and when the observer is in the *far field* (that is, far compared to the separation of the masses), then averaged over an orbit, the luminosity (cgs units) is

$$\dot{E}_{GW} = \frac{32G^4 m_1{}^2 m_2{}^2 (m_1 + m_2)}{5c^5\, a^5} f(e) = \frac{32G^4 \mu^2 M^3}{5c^5\, a^5} f(e) \tag{2.2}$$

where m_1 and m_2 are the individual component masses, the total mass is $M = m_1 + m_2$, and μ is the *reduced mass*,

$$\mu \equiv \frac{m_1 m_2}{M} \tag{2.3}$$

a is the semi-major axis of the elliptical orbit, G and c are the universal gravitational constant and the speed of light in a vacuum, respectively, and the function, $f(e)$, is a unitless function of the eccentricity, e (see A.5 of the *online material* for the properties of an ellipse),

$$f(e) = \frac{1 + \frac{73}{24}e^2 + \frac{37}{96}e^4}{(1 - e^2)^{7/2}} \tag{2.4}$$

For circular orbits, $e = 0$ and $f(e) = 1$. Notice how strong a dependence there is on the semi-major axis of the orbit ($\dot{E}_{GW} \propto 1/a^5$) in Eq. (2.2). Closer binaries will emit much stronger gravitational radiation than loosely bound orbits. Also, if the instantaneous value of $\dot{E}$ is considered, rather than an average over the orbit, it will be highest at periastron.

If we wish to rewrite Eq. (2.2) in more common astronomical notation and eliminate a in favour of the period, $\mathcal{T}$ (which is the directly measurable quantity) we can use Kepler's 3rd law[12] (Eq. 9.5) to obtain

$$\dot{E} = 3.0 \times 10^{33} \frac{\left[\frac{\mu}{M_\odot}\right]^2 \left[\frac{M}{M_\odot}\right]^{4/3}}{\left[\frac{\mathcal{T}}{\mathrm{hr}}\right]^{10/3}} f(e) \quad \mathrm{erg\ s^{-1}} \tag{2.5}$$

This equation helps us to see the luminosity of a GW at a glance. Suppose we have two, 2 $M_\odot$ objects with a one-hour circular orbital period, then $\mu = 1\ M_\odot$ and $\dot{E} \approx 2 \times 10^{34}$ erg s^{-1}. This is about 10 times the EM radiation from the Sun. It is straight-forward to see that shorter periods (closer binaries) will increase the luminosity of a GW significantly, as will increasing their masses. We will see in the next sections that there are objects that would produce much higher values of $\dot{E}$. This suggests that the GW sky is very 'bright'! If there were some way of 'tuning' our eyes to be sensitive to the luminosity of gravitational radiation, these signals could dominate the night sky, depending on how many such systems are present.

The energy loss from a binary system due to the generation of a GW has a significant effect on the binary. The only way that a binary can manifest this energy loss is through the *decay* of the orbit. As the semi-major axis becomes smaller, again by Kepler's 3rd law, the period decreases and orbital speeds increase (Problem 2.3). This means that the orbital frequency increases and so does f_{GW}. Eqs.(2.2) or (2.5) show that the rate of energy loss by gravitational radiation *increases* strongly when a and $\mathcal{T}$ decrease. The net result is that the two bodies *inspiral* towards each other, accelerating faster and faster until they merge. Figure 2.5 illustrates this behaviour.

When the objects are close to merging, relativistic effects become strong and must be carefully included. For example, the speeds of the bodies approach the speed of light and GWs are beamed into a cone in the forward direction (i.e. in the directions of the moving masses). This is similar to the beaming of EM synchrotron radiation to be discussed in Section 10.5. There is so much energy in this forward beam that it exerts a back reaction on the mass which slows the acceleration. For elliptical orbits, the instantaneous GW emission is highest at periastron so this back reaction has the effect of turning initially elliptical orbits into circles. Thus, circular orbits are expected during an advanced inspiral.

[12] Newtonian gravity is sufficient for the orbital parameters at this point. See also Section 2.6.1.

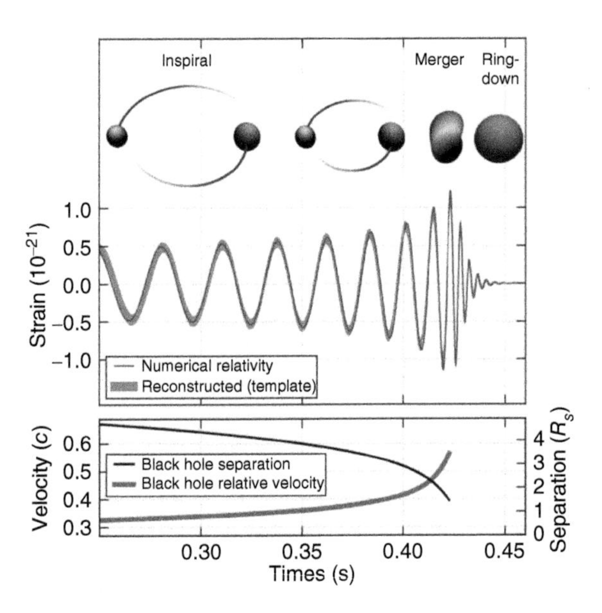

Figure 2.5 Example showing how the frequency and amplitude of GW150914 (the first GW ever detected) both increase with time as the two components (in this case, two black holes) inspiral towards each other. The top figure illustrates the two objects (top panel) and plots the strain as a function of time (bottom panel). The bottom figure shows the separation (labelled on the right) and the relative velocity (labelled on the left), where the former is in units of Schwarzschild radii (Section 9.1.3) and the latter in units of the speed of light. Source: Abbott et al. [1]. Licensed under CC BY 3.0.

The rapid increase in frequency towards the end of the inspiral (Figure 2.5, [2]) is referred to as a *chirp*, similar to a chirp heard in electronic devices when the frequency increases suddenly, and is loosely based on the chirping sound made by some birds. Once merger has occurred, there is a *ring-down* of the signal, consistent with a single final object. The precise shape of the waveform during an advanced chirp contains much information about the physics of the system. For example, the internal structure of neutron stars, currently not well known, can be examined for the first time in entirely new ways. Radii and densities can be probed and different waveforms result from rotating or non-rotating components so neutron star spins can be found. The ringing of a newly formed black hole has now been detected [17]. It may even be possible to test whether small departures from GR exist (so far there are none). Who knows what new surprises may be in store as this new window on the Universe is exploited? But first, let us examine what information can be extracted as the orbit evolves.

2.6 EVOLUTION OF A BINARY ORBIT

2.6.1 The Inspiral

Although GR is required to produce a gravitational wave, until the last few stable orbits at about the maximum amplitude of the GW signal, Newtonian mechanics can still be used to describe the *orbital motion* [20]. Departures occur when the orbital velocity starts to approach the speed of light, $v/c \approx 1$. Averaged over an orbit, the rate of change in semi-major axis and ellipticity are, respectively [29],

$$\dot{a} = -\frac{64G^3\mu M^2}{5c^5a^3}f(e) \tag{2.6}$$

$$\dot{e} = -\frac{304G^3 e\mu M^2}{15c^5a^4}g(e) \tag{2.7}$$

where

$$g(e) = \frac{1 + \frac{121}{304}e^2}{(1-e^2)^{5/2}} \tag{2.8}$$

Equation (2.6) shows that, as the binary becomes more compact (a decreases), then the *rate* at which the binary becomes more compact, *increases.* Equation (2.7) also shows that the rate at which the orbit circularizes with time is a strong function of a, so the orbit circularizes quickly as a decreases. The above equations lead to a relation between the ellipticity and semi-major axis,

$$\frac{da}{de} = \frac{12a}{19e}\left\{\frac{1+(73/24)e^2+(37/96)e^4}{(1-e^2)[1+(121/304)e^2]}\right\} \tag{2.9}$$

From an integration of Eqs. (2.6) and (2.7), we can find out how long it would take for a to reach zero from some initial semi-major axis of a_0, i.e. how long does it take for the two objects to merge? This is

$$\tau_m = \tau_{circ}k(e_0) = \frac{5c^5 {a_0}^4}{256G^3\mu M^2}k(e_0) \tag{2.10}$$

where τ_{circ} is the time-to-merge for a circular orbit, and $k(e)$ is a function of the initial eccentricity, e_0, whose approximate value is given by [6]

$$k(e_0) \approx (1 - {e_0}^2)^{3.689-0.243e_0-0.058{e_0}^2} \tag{2.11}$$

The function, $k(e_0)$, is less than 1, implying that binaries in elliptical orbits merge on a *shorter* timescale than binaries in circular orbits for the same initial semi-major axis, a_0. This is because the components spend more time closer together over an elliptical rather than a circular orbit. Clearly, longer merge times also correspond to larger separations.

In Figure 2.6, we plot more accurate relations between a, e, and time [15]. In the top figure, we give the ratio of initial to final semi-major axis, a_0/a_f, for a variety of initial and final ellipticity. In the bottom figure, we plot the ratio of time to the merge time for a circular orbit, t/τ_{circ}, where τ_{circ} is given in Eq. (2.10). The equations for these plots can be found in Section 2.2 of the *online material.* Notice that larger initial ellipticities generally give shorter timescales, as noted above. For example, if an orbit begins at $e_0 = 0.7$, and ends

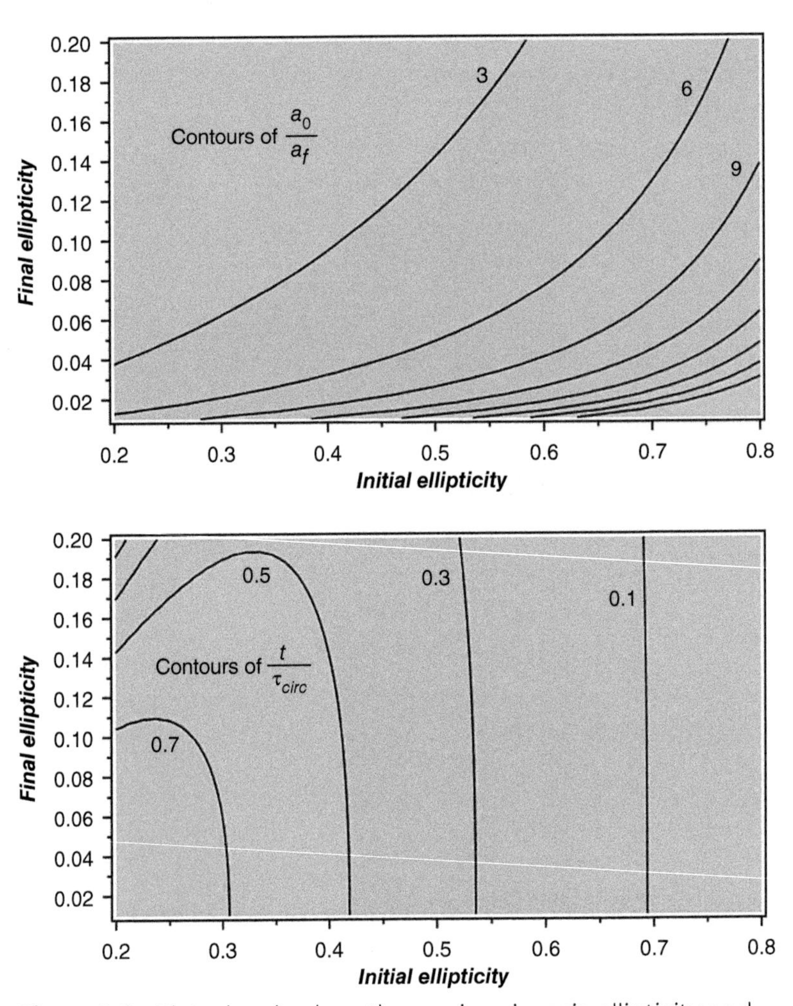

Figure 2.6 Plots showing how the semi-major axis, ellipticity, and time vary for a binary system as the orbit decays due to gravitational radiation. The *top* plot shows the ratio of initial to final semi-major axis, a_0/a_f, as ellipticity changes from e_0 to e_f. Curves are separated by 3 with the first three curves labelled. The *bottom* plot shows the ratio of time to the merge time for a circular orbit, t/τ_{circ}, as ellipticity changes. Curves are separated by 0.2.

at $e_f = 0.05$, then its semi-major axis is reduced to 8.4% of a_0 and the time it would take to reach e_f is $\approx 0.1\,\tau_{circ}$. Times can also be short if the initial and final ellipticities are close together as we can see at the top left of the bottom plot.

It is also useful to consider how rapidly the period is changing with time. The period in Kepler's 3rd law (Eq. 9.5) can be differentiated to find the rate at which the orbital period changes (Problem 2.4)

$$\dot{\mathcal{T}} = -\frac{192}{5}\frac{\pi}{c^5}\left(\frac{2\pi G\mathcal{M}_c}{\mathcal{T}}\right)^{5/3} f(e) \tag{2.12}$$

This equation is valid as long as the ellipticity, and therefore $f(e)$, is constant, or at least only slowly varying (Example 2.1). The quantity, $\mathcal{M}_c$, is called the *chirp mass*,

$$\mathcal{M}_c \equiv \frac{(m_1 m_2)^{3/5}}{M^{1/5}} \tag{2.13}$$

Both $\mathcal{T}$ and $\dot{\mathcal{T}}$ are measurable quantities, so if it could be shown that an orbit decayed and that it decayed according to the relation, $\dot{\mathcal{T}} \propto \mathcal{T}^{-5/3}$, then this would demonstrate the existence of GWs. We will see such a demonstration in Section 2.7.

2.6.2 The 'Death-Spiral'

As indicated earlier, when the binary comes closer together, the orbits *circularize*, and the frequency changes very rapidly, or 'chirps'. It is more convenient now to consider the change in frequency rather than period. Again, using Kepler's 3rd law and differentiating f_{GW} with respect to time, we find the change in frequency,

$$\dot{f}_{GW} = \left(\frac{96\pi^{8/3}G^{5/3}}{5c^5}\right) f_{GW}{}^{11/3}\mathcal{M}_c{}^{5/3} \tag{2.14}$$

where we have taken $f_{GW} = 2f_{orbit} = 2/\mathcal{T}$. Therefore, as the frequency increases (Figure 2.5), so does the rate at which the frequency increases. The fact that the orbits circularize is also apparent from observations since, if the orbits were elliptical, we would expect to see a modulation of the chirp signal at $f_{GW} = f_{orbit}$ and this is not (so far) observed [20].

Equation (2.14) can be rearranged to solve for the chirp mass

$$\mathcal{M}_c = \frac{c^3}{G}\left(\frac{5}{96}\pi^{-8/3} f_{GW}{}^{-11/3}\dot{f}_{GW}\right)^{3/5} \tag{2.15}$$

Equation (2.15) is an incredibly useful expression because the right hand side contains only values that can be *measured* as well as constants. The left hand side contains information about the masses of the two components. The frequency and frequency derivative change rapidly during the chirp, but their product must remain constant since the chirp mass is a constant.

As merging and ring-down are approached, relativistic effects must be taken into account for the orbital motion and higher order terms are needed to modify the Newtonian equations. If such terms can be measured, then the individual masses of the two components could also be found [10]. However, even without such information, it is still possible to place useful limits on the individual masses. For $m_1 \geq m_2$, it can be shown (Problem 2.6) that

$$m_1 \geq 2^{1/5}\mathcal{M}_c \quad \text{and} \quad m_2 \leq 2^{1/5}\mathcal{M}_c. \tag{2.16}$$

For example, suppose a chirp mass of 5 $M_\odot$ is determined from Eq. (2.15). Then, $m_1 \geq 5.7\ M_\odot$, implying that at least m_1 must be a black hole because this value exceeds 3 $M_\odot$ which is the approximate upper limit for the mass of a neutron star. Other constraints can be placed on the components by realizing that they must be separated by at least the sizes of the objects in order for there to be orbital motion at all.

The remaining and arguably most important relation involves the amplitude of the wave. Recall that all objects must respond to a passing GW by stretching and compressing as shown in Figure 2.3. In physics, we already have an expression that describes the stretching of a material and this is called the *strain*. Strain is the fractional change in a linear dimension, or for a GW [35],

$$h \equiv \frac{2\Delta l}{l} \tag{2.17}$$

where Δl is a change in the distance, l, between two particles[13]. We need to keep in mind that the stretching and compressing of an object by a GW means that the object itself is stationary within a stretching and compressing space. This is unlike the classical strain in which an object streches or compresses within space. The GW amplitude is not the same in every direction, as illustrated by Figure 2.4. Consequently, the measured strain would depend on the orientation of the detector with respect to the source geometry and it would also depend on time as the wave passes by. The amplitude of the wave for a circular binary orbit is (e.g. [27])

$$h_0 = \frac{4G^{5/3}}{c^4}(\pi f_{GW})^{2/3}\mathcal{M}_c^{\ 5/3}\frac{1}{D} \tag{2.18}$$

Here, D is the distance to the source[14]. So closer and more massive sources have a greater amplitude. Higher frequency binaries are also favoured and these are the ones that are closer together. Again, as Figure 2.5 shows, the amplitude increases as f_{GW} increases during the chirp. When converted to convenient units, Eq. (2.18) becomes

$$h_0 = 1.2 \times 10^{-19}\left[\frac{f_{GW}}{\mathrm{Hz}}\right]^{2/3}\left[\frac{\mathcal{M}_c}{M_\odot}\right]^{5/3}\left[\frac{D}{\mathrm{kpc}}\right]^{-1} \tag{2.19}$$

We again see the challenge in directly measuring a GW. A binary that is 1 kpc distant, has a chirp mass of 1 $M_\odot$, and a frequency of 1 Hz, would produce a fractional compression or stretching of space of only 10^{-19}. A source at a distance of 100 kpc (all else being equal) would reduce the constant to a stringent 10^{-21}! This seems practically impossible, yet it has been achieved recently, as we will see in Section 2.8. Thus, h_0 is a potentially measurable quantity.

An examination of Eqs. (2.15) and (2.18) leads us to a very important realization. Since h_0 is measurable and $\mathcal{M}_c$ is also determined from measured quantities, the only unknown is the distance, D. Thus, a direct measurement of GWs provides a new method of obtaining distances and is completely independent of traditional methods that rely on EM radiation. Objects whose EM luminosities are believed to be known are referred to as *standard*

[13] This simple equation applies, provided $l \ll \lambda$.
[14] This is the *luminosity distance*, which includes a redshift dependence as given in Eq. F.15 of the *online material*.

candles. By comparison, sources of GWs are now being referred to as *standard sirens*, consistent with the 'chirp' as an audible signal (Problem 2.13).

In the next two sections, we will see how these relations are useful in disentangling GW physics.

2.7 INDIRECT PROOF OF THE EXISTENCE OF GRAVITATIONAL WAVES

The first convincing verification of the existence of GWs came from the detection of light (Parts II–VI) coming from a pulsar. We will refer to it as the *Hulse and Taylor pulsar* because it was Hulse and Taylor who discovered this system [16]. The orbital motion of the observed pulsar was tracked over time as it, along with its unseen companion, moved around the centre of mass of the system. The orbit was clearly decaying. Moreover, the rate of decay beautifully followed the predicted relation of Eq. (2.12), showing unambiguously that the decay was associated with gravitational radiation [39, 41]. With additional information, it became clear that *both* components were neutron stars. Hulse and Taylor won the Nobel Prize in Physics in 1993 for this discovery.

Today, we know of many more binary systems in which both components are neutron stars and we present a subset of them in Table 2.1. The range of masses in this list is quite narrow, consistent with what is normally expected for the formation of a neutron star after a supernova explosion (Section 5.3.3). From this list, it can be seen that most systems have orbits that are close to circular ($e < 0.3$) but some are more eccentric.

Table 2.1 Binary neutron star systems.

System	$\mathcal{T}^a$(days)	$a_1 \sin i^b$ (*lt-sec*)	$\sin i^c$	e^d	$m_1{}^e$ ($M_\odot$)	$m_2{}^e$ ($M_\odot$)	D^f (kpc)	References[g]
B1534 + 12	0.421	3.7295	0.977	0.274	1.333	1.3454	1.05	
B1913 + 16[h]	0.323	2.3418	0.72	0.617	1.438	1.390	5.25	[16]
J1906 + 0746	0.166	1.4200	0.69	0.085	1.322	1.291	7.40	[40]
J0453 + 1559	4.07	14.467	0.97	0.113	1.560	1.174	0.52	[25]
J1756 − 2251	0.320	2.7565	0.93	0.181	1.341	1.230	0.73	
J1757 − 1854	0.183	2.2378	1.00	0.606	1.338	1.395	19.58	[8]
J1807 − 2500B	9.957	28.920	1.00	0.747	1.366	1.206	2.79	[22]

[a]Period.
[b]Semi-major axis of the orbit of m_1 times the sin of the inclination, in units of light-seconds.
[c]Sine of the orbital inclination.
[d]Orbital eccentricity.
[e]Masses of neutron stars 1 and 2.
[f]Distance to the binary.
[g]Reference, if the references below gave incomplete data.
[h]Hulse and Taylor pulsar [16].
Sources: Yang et al. [43]; Manchester et al. [23]; www.atnf.csiro.au/research/pulsar/psrcat/ and Hulse and Taylor [16].

An illustration as to what this kind of orbital decay can tell us about the binary system is presented in Example 2.1.

Example 2.1

For the Hulse & Taylor pulsar, find the following quantities and comment: $a, \dot{E}, \dot{a}, \dot{e}, \tau_m, \overline{v}_1, \mathcal{M}_c, f_{GW}, \lambda_{GW}, h_0$ *and* $\dot{\mathcal{T}}$.

a: From Table 2.1, $a_1 = (a_1 \sin i)/\sin i = 2.3418/0.72 = 3.25$ lt-sec. From the properties of binary elliptical orbits (Section 2.1 of the *online material*), $a_2 = (m_1/m_2)a_1 = 3.37$ lt-sec, so $a = a_1 + a_2 = 6.62$ lt-sec $= 1.99 \times 10^{11}$ cm. This is 1.4 × the diameter of the Sun. Since a typical neutron star might be 20 km across, the semi-major axis is $\approx 10^5 \times$ the diameter of either component.

$\dot{E}$: From Eq. (2.5), with $\mathcal{T} = 0.323$ days $= 7.75$ hours, $\mu = 0.707$ (Eq. (2.3)), and the function $f(0.617) = 11.8$ (Eq. (2.4)), we find $\dot{E} = 7.7 \times 10^{31}$ erg s^{-1} which is only 2% of the EM radiative power of the Sun. This energy loss rate is modest but is sufficient to result in the orbital decay of this system.

$\dot{a}, \dot{e}$: Using Eqs. (2.6) and (2.7), the rate at which the semi-major axis and ellipticity change are, with $g(0.617) = 3.82$ (Eq. (2.8)) and all other quantities known, $\dot{a} = -1.07 \times 10^{-5}$ cm s^{-1} $= -3.4$ meter yr^{-1} and $\dot{e} = -1.7 \times 10^{-17}$ s^{-1}, so very little change is seen at present.

τ_m: The time-to-merge correction for ellipticity is $k(e_0) = 0.185$ (Eq. (2.11)) giving a merger time from Eq. (2.10) of 1.03×10^{16} s $= 0.32$ Gyr[15]. This is about 2% of the age of the Universe.

$\overline{v}_1$: Since the orbits are elliptical, the speed of each component varies, but an average can be approximated by simply taking the perimeter, divided by the period. From Appendix A.5 of the *online material*, for an ellipse, $b_1 = a_1\sqrt{1-e^2} = 2.56$ lt-sec and the orbital period is 0.323 days, so $\overline{v}_1 = \left(2\pi\sqrt{\frac{a_1^2+b_1^2}{2}}\right)/\mathcal{T} = 1.98 \times 10^7$ cm s^{-1} $= 198$ km s^{-1}. This is much less than the speed of light, so the orbit is not relativistic.

$\mathcal{M}_c$: Using the masses of Table 2.1, $\mathcal{M}_c = 1.23\,M_\odot$. Notice that the chirp mass is less than either of the component masses in this case (but see Problem 2.6).

f_{GW}, λ_{GW}: From Eq. (2.1) and the relation between period and frequency, $f_{GW} = 2/\mathcal{T} = 7.2 \times 10^{-5}$ Hz $\Rightarrow \lambda_{GW} = c/f_{GW} = 4.2 \times 10^{14}$ cm which is 28 AU, so there would be no appreciable difference in amplitude measured anywhere on the surface of the Earth at a given time.

h_0: From Eq. (2.19) and the known distance from Table 2.1, we find $h_0 = 5.5 \times 10^{-23}$. This is currently outside of measurable limits (Figure 2.8).

$\dot{\mathcal{T}}$: From Eq. (2.12), $\dot{\mathcal{T}} = -2.4 \times 10^{-12}$ s s^{-1} $= -0.076$ ms yr^{-1}. This is a highly accurate measurement, facilitated by the fact that pulsars are highly accurate 'clocks'. See [39] for more details.

---

[15] A more accurate calculation gives 0.38 Gyr [5].

As Example 2.1 shows, it is not possible to measure the GW *directly* from the Hulse and Taylor pulsar directly, because h_0 is too small, and this is also true of the other binary pulsars in the list (Problem 2.7). Even though these objects imply that gravitational radiation must exist, what is really wanted is *direct* proof of their existence, and that means making a measurement of the wave itself.

2.8 DIRECT PROOF OF THE EXISTENCE OF GRAVITATIONAL WAVES

On 15 September 2015, two widely separated detectors, one in Hanford in Washington State and one in Livingston, Louisiana (Figure 2.7), directly detected the stretching and squeezing of space due to a gravitational wave. The source was called GW150914 (Table 2.2). The strain was miniscule (10^{-21}) yet both instruments measured the signal as the GW passed by. How could such a sensitive measurement be possible?

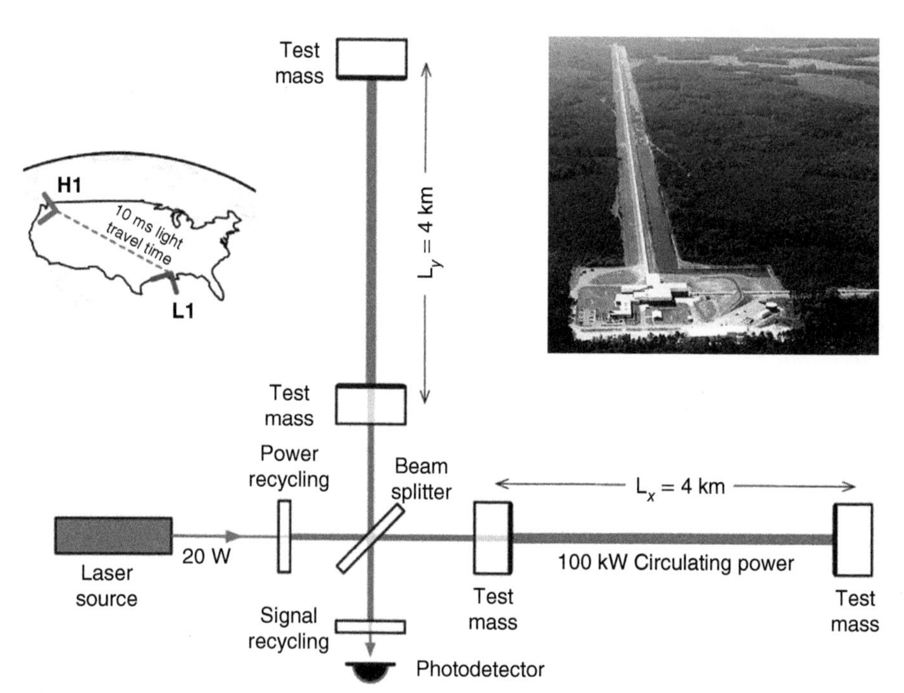

Figure 2.7 The Advanced Laser Interferometer Gravitational-Wave Observatory (aLIGO) showing its two 4 km long arms at right angles to each other, its laser source and test masses. There is one such interferometer at each of the two stations marked H1 and L1 on the map in the left inset. Source: Abbott et al. [1]. Licensed under CC BY 3.0. The right inset shows an aerial photograph of the L-shaped LIGO detector in Livingston, Louisiana. Source: Reproduced courtesy of LIGO, Livingston, Louisiana.

Table 2.2 Direct gravitational-wave detections[a]

System	$m_1{}^b(M_\odot)$	$m_2{}^b(M_\odot)$	$\mathcal{M}_c{}^c(M_\odot)$	$M_f{}^d(M_\odot)$	$E_{rad}{}^e(M_\odot c^2)$	$\dot{E}_{peak}{}^f$ ($\times 10^{56}$ erg s^{-1})	D^g(Mpc)
GW150914	35.6	30.6	28.6	63.1	3.1	3.6	430
GW151012	23.3	13.6	15.2	35.7	1.5	3.2	1060
GW151226	13.7	7.7	8.9	20.5	1.0	3.4	440
GW170104	31.0	20.1	21.5	49.1	2.2	3.3	960
GW170608	10.9	7.6	7.9	17.8	0.9	3.5	320
GW170729	50.6	34.3	35.7	80.3	4.8	4.2	2750
GW170809	35.2	23.8	25.0	56.4	2.7	3.5	990
GW170814	30.7	25.3	24.2	53.4	2.7	3.7	580
GW170817	1.46	1.27	1.186	$\leq$2.8	$\geq$0.04	$\geq$0.1	40
GW170818	35.5	26.8	26.7	59.8	2.7	3.4	1020
GW170823	39.6	29.4	29.3	65.6	3.3	3.6	1850

[a]See [3] for these quantities with their error bars. Some parameters are being refined with time. Detections are named as GW-year-month-day of discovery.
[b]Mass of the component.
[c]Chirp mass.
[d]Final mass.
[e]Radiated energy.
[f]Peak luminosity.
[g]Distance.
Source: Abbott et al. [3]. Licensed under CC BY 4.0.

The measurement was made with the advanced Laser Interferometer Gravitational-Wave Observatory (aLIGO). Figure 2.7 shows the observational set-up for one detector. Light at an infrared wavelength of $\lambda = 1064$ nm is shone from a laser through a beam splitter that allows some of the light to travel down one of the 4 km long arms and some of the light down the other 4 km arm at right angles to it. A high vacuum is maintained in each arm to limit interaction with air molecules along the way. The light passes through test masses which are actually partially reflective and partially transmissive mirrors (each 40 kg in mass) that are suspended by glass fibres in order to be isolated from man-made, seismic, or thermal vibrations. The laser light reflects back from the ends of the arms, returning to the beam splitter and the signal then combines and is detected at a photodetector shown at the bottom of Figure 2.7. Any difference in the path lengths between the two arms (due to space stretching in one direction and compressing in the other) will result in an *interference pattern* on the detector, allowing the difference in distance to be measured very accurately.

This is called a *Michelson interferometer*, a well-known instrument that has been used since the nineteenth century, but aLIGO is much larger and much more sensitive. An important sensitivity improvement comes from the fact that, as the laser light travels down each arm, it is reflected back and forth between the test masses within the arm 280 times, effectively making the lengths of the arms 1120 km – much larger than their physical length. This increases the denominator in Eq. (2.17) thereby lowering the detectable strain limit.

To detect a strain of 10^{-21}, a path difference of $h_0 l = 10^{-21} \times 1120$ km $\approx 10^{-13}$ cm is required. This is approximately the size of a proton! Yet aLIGO can measure to tolerances 10 000 times smaller than this. Sensitivity improvements come from highly amplified laser light ('power recycling' in the figure) so that the signal-to-noise ratio (see Section 4.5) is high and also from the ability to tune the signal output highly accurately ('signal recycling'). Ultimately tiny phase shifts (of order 10^{-8} rad [7]) can be measured. Future detectors will do even better.

A *characteristic strain* has also been defined that takes into account the integration of a signal over time near some frequency. For inspiralling sources, provided the chirp time is less than the observing time, this is [12]

$$h_c(f_{GW}) \equiv \sqrt{\frac{2 f_{GW}^{\,2}}{\dot{f}_{GW}}}\, h_0 \tag{2.20}$$

For the gravitational-wave signal, GW150914, the frequency increased from 35 Hz to a maximum amplitude frequency of 150 Hz. If the maximum amplitude corresponds to the final frequency at which Newtonian approximations can be used, then Eq. (2.14) and data from Table 2.2 lead to $\dot{f} = 1.48 \times 10^4$, and the characteristic strain is a factor of 1.7 higher than the instantaneous strain. This quantity has been used to compare signals from different instruments (Figure 2.8).

The first gravitational-wave signal, GW150914, was discovered very quickly (Figure 2.5), after the less sensitive earlier version of LIGO underwent an upgrade to aLIGO. It was detected at both the Hanford and Livingston interferometers[16]. This event (and subsequent discoveries, Table 2.2) has been carefully compared to waveform expectations from GR. During ring-down, for example, numerical GR models are employed to develop waveform templates for non-rotating and rotating black holes. In the case of neutron stars, GW waveforms provide the potential to probe the interior *equations of state*[17], which are currently not well known.

With the exception of GW170817, all sources in Table 2.2 have individual masses that exceed the neutron star upper mass limit which implies that they are black hole – black hole binaries. The final merged object is a more massive black hole. The individual black hole masses are high, but are consistent with what is expected from the end point of massive stellar evolution [4, 24, 37]. Notice that the final mass, M_F is *less than* the sum of the individual masses, $m_1 + m_2$. The difference is accounted for by the total energy radiated away via GWs, E_{rad}. For GW150914, $E_{rad} = 3.1\ M_\odot c^2$ (6×10^{54} ergs), equivalent to converting three Suns completely into GWs. As indicated in Section 2.5, our Universe is 'bright' in gravitational radiation.

---

[16] A third detector, *Virgo*, located in Cascina, Italy, was not operating at the time.

[17] An *equation of state* (EOS) is a relation between pressure, density, and temperature (e.g. Eq. 5.11), although temperature does not factor into the EOS of neutron stars.

2.9 EVEN NEWER WINDOWS

Inspiralling binaries are not the only objects that are expected to produce detectable GW signals, although they are the only sources that so far have well-modelled waveforms and have actually been detected. As indicated in Section 2.4, any non-symmetric accelerating mass can produce GWs so there are potentially many GW sources. Other possibilities include supernovae, rotating neutron stars that have some axial asymmetry, massive (10^4–10^7 $M_\odot$) black hole binaries, white dwarf binaries, active galactic nuclei, extreme mass ratio inspirals (e.g. stellar mass objects falling into supermassive black holes), an unknown *stochastic background* that likely includes a variety of sources, even relics from the early Universe or cosmic strings, should they exist.

The potential appears limitless, and an extraordinary effort is underway to extend the network of detectors both on the ground and in space. GW detectors currently cannot 'locate' their sources on the sky with any precision, but networks of detectors will improve on this situation. Future space-based detectors, such as the Laser Interferometer

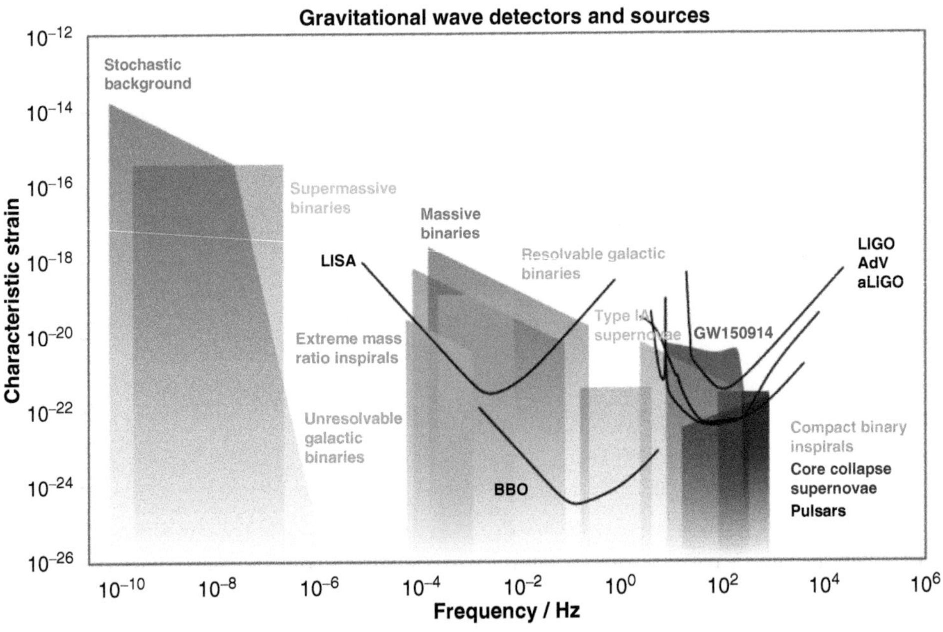

Figure 2.8 Characteristic strain as a function of GW frequency showing different types of astrophysical sources (colour). The sensitivity levels for some existing and planned GW detectors are marked with black curves. The ground-based detectors show curves with their highest sensitivity around 100 Hz. Space-based detectors (LISA, BBO) are sensitive at lower frequencies. The low frequency stochastic background must be probed using pulsar timing arrays which will be detecting EM radiation (see Chapter 9). Source: Moore et al. [26]. Licensed under CC BY 3.0. See http://gwplotter.com to explore more plot options and include more detectors.

Space Antenna (LISA), will have arms as long as a million kilometre, making it sensitive to GWs with wavelengths up to this value. In space, there are also no interfering seismic or man-made noise signals. A subset of current and future GW detectors is shown in Figure 2.8. This figure and its online counterpart (see caption) indicate the kinds of sources that could be detected at various GW wavelengths. In a sense, it is the GW analogue to the well-known electromagnetic spectrum (e.g. Figure 4.3 and Section 12.1.1).

Even with networks of GW detectors, positional precision and scientific understanding are improved significantly if the source can be identified and observed using traditional methods of detecting electromagnetic radiation. This has successfully occurred for the source, GW170817 (Table 2.2), a binary neutron GW source which was also detected in gamma-rays by the *Fermi* Gamma-ray Burst Monitor. Rapid follow-up observations showed transient emission at optical, UV, and near-infrared wavelengths, and X-ray and radio emission were also detected. The results both pin-pointed the exact location of the source in the galaxy, NGC 4993, as well as brought the full power of classical astronomical methods to bear on this GW source. The emerging picture is of two neutron stars merging followed by a short gamma-ray burst and a powerful nova whose light is powered by radioactive decay from products synthesized in the ejecta [21].

GW170817 truly represents a 'multi-messenger' astrophysical source. But in order to fully decode its message and the messages of other objects, we first need to understand what classical EM radiation can tell us – and that is the subject of the rest of this book.

PROBLEMS

2.1. (a) For circular orbits, determine the orbital period required of a binary that would generate a GW whose amplitude varies from maximum to minimum (1/2 a wavelength) over the diameter of the Earth. What is the corresponding GW frequency? Compare this result to the range of frequencies for Earth-based detectors (Figure 2.8) and comment.

(b) Suppose that a 2 $M_{\odot}$ and a 10 $M_{\odot}$ object are in orbit about their common centre of mass with the period computed in part (a). What would be the semi-major axis of such a system, assuming that Newtonian physics applies? Refer to Tables T.5–T.7 and determine whether either component of such a binary could be a star (i.e. in which nuclear burning is taking place in the interior). What types of object are possible?

2.2. Approximate the Solar System as a two-component binary consisting of its most massive components, the Sun and Jupiter, in circular orbits. Where is the centre of mass (the 'barycenter')? Is it within or outside of the Sun?

2.3. Beginning with Kepler's third law, find α in the proportionality, $\upsilon \propto a^{\alpha}$, where υ is the speed in a circular orbit, and a is the semi-major axis. If a decreased by a factor of 16, by what factor would the speed increase?

2.4. Derive Eq. (2.12).

2.5. Derive Eq. (2.14).

2.6. (a) Derive the expressions of Eq. (2.16). [Hint: Start by forming the ratio, $\mathcal{M}_c/m_1$, and simplify so that the right hand side contains terms in m_2/m_1.]

(b) For a chirp mass of 28.6 $M_\odot$, plot a graph of total mass, $M = m_1 + m_2$, as a function of m_1 over a range, $1 \leq m_1 \leq 100$, all in units of $M_\odot$. Find the minimum value of the total mass, M_{min}.

(c) Verify that $M_{min} = 4^{3/5}\mathcal{M}_c$ and verify that your results for m_1, m_2, and M_{min} are consistent with the values for GW150914 (Table 2.2).

2.7. Repeat Example 2.1 but for the binary neutron star system, J1906 + 0746.

2.8. (a) Determine a for the Hulse & Taylor binary system when the eccentricity has circularized to a value of $e = 0.2$. Call this value of the semi-major axis, a_p.

(b) Find τ_m from a starting value of $a_0 = a_p$. Compare this time to τ_m computed in Example 2.1 and comment.

2.9. Fill in two more columns of Table 2.1 labelled τ_{circ} and τ_m and comment on your findings. Which system has the smallest τ_{circ}? Which system will merge first?

2.10. Use Figure 2.6 to estimate a_f (lt-sec) and t (years) for a binary that starts at an initial ellipticity of $e_0 = 0.42$ and a final ellipticity of $e_f = 0.035$. Adopt values of $m_1 = m_2 = 1\ M_\odot$ and an initial semi-major axis of $a_0 = 3$ lt-sec.

2.11. A GW chirp is detected with the following properties: the maximum amplitude of the signal is $h_0 = 10^{-19}$, and the frequency and frequency derivative at this maximum are $f_{GW} = 57$ Hz and $\dot{f}_{GW} = 656\ s^{-2}$, respectively. Assume that $m_1 = m_2$ and that $\sin i \approx 1$. Find the following quantities in the units specified: $\mathcal{M}_c$ ($M_\odot$), m_1 ($M_\odot$), m_2 ($M_\odot$), M ($M_\odot$), $\mathcal{T}$ (s), $\dot{E}$ (erg s^{-1}), a (km), $\dot{a}$ (km s^{-1}), v (c), τ_m (sec), and D (Mpc). Is this source in our Milky Way galaxy?

2.12. (a) Find $\dot{E}$ for the GW generated by the Earth orbiting the Sun. Neglect any influences by the moon and other planets. Express your result in Watts and compare this energy to the EM radiative output of a 100 W light bulb.

(b) How long will it take for the Earth to merge with the Sun? Compare this time to the current age of the Universe (see Section 5.1).

(c) Suppose you were to set up a GW detector at the distance of Pluto which we will assume is in the far field of the Earth–Sun system. Find h_0, locate where it falls in Figure 2.8 and comment. [Note. There is essentially no change in the frequency during the observing time so Eq. (2.20) should not be used.]

JUST FOR FUN

2.13. Look up the range of hearing of the human ear and mark that range on Figure 2.8. What kinds of sources could you hear? Listen to some online material that mimics the chirp of a gravitational wave (see Ref. [32]).

2.14. For a rotating ellipsoid, Eqs. (2.2) or (2.5) must be modified to be

$$\dot{E} = \frac{32G}{5c^5} I^2 e^2 \left(\frac{2\pi}{\mathcal{T}}\right)^6 \tag{2.21}$$

where I is the moment of inertia, and the other quantities have their usual meanings. Estimate the GW luminosity of a spinning hard-boiled egg.

2.15. An asymmetric core-collapse supernova produces a GW of amplitude [19]

$$h = \frac{G M_{core}}{5Dc^4}\left(\frac{A_{\min}}{\Delta t}\right)^2 \tag{2.22}$$

where M_{core} is the mass of the collapsing core, D is the distance to the supernova, $A_{\min}$ is the minimum, major semi-axis of the core spheroid, and Δt is a characteristic time for a collapse. Typical parameters are $M_{core} = 0.24\ M_{\odot}$, $A_{\min} = 5.5$ km, and $\Delta t = 0.75$ ms. How close would a SN have to be, in order for the room that you are in right now to stretch by the width of a proton?

PART II

The EM Signal Observed

We now move to *electromagnetic (EM) radiation*. The radiation from an astronomical source can be thought of as a *signal* that provides us with information about it. In order to relate the signal received at the Earth to the physical conditions within an astronomical source, however, we first need ways to *describe* and *measure* light. This requires setting out the basic definitions for quantities involving light and the relationships between them. The definitions range from those associated with values measured at the Earth to those that are intrinsic to the source itself. We do this without regard (yet) for the processes that actually generate the light, an approach that is often followed in the study of Mechanics. For example, first one studies Kinematics which relates distances, velocities, and accelerations and the relations between them. Later, one considers Dynamics which deals with the forces that produce these motions. Measuring light involves a deep understanding of how the measurement process itself affects the signal and also how the Earth's atmosphere interferes. Our instrumentation imposes its own signature on an astronomical signal and it is important to account for this imposition. These steps are fundamental and lay the groundwork for turning the measurement of a weak glimmer of light into an understanding of what drives the most powerful objects in the Universe.

Chapter 3
Defining the Signal

... the distance of the invisible background [is] so immense that no ray from it has yet been able to reach us at all.

– Edgar Allan Poe (1848), Eureka: a prose poem. An essay on the material and spiritual universe, G. P. Putnam

3.1 THE POWER OF LIGHT – LUMINOSITY AND SPECTRAL POWER

The *luminosity*, L, of an object is the rate at which the object radiates away its energy (cgs units of erg s^{-1} or SI units of watts),

$$dE = L\,dt \tag{3.1}$$

This quantity has the same units as *power* and is simply the radiative power output from the object. It is an intrinsic quantity for a given object and does not depend on the observer's distance or viewing angle. If a star's luminosity is L_* at its surface, then at a distance r away, its luminosity is still L_*.

Any object that radiates, be it spherical or irregularly shaped, can be described by its luminosity. The Sun, for example, has a luminosity of $L_\odot = 3.85 \times 10^{33}$ erg s^{-1} (Table T.2), most of which is lost to space and not intercepted by the Earth (Example 3.1).

Example 3.1

Determine the fraction of the Sun's luminosity that is intercepted by the Earth. What luminosity does this correspond to?

At the distance of the Earth, the Sun's luminosity, $L_\odot$, is passing through the imaginary surface of a sphere of radius, $r_\oplus = 1$ AU. The Earth will be intercepting photons

Astrophysics: Decoding the Cosmos, Second Edition. Judith A. Irwin.

Companion website: www.wiley.com/go/irwin/astrophysics2e

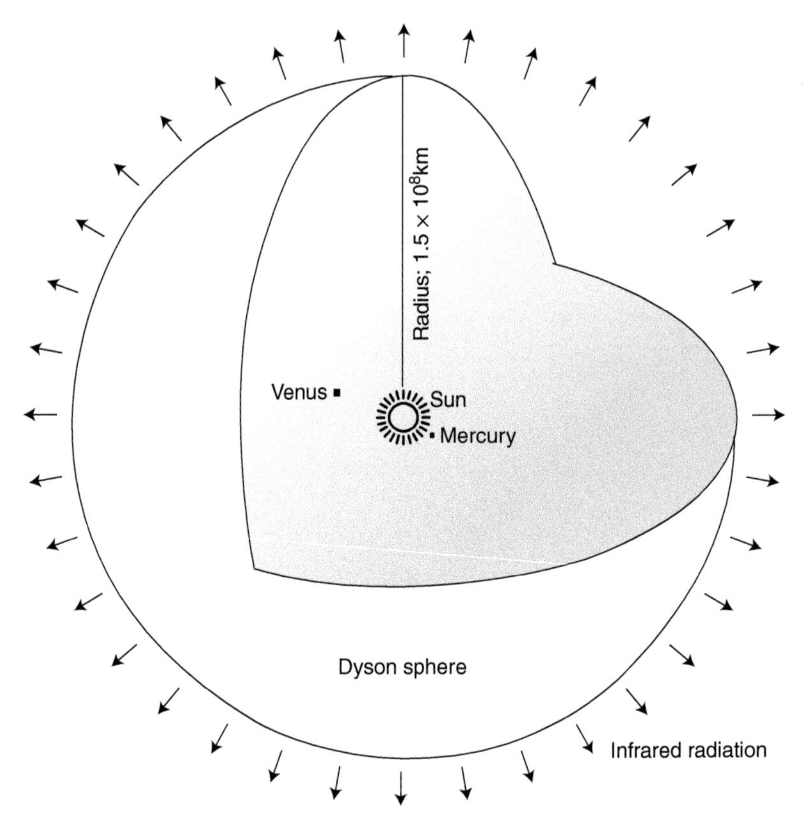

Figure 3.1 Illustration of a Dyson Sphere that could capture the entire luminous output from the Sun. Some have suggested that advanced civilizations, if they exist, would have discovered ways to build such spheres to harness all of the energy of their parent stars.

over only the cross-sectional area that is facing the Sun. This is because the Sun is so far away that incoming light rays are parallel. Thus, the fraction will be

$$\mathbf{f} = \frac{\pi R_{\oplus}^{\;2}}{4\pi r_{\oplus}^2} \tag{3.2}$$

where $R_{\oplus}$ is the radius of the Earth. Using the values of Table T.2, the fraction is $\mathbf{f} = 4.5 \times 10^{-10}$ and the intercepted luminosity is, therefore, $L_{int} = \mathbf{f} L_{\odot} = 1.73 \times 10^{24}$ erg s^{-1}. A hypothetical shell around a star that would allow a civilization to intercept *all* of its luminosity is called a *Dyson Sphere* (Figure 3.1).

When one refers to the luminosity of an object, it is the *bolometric* luminosity that is understood, i.e. the luminosity over all wavebands. However, it is not possible to determine this quantity easily since observations at different wavelengths require different techniques, different kinds of telescopes and, in some wavebands, the necessity of making measurements above the obscuring atmosphere of the Earth. Thus, it is common to

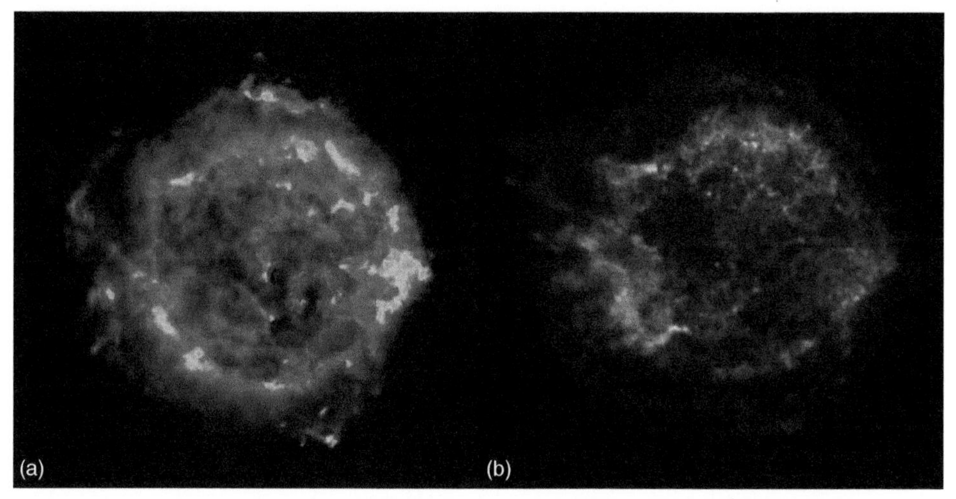

Figure 3.2 The supernova remnant, Cas A, at a distance of 3.4 kpc and with a linear diameter of ≈ 4 pc, was produced when a massive star exploded in the year 1680. It is currently expanding at a rate of 4000 km s^{-1} [18], and the proper motion (angular motion in the plane of the sky, see Section 9.1.1.1) of individual filaments has been observed. One side of the bipolar jet, emanating from the central object, can be seen at approximately 10 o'clock. (a) Radio image at λ 20 cm shown in *false colour* (see Section 4.6) from [2]. Source: Image courtesy of NRAO/AUI/NSF. (b) X-ray emission, with red, green and blue colours showing, respectively, the intensity of low, medium and high energy X-ray emission. Source: Reproduced courtesy of NASA/CXC/SAO.

specify the luminosity of an object for a given waveband (waveband ranges are given in Table I.1). For example, the *supernova remnant*, Cas A (Figure 3.2), has a radio luminosity (from $\nu_1 = 2 \times 10^7$ Hz to $\nu_2 = 2 \times 10^{10}$ Hz) of $L_{\text{radio}} = 3 \times 10^{35}$ erg s^{-1} [1] and an X-ray luminosity (from 0.3 to 10 keV) of $L_{\text{X-ray}} = 3 \times 10^{37}$ erg s^{-1} [6]. Its bolometric luminosity is the sum of these values plus the luminosities from all other bands over which it emits. It can be seen that the radio luminosity might justifiably be neglected when computing the total power output of Cas A.

Clearly, the source *spectrum* (the emission as a function of wavelength) is of some importance in understanding which wavebands, and which processes, are most important in terms of energy output. The spectrum may be represented mostly by *continuum emission* as implied here for Cas A (that is, emission that is continuous over some spectral region) or may include *spectral lines* (emission at discrete wavelengths, see Chapter 5, 7, or 11). Even very weak lines and weak continuum emission, however, can provide important clues about the processes that are occurring within an astronomical object and must not be neglected if a full understanding of the source is to be achieved.

In the optical region of the spectrum, various *passbands* (or filters) have been defined within which measurements are made (Figure 3.3). The Sun's luminosity, for example, could readily be approximated with measurements in V-band only, because 93% of

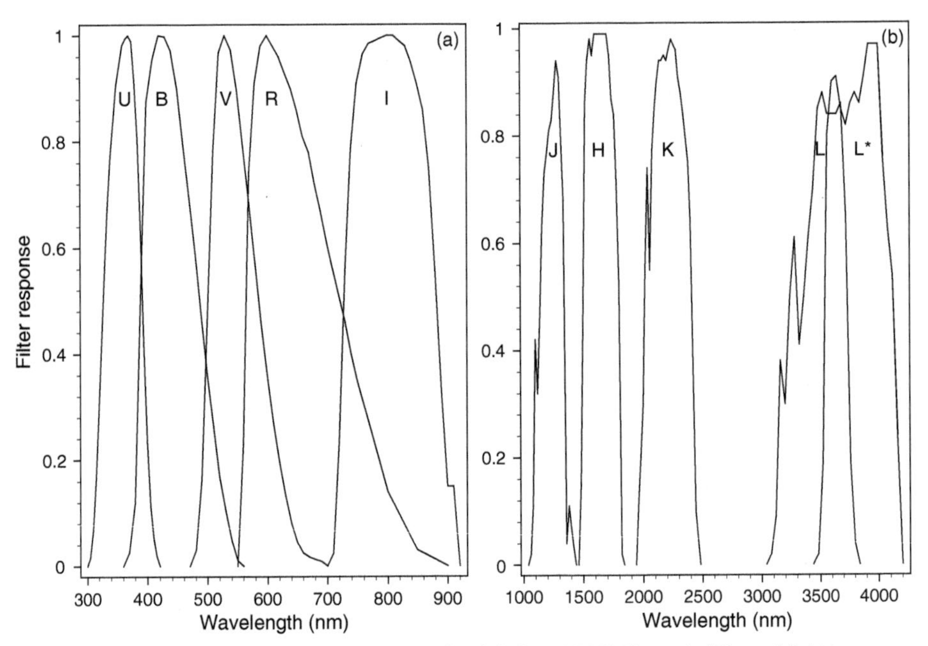

Figure 3.3 Filter bandpass responses for (a) the UBVRI bands [3] and (b) the JHKLL* bands [5]. (The U and B bands correspond to UX and BX of [3]). Corresponding data can be found in Table 3.1. Source: (a) Based on Bessell [3] and (b) modified from Bessell and Brett [5].

its bolometric luminosity is emitted in this band. However, crucial information is still contained in the other bands, revealing many secrets about our nearest stellar neighbour.

The *spectral luminosity* or *spectral power* is the luminosity per unit bandwidth and can be specified per unit wavelength, L_λ (cgs units of erg s^{-1} cm^{-1}) or per unit frequency, L_ν (erg s^{-1} Hz^{-1}),

$$dL = L_\lambda \, d\lambda = L_\nu \, d\nu \tag{3.3}$$

$$\text{so} \quad L = \int L_\lambda \, d\lambda = \int L_\nu \, d\nu \tag{3.4}$$

Note that, since $\lambda = \frac{c}{\nu}$,

$$d\lambda = -\frac{c}{\nu^2} \, d\nu \tag{3.5}$$

so the magnitudes of L_λ and L_ν will not be equal (Problem 3.1). The negative sign in Eq. (3.5) serves to indicate that, as wavelength increases, the frequency decreases. In equations like Eq. (3.4) in which the wavelength and frequency versions of a function are related to each other, this negative can be taken into account by ensuring that the lower limit to the integral is always the lower wavelength or frequency. Note that the cgs units of L_λ (erg s^{-1} cm^{-1}) are rarely used since 1 cm of bandwidth is exceedingly large (Table 3.1). Non-cgs units, such as erg s^{-1} Å^{-1}, are sometimes used instead.

It is important not to confuse the luminosity 'in a waveband' with the luminosity 'per unit waveband'. The former is an integration of the latter (Eq. (3.4)). An 'optical luminosity' (ergs s^{-1}) for example, results if a spectral luminosity (ergs s^{-1} Hz^{-1} or ergs s^{-1} cm^{-1}) has been integrated over *only* optical wavelengths. A true bolometric luminosity results if the integration is over *all* frequencies (or wavelengths) in which there is any significant emission.

Luminosity is a very important quantity because it is a basic parameter of the source and is directly related to energetics. Integrated over time, it provides a measure of the energy required to make the object shine over that timescale. For example, if we wish to know whether event A could power event B, we must certainly know the energies of both. However, luminosity is not a quantity that can be measured directly and must instead be derived from other measurable quantities that will shortly be described.

3.2 LIGHT THROUGH A SURFACE – FLUX AND FLUX DENSITY

The *flux* of a source, f (erg s^{-1} cm^{-2}), is the radiative energy per unit time passing through unit area,

$$dL = f\ dA \tag{3.6}$$

As with luminosity, we can define a flux in a given waveband or we can define it per unit spectral bandwidth. For example, the *spectral flux density*, or just *flux density*, (erg s^{-1} cm^{-2} Hz^{-1} or erg s^{-1} cm^{-2} cm^{-1})[1] is the flux per unit spectral bandwidth, either frequency or wavelength, respectively,

$$\begin{aligned} dL_\nu &= f_\nu dA \qquad & dL_\lambda &= f_\lambda dA \\ df &= f_\nu d\nu \qquad & df &= f_\lambda d\lambda \end{aligned} \tag{3.7}$$

A special unit for flux density, called the *Jansky* (Jy), is utilized in astronomy, most often in the infrared and radio parts of the spectrum,

$$1\ \text{Jy} = 10^{-26}\ \text{W m}^{-2}\ \text{Hz}^{-1} = 10^{-23}\ \text{erg s}^{-1}\ \text{cm}^{-2}\ \text{Hz}^{-1} \tag{3.8}$$

Radio sources that are greater than 1 Jy are considered to be strong sources by astronomical standards (Problem 3.3).

Equation (3.6) and the first line of Eq. (3.7) show the same relationships except for the wavelength or frequency dependence indicated by the subscripts. In the next equation, Eq. (3.9), we will include each version, but in future equations, we will give the relationships only for the unsubscripted quantities, to avoid repetition.

[1] The two 'cm' designations should remain separate. See the comments after Example 3.2.

The luminosity, L, of a source can be found from its flux via,

$$L = \int f dA = 4\pi r^2 f$$
$$L_\nu = \int f_\nu dA = 4\pi r^2 f_\nu, \qquad L_\lambda = \int f_\lambda dA = 4\pi r^2 f_\lambda \tag{3.9}$$

where r is the distance from the centre of the source to the position at which the flux has been determined. The $4\pi r^2$ in Eq. (3.9) is strictly only true for sources in which the photons that are generated can escape in all directions away from the source. This is usually assumed to be true, even if the source itself is irregular in shape (Figure 3.4). These photons pass through the imaginary surfaces of spheres as they travel outwards. The $\frac{1}{r^2}$ fall-off of flux is just due to the geometry of a sphere (Figure 3.5a). In principle, however, one could imagine other geometries. For example, the flux of a man-made laser beam would be constant with r if all emitted light rays are parallel and without losses en route

Figure 3.4 An image of the Centaurus A jet originating from an *active galactic nucleus* (AGN) at the centre of the galaxy which is located at the lower right of this image. Radio emission is shown in red and X-rays in blue. Source: Hardcastle et al. [12]. Even though gaseous material may be moving along the jet in a highly directional fashion, the RHS of Eq. (3.9) may still be used, *provided that* photons generated within the jet (such as at the knot at the centre of the arrows) escape outwards in all directions. Sometimes the light actually beams directionally, though, so assumptions need to be carefully examined (Problem 3.4).

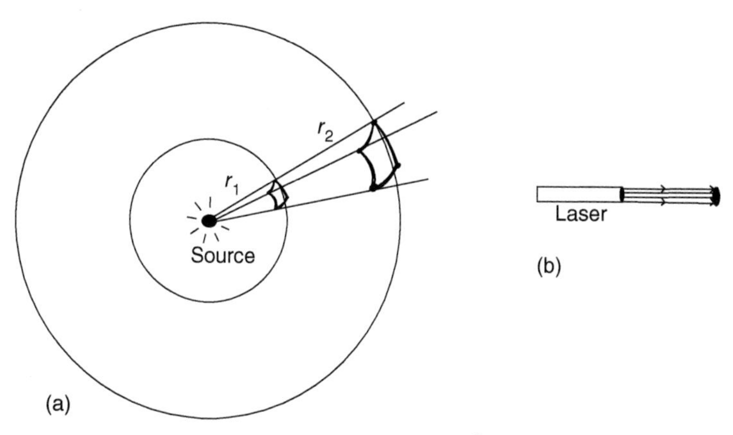

Figure 3.5 (a) Geometry illustrating the $\frac{1}{r^2}$ fall-off of flux with distance, r, from the source. The two spheres shown are imaginary surfaces. The same amount of energy per unit time is going through the two spheres shown. Because the area at r_2 is greater than the one at r_1, the energy per unit time per unit cm^2 is smaller at r_2 than r_1. Since measurements are made over size scales so much smaller than astronomical distances, the detector need not be curved as is shown here. (b) Geometry of an artificial laser. For a beam with no divergence, the flux does not change with distance.

(Figure 3.5b). Light that is beamed into a narrow cone, such as is occurring in pulsars[2] is an example of an intermediate case (Problem 3.4).

For *stars*, we now define the *astrophysical flux*, F, to be the flux at the surface of the star,

$$L_* = 4\pi R_*^2 F = 4\pi r^2 f \quad \Rightarrow f = \left(\frac{R_*}{r}\right)^2 F \tag{3.10}$$

where L_* is the star's luminosity and R_* is its radius. Notice that it is the luminosity that is the constant quantity, as indicated in Section 3.1.

Using values from Table T.2, the astrophysical flux of the Sun is $F_\odot = 6.33 \times 10^{10}$ erg s^{-1} cm^{-2} and the *Solar Constant*, denoted $S_\odot$, which is the flux of the Sun at a distance of 1 AU, is 1.367×10^6 erg s^{-1} cm^{-2}. The Solar Constant is of great importance since it is this flux that governs climate and life on Earth. Modern satellite data reveal that the solar 'constant' actually varies in magnitude, indicating that our Sun is a variable star (Figure 3.6). Earth-bound measurements failed to detect this variation since it is quite small and corrections for the atmosphere and other effects are large in comparison (e.g. Problem 3.5).

[2] Pulsars are rapidly spinning *neutron stars* with strong magnetic fields that emit their radiation in beamed cones. Neutron stars typically have masses of $\sim 1.5 \rightarrow 3$ $M_\odot$ in a diameter only tens of km across.

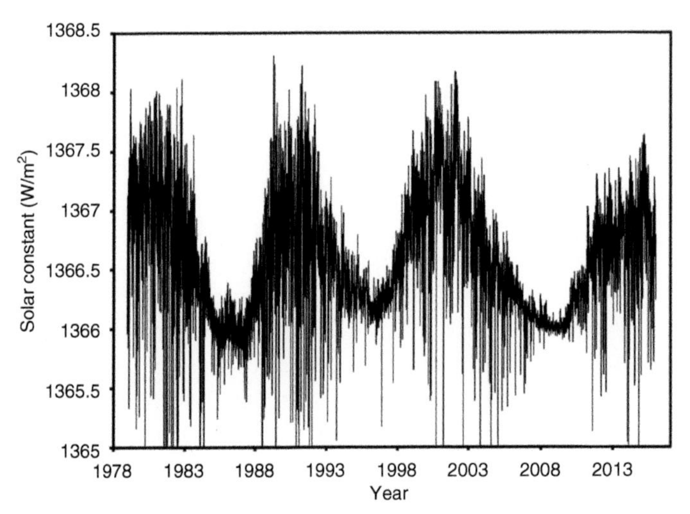

Figure 3.6 Plot of the Solar Constant as a function of time from satellite data. Data represent the corrected Total Solar Irradiance data (spot.colorado.edu/~koppg/TSI), adjusted to approximately 1 AU. The variation follows the 11-year sunspot cycle such that when there are more sunspots, the Sun, on average, is brighter. The peak-to-peak variation is no more than 0.1%. This plot provides definitive evidence that our Sun is a variable star.

The flux of a source in a given waveband is a quantity that is measurable, provided corrections are made for atmospheric and telescopic responses, as required (see Sections 4.2 and 4.3). If the distance to the source is known, its luminosity can then be calculated from Eq. (3.9) or Eq. (3.10).

3.3 THE BRIGHTNESS OF LIGHT – INTENSITY AND SPECIFIC INTENSITY

The *intensity*, I (erg s^{-1} cm^{-2} sr^{-1}), is the radiative energy per unit time per unit solid angle passing through a unit area that is perpendicular to the direction of the emission. The *specific intensity* (cgs units of erg s^{-1} cm^{-2} Hz^{-1} sr^{-1} or erg s^{-1} cm^{-2} cm^{-1} sr^{-1}) is the radiative energy per unit time per unit solid angle per unit spectral bandwidth (frequency or wavelength, respectively) passing through a unit area perpendicular to the direction of the emission. The intensity is related to the flux via,

$$df = I \cos\theta \, d\Omega \tag{3.11}$$

As before, Eq. (3.11) could also be written between the quantities per unit bandwidth, i.e. between the specific intensity and the flux density.

The specific intensity, I_ν, is the most basic of radiative quantities. Its formal definition is written,

$$dE = I_\nu \cos\theta d\nu d\Omega dA dt \tag{3.12}$$

Note that each differential quantity is independent of the others so when integrating, it does not matter in which order the integration is done.

The intensity isolates the emission that is within a given solid angle, at some angle from the perpendicular to the surface. The geometry is shown in Figure 3.7 for a situation (a) in which a real detector is receiving emission from a source in the sky and for a situation (b) in which an imaginary detector is placed on the surface of a star. In the first case, the source subtends some solid angle in the sky in a direction, θ, from the zenith. The factor $\cos\theta$ accounts for the foreshortening of the detector area as emission falls on it. Usually, a detector would be pointed directly at the source of interest in which case $\cos\theta = 1$. In the second case, the coordinate system has been placed at the surface of a star. At any position on the star's surface, radiation is emitted in all directions away from the surface. The intensity now refers to the emission in the direction θ radiating *into* solid angle, $d\Omega$. Example 3.2 indicates how the intensity relates to the flux for these two examples. Figure 3.7 also helps to illustrate the generality of these quantities. One could place the coordinate system at the centre of a star, in interstellar space, or wherever we wish to determine these radiative properties.

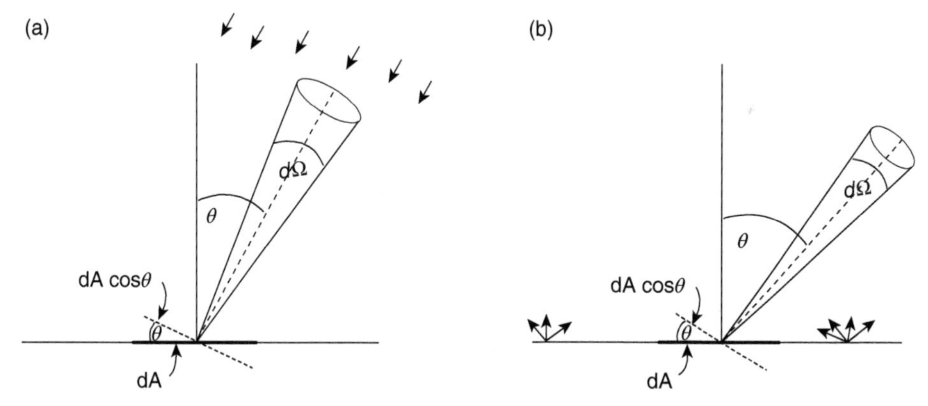

Figure 3.7 Diagrams showing intensity and its dependence on the direction and solid angle, using a spherical coordinate system such as described in Appendix B of the *online material*. (a) Here, dA would be an element of area of a detector on the Earth, the perpendicular upwards direction is towards the zenith, a source is in the sky in the direction, θ, and $d\Omega$ is an elemental solid angle on the source. The arrows show incoming rays from the *centre* of the source that flood the detector. (b) In this example, an imaginary detector is placed at the surface of a star. At each point on the surface, photons leave in all forward directions away from the surface. The intensity would be a measure of only those photons that pass through a given solid angle at a given angle, θ, from the vertical.

Example 3.2

Calculate the flux, given the intensity, for (a) a receiving case in which a detector is pointing directly at a source of small angular size, and (b) an emitting case in which the emission angle is large.

(a) If a detector is pointed directly at a uniform intensity source in the sky, the coordinate system shown in Figure 3.7a has essentially been tilted so that $\theta = 0$. Then for *small solid angle*, Ω, the flux would be,

$$f = \int_{\Omega} I \cos\theta \, d\Omega \approx I\Omega \tag{3.13}$$

(b) The astrophysical flux at the surface of an object (e.g. a star, Figure 3.7b) whose radiation is escaping freely at all angles outwards from the star must be calculated by integrating in spherical coordinates since the angle is no longer small (see Appendix B of the *online material*),

$$F = \int I \cos\theta d\Omega = \int_0^{2\pi} \int_0^{\frac{\pi}{2}} I \cos\theta \sin\theta d\theta d\phi = \pi I \tag{3.14}$$

Figure 3.8 shows a practical example as to how one might calculate the flux of a source for a case corresponding to Example 3.2a, but for which the intensity of the source varies with position. The image is typical of what is observed from a digital camera with the squares representing pixels. Each square subtends a solid angle of size, Ω_i, on the sky and the intensity in each square is I_i. As indicated in the caption, to find the flux, one needs to add up all the intensities times the solid angle of a square. As long as the source is not too large and the detector is pointing directly at it, then $\cos\theta \approx 1$ for all pixels.

The intensity in a given waveband is a measurable quantity, *provided a solid angle can also be measured.* If a source is so small or so far away that its angular size cannot be discerned (i.e. it is *unresolved*, see Sections 4.2.3, 4.2.4, 4.3.2), then the intensity cannot be found. In such cases, it is the flux (more typically the flux density) that is measured, as shown in Figure 3.9. The intensity can then only be determined if some other independent information about the angular size of the object is known.

Almost all stars other than the Sun are unresolved and therefore only the flux (or flux density) can be measured. However, there are a few large and/or nearby stars that can be resolved using special techniques (see Section 4.3.3), allowing for a crude determination of intensity. An example is shown in Figure 3.10.

Intensity (or specific intensity) is also referred to as *brightness* which has its intuitive meaning. A faint source has a lower value of specific intensity than a bright source. Note that it is possible for a source that is faint to have a larger flux density than a source that is bright if it subtends a larger solid angle in the sky (Problem 3.10).

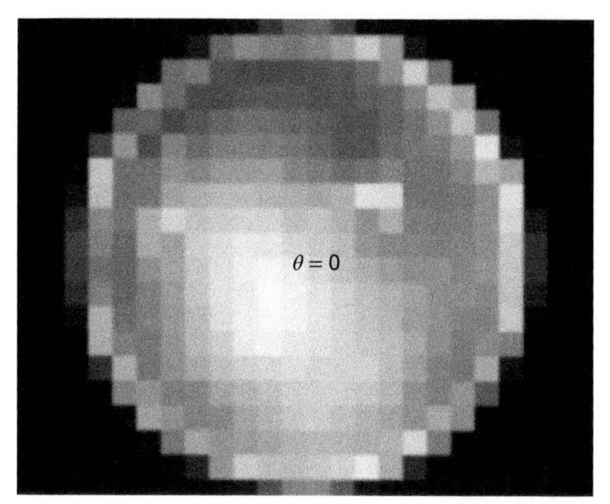

Figure 3.8 Looking directly at a hypothetical object in the sky corresponding to the situation shown in Figure 3.7 (a) but for $\theta = 0$ (i.e. the detector pointing directly at the source). The object subtends a total solid angle, Ω, which is small and therefore $\theta \approx 0$ at any location on the source. In this example, the object is of nonuniform brightness and Ω is split up into many small square solid angles, each of size, Ω_i, and within which the intensity is I_i. Then we can approximate $f = \int I \cos\theta \, d\Omega$ using $f \approx \sum I_i \, \Omega_i$. Basically, to find the flux, we add up the individual fluxes of all elements. If all squares are pixels of the same solid angle such that $\Omega_i = \Omega_p$, then $f \approx \Omega_p \sum I_i$.

Figure 3.9 In this case, a star has a very small angular size (left) and so, when detected in a square pixel of solid angle, Ω_p (right), which is determined by the properties of the detector, its light is 'smeared out' to fill that pixel. In such a case, it is impossible to determine the intensity of the surface of the star. However, the flux of the star, f_*, is preserved, i.e. $f_* = \overline{I_*}\Omega_* = \overline{I}\,\Omega_p$ (Eq. (3.13)) where $\overline{I_*}$ is the true intensity of the star, Ω_* is the true solid angle subtended by the star, and $\overline{I}$ is the mean intensity in the pixel. Thus, for an object of angular size smaller than can be resolved by the available instruments, we measure the flux (or flux density), but not intensity (or specific intensity) of the object.

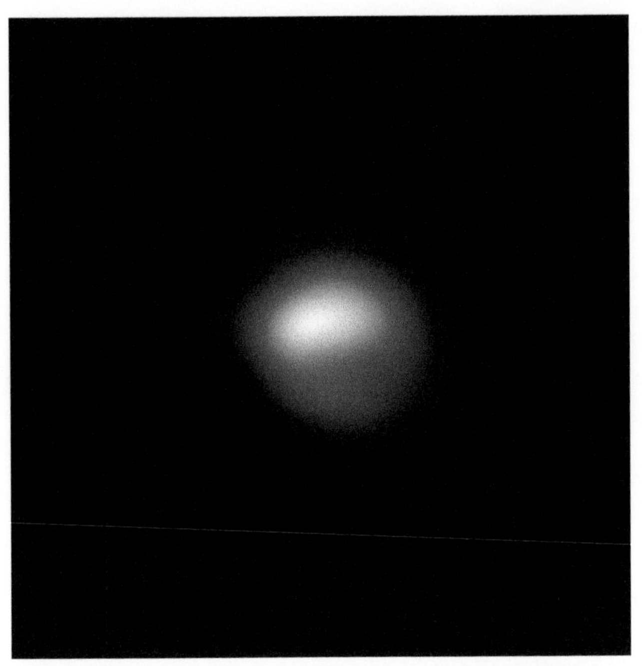

Figure 3.10 The bright supergiant star, Betelgeuse, marking the left (from our viewpoint) shoulder of Orion, was imaged using the *SPHERE* (Spectro-Polarimetric High-contrast Exoplanet REsearch) *adaptive optics* instrument on the Very Large Telescope operated by the European Southern Observatory (ESO). Betelgeuse normally brightens and dims by about 0.3 mag in the V-band but, surprisingly, in 2019, Betelgeuse began to dim suddenly. By February, 2020, its V magnitude was only 1.6 compared to a normal value of 0.45 (see Section 3.6 for a discussion of magnitudes). A change of 1.0 mag occurred between September 2019 and February 2020 [10]. Surface brightness features can be seen from the nonuniform brightness in the above photo. The diameter of Betelgeuse subtends about 45 milliarcseconds (mas). Source: ESO/M. Montargès et al., Reproduced under the Creative Commons Attribution 4.0 International Licence.

The intensity and specific intensity are *independent of distance* (constant with distance) in the absence of any intervening matter[3]. The easiest way to see this is via Eq. (3.13). Both

[3] More accurately, I/n^2 is independent of distance along a ray path, where n is the index of refraction, but the difference is negligible for our purposes.

f and Ω decline as $\frac{1}{r^2}$ (Eq. (3.10) and Eq. (I.10), respectively) and therefore I is constant with distance. The Sun, for example, has $I_{\odot} = F_{\odot}/\pi = 2.01 \times 10^{10}$ erg s^{-1} cm^{-2} sr^{-1} as viewed from any source at which the Sun subtends a small, measurable solid angle. The constancy of I with distance is general, however, applying to large angles as well. This is a very important result, since a measurement of I allows the determination of some properties of the source without having to know its distance (e.g. Section 6.1).

Intensity and specific intensity apply to light, but one could use the same concepts when speaking about particles as well. As indicated in Section 1.2.2, the quantity, $J(E)$, which describes the cosmic ray spectrum, is a kind of specific intensity for particles. Instead of units of erg s^{-1} cm^{-2} Hz^{-1} sr^{-1}, for example, the (non-cgs) units are number of particles s^{-1} m^{-2} GeV^{-1} sr^{-1}. Instead of the energy of a photon in erg, we have simply a number of particles, and instead of a quantity 'per Hz' of bandwidth, there is a quantity 'per GeV' of energy. See Example 3.3.

Example 3.3

Between the energies of 10^{16} and 10^{19} eV (above the knee), the spectrum of Figure 1.8 can be fitted by the function,

$$J(E) = 2.1 \times 10^{7} E^{-3.08} \text{ s}^{-1} \text{ m}^{-2} \text{ sr}^{-1} \text{ GeV}^{-1} \tag{3.15}$$

where E is in units of GeV [14]. What is the total rate of cosmic rays hitting the Earth's atmosphere in this energy range?

We first integrate over the energy range, $E_1 = 10^7$ GeV to $E_2 = 10^{10}$ GeV,

$$\int_{E_1}^{E_2} J(E)dE = \frac{2.1 \times 10^{7}}{-2.08}(E_2^{-2.08} - E_1^{-2.08}) = 2.78 \times 10^{-8} \text{ s}^{-1} \text{ m}^{-2} \text{ sr}^{-1} \tag{3.16}$$

Assuming that CRs are received from all forward angles in the direction of space but not from rearward angles in the direction of the Earth, the integration over solid angle amounts to a multiplication by π steradians (this is equivalent to the geometry of Example 3.2.b but for the receiving, rather than the emitting, case). The result is $f = 8.73 \times 10^{-8}$ s^{-1} m^{-2} at any position at the top of the atmosphere.

Finally, we integrate over the surface area of the Earth at the top of the atmosphere, which we take to be $4\pi R_{\oplus}^2$, because the radius of the Earth plus its atmosphere is approximately equal to $R_{\oplus}$ (6.371×10^6 m, Table T.2). The result is ≈ 45 million cosmic ray particles s^{-1} for this energy range.

It is straightforward to find the total amount of energy per second, rather than the total number of particles per second, by first changing the integral of Eq. (3.16) to read $\int E\,J(E)\,dE$. The result is 8.6×10^{14} GeV s^{-1}.

Since photons can also be thought of as particles (though massless), a similar approach can be taken for light, remembering that the energy of a photon is $E = h\nu$ (Problem 3.11).

3.4 LIGHT FROM ALL ANGLES – ENERGY DENSITY AND MEAN INTENSITY

The *energy density*, u (erg cm^{-3}), is the radiative energy per unit volume. It describes the energy content of radiation in a unit volume of space,

$$du = \frac{dE}{dV} \tag{3.17}$$

The *specific energy density* is the energy density per unit bandwidth and, as usual, $u = \int u_\nu \, d\nu = \int u_\lambda \, d\lambda$. The energy density is related to the intensity (see Figure 3.11, Eq. (3.12)) by,

$$u = \frac{1}{c}\int I d\Omega = \frac{4\pi}{c} J \tag{3.18}$$

where J is the *mean intensity*, defined by,

$$J \equiv \frac{1}{4\pi}\int I d\Omega \tag{3.19}$$

The mean intensity is therefore the intensity averaged over all directions. In an isotropic radiation field, $J = I$. In reality, radiation fields are generally not isotropic, but some are close to it or can be approximated as isotropic, for example, in the centres of stars or when considering the 2.7 K cosmic microwave background radiation (Section 5.1). In a nonisotropic radiation field, J is not constant with distance, even though I is. Example 3.4 provides a sample computation.

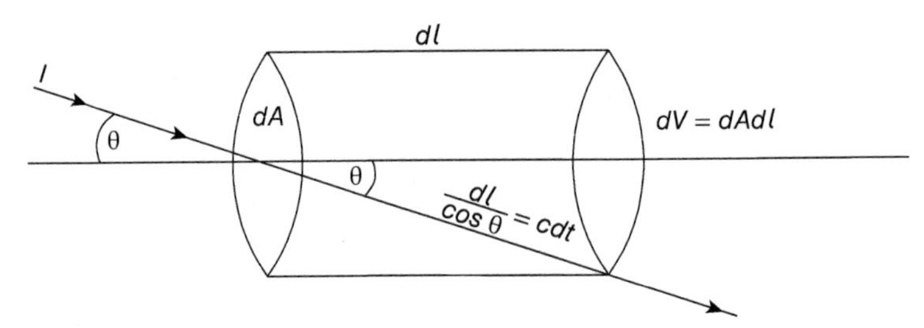

Figure 3.11 This diagram is helpful in relating the energy density (the radiative energy per unit volume) to the light intensity. An individual ray spends a time, $dt = dl/(c\cos\theta)$ in an infinitesimal cylindrical volume of size, $dV = dl dA$. Combined with Eq. (3.12), the result is Eq. (3.18).

Example 3.4

Compute the mean intensity and the energy density at the distance of Mars. Assume that the only important source of light is the Sun.

$$\begin{aligned} J &= \frac{1}{4\pi}\int_0^{4\pi} I d\Omega \\ &= \frac{1}{4\pi}\int_{\Omega_\odot} I_\odot d\Omega \approx \frac{I_\odot \Omega_\odot}{4\pi} = \frac{I_\odot}{4\pi}\frac{\pi\theta_\odot^{\,2}}{4} = \frac{I_\odot}{16}\left(\frac{2R_\odot}{r_{\mathrm{Mars}}}\right)^2 \end{aligned} \tag{3.20}$$

Figure 3.12 Why is the night sky dark? If the Universe is infinite and populated in all directions by stars, then eventually every sight line should intersect the surface of a star. Since I is constant with distance, the night sky should be as bright as the surface of a typical star. This is known as Olbers' Paradox, though Olbers was not the first to note this discrepancy.

where we have used Eq. (I.11) to express the solid angle in terms of the linear angle, and the small-angle formula (Eq. I.9) to express the linear angle in terms of the size of the Sun and the distance of Mars. Inserting $I_\odot = 2.01 \times 10^{10}$ erg s^{-1} cm^{-2} sr^{-1} (Section 3.3), $R_\odot = 6.96 \times 10^{10}$ cm, and $r_{\text{Mars}} = 2.28 \times 10^{13}$ cm (Tables T2, T.3), we find, $J = 4.7 \times 10^4$ erg s^{-1} cm^{-2} sr^{-1}. Then $u = \frac{4\pi}{c}(4.7 \times 10^4) = 2.0 \times 10^{-5}$erg cm^{-3}.

The radiation field (u or J) in interstellar space due to randomly distributed stars (Problem 3.12) must be computed over a solid angle of 4π steradians, given that starlight contributes from many directions in the sky. However, in that case, $J \neq I$ because there is no emission from directions between the stars. If there were so many stars that every line of sight eventually intersected the surface of a star of brightness, I_*, then $J = I_*$ and the entire sky would appear as bright as I_*. This would be true even if the stars were at great distances since I_*, being an intensity, is independent of distance. If this is the case, we would say that the stellar *covering factor* is unity.

A variant of this concept is called *Olbers' Paradox* after the German astronomer, Heinrich Wilhelm Olbers who popularized it in the nineteenth century. It was discussed as early as 1610, though, by the German astronomer, Johannes Kepler, and was based on the idea of an infinite starry Universe which had been propounded by the English astronomer and mathematician, Thomas Digges, around 1576. If the Universe is infinite and populated throughout with stars, then every line of sight should eventually intersect a star and the night sky as seen from Earth should be as bright as a typical stellar surface. Why, then, is the night sky dark? Kepler took the simple observation of a dark night sky as an argument for the finite extent of the Universe, or at least of its stars.

The modern explanation, however, lies with the intimate relation between time and space on cosmological scales. Because the speed of light is constant, as we look farther into space, we also look farther back in time. The observable Universe is not infinitely old but rather had a beginning (Section 5.1) and the formation of stars occurred afterwards. The required number of stars for a bright night sky is $\approx 10^{60}$ and the volume needed to contain this quantity of stars implies a distance of 10^{23} light years [13]. This means that we need to see stars at an epoch corresponding to 10^{23} years ago for the night sky to be bright. The Universe, however, is younger than this by 13 orders of magnitude (Section 5.1)! Thus, as we look out into space and back in time, our sight lines eventually reach an epoch before the formation of the first stars when the covering factor is still much less than unity. (Today, we refer to this epoch as *the dark ages.*) Remarkably, this solution was hinted at by Edgar Allan Poe in his prose–poem, *Eureka* in 1848 (see the prologue to this chapter).

3.5 HOW LIGHT PUSHES – RADIATION PRESSURE

Radiation pressure is the momentum (p) flux of radiation (the rate of momentum transfer due to photons, per unit area). It can also be thought of as the force, F, per unit area exerted by radiation and, since force is a vector, we will treat radiation pressure in this way as well[4]. Thus, the pressure can be separated into its normal, $P_\perp$, and tangential, $P_\parallel$, components with respect to the surface of a wall. The normal component of radiation pressure will be,

$$dP_\perp = \frac{dF_\perp}{dA} = \frac{dp}{dt dA}\cos\theta = \frac{dE}{c dt dA}\cos\theta \tag{3.21}$$

where we have expressed the momentum of a photon in terms of its energy (Table I.2). The angle, θ, is the angle between the incoming photon and the normal as shown in Figure 3.13. Using Eq. (3.12) we obtain,

$$dP_\perp = \left(\frac{1}{c}\right) I\cos^2\theta \, d\Omega \tag{3.22}$$

For the tangential component, we use the same development but take the sine of the incident angle, yielding,

$$dP_\parallel = \left(\frac{1}{c}\right) I\cos\theta \sin\theta \, d\Omega \tag{3.23}$$

[4] Pressure is actually a *tensor* which is described mathematically by a matrix, and a vector is a specific kind of tensor. We do not need a full mathematical treatment of pressure as a tensor, however, to appreciate the meaning of radiation pressure here.

Figure 3.13 An incoming photon exerts a pressure on a surface. For perfect absorption, the area will experience a force in the direction, $\vec{F}_{absorption}$. For perfect reflection, only the normal component of the force is effective and the resulting force will be in the direction, $\vec{F}_{reflection}$.

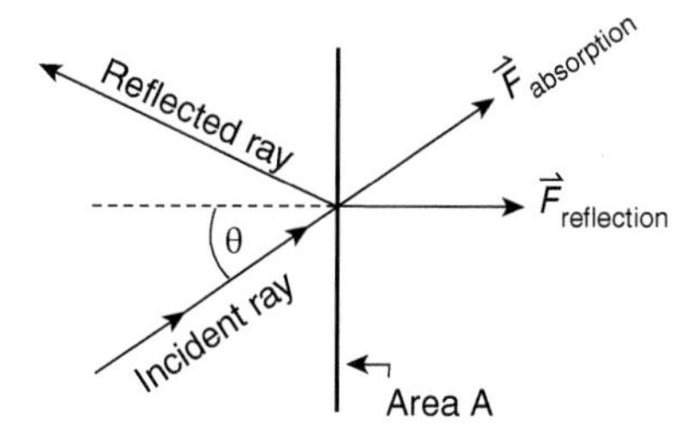

Suppose the radiation field is *isotropic*. We then need to integrate each component over 4π steradians.

$$P_{\perp} = \left(\frac{1}{c}\right)\int_{4\pi} I\cos^2\theta d\Omega = \frac{4\pi}{3c}I$$

$$P_{\parallel} = \left(\frac{1}{c}\right)\int_{4\pi} I\cos\theta\sin\theta d\Omega = 0$$

$$\text{therefore } P = \sqrt{{P_{\perp}}^2 + {P_{\parallel}}^2} = \frac{4\pi}{3c}I = \frac{1}{3}u \tag{3.24}$$

where we have used a spherical coordinate system for the integration (Appendix B of the *online material*), Eq. (3.18), and the fact that $J = I$ in an isotropic radiation field. Note that the units of pressure are equivalent to the units of energy density, as indicated in Table I.4. It is clear that the tangential component disappears, but the perpendicular component does not. Because photons carry momentum, the pressure is not zero in an isotropic radiation field. A surface placed within an isotropic radiation field will not experience a *net* force, however. This is similar to the pressure of particles in a thermal gas. There is no net force in one direction or another, but there is still a pressure associated with a 'photon gas'.

Now suppose the incoming radiation is in a *narrow beam from a fixed angle*, θ. This means that the solid angle subtended by that beam, Ω, is small. Then there is a force against the wall and the wall could accelerate if it is not fixed. However, the resulting acceleration depends on the kind of surface the photons are hitting. We consider two cases, illustrated in Figure 3.13: that in which the photon loses all of its energy to the wall (perfect absorption) and that in which the photon loses none of its energy to the wall (perfect reflection).

For *perfect absorption*, integrating Eqs. (3.22) and (3.23) with θ, Ω, constant, yields,

$$P_{\perp} = \left(\frac{1}{c}\right) I\Omega\cos^2\theta = \frac{f}{c}\cos^2\theta$$

$$P_{\parallel} = \left(\frac{1}{c}\right) I\Omega\cos\theta\sin\theta = \frac{f}{c}\cos\theta\sin\theta$$

$$P = \sqrt{{P_{\perp}}^2 + {P_{\parallel}}^2} = \frac{f}{c}\cos\theta \tag{3.25}$$

where we have used Eq. (3.13) with f the flux along the directed beam[5].

[5] For a narrow beam, this is equivalent to the Poynting flux (Table I.2).

For *perfect reflection*, only the normal component will have any effect against the wall (as if the surface were hit by a ball that bounces off). Also, because the momentum of the photon reverses direction upon reflection, the change in momentum is twice the value of the absorption case[6]. Thus, the situation can be described by Eq. (3.22) except for a factor of 2.

$$P = P_{\perp} = \left(\frac{2}{c}\right) I\Omega \cos^2\theta = \frac{2f}{c}\cos^2\theta \tag{3.26}$$

A comparison of Eqs. (3.25) and (3.26) shows that, provided the incident angle is not very large, a reflecting surface will experience a considerably greater radiation force than an absorbing surface. Moreover, as Figure 3.13 illustrates, the direction of the surface is *not* directly away from the source of radiation as it must be for the absorbing case. These principles are fundamental to the concept of a *Solar sail* (Figure 3.14).

The direction of motion depends on the angle between the radiation source and the normal to the surface, so it is possible to 'tack' a Solar sail by altering the angle of the sail, in a fashion similar to the way in which a sailboat tacks in the wind. Moreover, even if the acceleration is initially very small, it is *continuous* and thus high velocities could eventually be achieved for spacecraft powered by Solar sails (Problem 3.14). To date, the *Planetary Society* has achieved success with its LightSail 2 solar sail in orbit about the Earth[7].

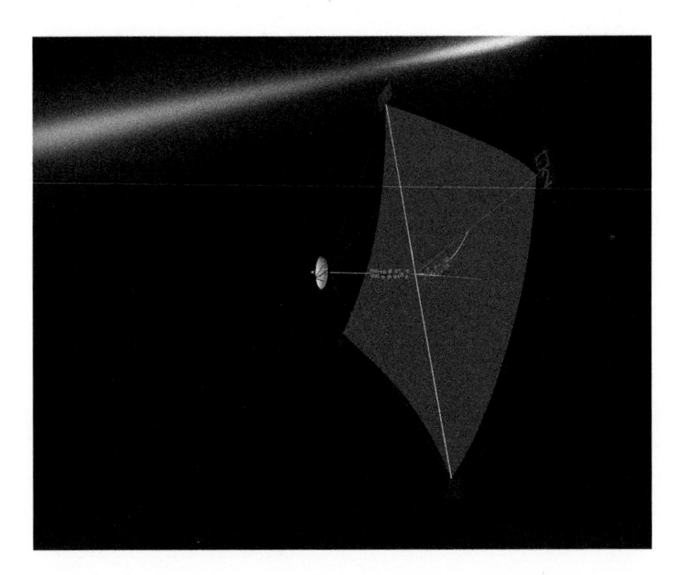

Figure 3.14 Artist's conception of a thin, reflective Solar sail, half a kilometre across. Source: Reproduced by permission of NASA/MSFC.

[6] A simple way of looking at this is when a particle hits a wall and bounces off. The particle exerts a force against the wall when the ball hits it and briefly 'sticks', and then the particle exerts another force against the wall when the particle jumps off again.
[7] See www.planetary.org, EXPLORE LightSail.

3.6 THE HUMAN PERCEPTION OF LIGHT – MAGNITUDES

Magnitudes are used to characterize light in the optical part of the spectrum, including the near IR and near UV. This is a logarithmic system for light, similar to decibels for sound, and is based on the fact that the response of the eye is logarithmic. It was first introduced in a rudimentary form by Hipparchus of Nicaea in about 150 BCE who labelled the brightest stars he could see by eye as 'first magnitude', the second brightest as 'second magnitude', and so on. Thus began a system in which brighter stars have *lower* numerical magnitudes, a sometimes confusing fact. As the human eye has been the dominant astronomical detector throughout most of history, a logarithmic system has been quite appropriate. Today, the need for such a system is less obvious since the detector of choice is the CCD (charge-coupled device, Section 4.2.2) whose response is linear. However, since magnitudes are entrenched in the astronomical literature, still widely used today, and well-characterized and calibrated, it is very important to understand this system.

3.6.1 Apparent Magnitude

The *apparent magnitude* and its corresponding flux density are values as measured above the Earth's atmosphere or, equivalently, as measured from the Earth's surface, corrected for the effects of the atmosphere,

$$\begin{aligned} m_\lambda - m_{\lambda_0} &= -2.5\log\left(\frac{f_\lambda}{f_{\lambda_0}}\right) \\ m_\nu - m_{\nu_0} &= -2.5\log\left(\frac{f_\nu}{f_{\nu_0}}\right) \end{aligned} \tag{3.27}$$

where the subscript, 0, refers to a standard calibrator used as a reference, m_λ and m_ν are apparent magnitudes in some waveband and f_λ, f_ν are flux densities in the same band. In Eq. (3.27) and all subsequent equations with magnitudes, 'log' refers to the base-10 logarithm. Note that Eq. (3.27) could also be written as a ratio of fluxes, since this would only require multiplying the flux density (numerator and denominator) by an effective bandwidth to make this conversion. This system is a *relative* one, such that the magnitude of the star of interest can be related to that of *any* other star in the same waveband via Eq. (3.27). For example, if a star has a flux density that is 100 times greater than a second star, then its magnitude will be 5 less than the second star. However, in order to assign a specific magnitude to a specific star, it is necessary to identify certain standard stars with known flux densities to which all others can be compared.

Several slightly different calibration systems have evolved over the years so, for careful and precise work, it is necessary to specify which system is being used when measuring or stating a magnitude. An example of such a system is the UBVRIJHKL Cousins–Glass–Johnson system for which parameters are provided in Table 3.1. The corresponding wavebands, U, B, V, etc., are illustrated in Figure 3.3. The apparent magnitude is

Table 3.1 Standard filters and magnitude calibration[a]

	U	B	V	R	I	J	H	K	L	L*
λ_{eff} [b]	0.366	0.438	0.545	0.641	0.798	1.22	1.63	2.19	3.45	3.80
$\Delta\lambda$ [c]	0.065	0.098	0.085	0.156	0.154	0.206	0.298	0.396	0.495	0.588
f_{ν_0} [d]	1.790	4.063	3.636	3.064	2.416	1.589	1.021	0.640	0.285	0.238
f_{λ_0} [e]	417.5	632	363.1	217.7	112.6	31.47	11.38	3.961	0.708	0.489
ZP_ν	0.770	−0.120	0.000	0.186	0.444	0.899	1.379	1.886	2.765	2.961
ZP_λ	−0.152	−0.601	0.000	0.555	1.271	2.655	3.760	4.906	6.775	7.177

[a] UBVRIJHKL Cousins–Glass–Johnson system [4]. The table values relate to a fictitious A0 star which has 0 magnitude in all bands. A star of flux density, f_ν, in units of 10^{-20} erg s^{-1} cm^{-2} Hz^{-1}, or f_λ, in units of 10^{-11} erg s^{-1} cm^{-2} Å^{-1}, will have a magnitude, $m_\nu = -2.5\log(f_\nu) - 48.598 - ZP_\nu$ or $m_\lambda = -2.5\log(f_\lambda) - 21.100 - ZP_\lambda$, respectively.
[b] The effective wavelength, in μm, is defined by $\lambda_{eff} = [\int \lambda f(\lambda)\, R_W(\lambda)\, d\lambda]/[\int f(\lambda)\, R_W(\lambda)\, d\lambda]$, where $f(\lambda)$ is the flux of the star at wavelength, λ, and $R_W(\lambda)$ is the response function of the filter in band W (see Figure 3.3). Thus, the effective wavelength varies with the spectrum of the star considered.
[c] Full width at half-maximum (FWHM) of the filters in μm.
[d] Units of 10^{-20} erg s^{-1} cm^{-2} Hz^{-1}.
[e] Units of 10^{-11} erg s^{-1} cm^{-2} Å^{-1}.
Source: Bessell et al. [4], © 1998, The European Southern Observatory.

commonly written in such a way as to specify these wavebands directly, e.g.

$$V - V_0 = -2.5\log\left(\frac{f_V}{f_{V_0}}\right)$$
$$B - B_0 = -2.5\log\left(\frac{f_B}{f_{B_0}}\right) \qquad \text{etc.} \tag{3.28}$$

where the flux densities can be expressed in either their λ-dependent or ν-dependent forms. The V-band ('visual') especially, since it corresponds to the waveband in which the eye is most sensitive (cf. Table T.4), has been widely and extensively used. B-band is 'blue' and R-band is 'red' to a good approximation. Some examples of apparent magnitudes are provided in Table 3.2.

The standard calibrator in most systems has historically been the star, Vega. Thus, Vega would have a magnitude of 0 in all wavebands (i.e. $U_0 = 0$, $B_0 = 0$, etc.) and its flux density in these bands would be tabulated. However, concerns over possible variability of this star, its possible IR excess, and the fact that it is not observable from the Southern hemisphere, has led to modified approaches in which the star Sirius is also taken as a calibrator and/or in which a model star is used instead. The latter approach has been taken in Table 3.1 which lists the reference flux densities for reference magnitudes of zero in all filters. The flux density and reference flux density must be in the same units. The above equations can be rewritten as,

$$m_\lambda = -2.5\log(f_\lambda) - 21.100 - ZP_\lambda$$
$$m_\nu = -2.5\log(f_\nu) - 48.598 - ZP_\nu \tag{3.29}$$

Table 3.2 Examples of apparent visual magnitudes[a]

Object or item	Visual magnitude	Comments
Sun	−26.8	
Approx. maximum of a supernova	−15	assuming V = 0 precursor
Full Moon	−12.7	
Venus	−3.8 to −4.5[b]	brightest planet
Jupiter	−1.7 to −2.5[b]	
Sirius	−1.44	brightest night-time star
Vega	0.03	star in constellation Lyra
Betelgeuse	0.45	star in Orion[c]
Spica	0.98	star in Virgo
Deneb	1.23	star in Cygnus
Aldebaran	1.54	star in Taurus
Polaris	1.97	the North Star[c]
Limiting magnitude[d]	3.0	major city
Ganymede	4.6	brightest moon of Jupiter
Uranus	5.7[e]	
Limiting magnitude[d]	6.5	dark clear sky
Ceres	6.8[e]	brightest asteroid
Pluto	13.8[e]	
Jupiter-like planet	26.5	at a distance of 10 pc
Limiting magnitude of HST[f]	28.8	1 h on A0V star
Limiting magnitude of ELT[g]	25.6[h]	39 m telescope

[a]From [11] (probable error at most 0.03 mag), and online sources.
[b]Typical range over a year.
[c]Variable stars. Betelgeuse has shown recent strong variability (see Figure 3.10). The value here is its 'normal' predimmed value.
[d]This is the faintest star that could be observed by eye without a telescope. It will vary with the individual and conditions.
[e]At or close to *opposition* (180° from the Sun as seen from the Earth).
[f]'Hubble Space Telescope', from Space Telescope Science Institute online documentation. The limiting magnitude varies with instrument used. The quoted value is the best case.
[g]'Extremely Large Telescope' refers to the European Southern Observatory's telescope in Chile.
[h]Assumes a three-hour exposure.
Source: Gupta [11]. © 2004, Royal Astronomical Society of Canada and a variety of online sources.

where f_λ is in units of erg cm^{-2} s^{-1} Å^{-1}, f_ν is in units of erg cm^{-2} s^{-1} Hz^{-1}, and ZP_λ, ZP_ν are called *zero point* values [4]. Example 3.5 provides a sample calculation.

Example 3.5

An apparent magnitude of B = 1.95 is measured at a particular time for the variable star, Betelgeuse. Determine its flux density in units of erg s^{-1} cm^{-2} Å^{-1}.

From Eq. (3.28) and the values from Table 3.1, we have

$$B - B_0 = 1.95 - 0 = -2.5\log\left(\frac{f_B}{632\times10^{-11}}\right) \tag{3.30}$$

Solving, this gives $f_B = 1.0\times10^{-9}$ erg cm^{-2} s^{-1} Å^{-1}, for Betelgeuse. Eq. (3.29) can also be used,

$$B = 1.95 = -2.5\log(f_B) - 21.100 + 0.601 \tag{3.31}$$

which, on solving, gives the same result.

3.6.2 Absolute Magnitude

Flux densities fall off as $\frac{1}{r^2}$, so measurements of apparent magnitude between stars do not provide a useful comparison of the *intrinsic* properties of stars without taking into account their various distances. Thus, the *absolute magnitude* has been introduced, either as a bolometric quantity, M_{bol}, or in some waveband (e.g. M_V, M_B, etc.). The absolute magnitude of a star is the magnitude that would be measured if the star were placed at a distance of 10 pc. Since the magnitude scale is relative, we can let the reference star in Eq. (3.27) be the *same* star as is being measured but placed at a distance of 10 pc,

$$m - M = -2.5\log\left(\frac{f}{f_{10\text{pc}}}\right) = -5 + 5\log\left(\frac{d}{\text{pc}}\right) \tag{3.32}$$

where we have dropped the subscripts for simplicity and we have used $f/f_{10pc} = (10/d)^2$ from Eq. (3.9). Here, d is the distance to the star in pc. Eq. (3.32) provides the relationship between the apparent and absolute magnitudes for any given star. The quantity, $m - M$, is called the *distance modulus.* Because this quantity is directly related to the distance, it is sometimes quoted as a proxy for distance. Writing a similar equation for a reference star and combining with Eq. (3.32), we find (e.g. Problem 3.15),

$$M - M_{b\odot} = -2.5\log\left(\frac{L}{L_\odot}\right) \tag{3.33}$$

where we have used the Sun for the reference star (Table T.2). Eq. (3.33) has been explicitly written with bolometric quantities but one could also isolate specific bands, as before, provided the corresponding reference values are used.

3.6.3 The Colour Index and Bolometric Correction

The *colour index* is the difference between two magnitudes in different bandpasses for the same star, for example,

$$B - V = -2.5\log\left(\frac{f_B}{f_V}\frac{f_{V_0}}{f_{B_0}}\right) = -2.5\log\left(\frac{f_B}{f_V}\right) - (ZP_B - ZP_V) \tag{3.34}$$

or between any other two bands (V – R, V – I, etc.). Eq. (3.34) is derivable from Eqs. (3.28) or (3.29). Various colour indices are provided for different kinds of stars[8] in Tables T.5–T.7. Since this quantity is basically a measure of the ratio of flux densities at two different wavelengths (with a correction for zero point), it is an indication of the *colour* of the star. A positive value for B – V, for example, means that the blue magnitude is higher than the V magnitude, and therefore, the flux density in the V band is *higher* than that in the B band; hence, the star will appear more 'yellow' than 'blue'.

The colour index, since it applies to a single star, is *independent of distance*. (To see this, note that converting the flux density to a distance-corrected luminosity would require the same factors in the numerator and denominator of Eq. (3.34)). Consequently, the colour index can be compared directly, star to star, without concern for the star's distance[9]. We will see in Section 6.1.3 that colour indices are a measure of the surface temperature of a star. This means that stellar temperatures can be determined without having to know their distances, a highly useful bit of information!

Because a colour index could be written between any two bands, one can also define an index between one band and *all* bands. This is called the *bolometric correction*, often defined for the V band,

$$\mathrm{BC} = m_{\mathrm{bol}} - \mathrm{V} = M_{\mathrm{bol}} - M_{\mathrm{V}} \tag{3.35}$$

For any given star, this quantity is a correction factor that allows one to convert from a V band measurement to the bolometric magnitude (Problem 3.19). Values of BC are provided for various stellar types in Tables T.5–T.7.

3.6.4 *Gaia* and the HR Diagram

The *Gaia* (Global Astrometric Interferometer for Astrophysics) satellite of the European Space Agency (ESA) was launched in 2013 with goals to accurately measure positions, distances, magnitudes and motions of stars. The concepts were standard, for example measurements of *parallax* which give stellar *distances* (see Section 3.1 of the *online material*) have been successfully carried out since the nineteenth century. However, the gain in positional accuracy was a factor of 50–100 over the best predecessor satellite (*Hipparcos*) and the increase in the number of measured stars was a factor of 10 000 [8]. In all, *Gaia* is measuring almost two *billion* stars, allowing for an accurate three-dimensional picture of our Milky Way neighbourhood to a variety of distances depending on the stellar magnitude.

Figure 3.15 shows over four million of these stars in one extremely important result. This is a plot of absolute magnitude (y-axis) as a function of colour index (x-axis) for stars in the sample that have low *extinction*, meaning that there is little obscuration by dust (Section 5.5.1). Such a plot is called a *Hertzsprung–Russell* (*HR*) *diagram* or a *colour–magnitude diagram* (*CMD*). To convert from the measured apparent magnitude to

[8] Stellar spectral types will be discussed in Section 5.4.7.
[9] This is not the case, however, if the stars are in different galaxies and the spectrum of one star is significantly *redshifted* (Section 9.1) compared to the other.

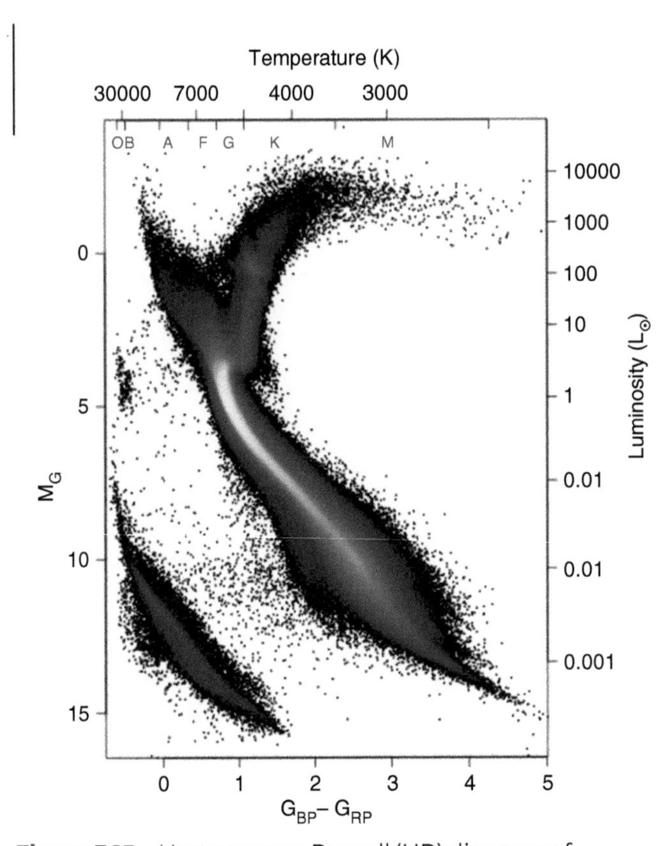

Figure 3.15 Hertzsprung–Russell (HR) diagram of 4 276 690 stars with low extinction observed with the *Gaia* satellite. Colours represent the numbers of stars, with yellow corresponding to the highest number. The absolute magnitude in the G-band (y-axis) is plotted against colour index, G_{BP}-G_{RP} (x-axis). Most stars fall on a region passing from upper left (hot, luminous stars) to lower right (dim, cool stars), called the *main sequence* and the parallel curve at the lower left represents *white dwarfs*. The main sequence is defined by stars that are burning hydrogen into helium in their cores whereas white dwarfs are remnants of low mass stars after all nuclear burning has ceased (Section 5.3.2). Luminosity is given at the right, and the temperature is at the top. Stellar spectral types for the main sequence are also given at the top. Source: Gaia Collaboration [9]. Licensed under CC BY 4.0.

Table 3.3 Gaia filter data[a]

	G	G_{BP}	G_{RP}
λ_m^b	639.74	516.47[c]	783.05
$\Delta\lambda^d$	330 → 1050	330 → 680	630 → 1050

[a]G filter information for Gaia.
[b]The mean wavelength, in nm, is defined by $\lambda_m = [\int \lambda p(\lambda)\, d\lambda]/[\int p(\lambda)\, d\lambda]$, where $p(\lambda)$ is the response function of the filter defined as the ratio of the recorded photo-electrons to the number of photons entering into the instrument, as a function of wavelength [17].
[c]The value is 511.78 nm for faint sources.
[d]Wavelength range of the filter in nm [7].

absolute magnitude requires that the distance be known (Eq. (3.32)), a feat that has been accomplished to unprecedented accuracy by *Gaia*. (A stylized HR diagram showing stellar spectral types can be seen in Figure 7.12.)

As indicated in Section 3.6.1, whenever magnitudes are specified, it is important to know which magnitude system is being used. The *Gaia* magnitudes of Figure 3.15 were measured in three filters (Table 3.3): a very broadband G filter that stretches over the entire visible spectrum, and two narrower 'blue' (G_{BP}) and 'red' (G_{RP}) filters. These blue and red filters do not exactly match the B and R filters of Table 3.1 so transformation equations are required to relate these systems. A subset of these transformations is

$$\begin{aligned}
G - V &= -0.02907 - 0.02385(B - V) - 0.2297(B - V)^2 - 0.001768(B - V)^3; \quad (\sigma = 0.063)\\
G - V &= -0.01746 + 0.008092(V - I) - 0.2810(V - I)^2 + 0.03655(V - I)^3; \quad (\sigma = 0.047)\\
G - V &= -0.01760 - 0.006860(G_{BP} - G_{RP}) - 0.1732(G_{BP} - G_{RP})^2; \quad (\sigma = 0.046)\\
G - I &= +0.02085 + 0.7419(G_{BP} - G_{RP}) - 0.09631(G_{BP} - G_{RP})^2; \quad (\sigma = 0.050)
\end{aligned} \tag{3.36}$$

where the uncertainties, σ, are provided in parentheses[10]. Example 3.6 provides a sample calculation.

Example 3.6

The star, Procyon, in the constellation, Canis Minor (the little dog), has an apparent magnitude of V = +0.34 and a colour index of B-V = +0.42. Find G and ($G_{BP} - G_{RP}$).

We use the first of Eqs. (3.36)

$$G - V = -0.02907 - 0.02385(0.42) - 0.2297(0.42)^2 - 0.001768(0.42)^3 = -0.0797$$

so G = −0.0797 + 0.34 = 0.26.

[10] Gaia Data Release 2 Documentation release 1.2, https://gea.esac.esa.int/archive/documentation/GDR2/ Part II *Gaia Data Processing*, section 5.3.7.

From the third of Eqs. (3.36)

$$-0.0797 = -0.01760 - 0.006860(G_{BP} - G_{RP}) - 0.1732(G_{BP} - G_{RP})^2$$

There are two real solutions to this equation: $G_{BP} - G_{RP} = +0.579$ (higher flux in the red filter) or $G_{BP} - G_{RP} = -0.619$ (higher flux in the blue filter). Remember that a lower numerical value for magnitude means a higher flux. Since we know that B-V = +0.42, this means that the higher flux belongs to the 'redder' V rather than the blue B. Then, the correct solution must be $G_{BP} - G_{RP} = +0.579$ for any normal stellar spectrum.

From the CMD, physical properties of stars can be found. The absolute magnitude can be converted into luminosity (Eq. (3.33)) when the distance is known. Also, the colour index, which is a measure of the ratio of flux densities (Eq. (3.34)), can be converted to a temperature (see, e.g., the calibration of Figure T.1). Therefore, the HR diagram can be plotted as luminosity (y-axis) against temperature (x-axis) which are more physically meaningful parameters for stars and are also plotted in Figure 3.15. The HR diagram shows that stars do *not* have arbitrary temperatures and luminosities but rather fall along well-defined regions in $L - T$ parameter space. This diagram contains a treasure trove of information about the physics and driving power of stars, including our Sun, along with their past and future evolutions, and will be discussed more fully in Section 5.3.2.

3.6.5 Magnitudes Beyond Stars

Magnitudes are widely used in optical astronomy and, though the system was developed to describe stars (for which specific intensities cannot be measured, Section 3.3), it can be applied to any object, extended or point-like, as an alternate description of the flux density. The one restriction is that magnitudes are generally only used for the optical band, which includes the near IR and near UV.

An example of a point-like object that is *not* a star is a *QSO*, or 'quasi-stellar object'. A QSO is the bright active nucleus of a very distant galaxy that looks star-like at optical wavelengths. QSOs that also emit strongly at radio wavelengths are called *quasars*, although sometimes the terms are used interchangeably. An apparent magnitude may be quoted for this or other point-like sources.

As for extended objects, galaxies are typical examples. It is common to express the brightness of a galaxy, not as a specific intensity, but rather in terms of magnitudes per unit solid angle (Problem 3.20).

3.7 LIGHT ALIGNED – POLARIZATION

The magnetic and electric field vectors of a wave are perpendicular to each other and to the direction of propagation (Figure I.1). A signal consists of many such waves travelling in the same direction in which case the electric field vectors are usually randomly oriented around the plane perpendicular to the propagation direction. However, if all of the electric field vectors are aligned (say all along the z-axis of Figure I.1) then the signal is said to be polarized. Partial polarization occurs if some of the waves are aligned but others randomly oriented. The *degree of polarization*, D_p is defined as the fraction of total intensity that is polarized (expressed as a percentage),

$$D_p \equiv \frac{I_{pol}}{I_{tot}} = \frac{I_{pol}}{I_{pol} + I_{unpol}} \tag{3.37}$$

Polarization can be generated internally by processes intrinsic to the energy generation mechanism (Section 10.5, for example), or polarization can result from the scattering of light from particles, be they electrons, atoms, or dust particles (Section 7.1 or Appendix D of the *online material*). When polarization is observed, D_p is usually of order only a few percent and 'strong' polarization, such as seen in radio jets (Figure 3.4), typically is $D_P \lesssim 15\%$ [16]. Values of D_p over 90%, however, have been detected in the jets of some pulsars [15] and in the low-frequency radio emission from planets (Section 10.5.1). For the Milky Way, $D_p = 2\%$ over distances 2–6 kpc from us due to scattering by dust [16]. In practical terms, this means that polarized emission is usually much fainter than unpolarized emission and requires greater effort to detect.

PROBLEMS

3.1. Assuming that Cas A (see Section 3.1 for data) has a spectral luminosity between $\nu_1 = 10^7$ Hz and $\nu_2 = 2 \times 10^{10}$ Hz of the form, $L_\nu = K\nu^{-0.7}$, where K is a constant, determine the value of K and specify its units. Also find L_ν and L_λ at $\nu = 10^9$ Hz.

3.2. Find, by comparison with exact trigonometry, the angle, θ (provide a numerical value in degrees), above which the small-angle formula (Eq. I.9), departs from the exact result by more than 1%. How does this compare with the relatively large angle subtended by the Sun?

3.3. Determine the flux density (in Jy) of a cell phone that emits 2 mW cm^{-2} at a frequency of 1900 MHz over a bandwidth of 30 kHz (assume a flat spectrum) and of the Sun, as measured at the Earth, at the same frequency. Equation 6.6 provides an expression for calculating the specific intensity of the Sun at that frequency. Compare these to the flux density of the supernova remnant, Cas A (~1900 Jy as measured at the Earth at 1900 MHz) which is the strongest radio source in the sky after the Sun. Comment on the potential of cell phones to interfere with the detection of astronomical signals.

3.4. (a) Consider a pulsar with radiation that is beamed uniformly into a circular cone of solid angle, Ω. Rewrite the right-hand side (RHS) of Eq. (3.9) for this case (see Eq. I.10).

(b) If $\Omega = 0.02$ sr, determine the error that would result in L if the RHS of Eq. (3.9) were used rather than the correct result from part (a).

3.5. Determine the percentage variation in the solar flux incident on the Earth due to its elliptical orbit. Compare this to the variation shown in Figure 3.6.

3.6. [*Online*] Determine the flux in a perfectly isotropic radiation field (see Section B.2).

3.7. (a) Determine the flux and intensity of the Sun (i) at its surface, (ii) at the mean distance of Mars, and (iii) at the mean distance of Pluto.
(b) How large (in arcmin) would the Sun appear in the sky at the distances of the two planets? Would it appear resolved or as a point source to the naked eye at these locations? That is, would the angular diameter of the Sun be larger than the resolution of the human eye (Table T.4) or smaller?

3.8. [*Online*] While standing near the Arctic Circle, you observe a 'coronal' aurora, which is an aurora that originates in the zenith and streaks down around you on all sides. Assume that the specific intensity of the aurora can be described by the function $I_\nu = I_{\nu 0} \cos(z)$, where z is the zenith angle, i.e. the angle from the zenith downwards. If the maximum zenith angle reached by the aurora is 45°, find a simple expression for the flux density, f_ν, in terms of $I_{\nu 0}$ (see Section B.2).

3.9. The radio spectrum of Cas A, whose image is shown in Figure 3.2, is given in Figure 10.15 in a log–log plot. The plotted specific intensity can be represented by $I_\nu = I_{\nu 0}(\nu/\nu_0)^\alpha$ in the part of the graph that is declining with frequency, where ν_0 is any reference frequency in this part of the plot, $I_{\nu 0}$ is the specific intensity measured at ν_0 and α can be measured from the slope. In Problem 3.1, we assumed that $\alpha = -0.7$. Now, instead, measure this value from the graph and determine, for the radio band from $\nu_1 = 2 \times 10^7$ to $\nu_2 = 2 \times 10^{10}$ Hz,
(a) the intensity of Cas A, I,
(b) the solid angle that it subtends in the sky, Ω,
(c) its flux, f,
(d) its radio luminosity, L_{rad} (erg s^{-1}). Confirm that this value is approximately equal to the value given in Section 3.1.

3.10. On average, the brightness of the Whirlpool Galaxy, M 51 (see Figure 5.8 or 11.8), which subtends an ellipse of major × minor axis, $11.2' \times 6.90'$ in the sky, is 2.1 times that of the Andromeda Galaxy, M 31 (subtending $190' \times 60'$). What is the ratio of their flux densities? (Eq. I.11 will be useful.)

3.11. The specific intensity of a hot ($T = 40\,000$ K) star is given by Eq. (6.6) when it is observed at infrared wavelengths (Table I.1). *For this band*, find the following:
(a) the intensity.
(b) the number of photons s^{-1} cm^{-2} sr^{-1}.
(c) the flux at the surface of the star.
(d) the luminosity, L_{IR}, if its radius is 10 $R_\odot$.
(e) the total number of photons s^{-1}.

3.12. Where does the Solar System end? There are different ways of defining the 'edge' of the Solar System[11], but suppose we take it to be the distance (in AU) at which the radiation energy density from the Sun is equivalent to the ambient mean energy density of interstellar space, the latter about 10^{-12} erg cm^{-3}. After more than 40 years of space travel, how far away is the Voyager 1 spacecraft (do an internet search for this value)? Is it out of the Solar System?

[11] For example, see https://spaceplace.nasa.gov/oort-cloud/en.

3.13. Suppose we approximate the entire non-solar CR energy spectrum shown in Figure 1.8 from 10^{10} to 10^{20} eV, with the function

$$J(E) = 5 \times 10^4 E^{-2.7} \tag{3.38}$$

where the units are the same as specified in Eq. (3.15). Find the energy density of non-Solar cosmic rays (eV cm^{-3}) in the vicinity of the Earth.

3.14. Consider a circular, perfectly reflecting solar sail that is initially at rest at a distance of 1 AU from the Sun and pointing directly at it. Assume that the initial location of the sail is far enough from the Earth that the Earth's gravitational field and any atmospheric or ionic drag are negligible. The sail is carrying a payload of 1000 kg (which dominates its mass), and its radius is $R_s = 500$ m.

(a) Derive an expression for the acceleration as a function of distance, $a(r)$, for this Solar sail. Include the Sun's gravity as well as its radiation pressure. (The constants may be evaluated to simplify the expression.)

(b) Manipulate and integrate this equation to find an expression for the velocity of the Solar sail as a function of distance, $v(r)$. Evaluate the expression to find the velocity of the Solar sail by the time it reaches the orbit of Mars.

(c) Finally, derive an expression for the time it would take for the sail to reach the orbit of Mars. Evaluate it to find the time. Express the time as seconds, months, or years, whatever is most appropriate.

3.15. Derive Eq. (3.33) (see Section 3.6.2).

3.16. Repeat Example 3.5 but expressing the flux density in its frequency-dependent form.

3.17. For both the U band and L* band filters, convert the value of f_λ read from Table 3.1 into f_ν and verify that your answer agrees with the corresponding value of f_ν in the table. Why might there be minor differences?

3.18. Refer to Table 3.2 for the following.

(a) Determine the range (ratio of maximum-to-minimum flux density) over which the unaided human eye can detect light from astronomical objects. Research the range of human hearing from 'barely audible' to the 'pain threshold' and compare the resulting range to your results for the eye.

(b) Find a web-based tool that will return the number of stars in the sky above a given magnitude limit[12]. What percentage of stars in the night sky would one lose by moving from a very dark country site into a nearby light-polluted city?

(c) The star, Betelgeuse, is at a distance of 130 pc. Determine how far away it would have to be before it would be invisible to the unaided eye, if it were to undergo a supernova explosion. Assume that Betelgeuse's magnitude has its 'normal' pre-dimmed value.

[12] An example is https://simbad.u-strasbg.fr/simbad/sim-fsam.

3.19. A star at a distance of 25 pc is measured to have an apparent magnitude of V = 7.5. This particular type of star is known to have a bolometric correction of BC = −0.18. Determine the following quantities:
   (a) the flux density, f_V in units of erg cm^{-2} s^{-1} Å^{-1},
   (b) the absolute V magnitude, M_V,
   (c) the distance modulus,
   (d) the bolometric apparent and absolute magnitudes, m_{bol}, and M_{bol}, respectively,
   (e) the luminosity, L in units of $L_\odot$.

3.20. A galaxy of uniform brightness at a distance of 16 Mpc appears elliptical on the sky with major and minor axis dimensions, $7.9' \times 1.4'$. It is observed in the radio band centred at 1.4 GHz (bandwidth = 600 MHz) to have a specific intensity of 4.8 mJy beam^{-1}, where the 'beam' is a circular solid angle of diameter, 15″ and uniform response. A measurement is then made in the optical B band of 22.8 magnitudes per pixel, where the pixel corresponds to a square on the sky which is one arcsecond on a side. Assume that the spectrum is approximately flat in each band and determine (all in cgs units) I_ν, f_ν, f, and L of the galaxy in each band. In which band is the source brighter? In which band is it more luminous?

3.21. The limiting magnitude of some instruments can be pushed fainter by taking extremely long exposures. Estimate the limiting magnitude of the Hubble Ultra-Deep Field from the information given in Figure 5.1.

3.22. The brightest star in the night sky is Sirius at a distance of d = 2.64 pc, which consists of two components: Sirius A (V = −1.47, I = −1.44), and the much fainter Sirius B (V = 8.528, I = 8.802). Place these two stars on the HR diagram of Figure 3.15 and comment on what kinds of stars they are.

Figure 3.16 The Tesla Roadster containing Starman with the Earth overhead. Source: Reproduced under the Creative Commons CCO 1.0 Universal Public Domain Dedication, Credit: SpaceX/CCO.

JUST FOR FUN

3.23. Suppose the Sun instantly changed into a white dwarf of the same mass. What would be the flux at the distance of the Earth? Would the white dwarf be resolved or would it look like a point source to the naked eye? Suppose you were on Mercury instead? Would it be more hospitable?

3.24. [*Online*] In 2018, SpaceX, under the leadership of Elon Musk, launched the 'Tesla Roadster' into space. This car contained a dummy, affectionately called 'Starman' (Figure 3.16). Suppose that you were Starman at a distance of 14 000 km from the centre of the Earth, and looking at a full-Earth directly overhead. Sketch a diagram showing the angle subtended by the Earth from your vantagepoint. Take the intensity of the Earth to be constant with an average value of $\bar{I}$ and find the flux in terms of this quantity (see Eq. (B.5)).

3.25. Find out where Starman is now[13]. What are the intensity, flux and angular diameter of the Sun, from his position?

[13] For example, www.whereisroadster.com.

Chapter 4

Measuring the Signal

All these facts were discovered ... with the aid of a spyglass which I devised, after first being illuminated by divine grace.

– Galileo Galilei in *The Starry Messenger*

4.1 SPECTRAL FILTERS AND THE PANCHROMATIC UNIVERSE

The first astronomical instrument was the human eye. From prehistoric times until today, human beings have surveyed the heavens with this most elegant and effective 'telescope'. The eye, like any other instrument, acts like a filter, accepting light at some wavelengths and filtering out others. *The spectral response function* of the human eye is shown in Figure 4.1 (see also Table T.4) and illustrates the relative ability of the eye to detect light at different wavelengths in the visual part of the electromagnetic spectrum[1]. If an incoming signal were uniformly bright across all wavelengths, the eye would not perceive it that way, but rather would see light in the range 500–550 nm (depending on conditions) as the brightest. Outside of this relatively narrow visual band, the eye is unable to detect a signal at all.

Attempts have been made to improve upon the human eye for at least 700 years. Early versions of eyeglasses, for example, date to 1284 or 1285 AD and possibly earlier in a more

[1] 'Visual' will be used to indicate the part of the spectrum to which the human eye is sensitive, where as 'optical' will refer to the part of the spectrum to which ground-based optical telescopes are sensitive; the latter includes the near-IR and the near-UV.

Astrophysics: Decoding the Cosmos, Second Edition. Judith A. Irwin.

Companion website: www.wiley.com/go/irwin/astrophysics2e

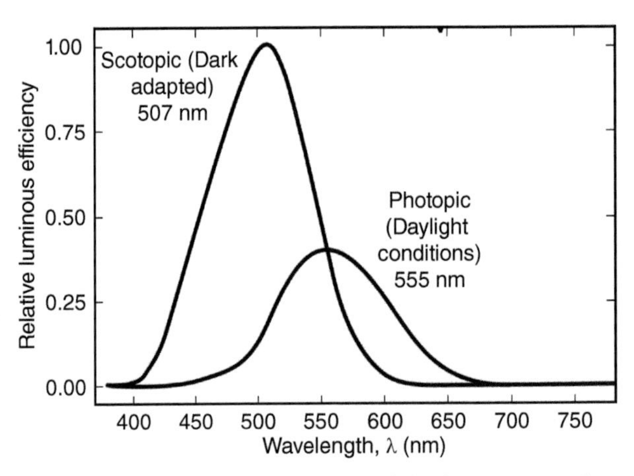

Figure 4.1 Response function of the human eye for photopic (daylight) and scotopic (dark-adapted) conditions. The wavelengths are indicated. The dark-adapted response is dominated by rods in the eye's retina, whereas daylight vision is dominated by the cones. Sensitivity to light is much higher under dark-adapted conditions, as shown, but the ability to distinguish colour is markedly diminished. Related information can be found in Table T.4.

primitive fashion. However, it was not until Galileo first turned a telescope on the Moon, the Sun, and planets in the early seventeenth Century that combinations of lenses were put towards the purpose of astronomical observation. Since Galileo, we have enjoyed 400 years of increasingly larger and more sophisticated optical telescopes. By contrast, other wavebands were not even known to exist until the discovery of the infrared by William Herschel in 1800. Once it was understood that there could be emission invisible to the human eye (see Table I.1), it was only a matter of time before a variety of developing technologies provided the means to detect and form images in non-visual wavebands. The technology to detect radio signals, for example, matured rapidly after World War II. In a sense, such technologies provide us with eyeglasses to enhance the response of the human eye 'sideways' into other wavebands.

One other important constraint inhibited the discovery of far-IR, UV, X-ray, and γ-ray emission from astronomical objects until the latter half of the twentieth Century – that of the Earth's atmosphere. The spectral response function of the atmosphere is shown in Figure 4.2 and reveals that only in the optical and radio wavebands is the atmosphere sufficiently transparent for ground-based observations. The 1957 launch of Sputnik by the former Soviet Union marked the dawn of the space age and, with it, a rich era of astronomical discovery. This era continues today, with space-based probes realizing increasingly sensitive and highly detailed images of astronomical objects at the previously inaccessible 'invisible' wavelengths.

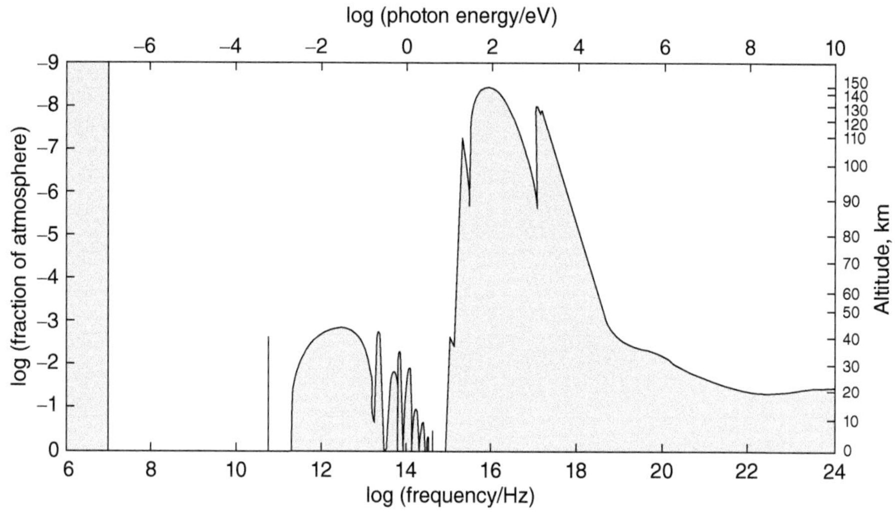

Figure 4.2 Atmospheric transparency curve for the Earth. The altitude (right-hand side) indicates how high one would have to put a telescope in order for the atmosphere to be transparent at that frequency. This is expressed as a fraction of the total atmosphere on the left-hand side. In this plot, wavelength decreases to the right in contrast to Figure 4.1. The two atmospheric windows are in the radio and optical parts of the spectrum. Most atmospheric absorption is due to water vapour (H_2O), then carbon dioxide (CO_2), and then ozone (O_3). The single spike around 60 GHz is due to oxygen (O_2). Source: Adapted from [14].

Thus, the history of astronomy has largely been driven by two spectral response functions: that of the human eye and that of the Earth's atmosphere. Our knowledge of the Universe is still based on information obtained through a strong optical and ground-based bias. In 2020, for example, there were 42 optical ground-based research telescopes greater than 3.0 m in diameter and several more with diameters in the tens-of-metres category were either under construction or planned (e.g. the ELT mentioned in Table 3.2). There were also seven operational optical telescopes in orbit[2]. We have come a long way since Galileo's 4-cm diameter 'spyglass' cracked the door open on the sky! The number and quality of non-optical and space-based data are rapidly increasing and are revealing aspects of our Universe never before seen – or even contemplated.

Why is such a concerted effort required over so many wavebands? Figure 4.3, which shows the spectrum of the galaxy, NGC 2903, provides a good illustration. This rather normal spiral galaxy (see Figure 4.11 for an optical image) emits over a wavelength range spanning 10 orders of magnitude, and probably more if the observations were available. A wealth of information is hidden in this plot with each band revealing new and different insights about the source. If an image were taken at each waveband, the images would not

[2] See updates at https://en.wikipedia.org/wiki/List_of_largest_optical_reflecting_telescopes and https://en.wikipedia.org/wiki/List_of_space_telescopes.

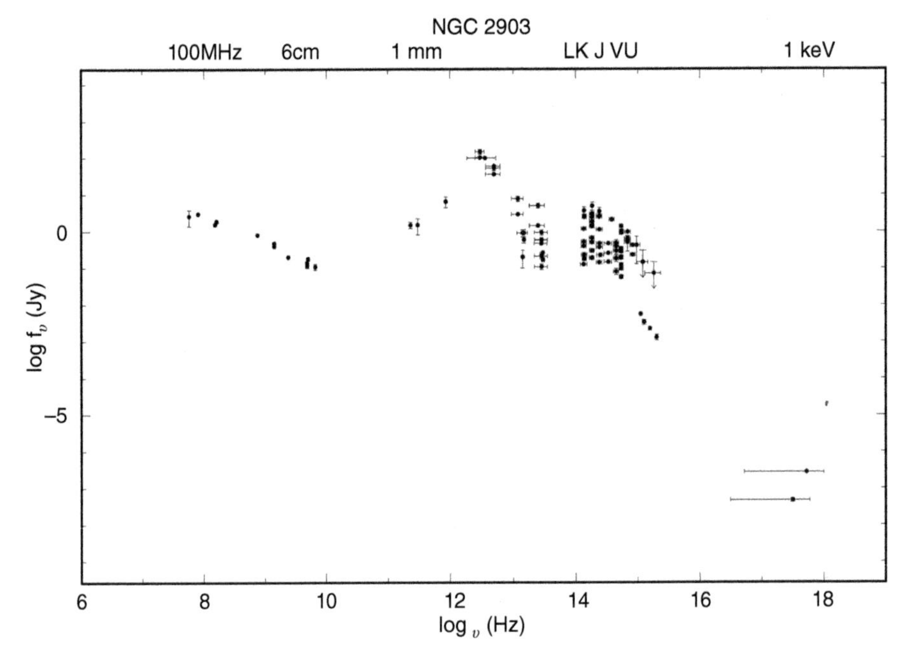

Figure 4.3 Spectrum of the galaxy, NGC 2903. The frequency is shown at the bottom, and some specific wavebands are labelled at the top. See Section 12.1.1 for a discussion of this spectrum. Source: Adapted from the NASA/IPAC Extra galactic Database [15].

be identical to each other. Further clues about the system are uncovered by this changing appearance with waveband. It is clearly impossible to truly understand the galaxy with observations in the optical band alone – an approach is required that is 'panchromatic', or spanning all wavebands. If we now think about the many different types of objects one could observe in the Universe besides normal galaxies, the imperative to widen our waveband horizons becomes even greater.

Astronomy is now in a discovery era, driven by technological improvements in the telescope and its concomitant instruments. It is therefore essential to have some understanding of how this technology, whether space-based or under the blanket of our atmosphere, collects, filters, distorts, and ultimately reveals the secrets of the astronomical sources that so intrigue us.

4.2 CATCHING THE SIGNAL – THE TELESCOPE

Although the technology required to collect light and the quality of the result vary enormously across wavebands, the basic principles are the same: the job of a telescope is to collect as much light as possible and to focus it on a detector[3], forming an image,

[3] Focussing is not possible, however, at the highest γ-ray energies (see Figure 4.7b).

if possible. For scientific purposes, there are two main ingredients in this process: a surface that collects and focusses the light called the *objective* (also called the primary lens, or mirror or antenna, depending on what is used), and a device (the *detector*) that detects and converts the received signal into some form, usually digital, for storage and analysis via computer. This is analogous to what the human eye does. The pupil is the opening through which light can pass, acting as the collector, focussing is achieved by the lens and the interior fluid, and the detector is the retina with its plethora of rods and cones. The eye's detector then converts the light into electrical signals that are sent to the brain to be analysed. A telescope, therefore, acts rather like a giant eye.

Figure 4.4 shows a simple telescope that uses a lens as the objective. However, the relationships given below are the same for a reflecting instrument. Since astronomical signals are at a great distance, the incoming wavefronts from a source at any angle in the sky will be plane parallel and are intercepted by the entire aperture, as shown here and in Figure 4.5. The focussed image is in the 'focal plane' a distance, f, from the objective behind (to the right of) the lens, and this is where the detector must therefore be placed. The geometry of the diagram shows how the angular size of any object on the sky, θ, is related to a linear size on the detector, l. The angles are small, so we can use the small angle formula (Eq. I.9)

$$l = f\theta \tag{4.1}$$

where θ is in radians and l and f have the same units. If θ is the angle of the galaxy on the sky, then l will be linear size of that galaxy on the detector. Telescopes with longer focal

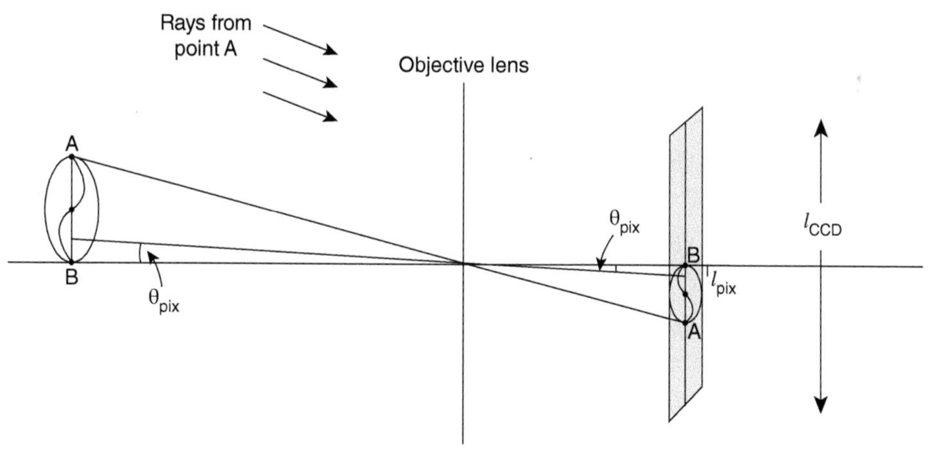

Figure 4.4 Diagram of a simple optical telescope, showing the relationships between the focal length, f, the linear scale on the detector, l, and the angular scale in the sky, θ. Subscripts on l indicate whether it is the full size of the CCD or just the size of a pixel (Section 4.2.2). Note that the angles have been exaggerated for clarity, but they are typically very small. The rays from a very distant point, A, all impinge upon the objective parallel to each other and at the same angle. The same is true of the rays from point B. The rays shown are those passing through the centre of the lens which experience no net bending due to lens symmetry.

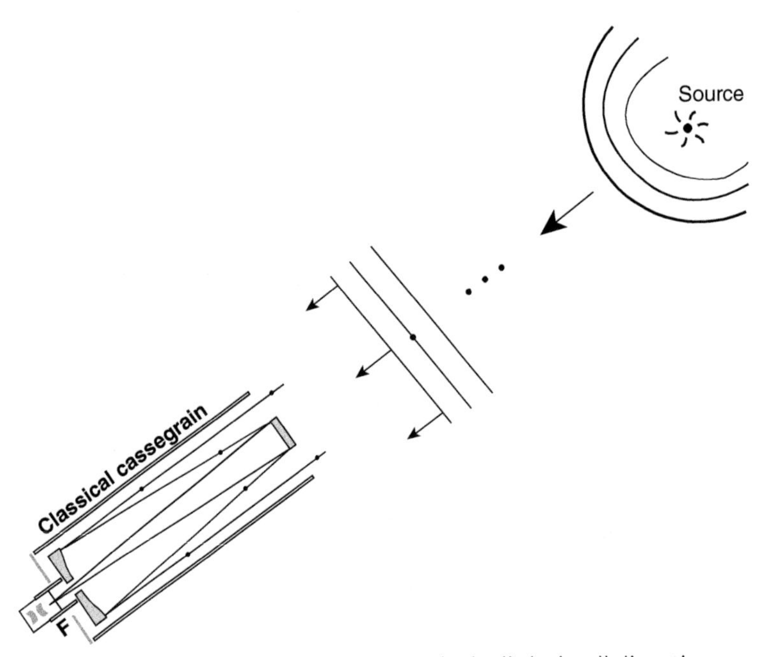

Figure 4.5 A distant point source emits its light in all directions, but the wavefront is plane parallel by the time it reaches the telescope. Rays denote the direction of the incoming plane parallel wavefront, in this case from a point source 'on-axis', i.e. at the centre of the field of view. This telescope shows a typical design in which curved mirrors collect and focus the light onto a detector behind the primary mirror. Such a design may have different names depending on the curvature of the mirrors and whether additional corrective lenses are used. To measure the focal length, start at the primary mirror, measure to the secondary mirror, and then add the length from the secondary mirror to the focal plane at F.

lengths result in larger images on the detector. Since f is a constant for any telescope, so is the ratio, θ/l. This ratio is called the *plate scale* for optical instruments, usually expressed in non-cgs units such as arcsec/mm.

Telescopes are designated by their diameters and *focal ratios*, the latter being the ratio of the focal length to the diameter of the objective. For example, a 3 m f/5 telescope would be 3 m in diameter and have a focal length of $f = 5 \times 3 = 15$ m. Bigger telescopes can detect fainter signals (Section 4.2.1). Bigger telescopes can also produce images that show finer detail, referred to as having good *resolution* (Section 4.2.4). However, resolution is subject to other constraints which will be outlined in Sections 4.2.3 and 4.3.2.

4.2.1 Collecting and Focussing the Signal

At optical wavelengths, collecting and focussing light are done via either a *lens* (refracting the light) or *curved mirror* (reflecting the light). For everything else equal, larger collecting areas result in brighter images on the detector because a larger area will intercept more of the source luminosity. Since astronomical sources are typically very faint, telescopes are built as large as costs and mechanical structures allow. Large optical telescopes used for research always use mirrors, since they require finishing only on one side and can be supported more easily when the mirror is at the bottom of the telescope tube. Moreover, since reflection affects all wavelengths of optical light the same way whereas refraction bends light differently for different wavelengths (e.g. Section 7.3) the use of mirrors avoids certain *aberrations* associated with lenses[4]. Since the light then reflects upwards, either a detector must be placed at the focal point high up in the tube, called the *prime focus*, or else another *secondary mirror* must be used to reflect the signal back down again to a detector near the bottom (Figure 4.5) or in some other direction. If the incoming light changes direction via mirrors, the light path is said to be *folded*. Modern detectors can be rather massive and so are often placed at or near the bottom of the telescope. A variety of light paths are possible, however, including those with secondary mirrors at positions that do not block the aperture and those with detectors on platforms adjacent to the telescope tube.

Single-dish radio telescopes, infrared, optical, and ultraviolet telescopes all operate similarly, with collection and focussing achieved via a primary reflector. Figure 4.6 shows a radio and optical telescope example. The FAST radio telescope in China (Left) was built on the same principle as the Arecibo Radio Telescope in Puerto Rico. The primary 'mirror' (reflector or 'dish') is 500 m across and fixed in position in a naturally occurring bowl (essentially a sinkhole) in a mountainous region of Guizhou Province, China. The prime focus is located in a cabin that hangs from cables 140 m above the dish. Rather than tilting the primary reflector in order to point at different astronomical objects, it is the prime focus cabin that is moved around on its cables above the surface. Radio sources that are at some angle from the zenith can be 'pointed at' by adjusting the position of this prime focus cabin. For example, a source whose light comes in from the top right of the picture would need the focal cabin shifted over to the left. This also means that the full aperture is not perpendicular to the source ($\cos\theta \neq 1$ in Eq. 3.13), reducing the 'effective aperture' (Section 4.2.4) from 500 m to something that is typically closer to 300 m – still a very large telescope indeed! The Gemini telescope on Mauna Kea in Hawaii (Right) is not the largest optical telescope in the world, but it is a nice example of folded optics, with a light path similar to that shown in Figure 4.5. A variety of detectors can be placed at the prime focus which is below the opening at the centre of the primary mirror.

---

[4] In particular, lenses are subject to *chromatic aberration* because of wavelength-dependent refraction, which puts the focal point of blue light closer to the objective lens than the focal point of red light.

Figure 4.6 (Left) The 500-m Spherical radio Telescope (FAST) in Guizhou Province, China, which operates at radio wavelengths from 10 cm to 4.3 m, has a primary reflector that is half a km in diameter. Suspended above the centre of the dish at its prime focus is a 10-m diameter feed cabin (visible as a small white dot) housing the detectors. Source: Xinhua/Ou Dongqu. (Right) A view of Gemini North, an 8.1 m diameter optical telescope situated on Mauna Kea in Hawaii. Light coming down the open tube reflects back from the primary mirror to a secondary mirror which reflects the light again through a hole in the primary mirror to the detectors below. The effective focal length is 128.12 m. Source: Reproduced by permission of the Gemini observatory/AURA.

To maintain the integrity of the signal, irregularities on the surface of the primary reflector should be a small fraction (e.g. 1/10th) of a wavelength. Thus, optical telescopes require finely engineered smooth mirrors whereas a wire mesh may suffice at radio wavelengths. At X-ray wavelengths, however, even very smooth mirrors have limited focussing capabilities because X-rays tend to be absorbed, rather than reflected, when hitting a mirror perpendicularly. Instead, X-rays are focussed via a series of glancing reflections (Figure 4.7 Left)[5]. At even higher energies, γ-rays cannot be focussed at all and techniques must be used that are more familiar to particle physicists. The Fermi telescope (Figure 4.7 Right), for example, contains 16 planes of thin tungsten foils. When an incoming γ-ray hits a foil, it produces electron–positron pairs. The resulting charged particles then pass through up to 36 planes of detectors that can track the direction from which the original photons have come [1].

[5] For X-ray telescopes, the ability of the telescope to focus onto a detector is what limits the spatial resolution.

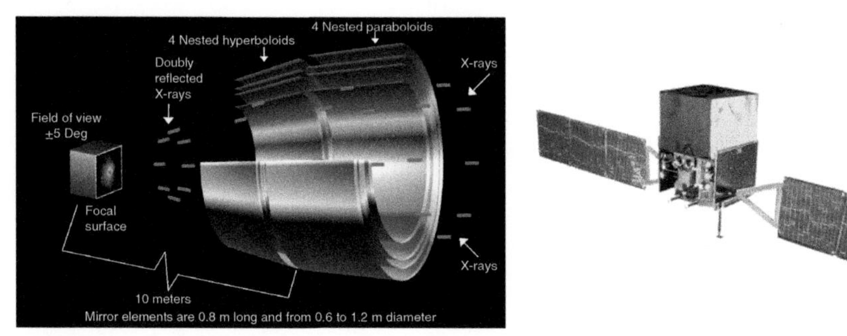

Figure 4.7 (Left) A schematic view of the space-based Chandra X-ray telescope (launched in 1999), showing how successive grazing reflections are used to focus the light onto a detector. Source: Reproduced by permission of NASA/CXC/SAO. (Right) Artist's drawing of the Fermi Gamma-ray Space Telescope (launched in 2008). This telescope is basically a large cube in space with no mirrors or tubes. The detector itself has built-in capabilities to measure the directions and energies of incident gamma rays. Fermi is funded by the US, France, Italy, Japan, and Sweden. Source: Reproduced by permission of NASA E/PO, Sonoma State University.

4.2.2 Detecting the Signal

The detector is a device for turning the collected light into another form that can be analysed. A wide variety of detectors are used depending on waveband and desired output: photographic films, radio receivers, microchannel plates[6], CCDs (see below) and many others.

The simplest detector is one that accepts a signal from a single position on the sky. The telescope is pointed at a position on the sky, and the detector records a value that represents source flux or intensity for only this one position. More often, a *spectrum* (emission as a function of wavelength or frequency) is obtained at this single position. If a map of an extended source is desired, the telescope must repoint to another position on the source, the detector makes a recording again, and so on until a picture of the source is built up consecutively. Some radio and mm-wave telescopes, which have complex receiver systems, still work on this principle. However, this process is slow and, in most cases, has been replaced by imaging detectors which consist of many 'picture elements' or *pixels*. In general, imaging detectors at the focus are called *focal plane arrays*.

For optical work, an example of a focal plane array is the *photographic plate* which has been widely used in the past and is still sometimes used today, especially when very large fields of view are desired or if attractive images, rather than numerical scientific results, are desired. However, the photograph has been virtually universally replaced for scientific work by the *Charge Coupled Device*, or *CCD* (Figure 4.8 Left), which is a semiconductor

[6] These are detection and amplifying devices used to convert a single high energy photon, such as an X-ray, into many electrons at the output.

Figure 4.8 (Left) A CCD camera like this one can be affixed to the back of an optical telescope to obtain images. Most of the weight of the camera is in supportive systems such as the cooling system. The CCD itself (upper left inset) is quite small. This particular CCD is 2.46 cm on a side with 1024 × 1024 pixels in the array, each of which is square and 24 μm on a side. Source: Reproduced by permission of Finger Lakes Instrumentation. (Right) The Arecibo L-Band Feed Array (ALFA) weighs 600 kg and fills a small laboratory. When placed at the focal plane of the telescope, its seven circular pixels of diameter 25 cm sample seven nearby regions on the sky with small gaps in between. Source: Reproduced by permission of CSIRO Australia. Credit: David Smyth.

detector (a 'chip') like those used in digital cameras. The CCD's high sensitivity to low light levels, its linear response to increasing light levels[7], and its direct interface to computers make it the ideal detector for most scientific purposes. Larger format CCDs are also now becoming available, making earlier limitations of small fields of view less problematic. In each pixel of the CCD, electrons are released when exposed to light in some energy range for which the CCD has been designed. This is called the *photoelectric effect*. The number of electrons released is proportional to the number of photons incident. The electrons remain at the location of the pixel during the exposure and are read out at the end, line by line, via an applied voltage. The information is then stored on the computer in a file that represents it as a two-dimensional numerical array with array locations corresponding to the location on the chip, higher numbers representing brighter light.

Multi-pixel detectors in the focal plane of the telescope are used across all wavebands from radio to X-rays, with the CCD replaced by whatever detector is required for the given band. At radio wavelengths, for example, the pixels consist of feedhorns which guide the signal into other electronic equipment. Compare the colossal seven pixel focal plane array which was used at the Arecibo Radio Telescope of Figure 4.8 Right, with the tiny CCD chip in the inset to Figure 4.8 Left!

[7] Any given pixel, however, will become *saturated* if exposed to too high a level.

4.2.3 Field of View and Pixel Resolution

Since any detector has a finite pixel size, l_{pix}, and a finite diameter, l_{det}, Eq. (4.1) indicates that these limits will impose corresponding limits on the angular resolution, θ_{pix}, also called the *pixel field of view* and the angular field of view, θ_{FOV}, respectively. Detectors with many small pixels have the potential to produce higher resolution images, provided there are no other limitations on the resolution (see Sections 4.2.4 and 4.3.2). For example, a photographic emulsion with tiny grains will show more detail than one with large grains. An illustration of an image on a CCD in which the resolution is limited by a large pixel size is shown in Figure 4.11c. A previous example, showing a single star smaller than the pixel field of view was given in Figure 3.9. As for the angular field of view, physically larger detectors are able to accept light from a larger range of angles, provided the telescope tube itself does not interfere with the light path.

4.2.4 Diffraction and Diffraction-limited Resolution

The telescope objective, however large it might be, collects only a portion of the light from any source that bathes the Earth. It is therefore like a circular aperture which accepts light that falls within it and rejects light that does not. This is analogous to a laboratory situation in which light passes through a small hole and is subsequently *diffracted*. Diffraction is the net result of interference *within* the aperture. Each point on any plane wavefront can be thought of as a new source of circular waves. This is known as *Huygen's Principle*. If there were no barrier (an infinite aperture), then the net result of these circular waves interfering with each other would still be a plane wave. However, with the barrier in place, only the points within the aperture interfere with each other and the net result is a wave that fans out in a circular fashion at the edges. The resulting diffraction pattern falling on a screen (or on a detector in the case of a telescope) is shown in Figure 4.9 for a narrow slit. For small angles and a circular aperture, the angular distance of the first null from the central point, θ_N, and the full width of the central peak at half-maximum intensity (FWHM), θ_{FWHM}, are, respectively,

$$\theta_N = 1.22\,\lambda/D \qquad \theta_{FWHM} = 1.02\,\lambda/D \tag{4.2}$$

where λ is the wavelength of the light and D is the diameter of the aperture. D could be somewhat less than the physical size of the aperture if the light path is partially blocked or otherwise imperfect. Similarly if the aperture is not perpendicular to the source (e.g. the FAST telescope, Section 4.2.1), the full diameter is not realized. In such a case, D must be replaced by the *effective aperture*, D_{eff}. It is clear from Eq. (4.2) that if we want a crisp small image of a point source, then a larger telescope (larger D) is required. It is also worth remembering that *larger* θ corresponds to *lower resolution*. Alternatively, saying that a telescope has *high resolution* means that θ is *small*, a sometimes confusing terminology.

A face-on view of the diffraction pattern in two dimensions for a circular aperture is shown in Figure 4.10a. This is the *spatial response function* of the telescope to a point

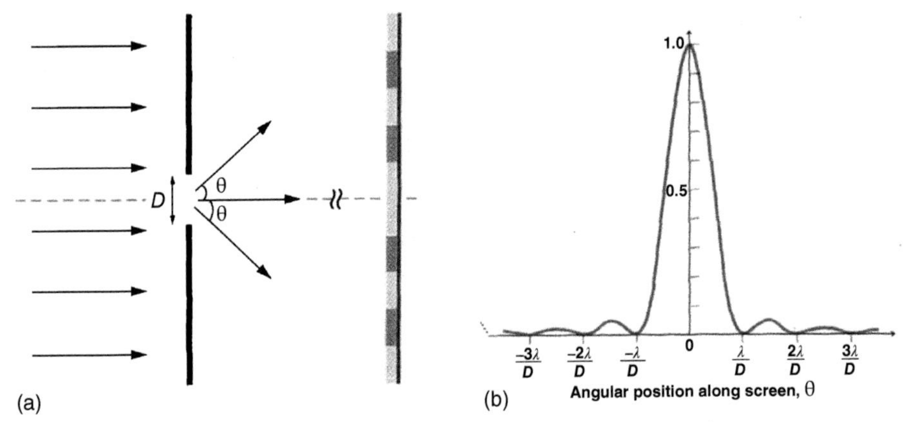

Figure 4.9 A laboratory aperture + screen is analogous to a telescope + detector. (a) Light is accepted only through a narrow slit and projected onto a screen some distance away. A diffraction pattern results from interference within the opening. (b) A plot of intensity as a function of angle, θ, across the screen yields this pattern.

source that is evenly illuminating the aperture. Most of the emission falls within the central bright region called the *Airy disc* in optical astronomy and called the *main lobe* or *main beam* in radio astronomy. However, a small amount of emission falls within the other surrounding peaks or diffraction rings (*sidelobes* in radio astronomy). For uniform aperture illumination, the Airy disc contains 84% of the power falling on the detector. The peak of the first diffraction ring occurs at a level 1.7% of the central peak and contains 8.6% of the power, while the values for the tenth ring are 10^{-3}% and 0.2%, respectively [20]. Thus, a point source in the sky will not form a point source on the detector, but rather a diffraction pattern in which the light is spread out as illustrated.

If a second star is now present, displaced from the first by some angle, θ, in the sky, then two sets of parallel light rays evenly illuminate the aperture, separated by this angle (see, e.g. points A and B in Figure 4.4). Two diffraction patterns will now be on the detector with their centres separated by θ. If these stars were separated from each other by smaller and smaller angles in the sky, the diffraction patterns on the detector would become progressively closer together, overlapping (as shown in Figure 4.10b) and finally merging. There is a minimum angle on the sky, θ_d, at which it is still possible to *just* distinguish that there are two sources rather than one. This limit is called the *diffraction-limited resolution* of the telescope, or just the *diffraction limit*, and provides a more technical definition of the concept of resolution. For a circular aperture, the resolution is often specified by the *Rayleigh criterion* which states that this will occur when the maximum of one image is placed at the first minimum (first null) of the other. In practise, separation by the FWHM is an adequate criterion and, as a rule of thumb, the diffraction-limited resolution of the telescope may be found from,

$$\theta_\mathrm{d} \approx \lambda/D \tag{4.3}$$

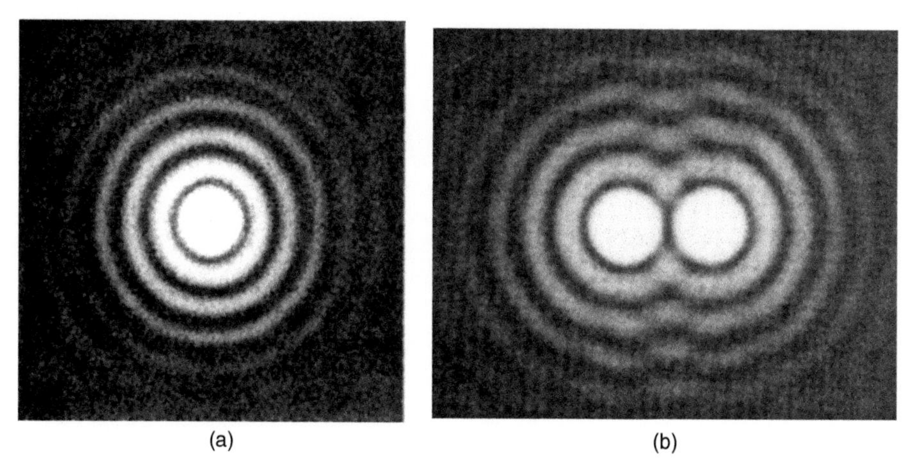

Figure 4.10 (a) The two-dimensional diffraction pattern from a circular aperture illuminated by a single point source at infinity. The bright central spot is called the *Airy disc*. (b) Two closely spaced point sources in the sky would create a pattern like this on the detector.

The brightness distribution of an astronomical source can be thought of as a set of many individual closely spaced point sources of different brightness. On the detector, each point in the image will have superimposed on it, the diffraction pattern of a point source. Mathematically, this can be represented as a *convolution* (see Appendix A.4 of the *online material*) of the brightness distribution of the source with the diffraction pattern of the telescope. The broader the central lobe of the diffraction pattern, the poorer, or lower, is the resolution, and the source will appear as if out of focus. Such a comparison is shown between Figure 4.11a and b. Since the light in the outer diffraction rings is much lower in intensity than the main lobe, these rings do not show up in Figure 4.11b.

Lower resolution, whether due to telescope diffraction or other limitations, has the effect of spreading out the signal and lowering its peak (note that the flux is preserved, cf. Figure 3.9). In the limit, if the resolution is very low, the angular size of the source, θ_S, becomes much less than θ_d and the source is said to be *unresolved* as if it were a point source (Problem 4.8). Its apparent angular size will then equal the angular size of the central lobe of the beam or Airy disc.

4.2.5 Weighting the Aperture – Interferometry

It is sometimes possible to modify the response function of the telescope. For example, the power going into the diffraction rings could be reduced by weighting the aperture so that it accepts more light towards the centre in comparison to the edges. However, such weighting has the effect of reducing the effective aperture diameter, D_{eff}, thereby worsening its resolution. The opposite is also true. Information at high-angular resolution can be selected by putting higher weight on the outer parts of the dish or mirror but only at the expense of increasing the power going into the diffraction rings.

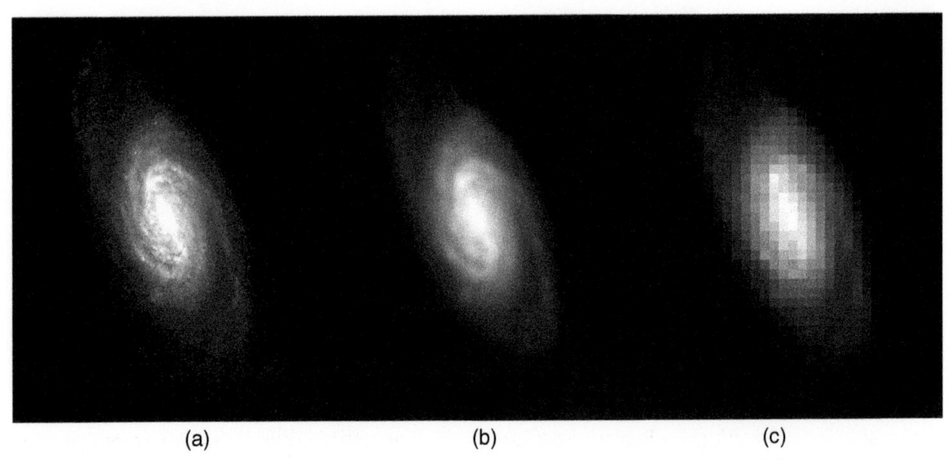

Figure 4.11 Illustration of resolution for the spiral galaxy, NGC 2903 (distance, $D = 8.6$ Mpc). The disc of this galaxy is circular, appearing as an ellipse because of its inclination, with the minor axis of the ellipse subtending the angle, 6.0′, and the major axis, 12.6′ on the sky. (a) A high-resolution image such as might result from using a large-aperture telescope. Source: Original image at left was taken from the NASA Extragalactic Database. (b) A low-resolution image such as from a telescope with a small aperture, or from a large-aperture telescope with poor seeing and (c) a low-resolution image due to large pixels on the detector. Source: ID: NGC 2903, http://nedwww.ipac.caltech.edu.

In the limit, one could weight the mirror to high resolution by simply applying a mask that blocks light from everywhere except the outer regions of the aperture. The weights, then, consist of zeros and ones – a zero where no light is accepted and a one where it is. This is a version of *optical interferometry*, a technique that was pioneered by the French astronomer, A. Labeyrie in the 1970s. When the mask only allows light through two widely separated openings, for example, it acts like a small interferometer, detecting the high-resolution information (referred to as small *spatial scales*) and masking out the low-resolution information (large spatial scales). As long as high-resolution information is all that is wanted (for example, a measurement of the size of the source only), then the mask effectively removes the unwanted noise (including atmospheric noise, Section 4.3.2) that does not contribute to the desired result. The downside is that the full collecting power of the mirror is lost as well as all the information that might be contained in larger spatial scales. One could instead apply more complicated masks (e.g. [12]) or a sequence of masks one after another and then piece together the results from the different observations. The mathematical transformation between an incoming signal from the sky and the signal that is actually measured at the detector is well known (technically a *Fourier Transform*), so an image can be reconstructed after the fact, by properly combining all the measurements.

But what about just using multiple telescopes instead of blocking apertures? The modern optical interferometers do just this, a good example being the four telescopes of the

European Southern Observatory's *Very Large Telescope* (VLT) shown in Figure 4.17 Right. The four main telescopes achieve a separation of 130 m and, together with some smaller auxiliary telescopes whose positions can be moved, a maximum separation (or *baseline*) of 200 m is possible. As the Earth rotates, the various baselines change with respect to the incoming wavefront from the source[8]. This has the effect of filling in many spacings of a large *synthesized* aperture, a technique, understandably, called *aperture synthesis.* The resolution of this synthesized aperture is the diameter of the maximum *baseline*, rather than the diameter of any individual mirror. Let us to return to Eq. (4.3) now. An individual mirror of the VLT has a diameter of $D = 8.2$ m. For an observation in H-band ($\lambda = 1.63\ \mu$m, Table 3.1), the telescope's diffraction-limited resolution is $\theta_d = 41$ mas. However, operated as an interferometer with a maximum baseline of $D = 200$ m, the resolution is now $\theta_d = 1.7$ mas. It is worth pausing to visualize what this means. At 1.7 mas, any object that is a mere 3 m in size (assuming that it is bright enough to be detected) could be resolved on the surface of the Moon!

Equation (4.3) shows that, to improve resolution, either D has to increase or λ has to decrease. If we want to probe the hidden secrets of astronomical objects at different wavelengths, then the only option is to increase the aperture size. This becomes problematic at the longer radio wavelengths because the telescope size becomes enormous (e.g. Figure 4.6 Left). Therefore, all high-resolution radio telescopes apply interferometric techniques with many individual dishes (antennae) at different stations, resulting in many baselines. Examples are the Karl G. Jansky *Very Large Array* in the United States and the Giant Metrewave Radio Telescope (GMRT) in India.

Each element is connected by cables to a *correlator* that cross-correlates all the signals. These detected signals can then be Fourier transformed to obtain an image. Although the longest baseline specifies the maximum spatial resolution, observations with many baselines that fill in the synthesized aperture allow for better image reconstruction. It was, in fact, at *radio* wavelengths that astronomical interferometry was originally pioneered in Cambridge, England. The first observations were made on the Sun in 1946 using a two-element (two-antenna or two-station) interferometer [18], the original goal being to eliminate unwanted extraneous noise.

A new international effort is now underway to build an interferometer whose collecting area is significantly larger than anything currently in use. Called the Square Kilometre Array (SKA), the increase in effective aperture will allow for fainter objects to be seen, as described in Section 4.2.1, while still maintaining good resolution. An image of one part of the SKA can be seen in Figure 10.13 along with more description in Section 10.5.1.

Suppose we want to make the baselines larger still? The limitation now is that the elements are so far apart that they cannot be physically connected by cables. The challenge then shifts to the need for accurate clocks (historically, *hydrogen maser clocks*) as well as accurate recordings. If a signal can be measured and recorded at each station, then the recordings can be brought together and correlated afterwards. As long as the

[8] The relevant baseline is the 'projected' baseline, i.e. the component of the baseline that is perpendicular to the incoming wavefront from the source. Earth rotation causes both the projected baseline and its orientation to change.

time is accurately known, the playback process can proceed just as if the antennae were physically connected. The first time that this Very Long Baseline Interferometry (VLBI) technique was successfully employed was in February of 1967 on a modest 200 m baseline in Ontario, Canada [3], followed soon after by a truly continental baseline of 3074 km between Ontario and British Columbia [4]. The quasars, 3C273 and 3C345, were observed and found to have sizes less than 0.02 arcsec.

Today, VLBI techniques are standard, but still impressive. Shorter wavelengths yield higher resolutions, but they are also more technically challenging. Arguably, the most impressive recent example is a $\lambda = 1.3$ mm VLBI observation by the *Event Horizon Telescope* (Figure 4.12). This telescope used baselines over almost the entire Earth, achieving a resolution of 20 μas, which was sufficient to probe the region near the supermassive ($M = 6.5 \times 10^9\ M_\odot$) black hole at the core of the galaxy, M 87, a distance 16.8 Mpc from us. The bright ring (inset) represents the gravitationally lensed photon sphere around the black hole; the dark region at the centre is where light is not escaping in the observer's direction. More will be said about the bending of light near a black hole in Section 9.2.

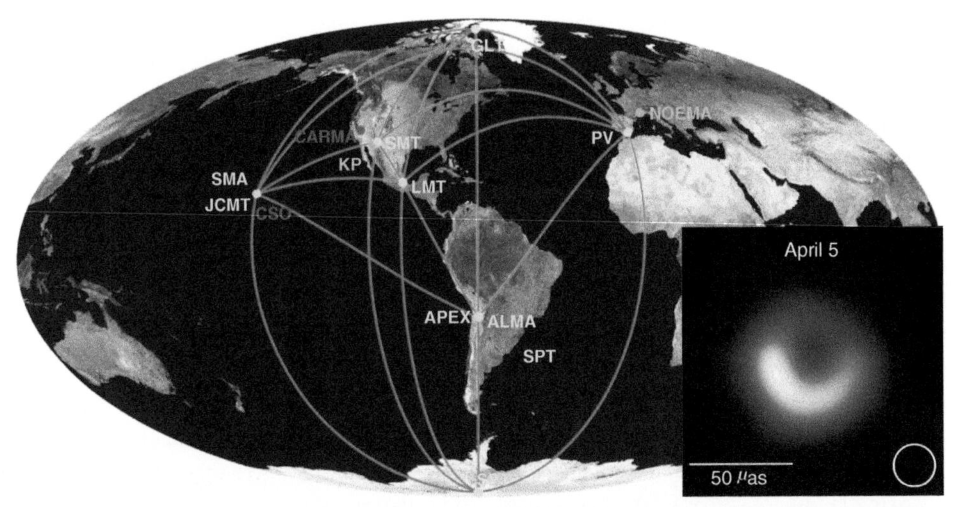

Figure 4.12 Stations of the Event Horizon Telescope (EHT) that were active in 2018 are labelled in yellow and red, and those being commissioned are labelled in green. Baselines are shown as connecting lines between stations. Source: The Event Horizon Telescope Collaboration [9]. Licensed under CC BY 3.0. *Inset:* The first image probing the region of the event horizon of a black hole in the galaxy, M 87. Source: The Event Horizon Telescope Collaboration [10]. Reproduced under the terms of the Creative Commons Attribution 3.0 licence. The wavelength was 1.3 mm, the resolution was 20 μas (circle at bottom right), and the black hole event horizon (Section 9.1.3) is 15 μas across.

4.3 THE CORRUPTED SIGNAL – THE ATMOSPHERE

Any observation from the Earth's surface must contend with the atmosphere. Not only does the signal pass through this scattering, absorbing and emitting layer, but the effect that the atmosphere has on a signal is strongly wavelength-dependent and is also variable with time and with local conditions. Details of scattering and absorption will be discussed more fully in Chapter 7, but here we consider the atmosphere in terms of its undesirable effects on observations.

The spectral response curve of the atmosphere (Figure 4.2) indicates that there are only two atmospheric windows for which ground-based observations are possible: the radio and the optical. Of these, the most transparent band is the radio. Radio wavelengths are large in comparison to the typical size of atmospheric particles (atoms, molecules, and dust particles), so the probability of interaction with these particles is small and atmospheric effects are mostly negligible. This means that observations in the radio band can be made day or night, during cloud cover or clear weather and from sea level, if desired.

The sharp cut-off at *very low radio frequency* ($\approx$ 10 MHz, Figure 4.2) is due to the *ionosphere* which reflects, rather than transmits, incoming waves. This is because the incoming waves have frequencies that are less than the *plasma frequency*, which will be discussed in Section 7.3. The ionosphere is an ionized layer ranging from a height of 50 –500 km with a maximum electron density (the number density of free electrons) of between 10^4 and 10^6 cm^{-3}, depending on conditions.

At the *high frequency end* of the radio band ($\nu \gtrsim 300$ GHz, $\lambda \lesssim 0.1$ cm), the atmosphere also becomes more important and radio telescopes operating in this range must be placed on high mountains and/or put in domes just like optical telescopes, as Figure 4.13 illustrates. Techniques such as *beam switching* or *chopping* are also employed at these as well as IR wavelengths. Chopping involves rapidly switching the light path back and forth from the source to a nearby region of blank sky. The image of the source is then corrected for time-variable atmospheric emission by subtracting off the nearby sky brightness distribution which closely corresponds in time.

The optical window (which includes the near-IR and near-UV) is strongly affected by the atmosphere, and therefore, optical research telescopes are placed at high altitudes above as much of the 'weather' as possible. Some specific effects of the atmosphere on an optical signal are discussed in Sections 4.3.1 and 4.3.2.

4.3.1 Atmospheric Refraction

The bending of light as it passes from one type of medium to another is described by *Snell's Law* (Table I.2). Such bending occurs as light passes from space into denser and denser regions of the atmosphere of the Earth. The bending increases systematically

Figure 4.13 The James Clerk Maxwell Telescope, operating at sub-mm wavelengths, is located on Mauna Kea in Hawaii at an elevation of 4200 m. Although this is a radio telescope, it is housed in a dome and, under normal operating conditions, the dome opening is covered by a protective cloth membrane that is transparent at the operating wavelengths. Source: Reproduced by permission of Robin Phillips.

with *zenith angle* which is the angle between the zenith and the source, making the source appear higher in the sky than it actually is. Extended objects that span a range of zenith angle will therefore also appear distorted, as does the Sun when it is close to the horizon at sunrise or sunset (Figure 4.14, Problem 4.2). Thus, models accounting for known systematic refraction must be built into telescope control software in order to accurately point the telescope to a source in the sky. Details of systematic atmospheric refraction at optical wavelengths are provided in Section 4.1 of the *online material* (see also [6]). A more general discussion of refraction is given in Section 7.3.

4.3.2 Seeing

The atmosphere is turbulent, and this turbulence can be approximated by 'cells' of gas that are constantly in motion. Many such cells with different sizes and slightly different temperatures, densities, and pressures may be present above the telescope aperture, even when the sky is perfectly clear. As a result, each cell has a slightly different index of refraction. This introduces a small random element of refraction superimposed on the systematic refraction discussed above. The net result is that an incoming wave that is originally

Figure 4.14 The Sun appears slightly flattened at sunrise or sunset because of refraction in the atmosphere. Refraction makes the Sun's position appear higher in the sky than its true location. Flattening occurs because the refraction of the lower limb is greater than the refraction of the upper limb. Source: Judith A. Irwin.

plane parallel will become distorted, and the angle, or tilt of the wavefront, results in a focal point on the detector that is shifted slightly from the non-tilted position. This can be thought of as a changing phase of the wavefront. If there are multiple cells above the aperture, then there will be multiple distorted images (each convolved with the telescope's diffraction pattern) on the detector (Figure 4.15). Each individual image is called a *speckle.* The larger the aperture, the more cells can be in front of it and the more speckles there will be on the detector. The largest effects are due to cells near the ground that are blown across the aperture by wind. The sizes of these atmospheric cells vary, but a typical value is $r_0 \approx 10$ cm in the visual, though it can be up to ≈ 20 cm at a high altitude where there is more atmospheric stability.

Since atmospheric cells are constantly in motion, the speckle pattern changes rapidly with time. The timescale over which significant changes in the pattern occur is of order the time it takes for a cell to move across the aperture,

$$t = a\frac{r_0}{v} \tag{4.4}$$

where v is the wind speed and the constant, a, has a value of ≈ 0.3. For a wind speed of $v = 300$ cm s^{-1} (11 km h^{-1}) and $r_0 = 10$ cm, therefore, this timescale is 10 ms. Thus, to see diffraction-limited images using a large-aperture telescope, it is necessary to take very rapid exposures shorter than 10 ms in duration.

This poses a problem, however, because such short timescales are generally not long enough to produce a significant response on the detector. A faint source requires an integration time that is much longer than 10 ms! During a longer exposure, then, the shifting

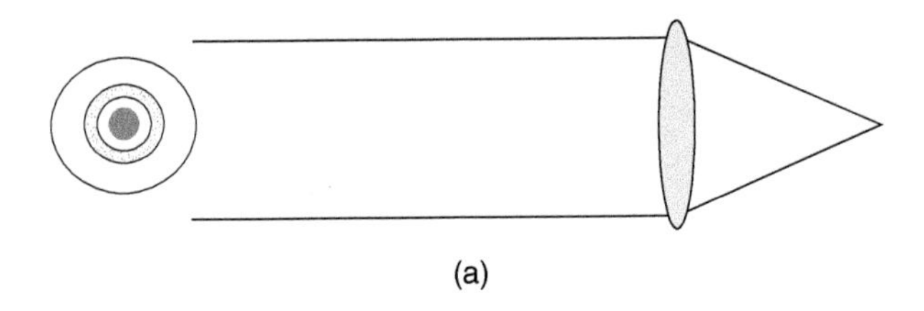

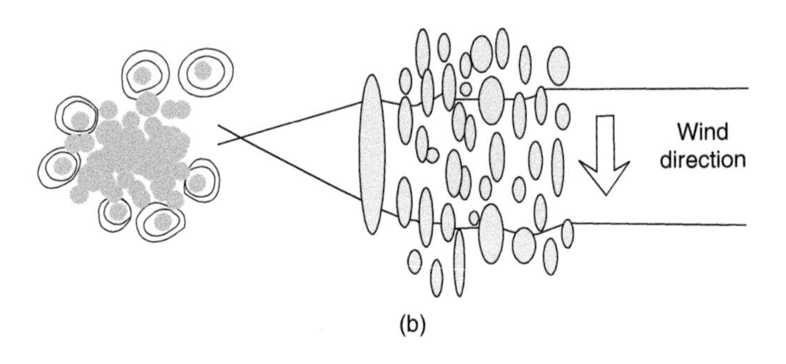

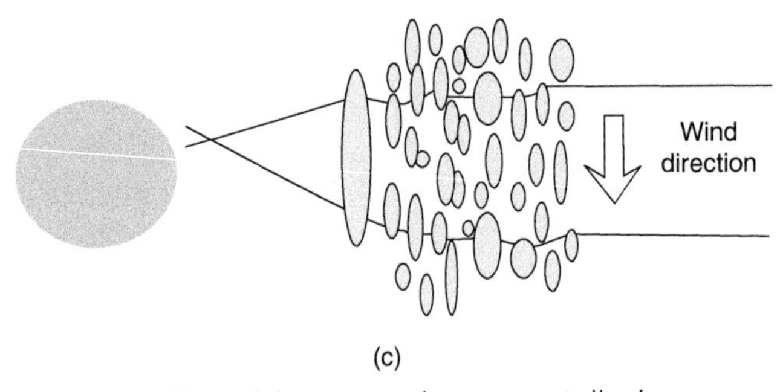

Figure 4.15 Effect of the atmosphere on a stellar image. (a) If the telescope were in space, the image of a point source would be the diffraction pattern of the telescope with the width of the central peak given by θ_d. (b) Individual atmospheric cells of typical size, 10 cm, move across the telescope aperture effectively breaking up the aperture into many small apertures about this size. Since the refraction is slightly different for the different cells, each one creates its own image, called a speckle, on the detector at a slightly shifted location. Each individual image has its own diffraction pattern, and the image changes rapidly with time as the atmospheric cells move across the aperture. (c) If a time exposure of the star is taken, the speckle pattern is seen as a blended spot, called the seeing disc.

speckles build up a signal on the detector, filling in the gaps and forming a smeared out image over a small region (Figure 4.15). The characteristic angular size of this region, θ_s, taken to be the FWHM, is called the *seeing disc* or just the *seeing*. Seeing, therefore, refers to the smeared out angular size of the signal due to the changing tilts of the wavefront with time. The apparent position of the star in the sky is at the centre of its seeing disc. Seeing varies with wavelength, and with altitude and geographic location. It also varies from night to night and can change over the course of the night.

Thus, the resolution of a large optical ground-based telescope has been limited, not by diffraction, θ_d, but by the seeing, θ_s, because over the timescale of a typical exposure, the stellar image will smear out over a size, θ_s. Large research telescopes have historically been designed to collect more light, allowing fainter sources to be detected. However, in the absence of corrective techniques (Section 4.3.3), their resolution has been limited by the size of an atmospheric cell. The resolution, $\theta_s = \lambda/r_0$, so the effective aperture of the telescope is equivalent to the size of the *atmospheric cell*, not the telescope mirror. If $r_0 = 10$ cm, then a 10 cm diameter telescope will have the same resolution as a 10 m telescope ($\approx 1''$, Eq. (4.3)). For the 10 cm diameter or smaller telescope, the resolution is *diffraction-limited* and for any larger telescope, it will be *seeing-limited*. Depending on the linear size of the pixels on the CCD, however, the resolution could also be *pixel-limited* (Example 4.1). The image of a point source on the detector, as actually measured, is called the *point spread function* (PSF). It is standard to determine the PSF during any observing session by pointing at an unresolved object (e.g. a star) and measuring its FWHM on the detector.

Seeing has been the curse of optical telescopes. Observatories at the tops of high mountains are not only escaping clouds, but are also searching for better seeing. Potential sites for new observatories may be monitored for a year first to measure the seeing before the site can be fully evaluated. Although somewhat of an exaggeration, seeing hinders our view of the night sky much like visualizing the world through a fish bowl. Wouldn't it be better to eliminate the bowl? In the next section, we will see how.

Example 4.1

An 8 in. f/10 telescope is outfitted with a 1024 × 1024 CCD with each pixel 9 μm on a side. Determine the field of view, the field of view of a pixel, and the resolution when this telescope is used, (a) in a back yard with seeing of $\theta_s = 2''$ and (b) on a high mountain with $\theta_s = 0.4''$. (c) Could this telescope achieve higher resolution at either of these locations?

In *cgs* units, the telescope diameter, $D = 20.32$ cm and pixel size $l_{pix} = 9 \times 10^{-4}$ cm. The CCD size is $l_{det} = 1024 \times l_{pix} = 0.92$ cm and the focal length, $f = 10 \times D = 203.2$ cm. From Eq. (4.1), the field of view is $\theta_{FOV} = l_{det}/f = 4.5 \times 10^{-3}$ rad, or 15.6′ square. By the same equation, the pixel field of view (converting to arcseconds) is $\theta_{pix} = l_{pix}/f = 0.91''$, also square. None of these results depend on location.

The diffraction limit of the telescope, from Eq. (4.3), adopting an observing wavelength of $\lambda = 507 \times 10^{-7}$ cm (Table T.4), is $\theta_d = \lambda/D = 2.5 \times 10^{-6}$ rad $= 0.5''$. This result is also independent of location.

(a) Since θ_s is larger than both the pixel resolution, θ_{pix}, and the diffraction-limited resolution, θ_d, the observations are seeing-limited and the resolution is 2″.
(b) In this case, $\theta_{pix} > \theta_d > \theta_s$ so the observations are pixel-limited and the resolution is 0.91″.
(c) No improvement can be achieved in the backyard case since the seeing cannot be changed. However, on the mountain, by changing the CCD to one which has much smaller pixels, the telescope could become diffraction-limited with a resolution of 0.5″. In fact, any telescope, large or small, that is pixel-limited displays poor planning on the part of the designers!

In this example, we have simply taken the largest value of θ to represent the resolution, be it pixel, diffraction, or seeing. A better approach would be to fold each together to approximate the FWHM of the PSF,

$$\theta_{PSF} = \sqrt{{\theta_d}^2 + {\theta_{pix}}^2 + {\theta_s}^2} \tag{4.5}$$

Of course, the best approach is to measure the PSF on a known unresolved source (e.g. a star) during the observing session. The size of the PSF is the actual resolution of the observations. If you then want to know whether another source on the map is resolved or unresolved, you could compare its FWHM size to that of the PSF. If the two sizes are equal, the source is unresolved.

4.3.3 Adaptive Optics

There are solutions to the problem of seeing. The most straightforward (but the most expensive) is to launch the telescope into orbit so that the observations do not have to contend with the atmosphere at all. This was done, for example, for the Hubble Space Telescope. Another is to improve the design of observatories to minimize local temperature gradients that can produce turbulence. It took some time to realize that much of the unwanted turbulence, and bad seeing was actually introduced by the dome structure itself. Open sides, such as shown in Figure 4.16, encourage smooth (laminar) airflow.

However, there is a very effective way of dealing with the problem of seeing – that of *adaptive optics* (AO). Now being routinely employed at major facilities, adaptive optics is a method of correcting for the seeing in real time by making rapid adjustments to a deformable mirror. An example of an adaptive optics system is shown in Figure 4.17 (Left).

After passing through the telescope aperture, the signal, along with a reference signal from a bright point source, is reflected from a deformable mirror. The reference signal is then split off to a 'wavefront sensor' which must sample the wavefront on millisecond timescales. The wavefront sensor breaks up the reference signal into many images via lenses. If the wavefront is planar, then the reference signal images are evenly spaced on the detector, but if the wavefront is distorted, then the images are unevenly spaced. After fast analysis of the spacing, a signal is sent to actuators at the rear of the deformable mirror which adjust the shape of the mirror to compensate for the wavefront distortion. The signal of astronomical interest is also reflecting from the same deformable mirror, so its image is therefore corrected as well.

Figure 4.16 The Gemini North Telescope on Mauna Kea, in Hawaii, showing the open sides which help to minimize local turbulence. Source: Reproduced by permission of Neelan Crawford, courtesy of Gemini Observatory/AURA.

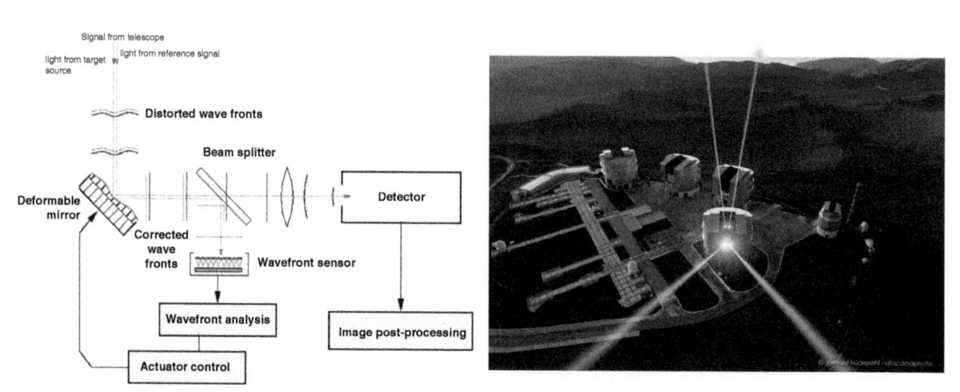

Figure 4.17 (Left) Simplified diagram of an adaptive optics system. Source: Adapted from Center for Adaptive Optics. http://cfao.ucolick.org/ao/how.php. (Right) Four different laser beams being emitted from one of the domes of the Very Large Telescope at the Paranal Observatory in the Atacama Desert of Chile. The laser beams create artificial guide stars for use in the adaptive optics system. The four telescopes in the four large domes of this picture operate as an optical interferometer. Source: European Southern Observatory / Gerhard Hudepohl (atacamaphoto.com). Reproduced under the Creative Commons Attribution 4.0 International Licence.

In order for this system to work, the reference signal must be bright and must be close enough in the sky (<1′ is desirable) to the astronomical source of interest that the two signals pass through the same atmosphere and same optical path. However, there are very few bright stars in the sky that are fortuitously placed close to any arbitrary astronomical source. The solution has been to generate a *laser guide star* (Figure 4.17 Right). A laser beam is emitted into the sky and excites or scatters off of particles in the atmosphere at an altitude of about 100 km, forming a point-like source as the reference beacon. A natural guide star must still be used for absolute positional information, but it need not be so bright or close to the target source.

The image quality that has been achieved by adaptive optics is remarkable, especially in the near-IR (Figure 4.18). At these longer optical wavelengths, it is easier to fine tune the system and the atmosphere is also more stable. With AO, a telescope can basically eliminate the seeing, as if it were above the atmosphere! Adaptive optics systems have been retrofitted to existing telescopes, but are now an integral part of designs for any new research-grade instrument [11]. As techniques continue to improve, new extremely

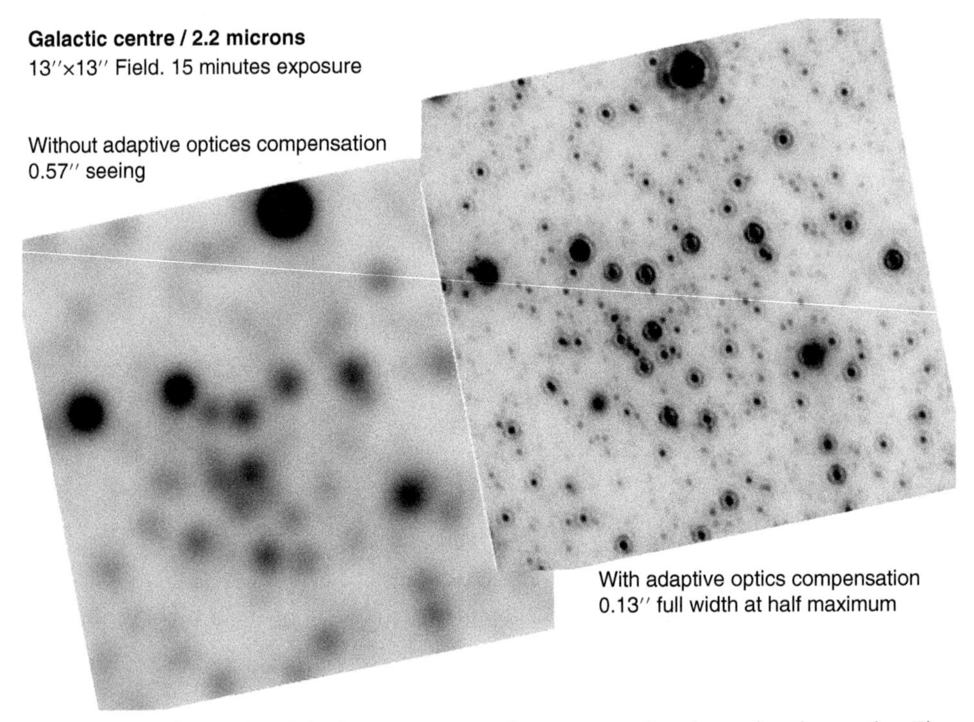

Figure 4.18 Example of the improvement that can result using adaptive optics. The image at left shows many stars crowded together in the region of the Galactic Centre. The image appears out of focus because of the seeing. At right is the same field after using adaptive optics. The star images now show the diffraction pattern of the telescope and the resolution approaches the diffraction limit. Source: Reproduced by permission of the Canada-France-Hawaii Telescope/1996

large telescopes will essentially 'subtract off' the atmosphere and probe the details of astronomical objects at the diffraction limit of the telescope (Problem 4.9). That atmospheric fish bowl is finally being eliminated. For a nice review, see [13].

4.3.4 Scintillation

Scintillation, commonly called 'twinkling', is the rapid change in *amplitude* of a wavefront with time due to the atmosphere (recall that seeing is the change in *phase* or wavefront tilt). It occurs for similar reasons as seeing, as atmospheric cells act like lenses focussing, defocussing, and shifting the signal. Scintillation, however, is primarily due to atmospheric fluctuations that are higher up, rather than near the ground. A characteristic timescale for scintillation is of order a ms, but variations are seen on a range of timescales from microseconds to seconds or longer [7]. The fluctuations in amplitude will be about a mean value which is the desired measurement, and since most exposure times are longer than the scintillation timescale, scintillation is not a serious problem for most astronomical measurements. However, it does introduce a kind of 'noise' into the optical path [16].

When viewed with the naked eye, a star, or other *point-like source*, appears to fluctuate in intensity ('scintillation') and also wander by a small amount (seeing). An *extended object* (like the moon) can be thought of as a combination of many point sources of different brightness, each 'blurred to' (convolved with) the size of the seeing disc, and each scintillating. If the eye could spatially resolve each of these points, it would see brightness fluctuations across an extended source. However, the resolution of the human eye ($\approx 1'$) is much poorer than the seeing ($\approx 1''$). Thus, each of the fluctuations across an extended source (some positive, some negative about the mean) are not seen individually by the eye but are instead integrated together. The result is that the eye perceives an extended source as steadily shining. The old adage 'stars twinkle and planets shine' is true for the planets visible by eye, but exceptions can occur (Problem 4.10).

4.3.5 Atmospheric Reddening

Reddening is the result of wavelength-dependent absorption and scattering. Shorter wavelength (blue) light is more readily scattered out of the line of sight than is light of longer (redder) wavelengths. Therefore longer, redder wavelengths will have greater intensity when viewed through an absorbing and scattering medium like the atmosphere. This is true at both the zenith and the horizon, but because viewing towards the horizon corresponds to viewing over a longer path length, the effect is more apparent, as is readily seen when viewing a red sunset. Reddening of starlight can also be noticed by eye by observing a given star when it is high in the sky and then again when the star is about to set. Molecules as well as dust and aerosols contribute to the reddening in the Earth's atmosphere. Absorption and scattering will be treated in greater detail in Chapter 7.

4.4 PROCESSING THE SIGNAL

It is not possible to detect a signal without affecting it. Both the atmosphere and the telescope and detector impose their spatial and spectral response patterns on the signal, as already seen. As light passes through other optical devices within the telescope (for example, through lenses, wavelength-dispersive media such as spectrographs, through filters or reflected from smaller mirrors), each 'interaction' also affects the signal. Moreover, other sources of emission which can pollute the image (such as detector noise or atmospheric emission) must be removed. The image that is detected, called the *raw image*, requires corrections for these effects, or at least a characterization of them, to know their limitations on the data. In addition, the corrected data, which consist of an array or arrays of numbers, must be converted to commonly understood units (e.g. magnitudes or Jy per unit solid angle) that are independent of telescope or location, i.e. the signal must be *calibrated*. The end result will be a *processed image* such as would be measured with 'perfect' instrumentation above the Earth's atmosphere.

4.4.1 Correcting the Signal

The steps involved in correcting the signal vary depending on waveband, the specifics of the telescope, and its instrumentation. For example, if the telescope is in orbit, corrections are required for variations in the signal due to variations in the orbit. If CCD images are taken rapidly, a new image may retain a 'memory', due to incomplete CCD readout from the previous image. Data from radio telescopes require radio interference to be excised and beam sidelobes to be carefully characterized or 'cleaned' out. Optical telescopes must have unwanted cosmic ray tracks or 'hits' removed and atmospheric emission lines either corrected for or avoided. Non-uniform responses across the field of view as well as in frequency must be 'flattened' and higher order distortions corrected if present. All of this sounds rather formidable, but it is possible to deal with it in a systematic way.

For work with an optical telescope and CCD, the minimum corrections would include the following steps. The CCD image would first have a constant positive offset subtracted from the image to account for a DC voltage level that exists in CCDs (the *bias correction*). The signal that results from thermal electrons in the detector, without any exposure to light, would then be subtracted (the thermal or dark correction). The non-uniform response of each pixel to light across the CCD must be corrected; this is done by dividing the image by another image that has been exposed to a source that is known to be evenly illuminated (the *flat-field correction*). Multiple frames of the same source, if they are present, are medianed (rather than averaged) together, a process that facilitates removing artefacts due to random cosmic rays hits. Finally, a background sky emission level, determined from a region on the CCD in which there is no source, would be subtracted from the image. Depending on the complexity of the instrumentation, it is usually more time-consuming to correct and characterize the data than to acquire the data in the first place!

4.4.2 Calibrating the Signal

The most common method of calibrating the signal is to compare it to a reference signal (the *calibrator*) of known flux density or brightness. There are often only a few *primary flux calibrators* at any wavelength, and all other observations are brought in line with them. The stars, Vega and Sirius, are good examples in the optical band, as are the quasars, 3C 286, and 3C 48, at radio wavelengths. Asteroids and planets are often used at sub-mm and/or IR wavelengths. Secondary calibrators may also be used. These would be sources whose fluxes are tied to those of the primary calibrators and are more widely dispersed in the sky so that one or more are accessible during any observing session.

Atmospheric attenuation is important at optical and sub-mm wavelengths and increases with zenith angle. At optical wavelengths, a calibrator must be monitored periodically at a variety of zenith angles over which the source of interest is also observed. The source data are then corrected according to the calibrator signal at the corresponding zenith angle. At sub-mm wavelengths, at least once per night a measurement of sky brightness as a function of zenith angle is made (called *sky dips*) and this information is combined with a model of the atmosphere to compute the attenuation.

Once the calibrators have been observed, it is then a straightforward matter of multiplying the source counts, in arbitrary units, by the factor determined from the calibrator (e.g. magnitudes/counts, Jy/counts, etc.). Because the calibrator and source are observed with the same equipment, the attenuation of the signal due to absorptions and scatterings in the instrumentation is automatically taken into account.

Clearly, the primary flux calibrators must be measured on an absolute scale for this system to work. This process is non-trivial, and considerable effort has been expended to make accurate measurements and bring them to a common system (e.g. [2, 17]). It consists of observing the source along with a man-made source of known brightness. At radio or sub-mm wavelengths, this might be a black absorbing vane (a black body, see Section 6.1 and Example 6.1) at fixed known temperature or, at optical wavelengths, a calibrated lamp. Alternatively, the flux densities of Sirius and Vega are also determined by calculating the expected values, using other known stellar parameters such as temperature, distance, and the shape of the spectrum. Even libraries of *synthetic* stellar spectra have been formed to help with calibration. Such libraries ensure that the full range of stellar parameters can be homogeneously and uniformly covered [5].

4.5 ANALYSING THE SIGNAL

The final result is a reduced, or processed, image, in units that are common to astronomy, ready for analysis, measurement, and other scientific scrutiny. Measurements of specific intensity and flux density can be made as described in Section 3.3 (e.g. Figure 3.8), and other derived parameters can then be determined, if possible, such as the flux, spectral power, or luminosity. It is therefore important to understand the characteristics of the processed images in order to place limits on the source parameters which are derived from them.

Once a calibrated image has been obtained (assuming no artefacts), it will be characterized by eight properties, three providing spatial information, three providing spectral information, and two related to the amplitude of the signal:

1. Position of map centre: The map centre is usually placed at the location of the target source in the sky and specified by some coordinates, Right Ascension and Declination (RA, DEC) being most common.
2. Field of view: The FOV is set by the detector size in the case of a simple system (Section 4.2.3), but it can be smaller for more complex, partially blocked light paths.
3. Spatial resolution: This is affected by telescope size, pixel size, and/or seeing, as described in Section 4.3.2.
4. Central wavelength or frequency: Different instrumentation is required for different wavebands but, within any given band, more accurate centring in wavelength or frequency can be achieved by tuning or using filters.
5. Bandwidth: Implicit in the setting of central wavelength or frequency is the fact that a range of wavelength (λ_1 to λ_2) or frequency (ν_1 to ν_2) will be detected.
6. Spectral resolution: If detailed spectral information is required, the waveband may be broken up into wavelength or frequency *channels* and each examined separately.
7. Noise level: A spatial region of the image that is devoid of emission (after all corrections and sky subtraction) will contain positive and negative fluctuating values about a mean of zero due to noise in the system. The noise, N, is normally measured as a *root mean square* (rms), i.e. $N = \sqrt{< x^2 >}$, where x is the calibrated pixel value and the angle brackets represent an average.
8. Signal-to-noise ratio (S/N): This is the ratio of the signal strength at any image location to the rms noise. In general, longer integration times (that is, longer exposure times) result in higher S/N images because N decreases with increasing integration time. In order for a signal to be considered a real detection, the S/N ratio must be high enough to be sure that it is not just a random peak in the noise level. A minimum value of $\approx$ 3–10 will often be used, although other factors can also play a part in this decision, such as the angular size of the signal on the detector or whether it is seen in the same location at other wavebands. Another way of characterizing the image as a whole is by the *dynamic range* which is the ratio of the peak signal to the minimum detectable signal on the map. This is therefore a measure of the 'stretch' in brightness that has been detected. A practical, though perhaps optimistic, measurement is to take the peak signal on the map over the rms noise. The S/N ratio may be different at every point in the map, but there is only one value of dynamic range for a map.

These properties must be kept in mind when determining and interpreting the scientific results that derive from the data. Example 4.2 provides an illustration of the kinds of questions one should ask prior to embarking on an observing run.

Example 4.2

A galaxy at a distance of $D = 25$ Mpc will be observed with a spatial resolution of 1″ at a frequency of 4.57×10^{14} Hz and bandwidth of $\Delta\nu = 2 \times 10^{12}$ Hz. The purpose is to detect the HII regions (these are ionized hydrogen regions around hot stars; more will be said about HII regions in Section 7.2.2) within this galaxy. The map noise expected for the integration time proposed is $N = 2 \times 10^{-17}$ erg s^{-1} cm^{-2} Hz^{-1} sr^{-1}. Could a spherical HII region of typical diameter, $d = 500$ pc, and luminosity, $L = 10^{39}$ erg s^{-1}, in this frequency band be detected from these observations? Would it be resolved?

At $D = 25$ Mpc $= 7.71 \times 10^{25}$ cm, the flux (Eq. 3.9) is $f = 1.34 \times 10^{-14}$ erg s^{-1} cm^{-2} and the mean flux density in a band of width, 2×10^{12} Hz, is (Eq. 3.7) $f_\nu = 6.69 \times 10^{-27}$ erg s^{-1} cm^{-2} Hz^{-1}. A diameter of 500 pc corresponds to an angle of 2×10^{-5} rad (Eq. I.9) and therefore a solid angle of $\Omega = 3.14 \times 10^{-10}$ sr (Eq. I.11) so the specific intensity is (Eq. 3.13) $I_\nu = 2.13 \times 10^{-17}$ erg s^{-1} cm^{-2} Hz^{-1} sr^{-1}. This is just at the rms noise level, so the HII region would not be at the minimum level for detection. A longer integration time should be chosen to increase the S/N. An angular resolution of 1″ corresponds (Eq. I.9) to a linear scale of 121 pc at the distance of this galaxy, so the HII region would indeed be resolved, if it were detected.

4.6 VISUALIZING THE SIGNAL

The final image can be displayed in a variety of ways. The galaxy, NGC 2903, has already been shown for different resolutions (Figure 4.11). Figure 4.19 shows the same galaxy, but using three different visualization schemes. In all three cases, the data (i.e. the numerical values in the 2-D array) are exactly the same. The first image shows the data as a linear greyscale (Figure 4.19a) similar to the linear response of the CCD with which it was obtained. This represents the data well, except that the range of brightness is very large and it is difficult to show the entire range on paper. Details near the centre of the galaxy are 'burned out' and the faint outer regions of the galaxy do not show up well. An alternative is Figure 4.19b which applies a logarithmic weighting to the data prior to displaying it. This has the effect of compressing the range of brightnesses so that both the bright and dim parts of the image are easily seen. The drawback is that there is less contrast in any part of the image, so the true range of brightness is not obvious. Figure 4.19c is a third possibility. Apart from being visually appealing, the use of colour, depending on how it is applied, allows details in both the bright and dim parts of the image to be highlighted. This is called *false colour* because the adopted colour scheme is chosen only for illustrative purposes and has no scientific meaning otherwise.

Figure 4.20 illustrates a representation of data when comparisons are desired between wavebands. Contours (Figure 4.20a) are most often used alone or in overlays like this when the spatial resolution is low or the image structure is not very complex. Otherwise, the contours become crowded. In Figure 4.20b, another optical image is shown. At first glance, this might be interpreted as being a single image for which colours have been applied

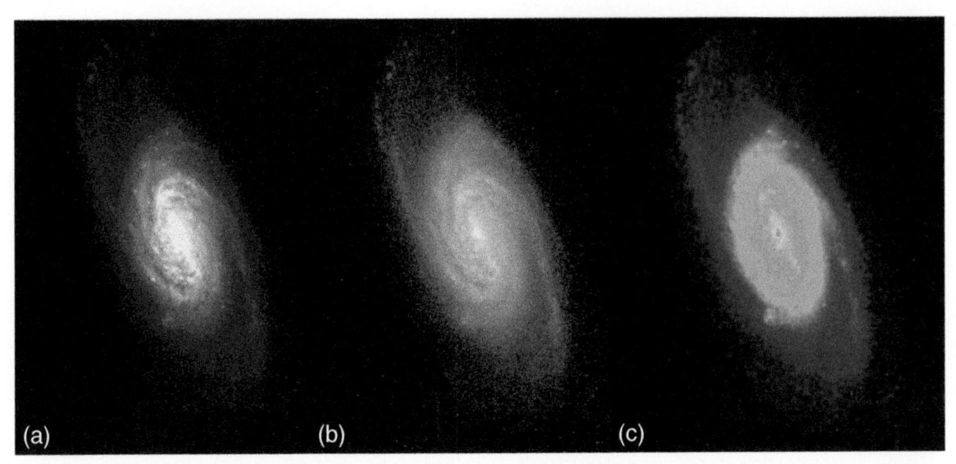

Figure 4.19 An optical image of the galaxy, NGC 2903 (distance, $D = 8.6$ Mpc), taken in a band centred at 500 nm, is represented in three different ways. (a) linear greyscale, (b) logarithmic greyscale, and (c) logarithmic false colour. Source: ID: NGC 2903, http://nedwww.ipac.caltech.edu.

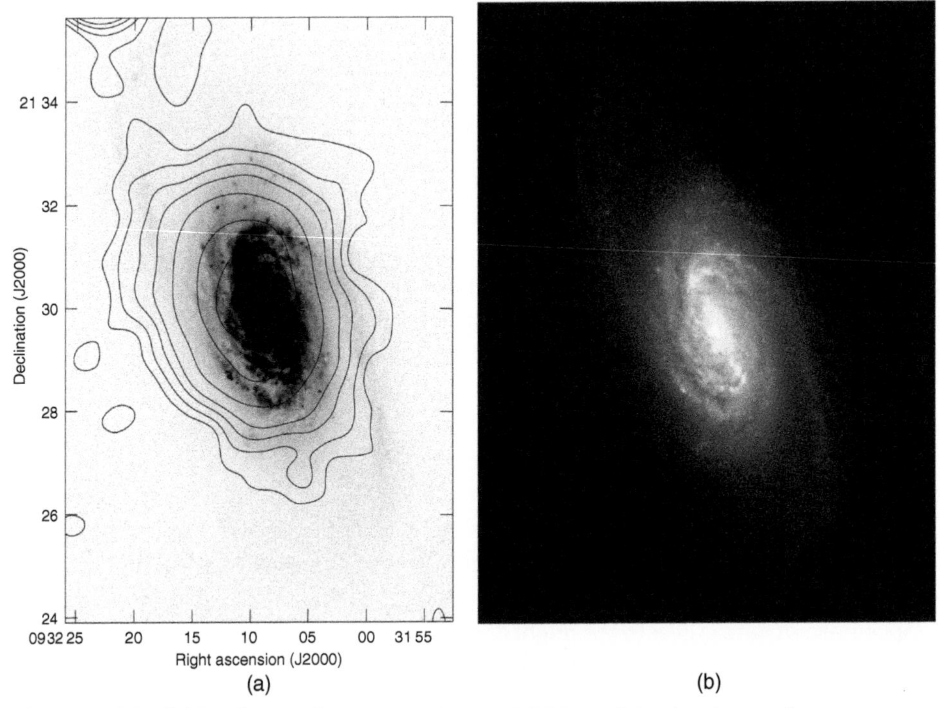

Figure 4.20 (a) Radio continuum contours at λ 20 cm (single-channel bandwidth = 0.66 MHz) overlaid on the same image of NGC 2903 as in Figure 4.19. The radio contours are at 0.6, 1.5, 3, 5, 10, 25, and 50 mJy beam^{-1}, the peak map level is 117 mJy beam^{-1}, the beam FWHM is 54.4″ (see Section 4.2.4), there are 1.7″ per pixel, and the rms noise is 0.5 mJy beam^{-1}. Here, 'beam' refers to the solid angle subtended by the beam (cf. Fig. 4.9 right). Tick marks on the y-axis are separated by 2 arcmin. Source: Judith A. Irwin. (b) Combination of three different images in three wavebands centred at λ 500 nm (shown as blue), λ 650 nm (green) and λ 820 nm (red). Source: ID: NGC 2903, http://nedwww.ipac.caltech.edu.

(i.e. false colour). In fact, it is a combination of three different images taken in three different wavebands for which different colours have been adopted. Colours corresponding to these wavebands were not intended to describe what the eye would see.

Any of the above approaches is valid depending on what information is meant to be conveyed. Since the 'true' colour of an astronomical object applies only to the narrow visual range of the electromagnetic spectrum to which the eye is sensitive, from a scientific point of view there is little motivation to reproduce true colour images of astronomical objects. However, the careful adoption of a visualization scheme can sometimes provide useful information 'at a glance'. For example, the compact objects seen in Figure 4.20 Right are green, which means that the compact objects emit most strongly in the λ 650 nm band (see caption). In fact, this band contains the Hα emission line[9] which suggests that these compact objects are HII regions (Section 7.2.2). Thus, while detailed scientific analysis requires that images in each band be analysed separately, a well-conceived image display can convey important scientific information [8]. Example 4.3 provides another illustration.

Example 4.3

Consider the strong background source just to the north of the HII region, IC 5146, visible in the radio contours of Figure 10.4. The location of this source is at RA = $21^h\,53^m\,35^s$, DEC = $47°\,21'$, *and the radio image is diffraction-limited, with the resolution specified by the size of the 'beam'. The radio telescope was pointed directly at the source when it obtained this image.*

(a) *What quantity is represented by the contours?*
(b) *Is this source resolved or unresolved?*
(c) *Estimate the flux density of this source (Jy).*
(d) *Express the peak specific intensity of this source in cgs units. (Assume that the beam response is uniform for this part of the question.)*

(a) The contour units are in milli-Janskys (defined in Eq. 3.8) per beam solid angle, (mJy beam^{-1}), so these are contours of specific intensity though not in cgs units.

(b) If the source is unresolved (source angular diameter, θ_{source}, $\ll$ the beam angular diameter, θ_b), then its observed size will be the *same* as the beam size because its emission will be spread out over the shape of the beam (Section 4.2.4). If θ_{source} approaches the size of the beam or exceeds it, then the observed size of the source will be broader than the beam. The source in the figure is the brightest source in the map, so its peak (see caption) is $I_{\nu\,\text{max}} = 318$ mJy beam^{-1}. The half-maximum contour is then $I_\nu = 159$ mJy beam^{-1}.

We now need to measure the full width of the half-maximum contour which we will do by using the tick marks on the Declination (ordinate or north–south) axis. *Do not use* the RA axis unless it is converted properly to minutes of arc as described in Section 4.2 of the *online* material. This measurement shows that the apparent source size, though somewhat oval in shape, is not significantly greater than the stated beam FWHM of $\approx 80''$ in any direction. Therefore, this source is unresolved, a conclusion that is reached by this quick comparison of its angular size at half-maximum with the beam size.

[9] Spectral lines will be discussed in Chapter 11. See Appendix C of the *online material* for the spectral lines of Hydrogen.

(c) The flux density of an unresolved source is preserved (e.g. Figure 3.9). Since the main beam has a Gaussian shape and the telescope is pointing directly at the source, the integral under the beam (of solid angle, Ω_b) can be approximated as (Eq. 3.13), $f_\nu = \int I_\nu \cos(\theta)\, d\Omega \approx I_{\nu max}\, \Omega_b = 318$ mJy. Basically, the flux density of an unresolved source is confined to a single beam, and therefore, its value can be read simply from the peak at its position for contours in 'per beam' units.

(d) This part requires more than a 'quick glance' at the map. From the definition of a milli-Jansky (Eq. 3.8), $I_\nu = 318 \times 10^{-26}$ erg $^{-1}$ cm^{-2} Hz^{-1} beam^{-1}. Converting the beam FWHM to radians, we find, $\theta_b = (80/60/60/180) * \pi = 3.88 \times 10^{-4}$ rad. Then, the beam solid angle is (Eq. I.11 for a beam that is approximately circular on the sky), $\Omega = \pi\, \theta_b{}^2/4 = 1.18 \times 10^{-7}$ sr. (This assumes that the beam has a uniform response. See Example 4.4 for a more realistic situation that accurately deals with the Gaussian response of the beam.) We finally find, $I_\nu = 318 \times 10^{-26}$ erg s^{-1} cm^{-2} Hz^{-1} beam^{-1} = 2.7×10^{-7} erg s^{-1} cm^{-2} Hz^{-1} sr^{-1} in cgs units.

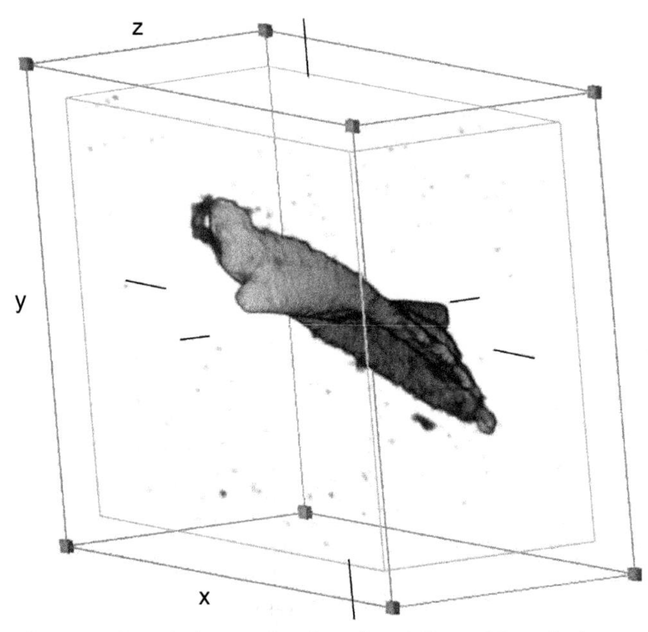

Figure 4.21 A *data cube* showing λ 21 cm HI emission from the galaxy, NGC 2903 in a three-dimensional 'space' with the *x*-axis being Right Ascension, the *y*-axis being Declination, and the *z*-axis being either wavelength, frequency, or velocity. For this galaxy, the emission on the *z*-axis extends from an observed wavelength of $\lambda_1 = 21.13215$ cm to $\lambda_2 = 21.15837$ cm, with the centre at $\lambda_{sys.} = 21.14526$ cm. An analysis of the emission in this cube can determine how the galaxy is rotating and how much mass is present (Section 9.1.1). Source: Judith A. Irwin

There is one possible element of an image that is still missing, however – that of the spectral dependence of the emission. The above images apply only to data obtained in a single spectral channel. However, if the frequency band has been split into many frequency channels (or equivalent in the wavelength regime), then an image is obtained in each of these channels. Rather than a two-dimensional image, we now have a three-dimensional *data cube*, the third axis being either frequency, or wavelength. If emission is observed in a spectral line (see Chapter 11), then the Doppler relation (Table I.2, Section 9.2.1) can be used to convert the third axis into a velocity. Figure 4.21 illustrates this for neutral hydrogen (HI) spectral line data (see footnote 9) obtained for NGC 2903. Data cubes like this contain a rich wealth of information about the source, its contents and dynamics.

4.7 COMPARING SIGNALS IN DISPARATE WAVEBANDS

Modern astronomy often requires comparisons between images at widely disparate wavebands. Gone are the days when one could be a 'radio astronomer' or an 'optical astronomer' or an 'X-ray astronomer'. Although the techniques that are used in the different bands do require a level of expertise that is band-specific, it is still necessary to compare information in different wavebands to extract the most useful information. It is common to do overlays as shown in Figure 4.20a, and even better, overlays in which the spatial resolutions are similar. A glance at Figure 4.20a shows that the optical image (greyscale) has a much higher resolution than the radio image (contours).

At optical or infrared wavelengths, for example, once the signal has been corrected for bias, dark signal and flat-fielded (Section 4.4.1) it is common to record some measure of flux density (cgs units or magnitudes) 'per pixel' on a CCD. The pixel response is uniform, that is, there is no weighting across the pixel. Knowledge of the plate scale (Section 4.2) provides a straightforward conversion to flux density per square arcsecond which is, of course, a (non-cgs) measure of specific intensity.

The situation is different for radio astronomy since the signal is commonly measured in Jy/beam as we saw in Example 4.3. Here, 'beam' refers to the diffraction pattern (e.g. Figure 4.9b) and it is often elliptical in RA and DEC on the sky. In Example 4.3, we happily converted from 'per beam' to 'per steradian' by simply converting the beam FHWM to radians and then converting to a solid angle. However, the radio beam is actually weighted by a Gaussian shape whose peak at the centre is normalized to 1.0, and therefore its response is not uniform, or 'flat', as was assumed in Example 4.3. Suppose we want to convert from 'per beam' to 'per pixel', where the pixel (or cell) size is smaller than the beam size. Now we have to deal with the Gaussian weighting of the beam. In a sense, we need to 'flat-field' the Gaussian beam. Example 4.4 provides an explanation.

Example 4.4

An IR image is measured in Jy/pixel, where the pixels are 1 arcsec on a side. A radio image is in units of Jy/beam, where the beam size is elliptical with a major axis FWHM of 15 arcsec (θ_1) and a minor axis FWHM of 12 arcsec (θ_2). The pixel size of the radio image has already been matched to the IR image. Convert the radio image so that it has the same units as the IR image.

From Eq. (I.11), we know how the 2-dimensional solid angle relates to the 1-dimensional FWHM when there is no weighting (uniform response),

$$\Omega = \frac{\pi\theta_1\theta_2}{4} \tag{4.6}$$

However, a Gaussian-weighted beam is defined by its 2-dimensional integral. This requires a weighting (Problem 4.14) as follows,

$$\Omega_{beam} = \frac{\pi\theta_1\theta_2}{4\ln(2)} \tag{4.7}$$

For the given conditions,

$$\Omega_{beam} = \frac{\pi(15)(12)}{4(2.063\times10^5 \text{arcsec/rad})^2 \ln(2)} = 4.79\times10^{-9}\ \text{sr}.$$

Now a pixel that is one arcsec on a side with a flat response will have a solid angle of

$$\Omega_{pixel} = \frac{(1)(1)}{(2.063\times10^5)^2} = 2.35\times10^{-11}\ \text{sr}.$$

Therefore, there are $\Omega_{beam}/\Omega_{pixel}$ [(sr/beam)/(sr/pixel)] = 204 pixels per beam. Finally, if the radio map is in units of Jy/beam and we instead want Jy/pixel, this requires us to divide every value in the radio map by a factor of 204.

In summary, any conversion of a Gaussian-weighted beam to a cgs solid angle in steradians requires Eq. (4.7). However, if the desire is to convert an elliptically or circularly shaped source in the sky into steradians, when the response function is 'flat', Eq. (4.6) is appropriate.

PROBLEMS

4.1. Using Figure 4.4 as a starting guide, (a) draw a diagram showing the rays that indicate the field of view on the sky, and (b) draw a diagram which illustrates why larger apertures should result in brighter images.

4.2. [*Online*] For the conditions given in Section 4.1 of the *online material*, determine the ellipticity (ratio of minor to major axis) of the sun when its lower limb appears to be at the horizon.

4.3. [*Online*]

(a) An observer, using the V-band filter (see Table 3.1), centres a star on his 256×256 pixel CCD detector. The star's observed zenith angle is 60°, the CCD pixel separation corresponds to 1″ on the sky, and the seeing is sub-arcsecond.

He then changes to a U-band filter followed by an I-band filter. Determine and sketch where the star will appear on the CCD in each of the filters, as well as its true position for the reference pressure and temperature given in Section 4.1 of the *online material*.

(b) Repeat part (a) but for a temperature of 0 °C.

(c) If the filters are removed and if the CCD could detect all wavelengths from the U-band to the I-band with uniform sensitivity, what would the image of the star look like on the CCD?

4.4. [*Online*] For observations at $\lambda 2.2\ \mu$m through atmospheric turbulence,

(a) Derive an expression for the variation in the index of refraction, δn, with temperature, δT. For a single layer within the turbulent region, write the variation in the refractive angle, δR, as a function of δT (see Eq. (4.2) of the *online material*).

(b) Evaluate δR (arcseconds) for a star at an observed zenith angle of 45° when the fluctuation in temperature, $\delta T = 0.3$ K, and for typical high altitude conditions, i.e. $P = 600 \times 10^2$ Pa; $T = 0$ °C.

(c) Assuming that the combination of different δR from different layers along the line of sight results in a seeing size of $\theta_s \approx 10\ \delta R$, determine the corresponding atmospheric cell size, r_0.

(d) For a wind speed of 20 km/h, how frequently must the deformable mirror be adjusted in an adaptive optics system to correct for seeing at this wavelength?

4.5. Determine whether the dark-adapted human eye is diffraction-limited, seeing-limited, or pixel-limited (Table T.4).

4.6. (a) Determine the spatial resolutions (arcsec) of the following instruments (assume that seeing for the ground-based instruments is 0.5″): (i) The James Webb Space Telescope (JWST, $D = 6.5$ m, $D_{eff} = 5.6$ m) at $\lambda 28\ \mu$m, and (ii) The James Clerk Maxwell Telescope (JCMT) ($D = 15$ m) using the 2nd Submillimetre Common-User Bolometer Array (SCUBA-2) detector ($D_{eff} = 12.4$ m) at $\lambda 450\ \mu$m.

(b) How large would a ground-based optical telescope operating at $\lambda 507$ nm have to be to match the resolution of the JWST? How large would a single-dish radio telescope operating at $\lambda 20$ cm have to be?

4.7. Using the information in the caption of Figure 3.2, determine the proper motion (in arcsec/year) of the filaments in Cas A. How long would it take for this proper motion to become measurable, using a radio telescope whose effective diameter is $D_{eff} = 32$ km?

4.8. A galaxy is face-on, circular in appearance, and subtends an angular diameter of 10″. In which of the following situations would the galaxy be like a point source, i.e. completely unresolved: (a) an amateur 'backyard' 2 in. diameter optical telescope with seeing of 2″, (b) the single-dish radio telescope in Arecibo, Puerto Rico, of diameter, 305 m, operating at $\lambda 20$ cm, (c) the 0.85 m Spitzer Space Telescope (launched 2003, but now retired from service) operating at the IR wavelength of $\lambda 60\ \mu$m?

4.9. A detector of the *European Extremely Large Telescope* (ELT, $D = 38.54$ m) will have a plate scale of 0.3016″/mm [19] and will use AO to eliminate the effects of atmospheric seeing.
(a) What is the focal length of the ELT? Will there be 'folded optics'?
(b) What is the spatial resolution at H-band (Table 3.1)?
(c) Estimate the expected number of resolution elements across the disc of Betelgeuse (Figure 3.10).
(d) If you had to choose the pixel size of the CCD for this instrument, what pixel size (μm) would you recommend?

4.10. On a dark clear night but with poor seeing ($\theta_s = 4''$), the two planets, Jupiter and Uranus (both near opposition), are viewed by eye. Determine, approximately, the number of independent seeing discs that are present across the face of these two planets and indicate whether they will twinkle or shine.

4.11. From the image and information provided for the radio continuum data in Figure 4.20a,
(a) list or estimate numerical values of the eight map parameters, as described in Section 4.5.
(b) estimate the following values (cgs units) for the galaxy in the λ20 cm band: (i) the flux density, (ii) the flux, (iii) the luminosity, and (iv) the spectral power.

4.12. A radio telescope operating at λ6 cm has a circular beam of $\theta_{FWHM} = 30''$. If the lowest possible rms noise on a map made using this telescope is 0.01 mJy beam^{-1}, could this telescope detect radio emission from the hot intracluster gas whose spectrum is plotted in Figure 10.3?

4.13. What is the dynamic range of stars in a major city (Table 3.2)? Repeat for a dark clear sky.

JUST FOR FUN

4.14. The plate scale of an African elephant's retina is 2.76°/mm, and there are 36.35 independent receptor cells (pixels) per mm. Assume a pupil size of 14 mm and compute the resolution of an elephant's eye. Is it pixel-limited or diffraction-limited? Is it better or worse than a human being?

4.15. Apparently, the Great Wall of China can be seen from space. But what about its width? Would the width of the wall (≈ 5 m) be resolved to the human eye or would it just appear like a line? Work this out for the International Space Station (ISS) and for the stratospheric platform from which Felix Baumgartner jumped in 2012.

4.16. Imagine that we had the entire Solar System available for a radio VLBI network working at $\lambda = 1.3$ mm, with stations from Mercury to Pluto. What would be the resolution? In Section 4.2.5, we indicated that the VLT could resolve a 3 m-sized object on the Moon. How far away could a 3 m-sized object be resolved using our Solar System VLBI (SSVLBI) network (assuming that it is bright enough)? Could you detect an Earth-sized planet in another galaxy?

PART III

Matter and Radiation Essentials

Radiation and matter can be thought of as different manifestations of energy. Consider, for example, $E = h\nu$ or $E = \mathrm{m}\, c^2$ (Table I.2). These different forms of energy interact with each other in very specific ways. Were it not for this fact, we could discern very little about the Universe around us. In astrophysics, it is radiation that we observe, but it is matter that we most often seek to understand. What is its origin? How does it behave? What is its energy source? By what process does it radiate? How does it evolve with time? Is the radiation that we see even a good indicator of the true content of the Universe?

These are challenging questions and not easily answered. We are not left without some tools, however. The radiation that we detect does indeed provide us with evidence as to the nature of this vast home in which we live. Sometimes, the evidence is strong from a single set of observations that are thorough and well executed. Sometimes it is strong because it comes from completely independent data sets which, when considered together, present a consistent and compelling picture. Sometimes we only have clues and hints as to the direction that the truth lies. Science is a process, not an endpoint, providing us with a snapshot of our natural world. We would always like that snapshot to be brighter, clearer, more detailed or larger, but to have a picture is infinitely better than being left in the dark. Even more encouraging is the fact that we continue to see the picture focussing and refocussing before our eyes.

To understand this process, or to be a part of it, there are some important concepts related to matter and radiation that should first be appreciated. In this section, therefore, we present some 'essentials' of matter and radiation in preparation for a discussion of the interaction between these two forms of energy. It is this matter/radiation interaction which is at the heart and soul of astrophysics.

Chapter 5
Matter Essentials

In truth, it was the greatest wonder that has ever shown itself in the whole of nature since the beginning of the world, or in any case as great as [when the] Sun was stopped by Joshua's prayers.
–Tycho Brahe, on the discovery of the supernova that bears his name

5.1 THE BIG BANG

The Universe, including space, time, energy in its various forms, virtual particles, and photons, visible and unseen matter, is believed to have begun in a primordial fireball called, prosaically, the *Big Bang*. It is important to think of the Big Bang, not as an event that has occurred at some point in time and whose result, the Universe, is now expanding into space. Space and time, as we know them, did not pre-exist, but rather came into being with the Big Bang. There is no lack of alternative ideas, such as oscillating universes, 'multiverses' (of which ours would be just one), and a Universe filled with 'strings' floating in higher dimensions, but so far, there is no convincing observational evidence to back them up.

By contrast, there is good evidence for the Big Bang that, collectively, present a compelling case. For example, a variety of observations suggests that the Universe has *evolved*. Since light travels at a finite speed, as we look farther into space, we also look back to earlier epochs in the history of the Universe. This early Universe was very different from what we see today, showing different kinds and admixtures of sources, and objects with properties that differ from those that are local (Figure 5.1). The expansion of the Universe itself (Section 9.1.2) is consistent with this view. The age of the Universe, derived from its cosmological properties, is 13.801 ± 0.024 Gyr [21] (although this error bar could be somewhat optimistic). The age of the oldest stars, determined by a variety of independent means, is comparable to the age of the Universe though with a larger error bar of $\gtrsim 2$ Gyr [3]. This suggests that stars began to form soon after the Big Bang.

The fact that the night sky is dark is also consistent with a Universe of finite age (Section 3.4). At the earliest epoch that it is currently possible to observe, we see the *cosmic microwave background* (CMB), an opaque glow at 2.7 K that fills the sky in all

Astrophysics: Decoding the Cosmos, Second Edition. Judith A. Irwin.

Companion website: www.wiley.com/go/irwin/astrophysics2e

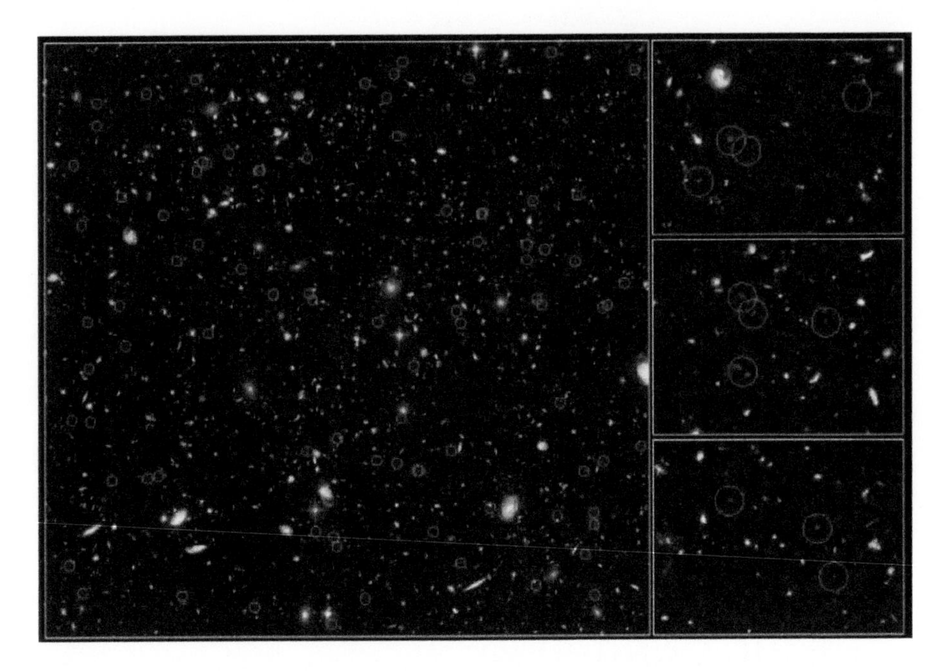

Figure 5.1 The Hubble Ultra-Deep Field, taken from the Hubble Space Telescope in a 10^6 s exposure. The red 'smudges', circled in these images (several blow-ups shown on right), are among the most distant galaxies observed optically. The faintest objects are 2.5×10^{-10} times as bright as what can be observed with the naked eye. These galaxies, or rather 'galaxy precursors', are seen as they were at a time when the Universe was only 5% of its current age. They are much smaller than the giant systems we see today and suggest that today's galaxies are formed by smaller systems building up into larger ones. (Source: Reproduced by permission of NASA, ESA, R. Windhorst [Arizona State University], and H. Yan [Spitzer Science Centre, Caltech]).

directions (Figure 5.2). The CMB is an imprint of the cooling primordial fireball. Tiny fluctuations in the CMB reveal details about the cosmological parameters that characterize the large-scale properties of the Universe. These fluctuations form the seeds from which galaxies later grow (see Figure 7.7). Finally, the abundances of the light elements, as predicted from nucleosynthesis in the first moments after the Big Bang, are in agreement with the observed values. Any theory that seeks to supplant this model would have to do a better job of addressing each of these points.

5.2 DARK AND LIGHT MATTER

What of the matter produced in the Big Bang? Strangely, most of it is invisible to us. The bulk of the material created in the immediate aftermath of the Big Bang is known as *dark*

Figure 5.2 Temperature map of the Cosmic Microwave Background (CMB) radiation as measured by the Wilkinson Microwave Anisotropy Probe (WMAP) satellite, launched in 2001. This is a *Hammer–Aitoff projection* in which the whole sky is shown in an elliptical projection. The CMB is remnant black body radiation (Section 6.1) from a time 380 000 years after the Big Bang when the temperature of the Universe cooled to about 3000 K, sufficient for electrons and protons to combine and form neutral atomic hydrogen. At this time, corresponding to a redshift of $z = 1000$ (Section 9.2.2), the Universe went from being opaque to being transparent. The mean observed temperature of the CMB, as we now measure it, is 2.725 K, but tiny fluctuations of temperature are shown in this image of order +200 μK (light) to −200 μK (dark) [2]. The temperature fluctuations are caused by acoustic (sound) waves. The angular scales over which perturbations can be seen provide information that constrains the cosmological parameters of our Universe. Source: Reproduced courtesy of The WMAP team and NASA.

matter because, to the limits of the available data, this material does not emit light at any wavelength. Its presence is inferred by the effect it has on neighbouring matter that can be seen. For example, galaxies rotate much faster in their outer regions than would be expected from the amount of mass that is inferred from the visible galaxy. This suggests that there is dark matter surrounding galaxies in a distribution that is more extended and more spherical than the visible material (see Figure 5.3). Similar arguments apply to clusters of galaxies. Adding up the mass from each member galaxy as well as any intracluster gas (e.g. Figure 10.8) results in a value that is far too low to bind the cluster together. These clusters are believed to be bound because the time it would take for the gas to escape and the galaxies to disperse is quite small if they are not. Moreover, a study of the bending of light from background sources, called *gravitational lensing* (see description in Section 9.2.1), confirms the presence of dark matter in clusters and elsewhere.

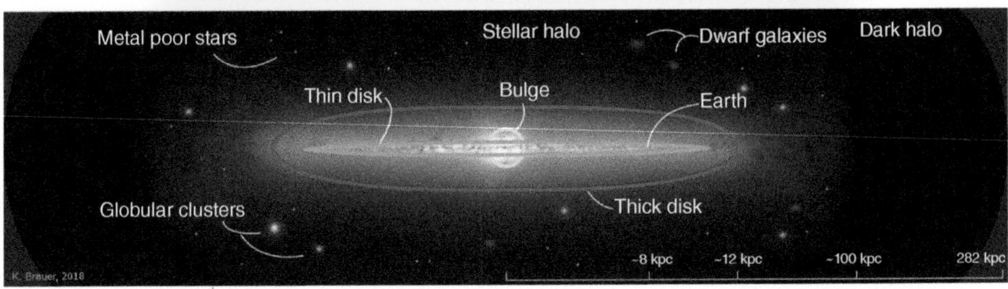

Figure 5.3 Sketches of the Milky Way galaxy face-on (top) and edge-on (bottom). The Milky Way is a barred spiral galaxy, about 35 kpc in diameter, and the Sun is located 8 kpc from the nucleus within the disk. The disk is thin in comparison with its diameter, only about 300 pc thick ('thin disk') surrounded by a 'thick disk' of uncertain size. The thin disk is active, containing stars, gas, and dust, and star formation occurs within it. The stellar distribution becomes more spherical, called the bulge, closer to the centre. Surrounding the disk is a halo of older metal poor stars and *globular clusters*, the latter containing typically 10^5 stars each, in a 'stellar halo'. An even larger distribution of dark matter (the 'dark halo') is believed to encompass the visible material. Source: Top image: © NASA/JPL-Caltech/R. Hurst (SSC/Caltech)). Bottom image: Courtesy of Kaley Brauer.

Many sensitive searches to detect dark matter have been carried out under the assumption that at least some of the material is *baryonic*. Baryonic matter is any matter containing baryons, which are protons or neutrons[1]. This includes all elements with their associated electrons as well as any object that is thought to have been baryonic in the past[2]. For example, if there are large spherical distributions of dark matter around galaxies, then such a distribution should also be present around our own galaxy, the Milky Way (Figure 5.3). Much observational effort has therefore been expended to measure the gravitational lensing from baryonic objects in the halo of the Milky Way, be they truly dark or just too dim to be directly detectable. The most likely candidates are objects that did not have sufficient mass to become stars (like Jupiter) or the remnants of stellar evolution such as white dwarfs, neutron stars, and black holes (see Sections 5.3.2 and 5.3.3 for a description of these objects). These have been collectively nicknamed *massive compact halo objects* (MACHOs). The result of such efforts (Section 9.2.2) is that the masses of all MACHOs fall far short of what is required to account for the dark matter.

The failure of observations to detect baryonic dark matter either directly or indirectly as well as other cosmological constraints has led to the current view that most dark matter consists of 'unusual' matter. Since the dark particles have been so elusive, they must not radiate or interact very easily with other matter, except via gravity. This rules out most known non-baryonic particles. Also, because dark matter does gravitate and has been present since the time the galaxies and stars were forming, it must also have played a part in determining the scale over which structures were established in the Universe. Non-baryonic particles such as *neutrinos* (Section 1.3), therefore, could not make up most of the dark matter because they move at relativistic speeds and the clusters of galaxies that form in a Universe dominated by such particles would be too large in comparison with what is observed. These criteria suggest that most of the dark matter must consist of particles not yet observed in Earth-based laboratories or accelerators; such particles are collectively referred to as *weakly interacting massive particles* (WIMPs). Some are predicted by well-established theory, and others are more ad hoc. But since classical telescopes have been unable to detect radiation from dark matter, the new 'telescopes' are now WIMP-seeking particle detectors in underground labs. For example, the Sudbury Neutrino Observatory, shown in Figure 1.9 Left, has expanded into *SnoLab* and is running a number of innovative and ground-breaking experiments in the hopes of detecting this elusive particle[3].

The search for dark matter is no less than a search for most of the mass of the Universe. Unless there is something seriously wrong with our understanding of gravity and current cosmological models, it is estimated that 83% of the *matter energy* in the Universe is dark and of unknown nature (Right column of Figure 5.4 under 'Matter'). Only 17%[4] is believed to be baryonic [16] and potentially visible via radiation (note that the energy density

[1] There are other types of baryons as well but they are unstable, with lifetimes of no more than about 10^{-10} s.

[2] An example of an object that is no longer baryonic is a black hole whose precursor was a massive star (Section 5.3.2).

[3] See https://www.snolab.ca/science/experiments.

[4] There is some variation in these quantities, depending on the specific observations.

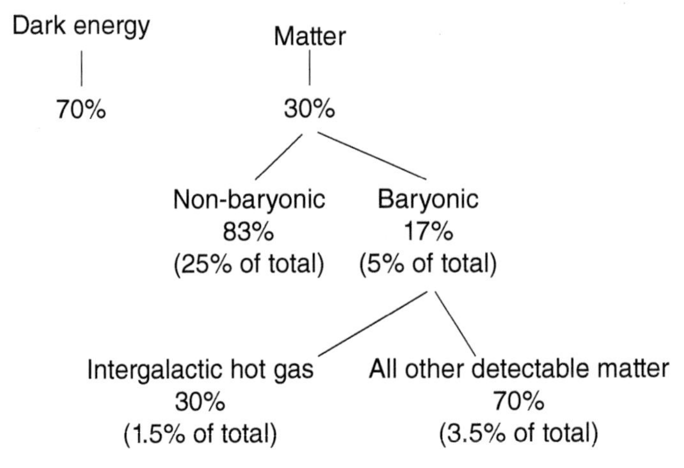

Figure 5.4 Estimates of the components of the energy density of the Universe from a variety of sources. 'Energy density' refers to all forms of energy. 'Dark Energy' is the component, as yet unknown, that causes the apparent acceleration of the Universe. 'Matter' includes all forms of matter, both light and dark, as well as all radiation, the latter being a small fraction at the present time. The values for dark energy and matter (first line) are thought to be known to within a few percentage points. The percentages of the total are referred to as Ω in Appendix F of *the online material.*

of radiation itself constitutes only a tiny fraction of the energy density of matter and is absorbed into the matter fractions quoted here). Of this 17%, a significant fraction (about 30%) has been 'missing' and not observed until recently. Historically, this has been called the *missing baryon problem.* These missing baryons have been difficult to observe because they are locked up in hot, low-density intergalactic gas that emits in the X-ray regime. Although X-ray emitting gas has been observed in and around galaxies for many decades, the missing component is weaker and more extended and requires sensitive observing techniques (e.g. the *Sunyaev–Zeldovich effect,* Section 10.6) to measure it [20, 26], see also [19].

Even more astonishing, however, are recent discoveries implying that the Universe is dominated, not by mass, but by a mysterious *dark energy* that causes the expansion of the Universe to *accelerate* with time. This is discussed further in Section 9.1.2. All mass, both light emitting matter and dark matter, make up only 30% of the total energy content of the Universe. The marvels that we actually *observe* in the Universe, from γ-ray bursts, to massive amounts of hot X-ray emitting gas in galaxy clusters, to the myriads of galaxies we see optically in the near and far Universe (Figure 5.1), to cold radio-emitting dust clouds, and all radiation – amount to a tiny 5% of this total energy (Figure 5.4). The revelation as to what dark energy and dark matter really are, when it finally comes, promises to revolutionize our understanding of the Universe and possibly of Physics itself.

5.3 ABUNDANCES OF THE ELEMENTS

5.3.1 Primordial Abundance

Of the small fraction of the Universe that makes up all of the stars, galaxies, and other objects that we see today, we still need to account for its constituents. The abundances of the elements can be quantified by their *number fraction* (Figure 5.9) or by their *mass fraction*. The latter is normally expressed as the mass fraction of hydrogen, X, helium, Y, and all heavier elements, Z, respectively,

$$X \equiv \frac{M_H}{M} \quad Y \equiv \frac{M_{He}}{M} \quad Z \equiv \frac{M_m}{M} \tag{5.1}$$

where M_H, M_{He}, and M_m are the total masses of hydrogen, helium, and all other elements, respectively, and M is the total mass of the object being considered. Thus,

$$X + Y + Z = 1 \tag{5.2}$$

Because the mass fraction of all elements other than hydrogen and helium tends to be quite low, these are collectively referred to as *metals* and Z is referred to as the *metallicity*.

In the immediate aftermath of the Big Bang, only hydrogen (H), helium (He), and trace amounts of the light elements, deuterium (D), helium-3 (^{3}He), and lithium-7 (^{7}Li) were formed[5]. Although the temperature was extremely high, the expansion of the Universe was so rapid that the density quickly became too low to drive the nuclear reactions required to create heavier elements. The abundances at this time are said to be *primordial* with values of [6]

$$X = 0.75 \quad Y = 0.25 \quad Z = \text{trace} \tag{5.3}$$

5.3.2 Stellar Evolution and ISM Enrichment

If the Big Bang could not produce the heavier elements, how then were they formed? The answer lies within stars. As the Universe expanded, inhomogeneities in the matter density resulted in the collapse and formation of the early protogalaxies and the stars within them. The process of star formation involves primarily a gravitational collapse of a molecular gas cloud, but the details of star formation are not entirely understood. Besides gravity, it is likely that turbulence and magnetic fields also play important roles [4]. As stars form, the collapse and compression of matter produce an increase in interior

[5] The numbers specify the atomic weight; for example, normal helium has two protons and two neutrons in the nucleus, whereas helium-3 has two protons and only one neutron. Deuterium is an isotope of hydrogen with one proton and one neutron in its nucleus. An isotope of an element has the same number of protons but different numbers of neutrons in the nucleus.

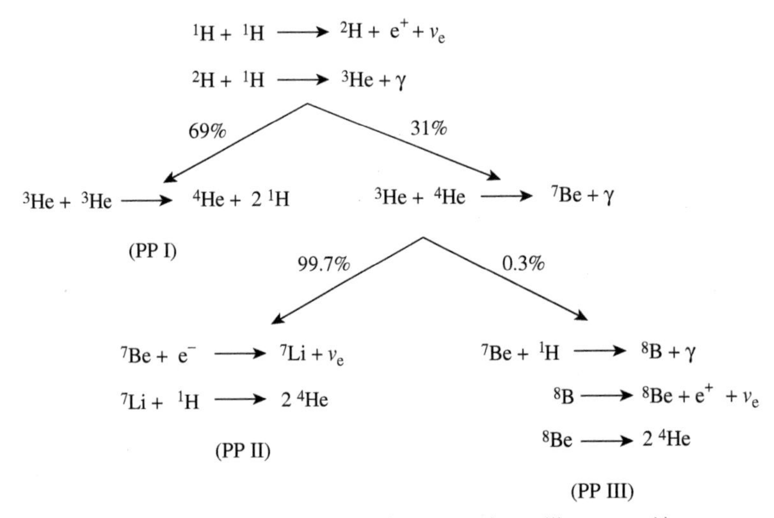

Figure 5.5 The *P–P* chain of nuclear reactions, illustrated here, powers all stars less than about 1.3 $M_\odot$. Note that the intermediate elements are used up again so the main product is ^{4}He. Two branching arrows indicate that two possible reactions can occur with the probabilities noted. However, the reaction in the PPI branch occurs more quickly than the precursor step to PPII and PPIII so, for the Sun, the production of ^{4}He ends via the PPI branch 86% of the time. Note that the first two reactions need to occur twice in order to drive the PPI reaction. In the Sun, the PPI chain can be summarized as $4H \rightarrow {}^4He + 2e^+ + 2\nu_e$ plus 26 MeV of energy. Of the non-nuclear photons or particles, γ represents a γ-ray photon, e^+ is a positron, and ν_e is an electron neutrino. The neutrino in the PPIII chain has historically been an important player in the so-called solar neutrino problem (Section 1.3.2.1).

temperature and density. If the mass of the object is less than $M_{*min} = 0.08\ M_\odot$, then the interior temperature does not rise sufficiently to ignite *nuclear burning* (*thermonuclear reactions*) in the core. Such objects are referred to as *brown dwarfs* – would-be stars that have never ignited. For higher mass objects, however, once the core temperature rises to a sufficiently high value (of order $\approx 10^7$ K), the first thermonuclear reactions can begin.

These reactions slowly convert the lighter elements into heavier ones through *fusion*. The energy released by this process produces the pressure required to halt any further collapse and also causes the star to shine. A current estimate of the star formation rate in the Milky Way is 1.7 $M_\odot$ per year [18].

Low mass stars (masses less than about 1.3 $M_\odot$) convert hydrogen into helium in their cores by a series of steps called the *P–P chain* (Figure 5.5) which ultimately turns four protons into a helium nucleus. This is the dominant energy generation mechanism in the Sun, for example, where it takes place in a core region of radius, $\approx 0.2\ R_\odot$. Because the mass of a He nucleus is slightly less than the mass of four free protons (4.0026 u_{amu}

compared to 4.0291 u_{amu}, Table T.1), there is a net mass loss that occurs in the process. This mass loss has an energy equivalent ($\Delta E = \Delta mc^2$), and it is this energy that powers the stars (Example 5.1). The energy, ΔE, is referred to as the *binding energy* of the nucleus since 'unbinding' the nucleus would require this amount of energy. The longest period of a star's stable lifetime is spent in core hydrogen burning, and therefore, the timescale set by this process sets the 'lifetime' of a star. It also *defines* the *main sequence* which can be seen as the region from upper left to lower right in the HR diagram shown in Figure 3.15. On the main sequence, more massive stars are both hotter and more luminous, placing them at the upper left of the figure. More massive stars on the main sequence also burn hydrogen into helium in their cores, but the dominant reactions are via a series of steps that are different from the P–P chain. The result, however, is the same. On the main sequence, all stars convert hydrogen into helium in their cores. Data for main sequence stars are given in Table T.5. Main sequences stars are also called *dwarfs* because a star on the main sequence is the smallest it will be compared to later stages in its evolution.

Example 5.1

Estimate the main sequence lifetime of the Sun by assuming that 15% of its mass is available for core hydrogen burning. Assume that the luminosity is constant with time.

The available mass is $M_a = 0.15\,X\,M_\odot$, where $X = 0.7154$ (Eqs. (5.1), (5.4)) is the mass fraction of hydrogen in the Sun, so the available number of protons is $N_a = 0.11\,M_\odot/m_p$, where m_p is the mass of a proton. The number of reactions is the number of available protons divided by the number of protons per reaction, $N_r = N_a/4 = 0.11\,M_\odot/(4\,m_p)$, and the total energy released is the number of reactions times the energy released per reaction, $E = N_r\,\Delta E = [0.11\,M_\odot/(4\,m_p)]\,(\Delta mc^2)$. Using $\Delta m = 4.4 \times 10^{-26}$ g and evaluating for the Sun (Table T.2), we find $E = 1.3 \times 10^{51}$ erg. Then, the lifetime for constant luminosity is $t = \frac{E}{L_\odot} = 3.4 \times 10^{17}$ s or about 10 Gyr.

The conversion of hydrogen into helium in the core of a star continues until the hydrogen fuel in the core is exhausted. A thick, hydrogen-burning shell then exists around an inert He core. The inert central core, having lost a source of pressure, begins to contract which in turn causes it to heat up. The heating of the core then pumps some extra energy into the hydrogen-burning shell around it. This additional energy is sufficient to produce greater pressure in the shell-burning region which then acts upon the non-burning outer layers of the star, causing them to expand. The resulting increase in the size of the star makes it more luminous, and the expansion produces cooling of the outer surface. The surface of the star is what is actually observed, so the star moves upwards and to the right, i.e. up the *red giant branch*, in the HR diagram as it becomes a *red giant*. This is the cause of the main branch from the main sequence towards the upper right in Figure 3.15. Data for giant stars are given in Table T.6, and a theoretical HR diagram with the evolutionary track of a one solar mass star is shown in Figure 5.6.

As the star becomes a red giant, the He core is compressing, its temperature rises, and the core density can become very high. For stars less massive than about 2.3 $M_\odot$,

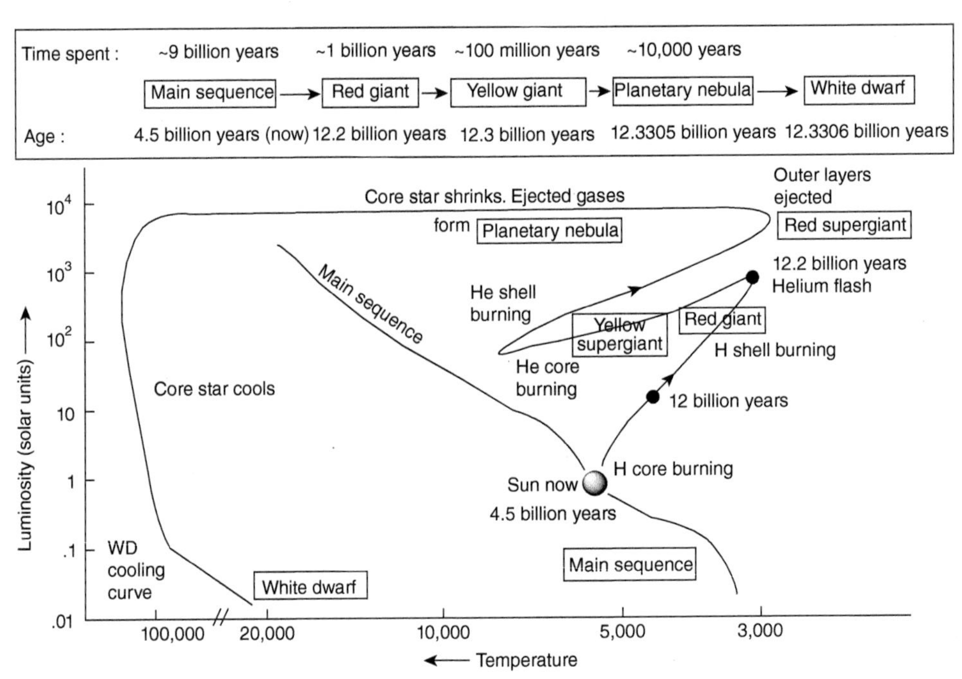

Figure 5.6 Evolutionary track of a 1 $M_{\odot}$ star in the HR diagram (see Figure 3.15 for an observational analogue). Source: Adapted from Stars and Nebulae, SDSS. http://skyserver.sdss.org/edr/en/astro/stars/stars.asp.

the dense He core becomes *electron degenerate*. This means that the electrons fill the available free energy states, starting from the lowest possible energies and filling each higher energy state sequentially. Only one electron can occupy any given quantum state (the Pauli Exclusion Principle, Appendix C.2 of the *online material*), and the occupation of these states becomes independent of temperature. If the atomic nuclei, and the electrons with them, are packed very closely together (small Δx) in the dense core, the Heisenberg Uncertainty Principle (Table I.2) states that the uncertainty in the momentum of the electrons is very high (large Δp). If the uncertainty in the momentum is high, then the momentum itself must be high and so must be the electron velocities. With sufficiently high electron velocities, the electrons themselves can supply enough pressure to hold up the core, dominating over the particle pressure produced by the random motion of nuclei (Section 5.4.1). A typical state of such material is one of extreme density, with the mass of the core containing up to 1.44 $M_{\odot}$ within a volume approximately equal to the size of the Earth.

For stars with masses between about 0.5 and 2.3 $M_{\odot}$, the core temperature eventually reaches a sufficiently high value (10^8 K) that He burning is ignited. He burning under degenerate conditions occurs quickly (within hours) and is called the *Helium flash*, although the rapid flash occurs within the star and therefore is not directly observable. The star, however, takes a sharp change of direction on the HR diagram (see Figure 5.6) as a result. The onset of He burning lifts the degeneracy, the core then stably burns helium, converting it to carbon[6], and the outer layers contract again. Hydrogen burning will still be occurring in a shell around the He-burning core so stars start to become differentiated, with carbon in the core, helium farther out, and hydrogen in the outer layers. A similar evolution results for stars of mass higher than 2.3 $M_{\odot}$ except that helium burning begins before the core becomes degenerate so there is no internal helium flash. For stars of mass lower than about 0.5 $M_{\odot}$, the temperature is not expected to become high enough to ignite helium burning. By the red giant phase, these stars have virtually reached the ends of their lives and should expel their outer layers as described below. However, the main sequence lifetimes of such low mass stars are greater than the age of the Universe (Problem 5.1)! Therefore, we have not yet seen the end products of stellar evolution for stars of such low mass.

The remaining evolution of stars depends on their mass. Stars like our Sun (Figure 5.6) eventually exhaust the helium in the core and evolve to the right again, i.e. up the *asymptotic giant branch* (uppermost curve going right in Figure 5.6) into a *red supergiant* (see Table T.7 for data on supergiant stars). The internal and external behaviour mimic the first hydrogen core exhaustion stage but now the carbon core becomes degenerate. The star will then go through a stage of mass loss via strong stellar winds, pulsating instability, and/or other forms of mass loss (Section 7.4.2). Since the greatly expanded outer layers are tenuous and weakly bound gravitationally, they are easily expelled, forming a nebula that can take on a rich variety of morphologies. These are called *planetary nebulae*[7], examples of which are shown in Figure 5.7. The complex nebular morphologies are not completely understood but seem to be related to different phases of mass loss. Circumstellar disks, magnetic fields, and the presence of companions also appear to play an important role [12].

What remains behind is the hot degenerate core at the centre of the nebula which is now referred to as a *white dwarf*. As can be seen in Figure 5.6, a white dwarf is very hot but has low luminosity because it is so small. No more nuclear reactions take place within it, and it slowly cools with time. The white dwarfs can be seen in a sequence below and to the left of the main sequence in the Hertzsprung–Russell diagram of Figure 3.15.

Stars more massive than the Sun can continue to go through a slow series of expansions and contractions in response to what is occurring inside them and what temperatures, and subsequent reactions, are achievable. This results in evolutionary tracks on the HR diagram that include repeated motions to the right or left in comparison with what is shown in Figure 5.6. Eventually, all stars with masses less than $\approx 8\ M_{\odot}$, which amounts to 95% of all stars that have evolved off the main sequence during the lifetime of the Milky Way [14] (Problem 5.3), go through the planetary nebula stage, leaving a white dwarf core behind. The white dwarf may contain both carbon and oxygen, but it will never be more massive

[6] This is called the triple alpha process because three He nuclei, also known as α particles, eventually become a carbon nucleus.

[7] The term 'planetary' is used for historical reasons. The first planetary nebulae that were discovered were of small angular size and showed a blue-greenish hue which was similar to the colour of the outer planets, Uranus and Neptune.

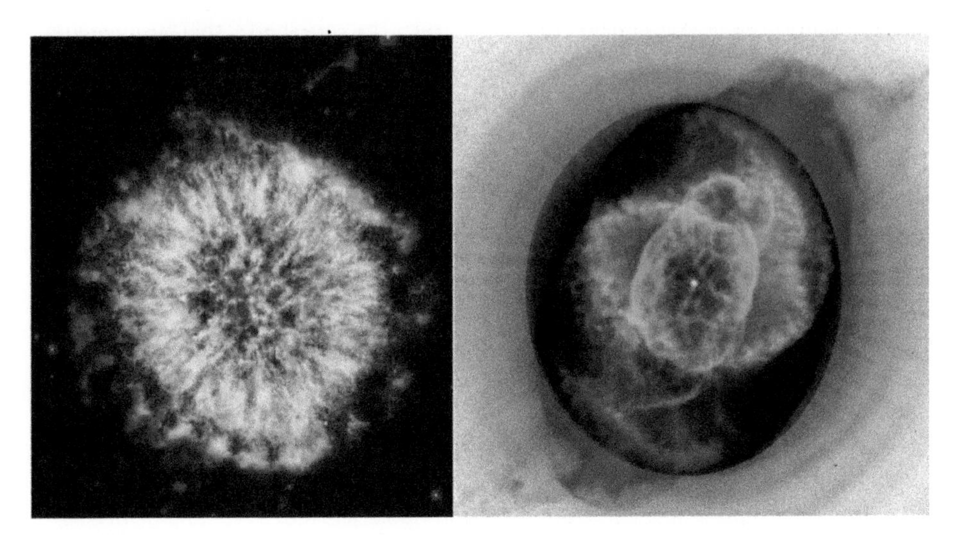

Figure 5.7 Examples of planetary nebulae, the end result of the evolution of low mass stars like our Sun. On the left is NGC 6751, and on the right is NGC 6543 (the Cat's Eye Nebula). The colour scheme is: red = [N II] (λ 6584 Å), green = [O III] (λ 5007 Å), and blue = V band. The square brackets mean that the spectral lines are 'forbidden' (Section 11.4.2). Source: For NGC 6751: Arsen R. Hajian, USNO, Bruce Balick, U Washington, Howard Bond, STScI, Nino Panagia, ESA, and Yervant Terzian, Cornell; For NGC 6543 : B. Balick, J. Wilson, and A. R. Hajian.

than 1.44 $M_\odot$, a value called the *Chandrasekhar limit* after the Indian astrophysicist who determined this maximum value (electron degeneracy pressure is insufficient to support a mass that is higher than this). As indicated earlier for the degenerate core of a star, a white dwarf is very dense, its mass having been compressed into a size that is only about the size of the Earth.

The material expelled as a planetary nebula will be enriched in heavy elements that have been formed in the interiors of the progenitor stars. From the enriched expelled material, new stars will form. Thus, as generations of stars are 'born' and 'die', the metallicity of the ISM (and of the Universe) increases with time.

5.3.3 Supernovae and Explosive Nucleosynthesis

A remaining event that contributes to metallicity and is responsible for elements heavier than iron (Fe) is a *supernova explosion*[8]. Stars of mass $\gtrsim 8\ M_\odot$ do not die out gradually as do lower mass stars, but rather explode, spewing their outer layers into the interstellar

[8] It has also become apparent that massive stars in a very large red giant phase can spew heavy elements more massive than Fe into the ISM via stellar winds. The heavy elements may dredge up from the interior.

medium (ISM) at speeds of tens of thousands of km s^{-1}. This is called a *Type II supernova*[9], one of the most energetic events in nature (Figure 5.8). A typical supernova kinetic energy is 10^{51} erg. Not only does the internal metal-rich material enter the ISM, but also the explosion itself is so energetic that it drives many more reactions, a process called *explosive nucleosynthesis.* It is in this explosive event that elements heavier than Fe are formed.

The reason for the extraordinarily different fate of high mass stars lies in the nature of intranuclear forces. The force that holds the nucleus together is called the *strong force,* and it acts on small ($\approx 10^{-13}$ cm) scales between *nucleons*[10], i.e. baryons that are in an atomic nucleus. For light elements with few nucleons, each 'feels' the force of every other nucleon and so as the number of nucleons grows, so does the binding force per nucleon.

Figure 5.8 The galaxy, M 51, before (left) and after (right) a Type II supernova went off in its disk. The supernova, named SN 2005cs, was discovered in June 2005 by Wolfgang Kloehr using an 8 in. diameter telescope and can be seen directly below the nucleus in the spiral arm closest to the nucleus. A supernova explosion takes only a fraction of a second but maximum light is achieved two to three weeks later. The brightness of a Type II may then plateau or decline in brightness over about 80 days followed by a continuing slow decline thereafter. This galaxy is 9.6 Mpc (3.3×10^6 ly) distant so the supernova detected in 2005 actually occurred 3.3 Myr earlier! Source: Reproduced by permission of R. Jay GaBany, Cosmotography.com.

[9] There are two main types of supernovae which occur in different environments and by different processes. We describe only Type II here. The Type I supernovae are more luminous and are used as distance indicators (Sect. 9.1.2). They also contribute to the abundances of the heavy metals, including those more massive than Fe.

[10] The strong force actually acts between *quarks* that are the constituents of nucleons.

The higher the binding force per nucleon, the more stable is the element and the greater is the mass deficit when it forms. Eventually, however, the nucleus becomes so large that Coulomb forces (in this case, the electrostatic repulsion between protons) which act on larger scales become important. Then as the nucleon number grows, the binding energy per nucleon decreases and the nucleus becomes less stable. The consequence of this behaviour is that light elements release energy when undergoing nuclear fusion, such as in stars, and heavy elements release energy during nuclear *fission*, such as currently occurs in nuclear reactors on Earth[11]. The element at which the binding energy per nucleon is highest and at which this transition occurs is iron[12].

For stars of mass $\gtrsim 8\ M_\odot$, the core temperature achieves a value high enough to initiate fusion reactions of Fe. However, rather than releasing energy during this reaction (called *exothermic*), the reactions now *require* energy (called *endothermic*). With no energy source in the core and the available energy going into endothermic reactions, the core pressure is removed and the core instantly implodes. Although the details of supernovae are not completely understood, it is believed that the core collapse generates a shock wave that triggers the explosion of the outer layers. The resulting material, enriched in heavy elements through explosive nucleosynthesis, contributes to the metallicity of the interstellar medium (ISM). The supernova itself can become extremely bright (Figure 5.8), sometimes outshining the light from the entire parent galaxy. Its light then declines in a characteristic exponential fashion with time, though changes in slope can also occur (see also Section 11.1.3). A plot of the variation of light with time is called a *light curve*. Eventually, the exploded outer layers become visible as a *supernova remnant*, such as the one shown in Figure 3.2.

What happens to the collapsing core? There are two possibilities.

If the core mass is not too high, the remnant is a *neutron star*, a star so dense that the mass of a Sun or more has been compressed into a volume only tens of km across. Typical values might be a mass of $1.5\ M_\odot$ and a radius of 10 km. In such a remnant, the density is so great that electrons have combined with protons to produce a 'sea' of neutrons plus a 'blast' of neutrinos. Modern models, in fact, *require* the blast of neutrinos in order to assist with the explosion of the outer layers in a supernova. We have already seen one example in this text, namely SN 1987A (Section 1.3.2.2). The density of a neutron star is approximately equal to the density of a single neutron, and the star is held up from further collapse by *neutron degeneracy*, the nuclear version of electron degeneracy. Neutron stars are almost always detected as *pulsars*. Pulsars are spinning neutron stars that emit EM radiation along magnetic poles that are at an angle to their rotational poles; as the pulsar spins, the radiation sweeps by our line of sight in pulses like a lighthouse.

---

[11] The first steps towards creating a *fusion* reactor have been taken in the establishment of ITER (Latin for 'the way' and an acronym for 'International Thermonuclear Experimental Reactor'), an international project located in France. Because only low pressures, in comparison with the centre of the Sun, are possible, ITER will operate at 1.5×10^8 K, an order of magnitude higher than the temperature at the centre of the Sun! If successful, future reactors will be like 'mini-Suns', available for power.

[12] The most tightly bound element is actually ^{62}Ni, but this element is not very abundant and therefore does not play as important a role in supernova explosions.

The second possibility, when the core mass is higher, is the formation of a *black hole*. There is some uncertainty about the upper limit to the mass of a stellar core that can produce a black hole, but a reasonable estimate is 3 $M_{\odot}$. A black hole is a region of space around a singularity within which nothing can escape, not even light. We have already seen how binary black holes can produce gravitational radiation (Section 2.8). More will be said about these astonishing objects in Chapter 9.

A remaining important type of supernova does not result from normal stellar evolution of massive stars, at all. These are the Type Ia supernovae (SNe Ia), and they start with a white dwarf in a binary star system. If matter from the companion falls onto the white dwarf, it can be pushed over its Chandrasekhar limit. The result is rapid collapse to a neutron star and concurrent explosion similar to the physics described above. Because the mass limit is so well defined (1.44 $M_{\odot}$), it means that the luminosity shows little variation from one supernova to another. Consequently, SNe Ia can be used as 'standard candles'. With the true luminosity known, the distance can be found (Eqs. 3.32 and 3.33). Supernovae are very bright, so they can be seen even in distant galaxies and are therefore used as calibrators when studying the large-scale expansion of our Universe (Section 9.1.2).

5.3.4 Abundances in the Milky Way, Its Star Formation History and the IMF

The elemental abundances of our Solar System, expressed as a number fraction, measured from Solar spectral lines and meteorites, are shown in Figure 5.9. This plot represents the abundances in the local Galactic neighbourhood at the time the Solar System formed. The values thus reflect the origin of the light elements in the Big Bang as well as subsequent evolution or other processes that could alter these relative abundances since that time. Enrichment of heavy elements by previous generations of stars is clearly the most important effect, the current supernova rate for the Milky Way being estimated at about two per century. Other processes are also involved. For example, high energy cosmic rays can break up heavier nuclei into smaller pieces by spallation, as introduced in Section 1.1. The stability of nuclei against radioactive decay will also affect these abundances.

Consistent with an origin in the Big Bang and in spite of all subsequent evolution, the abundance of H is still, by far, greater than all the other elements combined, constituting approximately 90%, by number, of the Universe. Normalized to a hydrogen abundance of 1, He constitutes only 8.5%, and Li is a mere 10^{-9}%[13]. Even the stable Fe (atomic number = 26) has an abundance only 3×10^{-3}% of hydrogen. The most abundant of the metals is oxygen at 6.8×10^{-2}%. Given the dominance of hydrogen in our Universe, it is important to understand this simple atom extremely well. A review of the properties of hydrogen is given in Appendix C of the *online material*.

[13] Li tends to be destroyed in stellar nuclear reactions.

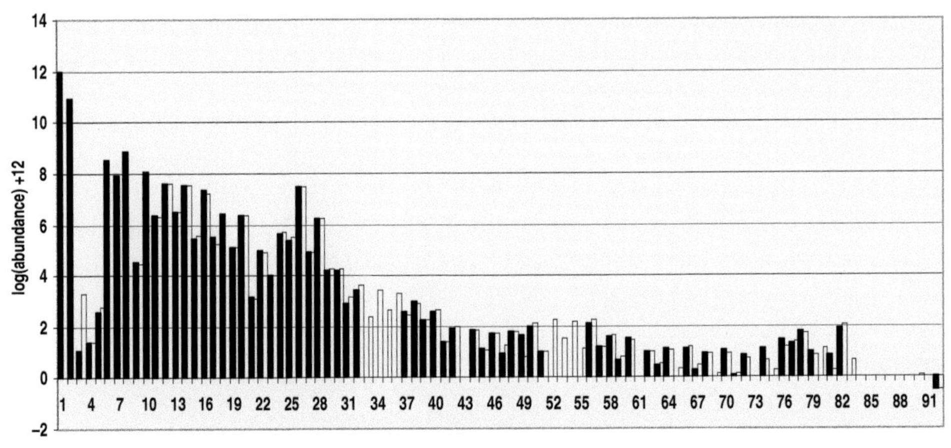

Figure 5.9 Logarithmic plot of elemental abundances in the Solar System as a function of atomic number. Black boxes represent values for the Sun, and clear box values are for meteorites. In several places, there are no data, for example a measurement for H in meteorites or for short-lived radioactive elements. All abundances have been normalized to hydrogen which has a value of 12 in this plot. For example, a value of 6 on the graph corresponds to an abundance of 10^{-6} with respect to hydrogen. Abundances refer to the numbers of particles, rather than their weights. Source: Data from Shore [23].

The mass fractions for the abundances, recommended by [1], referred to as *Solar abundance*, are

$$X = 0.7154 \qquad Y = 0.2703 \qquad Z = 0.0142 \tag{5.4}$$

From measurements of metallicities in interstellar clouds and other stars, Solar abundances appear to be typical of abundances elsewhere in the disk of our Milky Way.

As has been seen with the production of heavy elements within stars, there are many processes involved in the slow conversion of the primordial abundances of Eq. (5.3) to the Solar abundances of the Galactic disk. The increase in metallicity depends on the amount, rate, and metallicity of material expelled into the ISM by planetary nebulae and supernovae, and the heaviest elements originate in supernovae and the winds of massive red giants. Since the end points of stellar evolution and the enrichment of the ISM depend on stellar mass, it is important, then, to ask how many stars in different mass ranges are actually formed.

The admixture of stars of different masses, when first formed, is called the *initial mass function* (IMF). Various IMFs have been proposed in the astronomical literature, attesting to the difficulty in obtaining this function and its possible variation with environment. Examples are shown in Figure 5.10. The functional forms for the two curves, applicable to

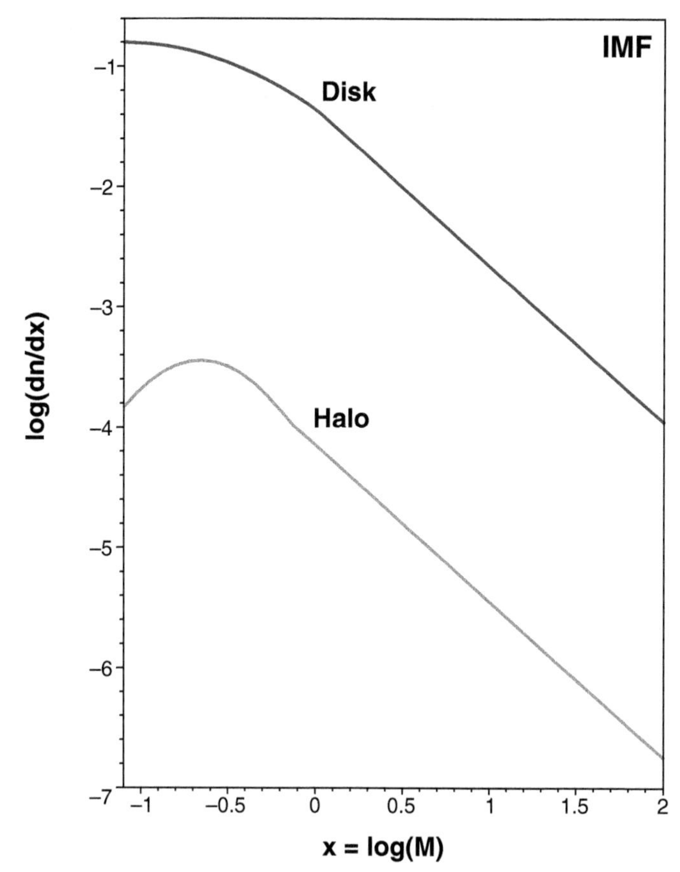

Figure 5.10 The initial mass function (IMF) for the disk and stellar halo of the Milky Way. Masses are in units of $M_\odot$. The IMF, given by $dn/d[\log(M)]$ (pc^{-3} $[\log(M_\odot)]^{-1}$), represents the number density of stars per logarithmic mass interval that have formed over the history of the region being studied. It is determined by counting the number density of stars in a given luminosity interval, applying corrections for observational biases, converting to the number density of stars in a given mass range, and then applying corrections for the number of stars that have evolved, i.e. that are no longer on the main sequence. The functional expressions are provided in Eqs. (5.5) and (5.6). There are some significant error bars (not shown) associated with the fitted parameters, but the plot provides a good idea as to how many stars of different masses have initially formed in the Milky Way. There are clearly many more stars per unit volume in the disk than in the halo and many more low mass stars in comparison with high mass stars.

the Milky Way's disk and halo, respectively, are,

$$\left[\frac{dn}{dx}\right]_{disk} = 0.158\ \exp\left[-\frac{1}{2}\left(\frac{x+1.10}{0.69}\right)^2\right] \qquad M \leq 1.0M_\odot$$

$$\left[\frac{dn}{dx}\right]_{disk} = 4.4\times10^{-2}\ M^{-1.3} \qquad M > 1.0M_\odot \qquad (5.5)$$

$$\left[\frac{dn}{dx}\right]_{halo} = (3.6\ \times10^{-4})\ \exp\left[-\frac{1}{2}\left(\frac{x+0.66}{0.33}\right)^2\right] \qquad M \leq 0.7\ M_\odot$$

$$\left[\frac{dn}{dx}\right]_{halo} = (7.1\times10^{-5})\,M^{-1.3} \qquad M > 0.7M_\odot \qquad (5.6)$$

where $x = \log(M)$ and M is the stellar mass in Solar mass units. The term, dn/dx, in units of $pc^{-3}\ [\log(M_\odot)]^{-1}$, gives the number density of stars per cubic parsec per logarithmic mass interval [4].

Figure 5.10 illustrates that far fewer high mass stars are formed in comparison with those of low mass (Example 5.2, Problem 5.2). High mass stars that end their lives as supernovae also live shorter lives since they burn their energy at a faster rate while undergoing normal hydrogen fusion. As long as the region of the ISM is rich in gas and able to form stars, this means that there will be many successive generations of massive stars forming and dying while low mass stars like our Sun continue to 'simmer' in the stable state of burning hydrogen to helium in their cores. Therefore, even though there are fewer high mass stars, the massive stars will have the greatest influence on the metallicity of the ISM.

Because high mass stars are short-lived (of order 10^6 year), they will also not stray far from their places of origin during their lifetimes. Therefore, the presence of high mass stars and any observational consequences of high mass stars, such as the formation of HII regions (which only exist around hot massive stars, Section 7.2.2) or supernovae, are taken to be proxies for active star formation. We say that hot high mass stars, HII regions, and supernovae are all *tracers* of massive star formation. It is important to note that, once some time has passed, the admixture of stars in a region will not be the mixture of Figure 5.10 because of the different rate of evolution of stars of different masses. Rather, the admixture of stars is determined by noting which stellar mix results in the observed HR diagram for a given region.

Example 5.2

How many disk stars per pc^3 with masses between 0.1 and 1 $M_\odot$ are initially formed in comparison with those between 10 and 100 $M_\odot$?

From Eq. (5.5) (see also Figure 5.10) with $x = \log M$, in the low mass range,

$$n = 0.158 \int_{x=-1}^{x=0} \exp\left(-\frac{1}{2}\left[\frac{x+1.10}{0.69}\right]^2\right) dx = 0.106\ pc^{-3}$$

In the high mass range,

$$n = 4.4\times10^{-2}\int_{x=1}^{x=2} M^{-1.3}dx = 4.4\times10^{-2}\int_{x=1}^{x=2} 10^{-1.3x}dx = 7.00\times10^{-4}pc^{-3}$$

Therefore, there are 151 times more stars formed per cubic parsec in the low mass range than the high mass range.

Abundances in our Milky Way are very important probes of its star formation history and, by implication, its global formation and evolution. For example the IMF, the quantity of gas present and knowledge of stellar evolution allow models to be built that predict the metallicity distribution in a given environment at the present time. The modelled result can then be compared to that observed. If one assumes an initially low metallicity and that the amount of gas in a given region of the Galaxy remains at its location without flowing out, or other gas flowing in (called the *closed box model*), then one finds that too few stars of low metallicity are observed in comparison with what the models predict[14]. This means that some modification is necessary in the assumptions of the model. For example, perhaps gas has accreted into the box (an *open box model*) so that subsequent generations of stars build up the heavier metals more quickly. Or perhaps there was some pre-enrichment so that the initial metallicity was not low to begin with. Clearly, the star formation rate (SFR) as a function of time is also important and modern models include this and other parameters in order to understand the observed metallicity distributions of stars.

When we look at the metallicity of stars in the disk in comparison with the halo, another important result emerges. Away from the gaseous, dusty spiral-like disk within which most of the visible material of our Galaxy is found, there is a more spherical halo of stars surrounding the disk in which there is very little gas and no cold dense gas from which stars can form (Figure 5.3 Bottom). The metallicities of halo stars vary, but a typical value is $Z = 0.0005$, less than 3% of the disk value. Clearly, the metallicity in the halo is very low in comparison with the disk and this provides a clue as to the comparative star formation history of these two regions.

The much lower density of stars, as shown by the IMF (Figure 5.10), also attests to the very different environment and history of the halo. Star formation in our Galaxy has mostly occurred and is still occurring in its disk but it is no longer occurring in the halo which now contains the oldest stars in the Galaxy. Although the metallicity in the halo is very low, it is not zero, indicating that some star formation has indeed occurred there in the past. In fact, similar results are found for other galaxies and imply that a period of rapid star formation and enrichment occurred very early in the galaxy formation process when smaller systems were joining to form larger ones (Figure 5.1). The thin galaxy disks that we see today with their beautiful spiral structure likely formed later when baryons settled into the central regions. Star formation then continued in the disk, forming a denser stellar environment there. More information on galaxy formation can be found in [9], and a simulation of the early structures that evolve into galaxies can be seen in Figure 7.7.

Dynamics, global structure, and stellar distribution all provide information about the formation of galaxies, but abundance is a key component. There is much that is still not fully understood about the formation process, and the fact that the dominant component is dark matter, of which we know little, is a strongly complicating factor. Nevertheless, even the tiniest of visible constituents and apparently insignificant observations such

[14] This is called the G dwarf problem.

as metallicity variations can offer just the insights required to piece together a coherent picture of Galactic history.

5.4 THE GASEOUS UNIVERSE

The two most abundant elements in our Universe, hydrogen and helium, are gases. Only under extreme pressure, such as at the centres of planets like Jupiter, can hydrogen exist as a metallic-like liquid. It follows, then, that most of the Universe is in a gaseous state. Indeed, most of the metals are also gaseous, given the pressures and temperatures under which they may be found, for example, silicon within the Sun, gaseous carbon in interstellar clouds, or iron atoms in hot tenuous gas. Only a tiny fraction of the observed Universe is in the solid state in the form of dust particles and the planets, asteroids, and comet nuclei that we see in our own Solar System or infer to be in others. Once established, the internal tensile strength of solids can keep them in this phase even in the low pressure environment of space. Ices of various molecules, such as water, carbon dioxide, and methane, are also a common constituent of solid Solar System material, especially in comets or on cold bodies, distant from the Sun. The liquid phase, however, appears to be rarest.

The Earth has a special place in our Solar System in that its temperature and surface pressure allow water to exist in all three phases on its surface. There is now mounting evidence that liquids can briefly flow on the surfaces of other planets and moons, for example, via volcanic flows. Liquid water may leak or be vented from under the icy surfaces of Europa, a moon of Jupiter, and Enceladus, a moon of Saturn. And it is possible that liquid water flowed on Mars in the past. However, conditions are not currently suitable for long-lived lakes or rivers of water on these objects.

In fact, conditions are not currently suitable for *any* substance to be in long-lived liquid form on extraterrestrial surfaces in the Solar system with one exception: Titan – a moon of Saturn. The pressure and temperature at the surface of this moon permit hydrocarbons, such as methane and ethane, to exist in stable liquid form on its surface. A beautiful example is Ontario Lacus, whose appearance and size resemble its namesake, Lake Ontario (Figure 5.11). Titan's atmosphere consists mostly of nitrogen (94%) and methane (CH_4, 5.7%). Its surface temperature is 94 K, and its surface pressure is higher than the Earth's at 1.5 atm. However, *methane rain* is present as part of a complex weather system on this moon[15].

The gases, of which the Universe is largely composed, can show a wide variety of properties, especially of temperature and density. A cold, dusty dense molecular cloud, for example, is quite different from the hot upper atmosphere or the deep interior of a star. Table 5.1 gives some typical densities and temperatures for various astrophysical gases, and images of a few different kinds of gaseous objects that can be found in the ISM are shown in Figures 3.2, 5.7, 5.13, and 5.17.

[15] The *triple point* of methane is at 90.67 K and 0.115 atm, close to the surface temperature and partial pressure of methane, of 94 K and 0.08 atm, respectively.

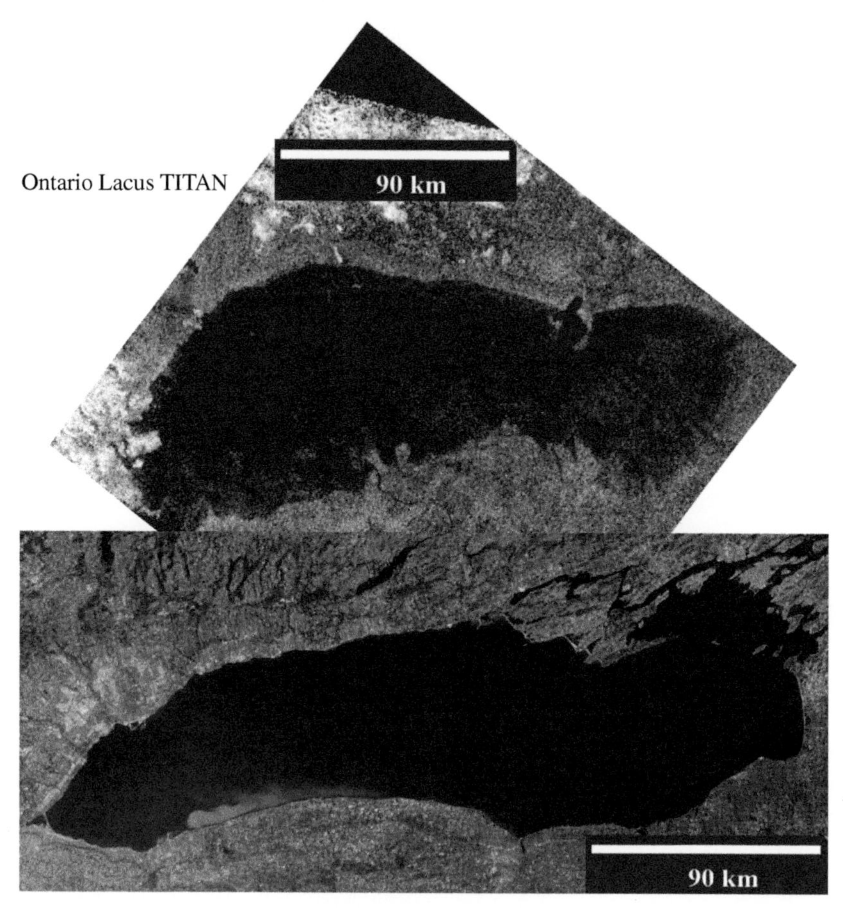

Figure 5.11 Top: Ontario Lacus ('Lake Ontario') looks like a giant footprint on the southern surface of Titan. This shallow lake is partially covered with hydrocarbons in liquid form (darker regions of lake) and is partially like a seasonal mudflat (lighter regions of lake) [7]. Notice the narrow winding river on the south side slightly left of centre. This view was obtained on 10 January 2010 from the Cassini spacecraft. NASA/JPL-Caltech. Bottom: satellite view of Lake Ontario in North America. Source: NASA.

Aside from density and temperature, another important difference is the *state of ionization*. Gases can be neutral or ionized or have some partial ionization. A neutral gas is one in which the atoms have their electrons still attached or *bound* to them so that there is no net charge on the atom. In gases with higher states of ionization (partially ionized), some atoms will have lost electrons, becoming *ions* with a net charge. An ionized gas is referred to as a *plasma* in astrophysics. Since hydrogen is the most abundant atom, a plasma that is highly ionized will have approximately equal numbers of free protons and free electrons, with some variation depending on the admixture of other elements and their state of ionization.

Table 5.1 Sample densities and temperatures in astrophysical gases.

Location	n_e[a](cm^{-3})	$n_m + n_H$[b](cm^{-3})	T (K)
Interplanetary space	$1\text{–}10^4$	≈ 0	$10^2 - 10^3$
Solar corona	$10^4\text{–}10^8$	≈ 0	$10^3 - 10^6$
Stellar atmosphere	10^{12}	≈ 0	10^4
Stellar interiors	10^{27}	≈ 0	$10^{7.5}$
Planetary nebulae	$10^3 - 10^5$	≈ 0	$10^3 - 10^4$
HII regions	$10^2 - 10^3$	≈ 0	$10^3 - 10^4$
Interstellar space[c]	$10^{-3} - 10$ (avg. ≈ 0.03)	$10^{-2} - 10^5$ (avg. ≈ 1)	10^2
Intergalactic space	$<10^{-5}$	≈ 0	$10^5 - 10^6$
Intergalactic HI clouds[d]		$10^{-6} - 10^{-3}$	$10^3 - 10^5$

[a]Electron density. For a pure hydrogen gas that is completely ionized, $n_e = n_p$ where n_p is the proton density.
[b]n_m is the molecular gas density (predominantly H_2), and n_H is the neutral atomic hydrogen (HI) gas density.
[c]The temperature quoted here is typical of interstellar HI clouds. However, there is a wide variety of densities, temperatures, and degrees of ionization in the interstellar medium. For example, molecular clouds tend to have low temperatures (5–200 K, typically 30 K) and high densities ($10^2 - 10^5$ cm^{-3}, typically 10^4 cm^{-3}). On the other hand, for ionized diffuse intercloud gas at $n_e \approx 0.03$ cm^{-3} the temperatures are much higher ($\approx 10^4$ K).
[d]Typical density range for the Ly α forest (see Figure 7.6 and [22]) for a redshift (Section 9.1.2) of $z = 3.1$. The ionized fraction is not easily determined. Values will vary with redshift.
Source: Lang [17].

The degree of ionization of any given element is specified by Roman numerals after the element name, where the number of electrons removed is given by the Roman numeral plus one. For example, neutral hydrogen is referred to as HI and ionized hydrogen is HII. For metals, there are many more electrons that could leave the atom. Thus one can have, for example, neutral carbon (CI) or triply-ionized carbon (CIV). Many such ions have been observed in various astronomical objects, including some with extremely high states of ionization. For example, neutral iron has 26 electrons, so iron with 25 electrons removed is Fe XXVI. This ion has been observed in the spectra of some highly energetic X-ray sources [5], including the Solar *corona*[16] (Figure 8.7).

Important goals in astrophysics are to determine the densities, temperatures, and ionization states of gases which include, of course, the atmospheres and interiors of stars. With sufficient information, the energetics involved in maintaining these gases in the observed states can then be determined. Armed with this knowledge, it may be possible to piece together a picture of the formation and evolution of the objects. Thus, there is considerable importance in understanding the physics of gases, especially of hydrogen.

[16] The solar corona is the faint, low density, outermost region of the Sun's atmosphere, visible to the eye during a total Solar eclipse. Coronal temperatures are typically 10^6 K but can be higher during times of high activity (Section 8.4.2).

5.4.1 Kinetic Temperature and the Maxwell–Boltzmann Velocity Distribution

Consider a gas in which energy has been exchanged between particles via *elastic collisions*. This means that only kinetic energy is exchanged; there are no radiative energy losses, ionizations, or other processes occurring between the particles. In such a case, a velocity distribution is set up such that the number density of particles in each velocity interval is described by the Maxwell–Boltzmann (M-B or simply 'Maxwellian') velocity distribution. The mathematical form of this distribution was given in Eq. (I.2) and is plotted in Figure 5.12.

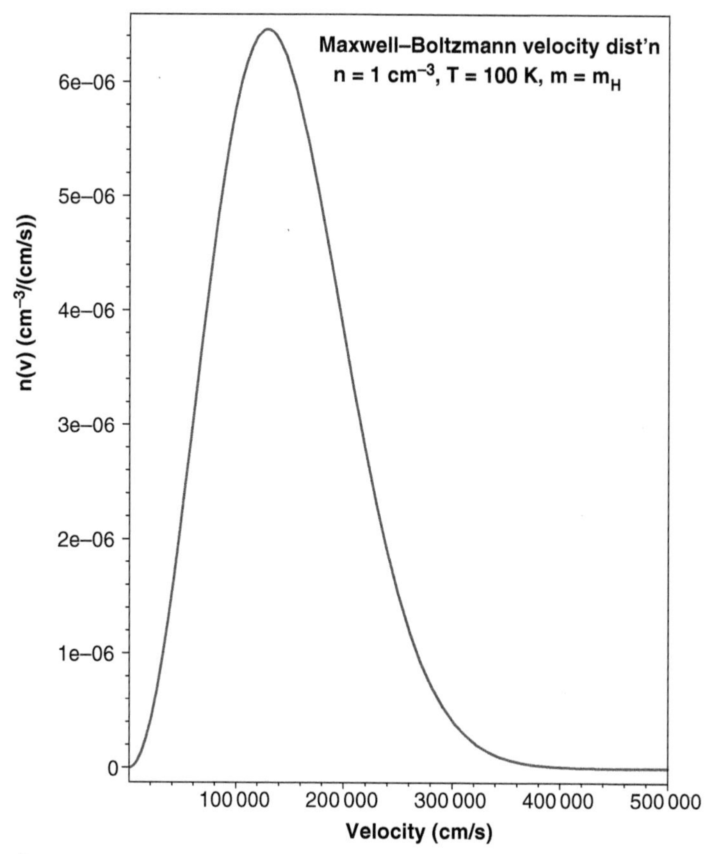

Figure 5.12 Maxwellian velocity distribution for neutral atomic hydrogen at $T = 100$ K and a total density of 1 cm^{-3}. The ordinate specifies the number of particles in each velocity interval, and the abscissa gives the velocity. The area under this curve returns the total density. See Eq. (I.2) for its functional form.

It is the Maxwellian velocity distribution that *defines* the gas *kinetic temperature*, written as T without a subscript. For particles of mass, m, T is related to the average of the squares of the particle speeds, or the *mean-square* particle speed, $\langle v^2 \rangle$,

$$\frac{1}{2}\mathrm{m}\langle v^2 \rangle = \frac{3}{2}kT \tag{5.7}$$

The *root-mean-square* particle speed is then $v_{\mathrm{rms}} = \sqrt{\langle v^2 \rangle}$ It can be shown (Problem 5.4) that the *most probable speed* is slightly lower,

$$v_{mp} = \sqrt{2\,k\,\mathrm{T}/\mathrm{m}} \tag{5.8}$$

and the *mean speed* is higher,

$$\langle v \rangle = \sqrt{8\,k\,\mathrm{T}/(\pi \mathrm{m})} \tag{5.9}$$

The left hand side of Eq. (5.7) represents the mean kinetic energy of the particles, $\langle E_k \rangle$, so T is a measure of the mean kinetic energy of particles in a gas. A gas that can be described by a single kinetic temperature is said to be in thermal *equilibrium*[17]. Such a result would occur if a hot gas and a cold gas were set beside each other in an insulated box. Eventually, both gases would reach a single warm temperature.

The Maxwell–Boltzmann distribution can be derived from *statistical mechanics* which applies statistics to the motions of large numbers of particles in a gas in order to predict its macroscopic properties. The resulting equation (Eq. I.2) can be seen to depend on a number of parameters, including the *Boltzmann factor*,

$$e^{-\chi/(k\mathrm{T})} \tag{5.10}$$

where, for the M-B distribution, $\chi = \frac{1}{2}\mathrm{m}\,v^2$ represents the kinetic energy of a particle. This factor indicates that there is a lower probability of finding a particle that has a very high kinetic energy compared to the mean kinetic energy of the gas. In addition, however, the Maxwellian velocity distribution also depends on the quantity, $4\,\pi\,v^2$ which represents the surface area of a shell about an origin in three-dimensional 'velocity-space'. As the velocity (or kinetic energy) of a particle increases, the velocity shells become larger, allowing more possible orientations of the velocity vector of a particle. This means that more 'states' are available to be occupied as the velocity increases. These competing effects result in the final Maxwellian distribution that first increases as v^2, reaches a peak, and then decreases exponentially (Figure 5.12). The constant multiplier in the equation is a normalization factor to ensure that the total number density of particles is correct.

If a gas can be described by a Maxwellian velocity distribution, then there must have been a sufficient number of elastic interactions that this thermal equilibrium could be established. How valid is this assumption for typical astrophysical conditions? For gases in which T is less than about 8×10^4 K, particle encounters are almost always elastic and there will therefore be many elastic collisions before an inelastic encounter occurs [24].

---

[17] The term, 'thermal equilibrium', is sometimes used differently from this definition, for example, to represent energy balance within stars.

The timescale required for establishing an equilibrium temperature, t_{therm}, can be determined by asking how long it would take for a faster test particle (the hotter gas particle) to slow down as it diffuses through the gas of slower particles (the colder gas particles)[18]. It can be shown that this timescale is quite short under astrophysical conditions and depends on the density of the dominant constituent of the gas. For example, in cool (T = 100 K) neutral atomic (HI) clouds in the interstellar medium, the timescale for establishing an equilibrium temperature via H–H collisions is $t_{therm} = 16/n_H$ year, where n_H is the hydrogen density. If $n_H = 1\ cm^{-3}$ (Table 5.1), then only 16 years are required, a very short time by astronomical standards. For a fully ionized hydrogen gas at $T = 10^4$ K and the same density, the equilibrium timescale for electron–proton 'collisions'[19] is only $t_{therm} = 8.2 \times 10^3$ s, that is, only a few hours. The timescale for electron–electron collisions is equivalent to that for electron–proton collisions, provided the electron and proton densities are equal in the gas. Clearly, higher densities facilitate the process of temperature equalization and Coulomb interactions shorten the timescale even more. In the low-density ISM, then, a Maxwellian velocity distribution is expected, and therefore, it should certainly be present in regions of higher density, such as stellar atmospheres (e.g. densities of order $10^{12}\ cm^{-3}$).

We therefore assume, in most cases, that astrophysical gases are in thermal equilibrium. Departures can occur, however, especially in gases of very low density. There may also be other physical effects that inhibit the process of temperature equalization. If the gases are physically prevented from diffusing together, thermal equilibrium will not be achieved (for example, magnetic fields can constrain the positions of charged particles in an ionized gas). Particles whose energetics are driven by processes that are intrinsically non-thermal in nature, such as cosmic rays (Section 1.3), will not achieve a Maxwellian distribution at all. Collisions of gas particles with dust particles require revisions from the above description as well. Figure 5.13 shows a situation common for an HII region in which dust is mixed with the ionized gas, yet the dust temperature may be two orders of magnitude lower than the kinetic temperature of the gas. Dust temperature, as quoted, is conceptually different from the gas temperature, however, since it does not refer to the random speeds of the dust particles but rather a temperature at which the dust grain radiates; this will be further elucidated in Section 6.2.

5.4.2 The Ideal Gas

An *ideal gas* is a gas that obeys the *ideal gas law* (or *perfect gas law*, see also Table I.4 and Problem I.3),

$$P_p = n\,k\,T = \frac{\rho}{\mu\, m_H} k\,T \tag{5.11}$$

where P_p is the particle pressure, n is the number density (cm^{-3}) of particles of any mass, k is Boltzmann's constant, T is the temperature, ρ is the mass density of the gas ($g\ cm^{-3}$), and

[18] The equilibrium timescale is one half of the slowing down time since the slower moving particles will also speed up. Such a calculation involves, as a starting point, determining the mean time between collisions, an example of which is provided in Section 5.4.3.

[19] For *coulomb interactions*, i.e. when the particles are charged, a 'collision' is considered to be an interaction that significantly affects the trajectory of a particle.

Figure 5.13 This image shows most of the *Lagoon Nebula* (also called Messier 8), which is an ionized hydrogen region (HII region) visible to the naked eye in the constellation of Sagittarius. The nebula is 1.6 kpc away with an average diameter of 14 pc (there is some uncertainty in the distance), and the complete nebula appears as large as the full Moon in angular size. The image shows a complexity of features with colours representing emission from different atoms (oxygen, silicon, and hydrogen) and dark bands and filaments representing obscuration by dust. The dust and gas are in close proximity yet the gas temperature is 7500 K, whereas dust temperatures are typically much lower, of order ~100 K or less. The mean density of electrons in the nebula is $n_e \approx 80\ \mathrm{cm}^{-3}$. Source: Reproduced by permission of Michael Sherick, Cabrillo Mesa Observatory, 2005.

$\mathrm{m_H}$ is the mass of the hydrogen atom. The quantity, μ, is called the *mean molecular weight*, defined by

$$\mu \equiv \frac{\langle \mathrm{m} \rangle}{\mathrm{m_H}} = \frac{1}{\mathrm{m_H} N} \sum_{i} \mathrm{m_i} \tag{5.12}$$

where $\mathrm{m_i}$ is the mass of the ith particle, and N is the total number of particles. The quantity, μ, gives the mean mass of free particles in units of $\mathrm{m_H}$ (Example 5.3). The mean molecular weight can also be expressed in terms of the abundances of the object.

For example, the mean molecular weights of a completely neutral and a completely ionized gas are (Problem 5.5), respectively,

$$\mu = \frac{1}{X + \frac{Y}{4} + \frac{Z}{\overline{A}_m}} \quad \text{(neutral)}$$

$$\mu = \frac{1}{2X + \frac{3Y}{4} + \frac{Z}{2}} \quad \text{(ionized)} \tag{5.13}$$

where $\overline{A}_m$ is the mean atomic weight of the metals in the gas. The atomic weight of particle i, A_i, is related to its mass, m_i, by $m_i = A_i\, m_H$. A derivation of the second expression makes use of the approximation, $\overline{A}_m/2 \gg 1$. Therefore, for a completely ionized gas with Solar abundance (Eq. (5.4)), $\mu = 0.61$. For the particle pressure of electrons alone, P_e, in a fully ionized gas, the number density of electrons, $n_e = \rho/(\mu_e\, m_H)$, can be used in Eq. (5.11) with

$$\mu_e = \frac{1}{X + \frac{Y}{2} + \frac{Z}{2}} \tag{5.14}$$

Example 5.3

Show that the mean molecular weight of a completely ionized pure hydrogen gas is 1/2.

In such a gas, half of the particles are electrons of mass, m_e, and half are protons of mass, m_p. From Eq. (5.12), $\mu = \frac{1}{m_H N}\left[\frac{N}{2}m_e + \frac{N}{2}m_p\right] \approx \frac{1}{m_H N}\left[\frac{N}{2}m_p\right] \approx \frac{1}{2}$.

The ideal gas law is derived under the assumption that, when energy is exchanged between particles, the interactions are elastic. Thus, if the particles were placed in a small box, the pressure against any side of the box would result from the collective motions of the particles against the wall, with no corrections required for particle–particle interactions. The more closely spaced the particles are, the more likely it will be that non-elastic interactions may occur. Therefore, a common, though not fool proof, way to decide whether such interactions are negligible is to determine the mean separation between particles, $\overline{r}$, and compare this to the particle radius[20], R_p. The mean separation between particles can be found from $nV = N$, where n is the gas density, V its volume, and N the total number of particles, and setting the total number of particles to 1,

$$\overline{r} = \frac{1}{n^{1/3}} \tag{5.15}$$

[20] Comparing to the particle diameter is more accurate, though factors of two are not that important for most comparisons, as Example 5.4 illustrates.

If $\bar{r} \gg R_p$, then the gas may be considered ideal. The caution to this rule of thumb is that other effects may be important, for example, Coulomb interactions between charged particles, or the fact that the effective volume within which a particle can move may be reduced due to a finite particle size.

The perfect gas law is an example of an *equation of state*, that is, an equation that describes how the physical properties of the gas (e.g. pressure, density, and temperature) are related at any location. It is important to note that a complete description of pressure includes all contributors, including particle pressure (due to the thermal motion of particles and also free thermal electrons, if present) as well as radiation pressure (Section 3.5),

$$P = P_p + P_{rad} \tag{5.16}$$

If electron degeneracy (Section 5.3.2) is an important contributor to the pressure, then it must be added as well and, if these other terms are significant in comparison with P_p, then the gas is non-ideal.

5.4.3 The Mean Free Path and Collision Rate

If a small test particle enters a gas containing larger field particles (the latter initially considered at rest), there will be some probability that it will collide with a field particle. The average distance that the particle travels before such an encounter occurs is called the *mean free path*, $\bar{l}$. The mean free path can be determined by considering a small volume, V, of the gas, as shown in Figure 5.14. The probability that the test particle will encounter a field particle will depend on the projected area presented by all field particles, A_p, divided by the total area presented by the region of space, A. If we are to ensure that an interaction does occur, then the probability is 1 so,

$$1 = \frac{A_p}{A} = \frac{N\sigma}{\pi R^2} = \frac{nV\sigma}{\pi R^2} = \frac{n\pi R^2 \bar{l}\sigma}{\pi R^2} = n\sigma\bar{l} \tag{5.17}$$

Here, N is the total number of field particles, σ is the cross-sectional area of each field particle, and the volume is $V = \pi R^2 \bar{l}$ as shown in the figure. Thus, the mean free path is given by,

$$\bar{l} = \frac{1}{n\sigma} \tag{5.18}$$

If the test particle has a velocity, υ, this distance can be expressed as $\bar{l} = \upsilon\bar{t}$, where $\bar{t}$ is mean time between collisions, thus,

$$\bar{t} = \frac{1}{n\,\upsilon\,\sigma} \tag{5.19}$$

The *collision rate* for this test particle is just the inverse of the mean time between collisions,

$$\mathcal{R} = n\,\upsilon\,\sigma \tag{5.20}$$

For a more general case with many test particles and field particles, υ is the relative velocity between the test and field particles. For identical particles in a one-temperature

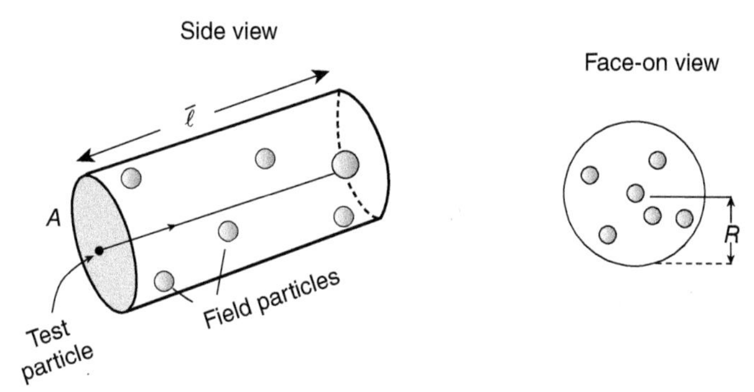

Figure 5.14 A test particle entering a gas containing field particles travels a mean free path (depth of the volume, left) before encountering a field particle. The probability of an encounter depends on the area presented by the field particles over the total area (right). If the entire cross-sectional area of the cylinder is occupied by particles, then the test particle will certainly encounter a field particle.

gas, the relevant velocity is the mean particle velocity, $\langle v \rangle$, as given in Eq. (5.9). If the test particle is of very low mass in comparison with the field particles, such as collisions between free electrons and protons in an ionized gas, then v is just the mean particle speed of the electrons since the protons can be thought of as 'at rest' in comparison. Also, the collision rate, $\mathcal{R}$ in Eq. (5.20) refers to the rate at which a single test particle will encounter another field particle in a gas in which the field particle density is n. If we wish to calculate the total collision rate in a gas due to collisions by all test particles, then we need to multiply $\mathcal{R}$ by the total number of test particles, $N_t = n_t V$, where n_t is the test particle density. The total collision rate per unit volume, ν_{tot} (s^{-1} cm^{-3}), is then,

$$\nu_{tot} = n \, n_t \, v \, \sigma \tag{5.21}$$

It does not matter which particle is considered the test particle and which is considered the field particle as long as the correct values of v and σ are used.

If all particles are neutral and can be approximated as elastically interacting spheres, then σ is just the geometrical cross-sectional area of the particles. However, the cross-sectional area may have to be modified in some circumstances, in which case it is called the *effective cross section*, or σ_{eff}. For example, σ_{eff} will differ from the geometric cross section if one or both sets of colliding particles are charged (such as electrons with protons or electrons with electrons) and σ may itself be a function of velocity, $\sigma(v)$. Thus, σ in the above equations must be replaced with the appropriate value of σ_{eff} which depends on the type of colliding particles and velocities under consideration.

Often, the quantity, γ_{coll}, where coll refers to the type of collision, is the measured value. This is the mean collision rate coefficient which folds together the quantities, v and σ with appropriate weighting. For example, for collisions between neutral hydrogen

Table 5.2 Sample collision parameters.

Temperature (K)		
	HI – HI[a]	γ_{H-H} (cm^3 s^{-1})
30		5.1×10^{-10}
100		7.4×10^{-10}
300		10.2×10^{-10}
1000		13.6×10^{-10}
	HI – HI de-excitation[b]	$\gamma_{21\,\text{cm line}}$ (cm^3 s^{-1})
30		3.0×10^{-11}
100		9.5×10^{-11}
300		16×10^{-11}
1000		25×10^{-11}
	Electron–proton with recomb.[c]	α_r (cm^3 s^{-1})
5000		4.54×10^{-13}
10 000		2.59×10^{-13}
20 000		2.52×10^{-13}
	Electron–proton without recomb.[d]	σ_{eff} (cm^2)
10^4		1.4×10^{-15}
10^5		1.4×10^{-17}
10^6		1.4×10^{-19}
	Electron–HI de-excitation[e]	$\gamma_{Ly\alpha}$ (2p → 1 s) (cm^3 s^{-1})
5000		6.0×10^{-9}
10 000		6.8×10^{-9}
20 000		8.4×10^{-9}

[a] Collisions between neutral H atoms. Here $\gamma_{H\text{-}H} = \langle v \rangle \sigma_{eff}$ [24].
[b] Collisions between neutral H atoms that result in the de-excitation of the hyperfine ground state (see Appendix C.4 of the *online material*).
[c] Collisions between an electron and proton that result in recombination to HI. In this case, γ_{coll} is referred to as α_r, which is the sum of $\alpha_{n',L}$ over all quantum levels, n and L (Appendix C of the *online material*). Case B recombination (see Footnote 6 in Chapter 7 or Section 11.4.2) is assumed.
[d] Collision *without* recombination. A collision is taken to be an interaction that significantly affects the trajectory of the electron. Taking $e^2/r_{eff} = m_e\, v^2$ to be the condition for a significant deflection [24], we find, $\sigma_{eff} = \pi\, r_{eff}{}^2 = \pi\, [e^2/(m_e v^2)]^2$. Using Eq. (5.9), this leads to $\sigma_{eff} = 1.4 \times 10^{-7}/T^2$. A more accurate derivation would include the cumulative effects of other particles in the gas and would also use the Maxwellian distribution instead of Eq. (5.9). Therefore, these results are meant to be 'indicative' only.
[e] Collisions that result in the de-excitation of the Ly α line in the permitted transition (Appendix C.2 of the *online material*) from $2p$ to $1s$ only. This has been computed from Eq. (4–11) in [24] using data from www.astronomy.ohio-state.edu/~pradhan/atomic.html. Note that the value for excitation will not be the same as for de-excitation nor will it be the same for the forbidden transition from $2s$ to $1s$.

atoms, $\gamma_{H\text{-}H} = \langle v \rangle \sigma_{eff}$ cm^3 s^{-1}. Then, the quantities of Eqs. (5.19) – (5.21) depend only on the collision rate coefficient and density. Some sample values of γ_{coll} and σ_{eff} for collisions between various types of particles are given in Table 5.2.

The above expressions are put to use in many astronomical situations. Example 5.4 provides a sample calculation for interstellar HI. Later (Example 7.4), we will provide an example for electron–proton collisions that result in recombination to HI. A final important point is that the 'test particle' could be a photon, a point we return to in Example 5.5 and Chapter 7.

Example 5.4

For an interstellar cloud of pure HI at $T = 100$ K and $n = 1\ \text{cm}^{-3}$, (a) calculate the geometric cross section, σ_g, for H–H collisions. What is the ratio of σ_{eff}/σ_g and σ_{eff}? (b) Determine the mean free path, mean time between collisions, and collision rate for an H atom. (c) What is the mean separation between particles? Is this an ideal gas?

(a) Using the Bohr radius, $a_0 = 5.29 \times 10^{-9}$ cm (Table T.1), $\sigma_g = \pi a_0^2 = 8.79 \times 10^{-17}\ \text{cm}^2$. By Eq. (5.9) and setting m to the mass of the hydrogen atom, $\langle v \rangle = 1.45 \times 10^5\ \text{cm s}^{-1}$. Then $\langle v \rangle \sigma_g = 1.27 \times 10^{-11}\ \text{cm}^3\ \text{s}^{-1}$. From Table 5.2, $\gamma_{H-H} = \langle v \rangle \sigma_{eff} = 7.4 \times 10^{-10}\ \text{cm}^3\ \text{s}^{-1}$, so $\sigma_{eff}/\sigma_g = (\langle v \rangle \sigma_{eff})/(\langle v \rangle \sigma_g) = 58.0$. Then $\sigma_{eff} = 58\sigma_g = 5.1 \times 10^{-15}\ \text{cm}^2$.

(b) All quantities are now known for the required calculations for which we will use either the effective cross section, σ_{eff} or γ_{H-H}. From Eq. (5.18), we find $\bar{l} = 2.0 \times 10^{14}$ cm (13 AU), from Eq. (5.19), $\bar{t} = 1.35 \times 10^9$ s (43 years) and from Eq. (5.20), $\mathcal{R} = 7.4 \times 10^{-10}\ \text{s}^{-1}$.

(c) From Eq. (5.15), $\bar{r} = 1$ cm. The 'effective radius' of the particle is $R_{eff} = \sqrt{\sigma_{eff}/\pi} = 4.0 \times 10^{-8}$ cm which is $\ll \bar{r}$ so this is an ideal gas.

5.4.4 Statistical Equilibrium, Thermodynamic Equilibrium, and Local Thermodynamic Equilibrium

Even under conditions in which most interactions between particles in a gas are elastic and a Maxwellian velocity distribution exists, there will be a minority of interactions in which kinetic energy is not perfectly conserved and some energy is exchanged in other ways. A good example is when collisions cause the excitation or de-excitation of atoms, a process that involves the transition of electrons from one quantum state to another (see Appendix C of the *online material*). For an excitation, a collision must exchange an energy that corresponds to the energy difference between levels. The probability of such a collisional energy exchange will be different for each transition. Thus, in a Maxwellian velocity distribution, a single value for $v\,\sigma_{eff}$ that would normally be used in Eqs. (5.19) and (5.20) does not describe the probability that a *specific* exciting or de-exciting interaction will occur. For transitions between states, i and j, $v\,\sigma_{eff}$ should be set to $v\,\sigma_{i,j}$ whose value is less than $v\,\sigma_{eff}$ since only a fraction of all collisions will result in a specific transition. We can see, then, that collisions between particles can result in changes in the level of excitation of an atom. They can also result in changes in the ionization of a gas if, for example, the collision imparts sufficient energy to eject an electron from an atom.

However, collisions are not the only way that excitations, de-excitations, or ionizations can occur. Other processes such as the absorption of a photon could result in an excitation or ionization, depending on photon energy. And de-excitation can also occur via photon interaction or spontaneously. In general, the populations of energy levels in atoms must be determined by including all processes that both populate and de-populate any given level. These processes include de-excitations from *all* higher levels (including free states) and excitations from *all* lower levels – collisional, radiative, and spontaneous.

In a steady state, the transition rate into any level equals the rate out, and this should be true for all levels, a situation that is called *statistical equilibrium*. Equations of statistical equilibrium are set up for each level and involve the density of the particles, the energy density of the radiation field, and coefficients describing collisional, radiative, and spontaneous transition probabilities. The coefficients may themselves be functions of other quantities such as quantum mechanical parameters or temperature. The equations are then solved to determine the populations of the levels for given conditions. This is a complex problem and is not always easily solved, nor will it be pursued further here.

However, not every term in the statistical equilibrium equations is always necessary. For example, if the gas is completely neutral, then there will be no recombinations of electrons from free states so those terms in the statistical equilibrium equations will drop off. There are, in fact, certain conditions under which the equations of statistical equilibrium can be greatly simplified (see Section 5.4.5 for a mathematical description of one case). These are when the gas is in *thermodynamic equilibrium* (TE) or in *local thermodynamic equilibrium* (LTE), so we need to understand what each of these mean.

If a gas is in *TE*, then the energy in the *radiation* field is in equilibrium with the kinetic energy of the *particles*. Such a situation would be set up if a gas were isolated, confined, and no radiation escaped or entered the region. Eventually, the radiation and particles would achieve an equilibrium in which energy is exchanged between photons and particles with no net energy gain or loss overall. The radiation field can be characterized by a *radiation temperature*, T_R. In general, T_R is a temperature that is used in an equation to calculate parameters of the radiation field such as the specific intensity or flux density (Eqs. 6.1 and 6.2 are examples). In TE, therefore, $T_R = T$ (see also, Section 6.1). There is no additional radiative source (which could make $T_R > T$) and no 'sink' to let photons escape (allowing $T_R < T$). The photons that are produced in the gas (represented via T_R) therefore must be tightly coupled to the random motions of the particles (represented by the kinetic temperature, T). Notice that *thermodynamic equilibrium* is a stronger statement than *thermal equilibrium*. The former says that both the kinetic and radiation temperatures are the same everywhere in the gas, whereas the latter just says that a single kinetic temperature applies to the gas and it is possible that the radiation temperature is different. An example of $T_R > T$ is a warm spring day on the ski slope. The air temperature (T) is still very cold but the Sun's radiation (T_R) is warm.

It is difficult to find an example in nature in which true TE holds and $T_R = T$, since energy will leave all objects eventually (the object will cool), even if very slowly. Probably the best example is the 2.7 K cosmic microwave background (CMB) radiation (Section 5.1

and Figure 5.2). This radiation filled the Universe in the past[21] (no loss of gas or photons from the Universe is possible), and the particles and radiation were in equilibrium. Small variations in temperature are observed with *position*, however, as shown in Figure 5.2.

A more common situation is one of *LTE*, that is, the gas has the properties of TE, but only *locally*. A good example is one in which some radiation can escape or leak out of the gas as, for example, in stars. The interiors of stars have densities that are high enough for the radiation and matter to be considered 'trapped together' locally and are therefore in equilibrium at any given location, yet there is a large-scale temperature gradient and a net flux of radiation outwards to the surface. Thus, the radiation temperature is close to the kinetic temperature *over some relevant size scale*. Example 5.5 provides a quantification of this condition for the Sun.

Example 5.5

Provide an argument showing that the interior of the Sun is in LTE.

We start by computing the mean free path of a photon in the Sun's interior by assuming that the gas is fully ionized, and adopt a cross section for interaction to be the Thomson scattering cross section[22] for which $\sigma = 6.65 \times 10^{-25}$ cm^2 (Table T.1). The mean density of the Sun (data from Table T.2) is $\overline{\rho} = M_\odot / \left(\frac{4}{3}\pi R_\odot{}^3\right) = 1.4\ \mathrm{cm}^{-3}$. The corresponding number density is then $n = \overline{\rho}/\mu m_H = 1.4 \times 10^{24} \mathrm{cm}^{-3}$, where we have used $\mu = 0.61$ for a completely ionized gas with Solar abundance (Section 5.4.2). Then, (Eq. (5.18)) $\overline{l} = 1/(n\sigma) \approx 1$ cm with these values. How much does the temperature change over this distance? The central temperature of the Sun is $T_c = 1.6 \times 10^7$ K and the surface temperature is $T_s = 5781$ K (Table T.2) so the global temperature gradient in the Sun is $\nabla T = \frac{(T_c - T_s)}{R_\odot} = \frac{1.6\mathrm{x}10^7}{6.96\mathrm{x}10^{10}} = 2.3 \times 10^{-4}\ \mathrm{K\,cm}^{-1}$. Thus, the average temperature change is only 2.3×10^{-4} K over the distance that a photon would travel before being reabsorbed. Since a photon that is released will be absorbed again in a region that is essentially at the same temperature, an LTE condition is satisfied – at least to an accuracy of $\sim 10^{-4}$ K. A more accurate derivation would have to take into account the true temperature gradient corresponding to the adopted density at each interior point, as well as regions that are not fully ionized. Nevertheless, the conclusion would be the same.

There are many examples in astrophysics for which LTE (or close to LTE) conditions hold. From an astrophysical point of view, the condition, $T_R \approx T$, is very useful, not only because of the simplification of the statistical equilibrium equations, but also because, by observing the *radiation*, we can infer the *kinetic temperature* of the gas. We will see this more explicitly in future chapters.

For stellar interiors, the fact that radiation is 'locally trapped' has implications for the transport of energy out of stars. The energy generated in the core does not travel

[21] The temperature at the time this occurred was much higher than 2.7 K but we see the lower temperature because of the expansion of the Universe (Section 9.1).
[22] Absorption processes that occur in an ionized gas, such as will be discussed in Section 7.4.1, are ignored, but these will lower the mean free path even more than we calculate in this example.

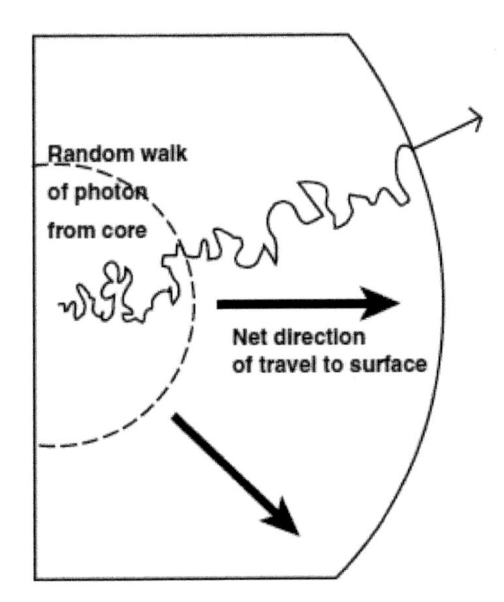

Figure 5.15 In the interiors of stars, photons travel a mean free path before interacting with a particle. The net result is a random walk to the surface.

straight to the surface to be radiated away. Rather, the photon will travel a mean free path and then either be absorbed, followed by re-emission of another photon in a random direction, or else be scattered into another random direction. Thus, there are many such absorptions, re-emissions, and scatterings as the radiation makes its way from the core to the surface, a process called *diffusion*. As this diffusion proceeds, the mean photon energy also degrades from γ-ray or X-ray energies in the core to UV, optical, or IR photons at the surface, depending on the type of star.

Although the photons near the surface are not the same ones as are originally generated in the core, the process can nevertheless be thought of as a single photon taking a *random walk* from the core to the surface[23] (Figure 5.15) and a diffusion timescale can be determined. The mean free path of a photon will vary with radius within the star, but we will now consider $\bar{l}$ to be an approximate value over all interior radii. For a random walk, it can be shown that, after taking N random steps of unit length, the straight-line distance travelled is only $\sqrt{N}$. Therefore, for a photon whose mean free path is $\bar{l}$ that takes a random walk towards the surface of a star of radius, R_*, we have $R_* = \sqrt{N}\bar{l}$. If the interaction time for scattering or absorption is small in comparison with the time of flight of the photon during each step, then the time for a photon to diffuse from the core to the surface is the total distance the photon actually travels divided by its speed, c,

$$t_{\mathrm{d}} = \frac{N\bar{l}}{c} = \left(\frac{R_*}{\bar{l}}\right)^2 \frac{\bar{l}}{c} = \frac{R_*^2}{\bar{l}c} \qquad (5.22)$$

However, the duration of the interaction time itself should also be taken into account. For the Sun, the diffusion time is of order, 10^7 year when this duration is taken included

[23] We ignore regions of the star that may be *convective*, that is, regions in which the gas itself has bulk motion up and down, carrying energy with it.

in the calculation[24] (Problem 5.7). This means that if there is a change in the rate at which energy is generated in the core of the Sun, it would take about 10^7 year before we even knew about it, as judged by the luminous output at the surface! Neutrinos, on the other hand, escape freely from the core, so a neutrino observatory (Figure 1.9) could detect a core change in about eight minutes, the time it takes for a particle moving at speed, *c*, to reach the Earth.

5.4.5 Excitation and the Boltzmann Equation

Let us now assume that our gas is in LTE and return to the issue of the populations of energy levels in atoms. Since there is an equilibrium between particles and radiation ($T = T_R$), there must be many interactions between the particles and photons in the radiation field in order for this equilibrium to be achieved. As indicated in Example 5.5, any emitted photon is reabsorbed or scattered in a region whose temperature has not significantly changed in the Sun. Consider the alternative (and unrealistic) *non-LTE* case for a moment. Suppose a photon were emitted from the core where $T_R \approx 10^7$ K but not absorbed again until it is close to the surface where $T_R \approx 10^4$ K. Clearly, this could only happen in a low-density gas in which the collision rate between particles is very low.

In the *LTE case*, then, the opposite is true. For bound states, excitations and de-excitations are *collisionally dominated.* That is, collision timescales that result in excitations and de-excitations are *shorter* than the timescales associated with photon-induced excitations and de-excitations. It is this property that leads to a simplification of the equations of statistical equilibrium. A short collisional timescale is more likely to occur when the gas is dense and/or the particle cross section to the interaction is large (see Eq. (5.19)).

In LTE, then, the equations of statistical equilibrium are much simplified and the population of states is given by the *Boltzmann equation*,

$$\frac{N_n}{N_1} = \frac{g_n}{g_1} e^{-\left(\frac{\Delta E_n}{kT}\right)} \tag{5.23}$$

Here, N_n is the number of atoms in which electrons are in an energy level, n, and N_1 is the number of atoms with electrons in the ground state, where n is the principal quantum number. ΔE_n is the energy difference between principle quantum state, n, and the ground state and g_n is the *statistical weight* (see Appendix C.2 of the *online material*) which specifies the number of available states at an energy level. For hydrogen, $g_n = 2n^2$. The Boltzmann equation can also be written to compare the populations between any two states, rather than a comparison with the ground state only. If the Boltzmann equation holds, then LTE conditions must be present.

The Boltzmann equation shows some similarity to the Maxwell–Boltzmann velocity distribution equation (Section 5.4.1), which should not be surprising given the discussion in Section 5.4.4. The populations of states depend on the Boltzmann factor, $e^{-\chi/(kT)}$,

[24] The duration of the interaction is only important for absorption/re-emission interactions; scattering should be almost instantaneous.

χ representing the excitation energy, as well as a quantity, g_n, that indicates the number of possible states in which a particle could be found at that energy. Thus, Eq. (5.23) is like a discrete, or quantum mechanical analogue to the Maxwellian velocity distribution[25].

There is a difficulty with Eq. (5.23) in that as $n \to \infty$, $g_n \to \infty$ yet $e^{-\left(\frac{\Delta E_n}{kT}\right)} \to$ constant because, at very high excitation, ΔE_n just becomes the ionization energy of the atom. Thus, the Boltzmann equation diverges at very high n. Physically, however, before n reaches values at which the Boltzmann equation diverges, other effects become important that effectively impose an upper limit to the principal quantum number, n_{max}. For example, the electrons at high n may become stripped off by the effects of neighbouring atoms. Coulomb interactions between the free charged particles and their distribution in the gas can also affect n_{max}.

Rather than considering how many particles are in an excited state in comparison with the ground state, a more general form of the Boltzmann equation is one which compares the number of particles in any excited state, N_n, to the *total* number of neutral particles, $N = N_1 + N_2 + \ldots + N_{n_{max}}$. From Eq. (5.23),

$$\frac{N_n}{N} = \frac{g_n e^{-\left(\frac{\Delta E_n}{kT}\right)}}{g_1 e^{-\left(\frac{\Delta E_1}{kT}\right)} + g_2 e^{-\left(\frac{\Delta E_2}{kT}\right)} + \ldots + g_{n_{max}} e^{-\left(\frac{\Delta E_{n_{max}}}{kT}\right)}} = \frac{g_n}{U} e^{-\left(\frac{\Delta E_n}{kT}\right)} \tag{5.24}$$

where $U \equiv \sum_{n=1}^{n_{max}} g_n e^{-\left(\frac{\Delta E_n}{kT}\right)}$ is called the *partition function*.

At temperatures below 3500 K, the partition function for hydrogen reverts to the statistical weight of the ground state, i.e. $g_1 e^{-\left(\frac{0}{kT}\right)} = g_1 = 2(1)^2$ because the remaining exponentials in the denominator are very small for low T. For hydrogen at higher temperatures such as in the atmospheres of stars, the partition function has been parameterized by,

$$\log(U) = 0.3013 - 0.00001\ \log(5040/T) \tag{5.25}$$

which is valid over the temperature range, 3500–20 000 K [11][26]. Evaluating this equation shows that, again, U = 2 to within 0.1% over this temperature range. Thus, for HI, Eq. (5.24) is essentially equivalent to Eq. (5.23) for temperatures up to 20 000 K. Example 5.6 shows that very high temperatures are required before there are significant numbers of particles in even the first excited state. Such high temperatures, as we shall see in the next section, will ionize the gas.

[25] Actually, the M-B velocity distribution is derived by discretizing free space so there is a quantum nature to M-B as well, but no gaps in the energy distribution as is the case for the Boltzmann Eqn.
[26] Pressure effects may increase the error at the highest temperatures.

Example 5.6

Determine the temperature required for the number of particles in the first excited state of hydrogen to be 10% of the number in the ground state.

By the Boltzmann equation (Eq. (5.23)), the equation for statistical weight, and ΔE (Eqs. C.6 and C.11 of the *online material*) expressed in erg with n' = 2 and n = 1, we have,

$$\frac{N_2}{N_1} = \frac{2(2^2)}{2(1^2)} \exp\left(-\frac{-2.18 \times 10^{-11}\left(\frac{1}{2^2} - \frac{1}{1^2}\right)}{kT}\right).$$

Setting the LHS to 0.1 and solving, we obtain $T = 3.2 \times 10^4$ K. Thus, the temperature must exceed about 30 000 K in order for there to be a significant number of particles in the first excited state.

When the gas density is low, such as in the ISM, LTE often no longer holds so the Boltzmann equation, as stated in Eq. (5.23), cannot be used. The rate at which an electron will spontaneously de-excite is determined by the *Einstein A coefficient* which, for de-excitation from the first excited state to the ground state ($L_y\,\alpha$) is $A_{2,1} = 6.3 \times 10^8$ s^{-1} (Appendix C plus notes to Table C.1 of the *online material*). The collision rate, however, is of order, $10^{-6}\ n_e$, where n_e is the number density of free electrons, i.e. the *electron density* [24]. Any transition to the first excited state, whether radiatively or collisionally induced, will therefore result in an immediate spontaneous de-excitation for the range of densities observed in interstellar space (Table 5.1). Similar arguments hold for the other excited states of hydrogen, indicating that hydrogen in the ISM is not in LTE. Moreover, this implies that *neutral hydrogen in the ISM is in the ground state.*

The Boltzmann equation (Eq. (5.23)) *defines the excitation temperature*, T_{ex}, i.e. T_{ex} is *whatever temperature, when put into the Boltzmann equation, results in the observed ratio of* $\frac{N_n}{N_1}$ (or whatever two levels are being compared). For a gas in LTE, *all* levels in the atom can be described by the same value of T_{ex}, in which case the excitation temperature is just the gas kinetic temperature. In non-LTE cases, the Boltzmann equation may still be used provided the excitation temperature, rather than the kinetic temperature, is inserted into Eq. (5.23). There could then be a different value of T_{ex} between every pair of energy levels in the atom. There are even cases in which T_{ex} could be a negative value if there are more particles in an excited state in comparison with a lower state. A 'negative temperature' only has meaning in the context of this equation. An example of such a situation is an *astrophysical maser*, as illustrated in Figure 5.16.[27]

ISM HI clouds are particularly interesting because, although all particles are in the ground state, this level is split into two hyperfine states due to the two possible orientations of the proton and electron spins (Appendix C.4 of the *online material*). A transition

[27] Maser stands for 'microwave amplification by stimulated emission of radiation'. Masers are the microwave version of the more common 'lasers' ('l' for 'light') but the principle is the same, i.e. particles are pumped up to a higher energy level that is 'metastable', meaning that the particle can remain in the higher state for a relatively long time so that particles accumulate in the higher state. The particles then cascade downwards to a lower state emitting a very strong signal at a single wavelength (coherent emission).

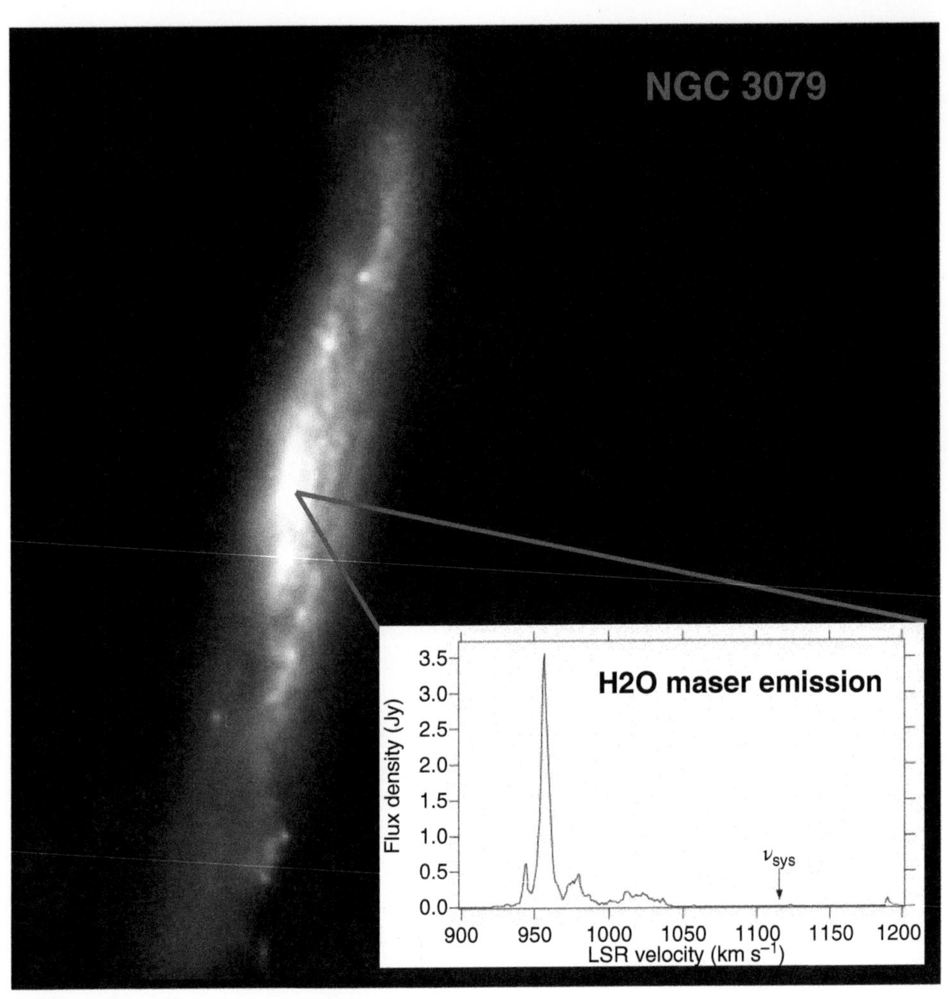

Figure 5.16 The nucleus of the edge-on spiral galaxy, NGC 3079 [10], harbours the most luminous H_2O maser known. The emission is at a rest frequency of 22 GHz, but the x-axis in the inset spectrum [27] has been converted to a velocity via the Doppler formula (Table I.2). The excitation temperature of this line is *negative* since, at any time, there are more particles in the upper energy level in comparison with the lower energy level. In astronomical sources, masers are usually associated with molecules such as H_2O or OH which are near sources of pumping such as hot stars or the nuclei of galaxies. Source: NASA [27].

from the upper to lower energy state results in the emission of radiation at a wavelength of λ 21 cm. The excitation temperature of this line is therefore referred to as the *spin temperature*, T_S. Because the spontaneous de-excitation rate of the upper hyperfine state is only $A_{1,1} = 3 \times 10^{-15}$ s^{-1} (unlike the case for Ly α) spontaneous de-excitations are very rare and, for ISM densities, this transition is *collisionally induced*. Since λ 21 cm radiation depends on collisions with other neutral particles that have random motions described by the kinetic temperature, then for this transition, $T_{ex} = T_S \approx T$. Thus, we can say that Eq. (5.23) holds *for this particular transition* and the HI spin temperature is about equal to the kinetic temperature of the gas. When $T_{ex} \approx T$ (for the λ 21 cm line or any other line), we say that the line is *thermalized.* This is a very useful result that will allow us to compute the temperatures of interstellar HI clouds from the λ 21 cm line (see Section 8.4.3).

5.4.6 Ionization and the Saha Equation

The ionization state of a gas in LTE can also be expressed in a fashion similar to the Boltzmann equation. For a gas of temperature, T, the number of particles that are in the $K+1$ state of ionization compared to the number of particles that are in the Kth state is given by the *Saha Equation,*

$$\frac{N_{K+1}}{N_K} = \frac{2U_{K+1}}{n_e U_K}\left(\frac{2\pi m_e kT}{h^2}\right)^{3/2} e^{-\frac{\chi_K}{kT}} \tag{5.26}$$

where U_{K+1}, U_K are the partition functions of the $K+1$ and Kth ionization states, respectively, n_e is the electron density, m_e is the electron mass, χ_K is the energy required to remove an electron from the ground state of the Kth ionization state, and the constants have their usual meanings. Like the Boltzmann equation, higher temperatures result in a higher excitation state – in this case, a higher fraction of particles that are ionized. However, unlike the Boltzmann equation, the Saha equation has a dependence on electron density. If there is a higher density of free electrons, then there is a greater probability that an electron will recombine with the atom, lowering the ionization state of the gas[28]. Regardless of which atom is being considered in the Saha Equation, the electron density must include contributions of free electrons from all atoms. Heavier atoms, for example, may contribute to n_e in larger proportion than their abundances (Figure 5.9) would suggest because they have more electrons and also their ionization energies may be lower[29].

Hydrogen only has one electron to be removed and can only exist in the singly ionized or neutral states, so Eq. (5.26) reduces to,

$$\frac{N_{HII}}{N_{HI}} = 2.41 \times 10^{15} \frac{T^{3/2}}{n_e} e^{-\frac{1.58\times10^5}{T}} \tag{5.27}$$

where we have used $\chi_{HI} = 13.6$ eV, $U_{HI} = 2$ (Section 5.4.5), and $U_{HII} = 1$ since the ionized hydrogen atom is just a free proton and only exists in a single state. It is now straightforward to show that a hydrogen atom tends to become ionized before there are many particles in even the first excited state (Example 5.7).

[28] An exception is if the densities are so high that electron degeneracy pressure (Section 5.3.2) starts to become important in which case the ionization state remains high even though n_e is also high.
[29] It is generally easier to strip off an outer electron from an atom that has other electrons around it since the effective nuclear potential is weakened.

Example 5.7

At the surface of a certain hot star, the conditions are as follows: T = 26 729 K and $n_e = 9.12 \times 10^{11}$ cm^{-3}. Assuming that LTE holds, determine the fraction of all hydrogen atoms that are ionized and compare this to the result of Example 5.6.

Using Eq. (5.27),

$$\frac{N_{\text{HII}}}{N_{\text{tot}}} = \frac{N_{\text{HII}}}{N_{\text{HI}} + N_{\text{HII}}} = \frac{\frac{N_{\text{HII}}}{N_{\text{HI}}}}{1 + \frac{N_{\text{HII}}}{N_{\text{HI}}}} = \frac{3.1 \times 10^7}{1 + 3.1 \times 10^7} \approx 1$$

Example 5.6 showed that a temperature of 32 000 K was required to place 10% of all neutral hydrogen atoms in the first excited state. Yet, from this example, it is clear that at an even lower temperature, virtually all hydrogen is ionized. Thus, a hydrogen gas starts to become appreciably ionized at temperatures lower than those required to excite the neutral atom. This is because there are many more possible states available for a free electron, than for a bound electron in the first excited state.

We conclude that, if a hydrogen gas in LTE is neutral, virtually all particles will be in the ground state. Since we have already seen that neutral hydrogen that is not in LTE will also be in the ground state, our conclusion is even firmer.

5.4.7 Probing the Gas

In order to probe the nature and properties of a gas, it is necessary to observe the gas in some tracer that is appropriate to its ionization and excitation state. From the previous sections, we know that hydrogen that is neutral will have its electrons in the ground state, and therefore, no emission lines, such as from the Lyman, Balmer, and Paschen series (Appendix C of the *online material*) would be seen because few electrons are in excited states. Thus, to obtain information on HI such as, for example, the shell shown in Figure 5.17, we require some probe of the ground state itself. This is provided by the λ 21 cm line that was introduced in Section 5.4.5 will be discussed further in Section 8.4.3. Another possibility is if a background signal excites the electron from the ground state upwards. The upwards pumping will produce absorption lines since the background photon that would normally pass through the gas is removed from the line of sight (Section 8.4.2).

Ionized hydrogen (HII) exists wherever there is a source of radiation or collisions energetic enough to eject the electron from the H atom. Examples are HII regions like the ones shown in Figures 5.13 and 10.4 in which an embedded hot star (or stars) provides the radiation field, or planetary nebulae, as in Figure 5.7, in which the radiation field is supplied by the central white dwarf. If the source of ionization were removed, HII would recombine to HI very quickly.

There are several ways to probe ionized gas. One is to observe the emission that occurs when the free electrons in these nebulae accelerate near positively charged nuclei

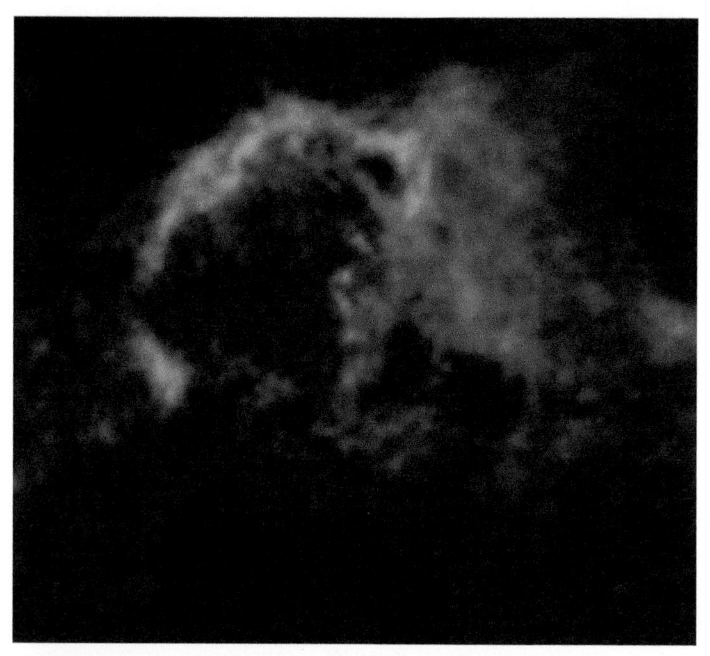

Figure 5.17 An expanding 1.8×10^5 $M_\odot$ HI shell in the low-density outer region of the Milky Way galaxy (distance from the Galactic Centre = 23.6 kpc), observed in the λ 21 cm line of neutral hydrogen. This shell is believed to represent the ISM HI that has been pushed outwards after a supernova exploded 4.3 Myr ago and after the emission from the supernova itself (e.g. similar to that shown in Figure 3.2) has faded from view. The radius of the shell is 180 pc, the expansion velocity is 11.8 kms^{-1}, and the temperature is 230 K [25]. Source: Composed for the Canadian Galactic Plane Survey by Jayanne English with the support of A. R. Taylor. Reproduced by permission of J. English.

(see Section 10.2). Another results from electrons recombining with nuclei. Electrons always have some probability of recombining with positively charged nuclei, so continuous ionizations and recombinations occur in a steady state in an ionized gas. This means that emission lines of hydrogen such as the Lyman and Balmer series lines can be observed as electrons cascade down various energy levels, emitting photons in the process. Even though the gas overall remains highly ionized, these transitions between bound states in atoms (meaning the atom is momentarily neutral) can be observed. Such emission lines are called *recombination lines* (see also Section 11.4). Wherever hydrogen emission lines are seen in the ISM, the gas from which these lines originate *must* be ionized because the electron of HI is in the ground state. This is a very important clue. If, say, an Hβ line is observed, then the gas must be ionized (HII); if the gas is ionized, then there must be a source of ionization such as a hot star;[30] if there is a hot star which we know live very short lives (Section 5.3.4), then there must be star formation in the vicinity. We can start to see how a straightforward observation and some knowledge of astrophysics allow us to piece together a story about the object of interest.

Gases that are partially ionized are also present in a number of astrophysical conditions such as stellar atmospheres or interiors, or diffuse interstellar regions farther from sources of ionizing radiation than the immediate HII region. Such gases may also be detected by their emission lines, depending on the degree of ionization and the number of particles that are present. For stellar atmospheres, however, an important probe of the conditions is provided by absorption lines. Absorption lines are formed when the temperature of the background stellar 'surface'[31] is hotter than the partially transparent stellar atmosphere that surrounds it. The background radiation excites the atmospheric atoms, and absorption lines are seen (see also Section 8.4.2).

Historically, the most important hydrogen absorption lines are those of the Balmer series because they occur in the optical part of the spectrum. However, observing a Balmer absorption line requires that there be a sufficient number of hydrogen atoms in which electrons are in the energy level, $n = 2$. As we know, few particles are in that state, but a small fraction can still produce an observable line. By considering the Boltzmann and Saha equations together, it can be shown that the fraction, N_2/N_{tot}, reaches a maximum at a temperature around 10^4 K (see Figure 5.18). At lower temperatures, there are insufficient numbers of particles with electrons in $n = 2$, and at higher temperatures, most of the particles have been ionized. Thus, the strength of the Balmer absorption lines seen in stars reaches a maximum around 10^4 K. The strength of these and other absorption lines in stellar atmospheres forms the basis of the *stellar classification system* in use today. The properties of the various stellar spectral types, which are ordered by temperature, are listed in Tables T.5 – T.7. Sample spectra, showing how various absorption lines change with spectral type, are illustrated in Figure 8.8. And spectral types are located on the HR diagram in Figure 7.12. Since stars are at 10^4 K for spectral type, A0V, this is the spectral type for which the Balmer absorption lines are strongest.

5.5 THE DUSTY UNIVERSE

Much has already been said about solid material, most of it being small, dust-like particles, in Section 1.1. Understanding nearby dusty or rocky material benefits from the fact that we can reach the material by artificial probes and/or retrieve extraterrestrial dust or meteorites for laboratory analysis on Earth. The resulting data have led to a wealth of understanding about the solid components in our Solar System.

Yet, it is not clear to what extent local material resembles the *interstellar dust* that is scattered throughout the Galaxy. Therefore, our knowledge of interstellar medium (ISM)

---

[30] A white dwarf is also possible, as previously mentioned.

[31] The surface occurs at the radius at which the star becomes opaque. See comments in Section 8.4.2 and Footnote 12 in Chapter 8.

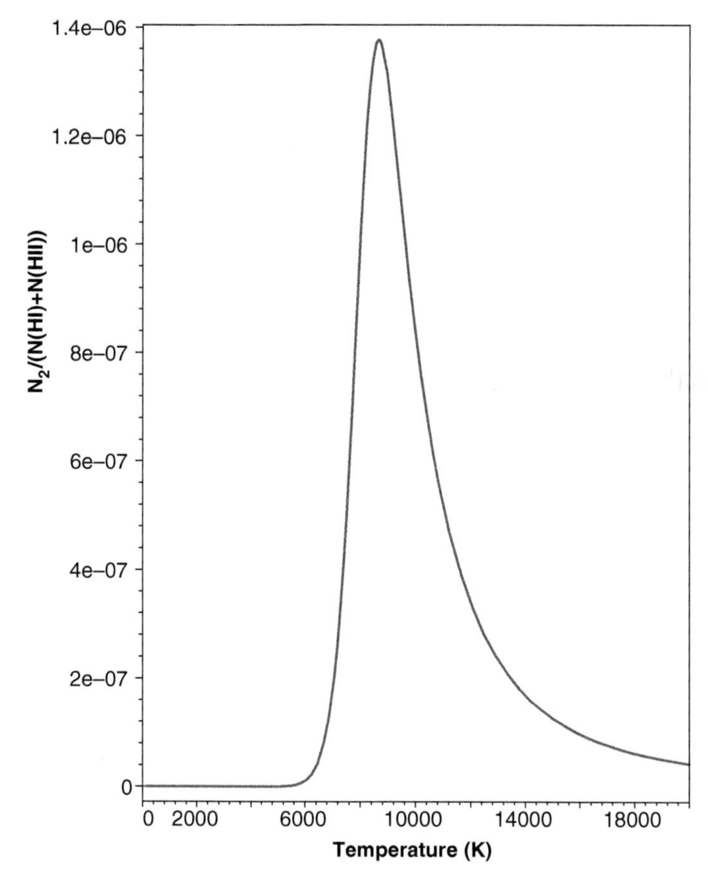

Figure 5.18 Fraction of hydrogen atoms that are in the first excited state, N_2, in comparison with the total number of hydrogen atoms, $N_{HI} + N_{HII}$. The adopted electron density at all temperatures has been taken to be constant at $n_e = 10^{13}\ cm^{-3}$ for this plot. This distribution indicates how the strength of the Balmer absorption lines varies with stellar surface temperature. Note that the fraction is still quite low, even at the peak.

dust properties must still be deciphered from the light that is emitted, absorbed, scattered, and polarized by this important component of the ISM. In this section, we will look at some of the observational effects of dust and discuss dust properties. Details of the interaction of light with dust will be dealt with in Section 7.5.

5.5.1 Observational Effects of Dust

Almost everywhere we look in the sky, we see some evidence for dust. Dark bands that cut a swath across galaxies (Figure 5.21), thin curves that follow spiral arms (Figure 5.8),

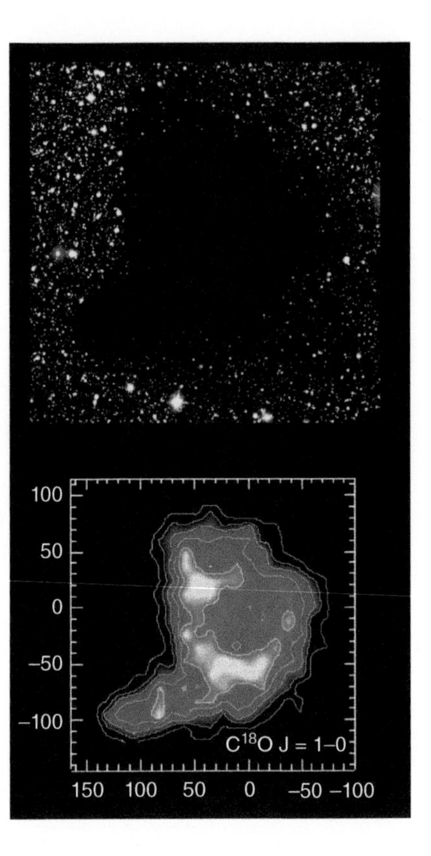

Figure 5.19 *Top:* At one time, dark regions like this in the sky were thought to be 'holes' – regions in which there were no stars. We now realize that these dark nebulae are dense molecular clouds. The dust makes up only about 1% of the mass of the cloud yet it is the dust that obscures the background starlight from view. This cloud, called Barnard 68, is quite nearby, about 150 pc away and 0.15 pc across. *Bottom:* The same cloud is shown by the emission of a spectral line of the carbon monoxide isotope, $C^{18}O$. Most of the mass of the cloud is in molecular hydrogen, H_2, but various other molecules and their isotopes are present, such as CS, N_2H^+, NH_3, H_2CO, and C_3H_2, among others [15]. Source: Reproduced by permission of the AAS and J. Alves.

filamentary patterns in HII regions (Figure 5.13), patchy dimming of starlight in the Milky Way (Figure 5.19), diffuse glows from scattered light around individual stars (Figure 7.3), and stellar disks hinting of hidden planetary bodies (Figure 6.13) all attest to the presence of small particulate (solid) matter in a wide variety of environments.

Although dust makes up only $\approx 1\%$ of the ISM, by mass, its effects are extremely important, especially in the optical and UV where dust strongly scatters and absorbs, obscuring the light from behind. Dust prevents us from seeing the nucleus of our Milky Way galaxy at optical wavelengths (Figure 5.3), and we must rely, instead, on other wavebands to penetrate through this barrier. Indeed, the depth of our view along the plane of the Milky Way is quite shallow, being restricted to about a kpc, on average, though there are some lines of sight that are clearer. What is worse, the distribution of dust is so patchy that it cannot be easily modelled or corrected for, except on an object-by-object or location-by-location basis. In the history of our understanding of the Milky Way, dust has been a major impediment to obtaining distances and sizes of various objects, including the large-scale structure of the Galaxy itself.

For example, if the absolute magnitude, M, is believed to be known for some object, say via a calibration of other similar objects of known distance, then a measurement of the apparent magnitude, m, leads to a measurement of distance via Eq. (3.32). If there is dust along a line of sight, however, the object will look dimmer than it otherwise would be, leading to a distance that is erroneously large. Early estimates of the size of the Milky Way were high by at least a factor of two for just this reason. In the presence of dust obscuration, therefore, Eq. (3.32) must be re-written to take this into account,

$$\mathrm{V} - M_{\mathrm{V}} - \mathrm{A}_{\mathrm{V}} = 5 \log d - 5 \tag{5.28}$$

The quantity, A_{V}, is the *total extinction* or just *extinction* in the V band which gives the amount of dimming, in magnitudes, as a result of scattering and absorption along the line of sight, that is (Eq. 3.28),

$$\mathrm{A}_{\mathrm{V}} = -2.5 \log \left(\frac{f_{\mathrm{V}}}{f_{\mathrm{V}_0}} \right) \tag{5.29}$$

where f_{V} is the flux density of the object with extinction and f_{V_0} is the flux density in the absence of extinction. Similar equations could be written for the other filter bands. Since the V band extinction per unit distance along the line of sight in the plane is about 1 mag kpc^{-1} [28], large errors in distance will occur if the extinction is ignored. For example, an object at a true distance of 1 kpc would inaccurately be placed at a distance of 1.6 kpc if dust extinction is not included (Problem 5.11).

Reddening was introduced in Section 4.3.5 in the context of the Earth's atmosphere but the process and effect are similar outside of the Earth's atmosphere. Reddening is therefore seen in the interstellar medium or in any dusty astrophysical environment. It is due to the fact that dust extinction is more effective at short wavelengths than at long wavelengths, so any object viewed at optical wavelengths will appear redder if seen through a veil of dust. Given this wavelength dependence of extinction, we can define a *selective extinction*, which is a measure of reddening at some wavelength, λ,

$$E_{\lambda-\mathrm{V}} = \mathrm{A}_{\lambda} - \mathrm{A}_{\mathrm{V}} \tag{5.30}$$

It is possible to measure the reddening of an astronomical object via a comparison with another object (a calibrator) that has negligible reddening. For example, if two stars of the same intrinsic type, and therefore the same absolute magnitude, are observed in the V and B bands, the apparent magnitudes of the star of interest and the calibrator, respectively, will be,

$$\begin{aligned} \mathrm{V} &= M_{\mathrm{V}} + 5\log(d) + \mathrm{A}_{\mathrm{V}} - 5 \qquad \text{(star)} \\ \mathrm{B} &= M_{\mathrm{B}} + 5\log(d) + \mathrm{A}_{\mathrm{B}} - 5 \qquad \text{(star)} \end{aligned} \tag{5.31}$$

$$\begin{aligned} \mathrm{V}_0 &= M_{\mathrm{V}} + 5\log(d_0) - 5 \qquad \text{(calibrator)} \\ \mathrm{B}_0 &= M_B + 5\log(d_0) - 5 \qquad \text{(calibrator)} \end{aligned} \tag{5.32}$$

where d and d_0 are the distances to the star of interest and the calibrator, respectively. Combining and rearranging these equations, and expressing $(\mathrm{B}_0 - \mathrm{V}_0)$ as $(\mathrm{B} - \mathrm{V})_0$, yields,

$$E_{\mathrm{B-V}} = \mathrm{A}_{\mathrm{B}} - \mathrm{A}_{\mathrm{V}} = (\mathrm{B} - \mathrm{V}) - (\mathrm{B} - \mathrm{V})_0 \tag{5.33}$$

This is an important result because it does not depend on knowing distances which are often difficult to obtain. One need only compare the quantities, B – V, for the star of interest and the calibrator star. Differences of magnitudes are also more accurately determined than the magnitudes themselves, in the event that there is some error in the absolute scale being used. It is straightforward, then, to determine $E_{\lambda-V}$ for any λ, provided unreddened (or minimally reddened) calibrators can be found. Note that E_{B-V} is almost always a positive quantity because blue light is more heavily extincted (will have a higher numerical value of the magnitude) than visual light. Thus, by Eq. (5.33), we have a way of determining the selective extinction of an object.

It is less straightforward to determine the total extinction, A_V. However, much effort has gone into determining this value for objects of known distance. As a result, it has been found that the ratio of total to selective extinction is very nearly constant in the Milky Way. This is basically a statement that the same dust that is producing the total extinction is also producing the reddening. For the diffuse ISM, the two values are related via [28],

$$R_V = \frac{A_V}{E_{B-V}} = 3.05 \pm 0.15 \tag{5.34}$$

In the absence of other information, if the selective extinction is measured, this relation may be used to find the total extinction which can then be used in Eq. (5.28) as required Example 5.8). However, there may be a larger variation in R_V than is suggested by the error bar of Eq. (5.34) depending on environment. The value may be higher in denser regions where the properties of grains may differ. For example, [13] find $R_V = 3.6$.

Example 5.8

An MOIII star in the diffuse ISM is measured to have blue and visual magnitudes of B = 8.76 and V = 7.09. Estimate the distance to this star.

From Table T.6, $(B-V)_0 = 1.57$ for a MOIII star. Therefore, the selective extinction of this star is (Eq. (5.33)) $E_{B-V} = (B-V) - (B-V)_0 = (8.76–7.09) - 1.57 = 1.67–1.57 = 0.10$. From Eq. (5.34), $A_V = 0.305$. Using $M_V = -0.2$ for this kind of star (Table T.6), Eq. (5.28) yields a distance of 249 pc. Notice that we knew what kind of star this is and therefore knew its absolute magnitude. If we had ignored dust and used Eq. (3.32) without any dust correction, the V magnitude measurement would have given an incorrect distance of 287 pc.

Historically, most observational attention has been paid to the optical waveband, so the optical extinction properties of dust have dominated the attention of astronomers for the purposes of distance or flux corrections. In other words, the dust has been considered 'in the way'. However, for dust, it is certainly true that 'one man's trash is another man's treasure'. The same dust that blocks our optical view of the cosmos is a crucial component in the complex chemistry of the ISM, and it traces magnetic fields, plays an important role in ISM dynamics, must be taken into account in abundance determinations, and contributes to driving stellar evolution. Thus, the study of dust is a rich and important sub-field of astronomy in its own right.

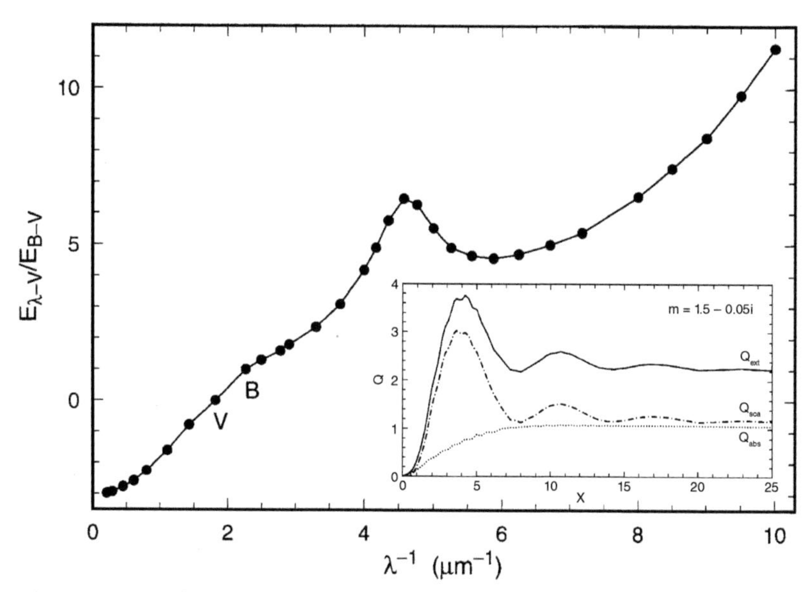

Figure 5.20 The mean extinction curve for the ISM in the optical band from $0.2 \rightarrow 10\,\mu m^{-1}$ ($\lambda\,5000 \rightarrow \lambda\,100$ nm). Frequency increases to the right and the V and B bands are marked. Note that larger $\frac{E_{\lambda-V}}{E_{B-V}}$ corresponds to more extinction, so background interstellar light will be more heavily extinguished at these wavelengths [28]. *Inset:* Absorption, scattering, and extinction efficiency factors as a function of the parameter, $x = 2\pi\,a/\lambda$ (a is the grain radius), for a weakly absorbing grain with index of refraction, m. When Q is multiplied by the geometric cross section, the result is the effective interaction cross section of the dust particle to the incident light (see Appendix D.3 of the *online material*). Source: Whittet [28].

Observational data that allow us to probe the characteristics of dust are therefore very important. An example is Figure 5.20 which shows the mean *extinction curve* for the Milky Way in the optical band. The extinction curve is a plot of the selective extinction, $E_{\lambda-V}$ (normalized by E_{B-V}) as a function of λ^{-1}. Since $E_{\lambda-V} = A_\lambda - A_V$, this plot shows how the extinction varies with wavelength. As the frequency ($\propto \lambda^{-1}$) increases, so does the extinction, so extinction is worse in the ultraviolet but not so bad in the infrared.

Structure in the extinction curve provides clues as to the make-up of interstellar dust. For example, there are several changes in slope and an obvious peak (excess extinction) in the ultraviolet at 4.6 μm^{-1} (λ2175 Å). Such spectral features can be compared to laboratory spectra of vibrational modes within solids in an attempt to determine the type of material responsible. For the λ2175 Å peak, the best candidate is graphite (Figure 5.22) or a related particle [28]. Overall, the underlying spectrum in the optical region shown in Figure 5.20 can be roughly approximated by $E_{\lambda-V} \propto 1/\lambda$ (ignoring the peak at 4.6 μm^{-1}).

See Example 7.8 for further details on the link between the extinction curve and the optical properties of grains.

In reality, because a complete description of dust requires many parameters, it is difficult to find a unique result using only the extinction curve. This is compounded by the fact that there may be different types and admixtures of grains in different interstellar environments. A good example is our own Solar System in which solid particles (and planets) become icier farther from the Sun. Fortunately, there are other ways of obtaining additional information about interstellar dust. These include the study of dust scattering and polarization characteristics (Section 7.5). In addition, a rich area of study is the *emission* properties of grains viewed in the infrared part of the spectrum. The same dust that causes extinction in the optical band will *radiate* in the IR (see Section 6.2.1). This is beautifully illustrated by the dust lane of the Sombrero Galaxy shown in Figure 5.21. With the advent of space-based infrared telescopes such as the Infrared Astronomical Satellite (IRAS) launched in the 1980s, the Spitzer Space Telescope (launched in 2003), the Herschel Space Observatory (launched in 2009), and soon NASA's James Webb Space Telescope, the study of dust has become a mature sub-discipline in astronomy. Put together, we have an emerging view of dust characteristics, as described in Sections 5.5.2 and 5.5.3.

5.5.2 Structure and Composition of Dust

Large grains, also called *classical grains*, are of order, 0.1 μm in size which is about the size of particles in smoke. *Very small grains* (VSGs) are of order, 0.01 μm which is about the size of a small virus. However, a continuum of particle sizes is likely present in a power law distribution such that there are more smaller grains and fewer large grains, i.e. $n(a) \propto a^{-q}$, where $n(a)$ is the number density per unit size range (cm^{-3} cm^{-1}) of grains of radius, a, and q is the power law index that must be fitted to the data. Ignoring some complexities, a function that fits the data reasonably well over the range, $a = 0.005$ μm to $a = 0.25$ μm is [28],

$$n(a) \propto a^{-3.5} \tag{5.35}$$

Grain composition is more difficult to ascertain because few spectral features are observed from these solid particles. Carbon appears to play an important role and, given its propensity to join together with other carbon atoms, this element can combine in a wide variety of ways, including highly ordered (*crystalline*) structures such as nanodiamonds, graphite, and fullerenes (Figure 5.22) as well as *amorphous* (irregular) structures. Carbon can also easily form hydrogenated rings, called *polycyclic aromatic hydrocarbons* (PAHs, Figure 5.23), commonly seen on Earth in the soot from automobiles. Some version of PAHs, though possibly not exactly in the same formations, may be in the interstellar medium, as suggested by observed mid-infrared spectral features. PAHs are planar, rather than three dimensional, and so could be thought of as large molecules, containing of order 50 atoms, rather than the three-dimensional structures that we identify with solid dust grains. PAHs are therefore at the transition between molecules and dust. Another component of dust is likely *silicates*, a class covering a wide variety of chemical compositions

Figure 5.21 The Sombrero Galaxy (M 104) in optical and infrared light. (a) A Hubble Space Telescope image reconstructed from a series of images taken in red, green, and blue filters. The dust lane *obscures* the light from the stars behind it. (Credit: Hubble Space Telescope/Hubble Heritage Team) (b) False colour image showing λ 3.6 mm data in blue, λ 4.5 mm data in green, λ 5.8 mm data in orange, and λ 8.0 μm data in red. In this image, blue/green represents starlight primarily from stars cooler than the Sun, while red/orange represents thermal emission from dust at its equilibrium temperature. Now the dust lane is *emitting*, rather than obscuring. Source: Reproduced by permission of NASA/JPL-Caltech/R. Kennicutt (University of Arizona), and the SINGS Team.

involving SiO_4. These appear to be in amorphous, rather than crystalline, structures, and also likely contain magnesium (Mg) and iron (Fe). A third category is ices, such as the solid phases of water (H_2O), methane (CH_4), and ammonia (NH_3).

The structure of the grains is less certain. 'Sooty' (carbon rich) mantles over silicate cores are one possibility. Ices may also form mantles about cores since they move more

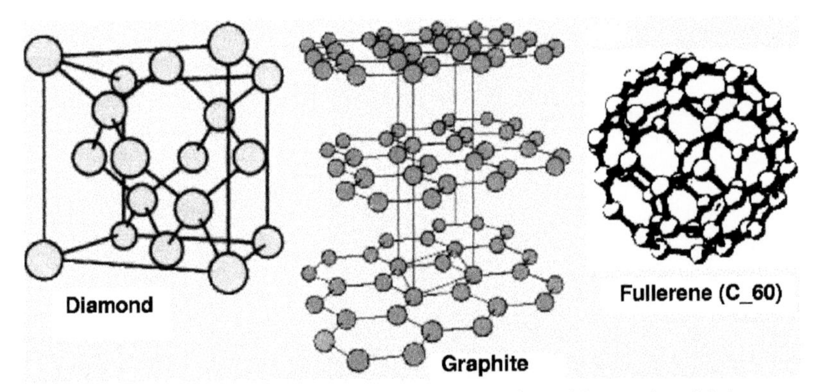

Figure 5.22 Sketch of the three different ordered forms in which carbon can exist in its solid phase. Source: Whittet [28]. © Douglas Whittet.

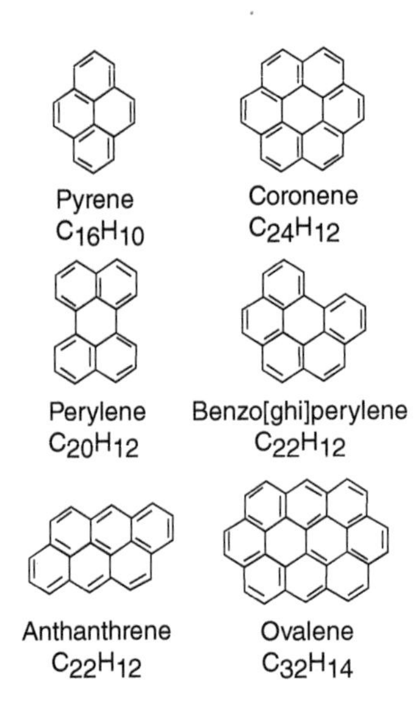

Figure 5.23 Sample configurations of polycyclic aromatic hydrocarbons (PAHs) which may make up the largest molecules in the ISM. Source: Data from http://pubchem.ncbi.nlm.nih.gov.

easily between the gas phase and the solid phase. As Figure 5.22 illustrates, dust grains are unlikely to be uniform spheres. It is clear that at least some fraction of grains are elongated when it is scattered from dust, because asymmetry is needed to explain polarized light, as shown in Figure D.6 of the *online material*.

5.5.3 The Origin of Dust

Where do these grains come from? It is well known that the envelopes of stars contain dust. The cooler atmospheres of evolved stars appear to be a source of dust since their atmospheres are rich in carbon and oxygen (Section 5.3.2), and dust can easily be driven from the extended envelopes of these stars into the ISM via radiation pressure (Section 3.5). However, it appears that 'stardust', although important, cannot be the sole source of interstellar dust, because more dust is present in the ISM than can be accounted for by stellar injection alone [28]. An additional source of dust, though, appears to be the metal-rich ejecta of supernovae. For example, the supernova remnant, Cas A, shown in Figure 3.2, contains 2–4 $M_\odot$ of dust, which is about 100 times the amount of dust that could have been swept up from the ISM during its expansion. This implies that dust was formed during the supernova event, itself [8].

Observationally, we know that the dust distribution in the Galaxy follows the dense, molecular gas distribution (Figure 5.19). There appear to be complex processes in these regions involving gas-phase chemical reactions, the absorption of particles onto grains, migration and reactions on grain surfaces, destruction of grains in shock waves, and/or growth of grains via coagulation or deposition onto mantles. These processes are not yet fully understood, but the careful study of light associated with dust-rich regions is slowly revealing the properties and nature of this important component of our Universe. The intimate link between dust grains and large molecules, especially organic molecules and carbon-based molecules such as PAHs, and related information such as the presence of amino acids in meteorites, hint that this branch of astrophysics may provide important clues about the origin of life itself.

PROBLEMS

5.1. (a) Write an equation for the estimated lifetime of a star on the main sequence in terms of stellar mass and luminosity, making the same assumptions as in Example 5.1. Evaluate known quantities and express the result in years with the stellar mass and luminosity in units of $M_\odot$ and $L_\odot$, respectively.

(b) Refer to Table T.5 and find the bolometric magnitude and luminosity of a 17 $M_\odot$ star. Evaluate the main sequence lifetime of this star, assuming that the formation time of a star and all subsequent evolution off the main sequence are small fractions of the main sequence lifetime. If the star formation is continuous, determine how many generations of 17 $M_\odot$ stars could occur during a single Solar lifetime.

(c) Evaluate the main sequence lifetime of a 0.34 $M_\odot$ star with the same assumptions as in part (b) and compare it to the age of the Universe.

5.2. Approximating stellar mass distributions by the IMFs shown in Figure 5.10 and provided in Eqs. (5.5) and (5.6),

(a) What is the number density of all stars formed from 0.1 to 100 $M_\odot$ in the disk and in the halo?

(b) Determine the mean separation between stars in both regions of the Galaxy.

(c) Express the mean separations of part (b) in terms of a 'typical' stellar diameter, D_t, for each region.

(d) Stars in the disk of the Galaxy are in orbit about its centre, but there is also a random component to stellar velocities of about 10 km s^{-1}. Suppose we treat this random component to be like the motions of particles in a gas. Assume that the effective diameter for star–star collisions is $10D_{\rm t}$ and compute the mean time between star–star collisions in the disk. What can you conclude about the probability of such collisions in the Galaxy, excluding stars in dense clusters?

5.3. For any given collection of stars that are formed in the disk of the Milky Way, determine the fraction of stars that will become Type II supernovae. If star formation continues in the same region (no gas entering or leaving) until the gas is used up, should this fraction increase, decrease, or stay the same with time? Explain.

5.4. From the M-B velocity distribution (Eq. I.2), derive Eqs. (5.8) and (5.9).

5.5. Show that Eqs. (5.13) result from Eq. (5.1) and Eq. (5.12).

5.6. For the HI shell shown in Figure 5.17, assume that the initial ISM had Solar abundance, uniform density, and $T = 200$ K before the supernova swept up the gas.
   (a) Calculate the average density, n, of particles in the pre-supernova (pre-SN) ISM. (You may assume that the mean atomic weight of the metals can be approximated by that of oxygen.)
   (b) Determine, for the pre-SN ISM, $\bar{l}$ (AU), $\bar{t}$ (year), $\mathcal{R}$ (s^{-1}), and $\bar{r}$ (cm). Is this an ideal gas?

5.7. (a) Calculate the time it would take a photon to travel from the core to the surface of the Sun if there were no interactions en route.
   (b) Adopting a mean free path of $\bar{l} = 0.3$ cm, determine the time (year) it takes for a photon to diffuse from the core to the surface of the Sun (see Eq. (5.22)) when the interaction time is negligible.
   (c) Write an expression for the diffusion time of a photon assuming each interaction takes 10^{-8} s and re-evaluate this time for the Sun.

5.8. For the following two ionized regions, determine the mean separation between particles, $\bar{r}$, and the mean free path for a photon, $\bar{l}$. Indicate whether the region should be considered an ideal gas and also whether it is in LTE. For simplicity, assume that Eq. (5.4) holds for both regions and that the most likely interaction for the photon is scattering from a free electron[32].
   (a) The solar region core at $R/R_\odot = 0.025$ (Table T.8).
   (b) A diffuse region in the Galaxy of diameter, $D = 1$ pc ($T = 10^5$ K, $n_e = 0.001$ cm^{-3}).

5.9. In Section 1.3, we indicated that a neutrino would travel through almost all matter without interacting, as if the matter were not even there. The cross section for neutrino scattering is 10^{-38} cm^2 at an energy of 1 GeV. Compute the mean free path for (a) the density of the interstellar medium, (b) the density of the Solar core, and (c) the density of a neutron star. For each case, comment on the ability of a neutrino to traverse these regions.

[32] Note, however, that other types of interactions may be important depending on frequency (see Section 7.1.1.2).

5.10. For this question, assume that LTE holds.

(a) Write an expression for the fraction of hydrogen that is in the neutral state, N_{HI}/N_{tot}. Evaluate all constants and express the result as a function of n_e and T.

(b) Consult Table T.9 to determine n_e and T applicable to the Sun's surface (i.e. letting $x = 0$ km) and evaluate N_{HI}/N_{tot}.

(c) What can you conclude about the ionization state of hydrogen at the surface of the Sun?

5.11. (a) Write an expression for the temperature, T as a function of energy level, n, for an LTE situation in which the number of hydrogen atoms with electrons in level n is equivalent to the number in the ground state.

(b) Using a spreadsheet or computer algebra software, compute T for the first 300 energy levels of hydrogen. Plot T(n).

(c) Briefly explain or comment on the behaviour of the plot.

5.12. (a) Find the factor by which the distance will be in error if dust extinction is ignored, i.e. find f, such that $d_{true} = d_{err}\, f$, where d_{true} is the true distance to the object, and d_{err} is the distance calculated without taking A_V into account.

(b) Suppose a F0Ib star is measured to have an apparent magnitude of $V = 4.5$. If $A_V = 0.8$, calculate the distance (pc) to the star, with and without the assumption of dust extinction.

5.13. (a) Determine the ratio of the number of large classical ($0.10 \le a \le 0.25\ \mu$m) dust grains per unit volume in comparison with VSGs ($0.005 \le a \le 0.01\ \mu$m).

(b) If all of these grains have, individually, the same shape and mass density, determine the ratio of the mass of classical grains in comparison with VSGs in some volume of space.

JUST FOR FUN

5.14. Suppose that the particles of an ideal gas were confined to move on a large frictionless surface, rather than freely in a volume. Let every particle be a cylinder of radius, r_p, with the cross section of the cylinder parallel to the surface, like disks sliding on a Crokinole board or rocks in the game of Curling.

(a) Rewrite Eq. (5.18) to find a mean free path, $\bar{l}$, for the particles in this two-dimensional gas.

(b) Estimate the number of skaters per unit area, $\mathcal{N}$, in the Rideau Canal Skateway shown in Figure 5.24. If all skaters were moving randomly, what would be the mean free path of a skater? What would be the mean time between collisions? What do you conclude about the motion of the skaters?

5.15. A fictitious alien race of spherical creatures lives inside the atmosphere of Jupiter (approx. 90% H_2 and 10% He) where the pressure and temperature are 0.45 bars and 128 K, respectively. In order for the Jovians to be neutrally buoyant, their density must be equal to the density of the atmosphere[33]. If the mass of a typical Jovian is 100 kg, what is his diameter?

Figure 5.24 Skaters on the Rideau Canal Skateway in Ottawa, Canada. The skateway is 7.9 m wide and 7.8 km long. Source: Saffron Blaze, Saffron Blaze Photography, http://www.mackenzie.co.

[33] This is a statement of *Archimedes' principle.*

Chapter 6

Radiation Essentials

We can hold one hand up for tomorrow – we can hold one hand up to the stars
– from *One Hand Up*, by the Strumbellas

6.1 BLACK BODY RADIATION

When a gas is in thermodynamic equilibrium (TE), the absorption and emission rates in the gas are in balance. Such a situation could be set up if the interior walls of a closed box were maintained at a single temperature, T. If no matter or radiation were permitted to escape from the box, then the gas particles and all radiation within it would eventually reach an equilibrium temperature, independent of the kind of material in the box or its shape. It can be shown that the resulting radiation field will be isotropic and that its spectrum (the emission as a function of frequency or wavelength) depends only upon T. Such radiation is referred to as *black body radiation*, historically called 'cavity radiation' because of the history of studying such radiation by setting up a 'thermal bath' within a cavity). The resulting specific intensity is described by a particular function called the *Planck function* or *Planck curve* (see Section 6.1 of the *online material* for a derivation) which, in frequency-dependent and wavelength-dependent forms, respectively, are given by,

$$B_\nu(\mathrm{T}) = \frac{2h\nu^3}{c^2}\frac{1}{\left(e^{\frac{h\nu}{kT}} - 1\right)} \quad \text{erg s}^{-1}\text{ cm}^{-2}\text{ Hz}^{-1}\text{ sr}^{-1} \tag{6.1}$$

$$B_\lambda(\mathrm{T}) = \frac{2hc^2}{\lambda^5}\frac{1}{\left(e^{\frac{hc}{\lambda kT}} - 1\right)} \quad \text{erg s}^{-1}\text{ cm}^{-2}\text{ cm}^{-1}\text{ sr}^{-1} \tag{6.2}$$

Astrophysics: Decoding the Cosmos, Second Edition. Judith A. Irwin.

Companion website: www.wiley.com/go/irwin/astrophysics2e

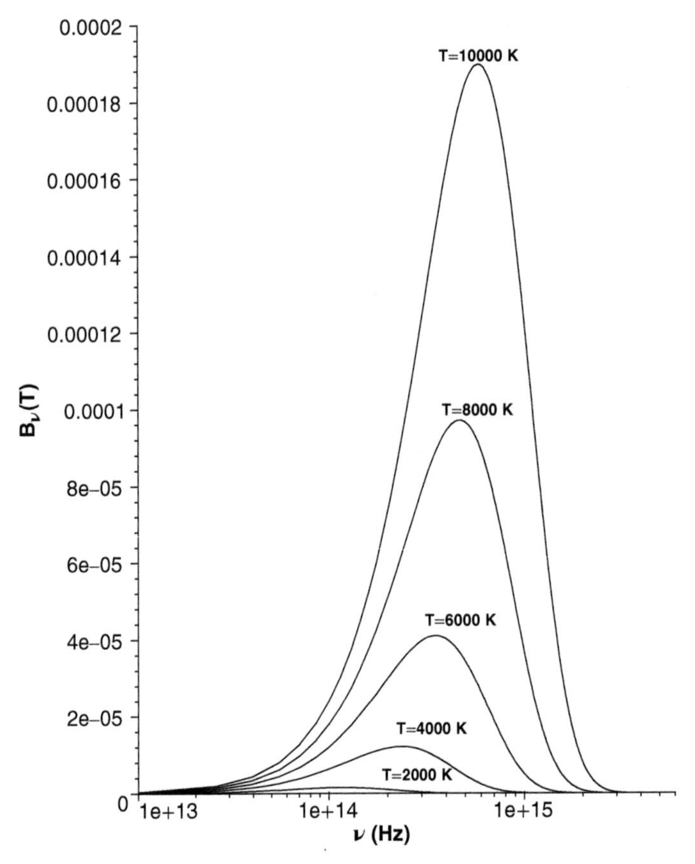

Figure 6.1 Planck curves for temperatures from 2000 to 10 000 K. The ordinate axis is in units of erg s^{-1} cm^{-2} Hz^{-1} sr^{-1}. Note that the peaks of the curves shift to higher frequency (smaller wavelength) as the temperature increases and also that hotter objects have higher specific intensities than cooler objects at *all* frequencies.

where cgs units have been specified. Plots of $B_\nu(T)$ for different temperatures are shown in Figure 6.1 and illustrate that, at higher temperatures, the specific intensity is higher at all frequencies. The two functions are related to each other via,

$$B_\nu d\nu = B_\lambda d\lambda \tag{6.3}$$

We have seen this kind of relation before (i.e. Eq. 3.3), but Eq. (6.3) applies particularly to the Planck function. The integrals under the two curves, from zero to infinity, are equal, but the functions themselves are different. They have different maxima (Section 6.1.3); and Eq. (3.5) must be used to convert from one form to the other. Since $B_\nu(T)$ and $B_\lambda(T)$ are specific intensities, the relations given in Chapter 3 involving specific intensity also apply to the Planck function. In this chapter, we will present Planck curves in both wavelength-dependent and frequency-dependent forms.

Because T in Eqs. (6.1) and (6.2) is a value that gives the specific intensity of the *radiation*, it is an example of a *radiation temperature*. However, we noted in Section 5.4.4 that for TE, or LTE 'locally', the radiation temperature is equal to the kinetic temperature of the gas. Thus, T in the above equations is written without subscript, indicating that it also represents kinetic temperature. Since there must be sufficient interactions between radiation and matter for an equilibrium temperature to be established, this implies that the gas must be *opaque*, as illustrated in Figure 6.2. For an object to be opaque, the mean free path of a photon must be less than the size of the object. The observer sees into the object only over a distance about equal to the mean free path. We thus come to the more formal definition of a black body, that is, a black body is an object that is a *perfect absorber*. This means that a black body absorbs every photon that hits it, regardless of the photon's frequency. The photon will neither be reflected back nor pass through unimpeded. The photon will be absorbed, implying that its mean free path is less than the object's size. Any opaque, non-reflecting object at a single temperature is therefore a black body. This could include solids, provided the solid material does not reflect light. For an object to remain at a single temperature, absorptions and emissions must be in balance. Thus, a black body must also be a *perfect emitter*. Black bodies are therefore *not* black in colour nor does the word 'black' imply that there is no emission. On the contrary, a black body emits the continuous Planck spectrum whose shape is dictated by its temperature.

As we found for thermodynamic equilibrium (Section 5.4.4) identifying a perfect black body in nature is difficult. However, there is one excellent example: that of the cosmic microwave background (CMB) radiation that can be seen at every position over the entire sky as shown in Figure 5.2. At every position on this map, a Planck spectrum is measured. The mottled appearance is due to the fact that there are temperature fluctuations of order 0.4 mK over the sky (see caption). Consequently, the Planck curves shift slightly in frequency and magnitude just as Figure 6.3 illustrates, although the latter figure shows much larger temperature variations. The global mean temperature for the data shown in

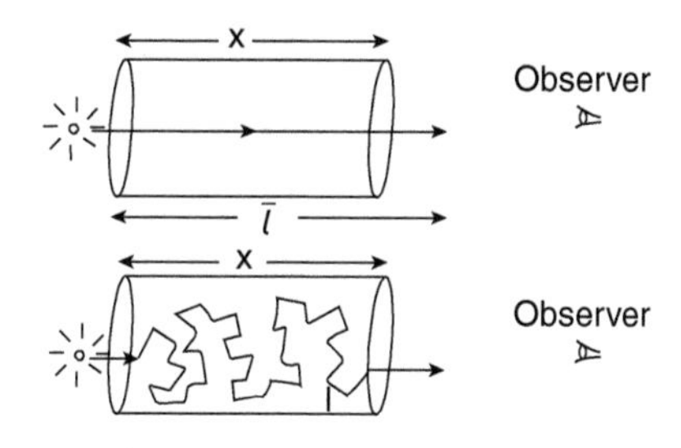

Figure 6.2 In these figures, a single source is shining just behind a gas cloud of temperature, T, as seen by an observer on the right. In the *top* picture, the mean free path of a photon, $\bar{l}$, is greater than the thickness, x, of the cloud. In such a case, the observer can see through the cloud to the source. In the bottom picture, the mean free path of a photon is much smaller than the cloud thickness. The photon diffuses to the other side (similar to Figure 5.15), coming into equilibrium with the temperature of the cloud. The observer will see into the cloud only as far as a mean free path, i.e. only into the outer 'skin' of the cloud, and will measure a Planck spectrum at the temperature of this skin.

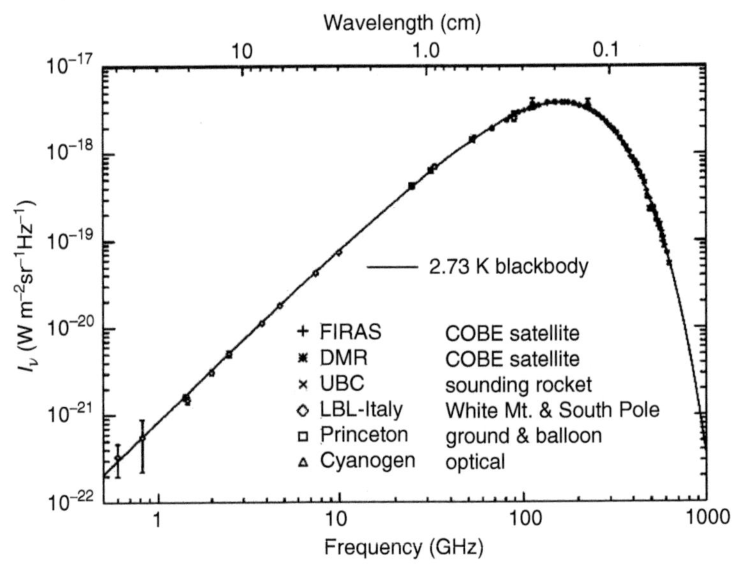

Figure 6.3 Data points for this curve are measurements of the cosmic microwave background (CMB) radiation from a variety of sources that are listed on the plot. The solid curve is the best fit black body spectrum, corresponding to a temperature of 2.73 K. Source: Courtesy G. F. Smoot and D. Scott in Hagiwara et al. [7].

Figure 6.3 is 2.73 K. The CMB is the best black body known in nature. According to [5], its temperature can be pinned down to $T_{CMB} = 2.72548 \pm 0.00057$ K!

Stars are also very good examples of black bodies, although some departures from a perfect curve do occur, as shown for the Sun in Figure 6.4. Even though the entire star is not at a single temperature, the core being much hotter than the surface, we cannot see into the interior (Figure 6.2) and, over the depth of the outer skin that *can* be viewed, the temperature is approximately constant. Since stars have temperatures that correspond to Planck curves that have peaks at or near the *optical* part of the spectrum, most of the astrophysical radiation that we see with our eyes, whether naked eye or through optical telescopes, is from *stars*. This includes the Sun, of course, and it is unlikely to be an accident that our eyes have the greatest sensitivity at just the frequency regime over which the Solar Planck curve peaks (Figure 6.5). When we look out into the vast depths of space, it is starlight that we see, from our own Galaxy, as shown in Figure 5.3, to nearby galaxies like those of Figures 5.8 and 4.19, to distant galaxy clusters like the one of Figure 9.6. Of all the emission mechanisms that we will discuss (see Part V), it is black body radiation, by far, that is most relevant to our visual appreciation of the sky.

If we can measure the specific intensity of an object and *know* it to be a black body, then by Eq. (6.1) or (6.2), we can determine its temperature. Recall that specific intensity is independent of distance in the absence of intervening matter (Section 3.3), which means that we do not need to know anything else about the object, not even its distance, to find T. Moreover, in principle, only a single measurement at one frequency is required to

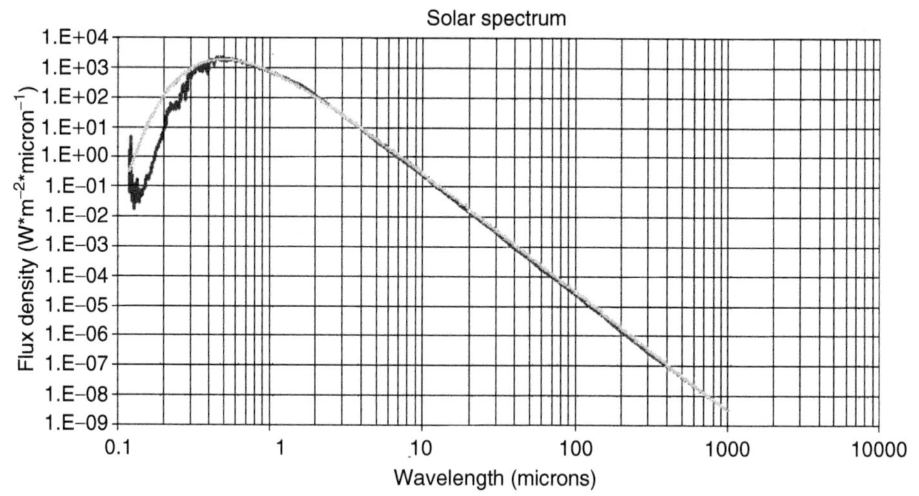

Figure 6.4 The Solar spectrum shown as a function of increasing wavelength in this logarithmic plot. The plot is of flux density as measured from the distance of the Earth; to convert to specific intensity, a division by $\Omega_{\odot}$ is required (see Eq. 3.13). The spectrum (black curve) is well fitted by a black body at T = 5781 K (grey shaded curve) but starts to depart significantly in the ultraviolet ($\lambda \lesssim 0.3\,\mu$m) due to numerous absorption lines. There are also significant departures in the X-ray and radio regions of the spectrum (not shown) due to solar activity. These data are consistent with a Solar Constant (cf. Figure 3.6) of 1366.1 Wm^{-2}.

give this result, although in practise, more measurements are usually needed to ensure that the spectrum is indeed Planckian. The flip side of this coin is that, because the Planck curve depends *only* on temperature, we can discern nothing about the kind or amount of material that emits this radiation by studying its Planck curve alone. Whether the black body is the glowing filament of an incandescent light bulb, 'super black' man-made paint that reflects less than 1% of the light that falls on it, or the interior of a kitchen oven after an equilibrium temperature has been achieved, the same Planck spectrum, dependent on T alone, will be measured.

6.1.1 The Brightness Temperature

Other than the CMB, most objects that we refer to as black bodies have emission spectra that, like stars, are approximations to the Planck curve. To account for departures from the Planck curve, we define the *brightness temperature* T_{B_ν}, at the observing frequency, ν, to be *the temperature which, when put into the Planck formula, results in the specific intensity that is actually measured at that frequency,*

$$I_\nu = \frac{2h\nu^3}{c^2}\frac{1}{\left(e^{\frac{h\nu}{kT_{B_\nu}}} - 1\right)} \tag{6.4}$$

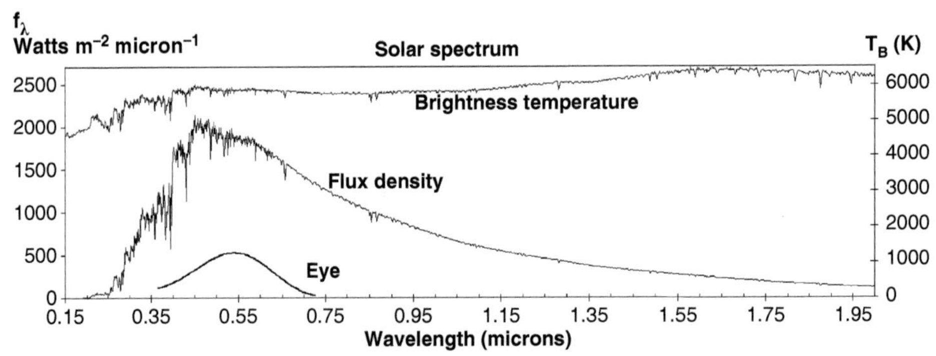

Figure 6.5 The Solar spectrum is again shown (middle curve, left scale) using the same data as in Figure 6.4 except over a more restricted wavelength range and on a linear scale. The brightness temperature (top curve, right scale) is also shown, derived by dividing the flux density by $\Omega_\odot$ (Eq. 3.13) and then using the Planck formula (the wavelength form of Eq. (6.4)) to recover T_B. The bottom curve, in arbitrary units, is the daylight photon flux response of the human eye smoothed to 30 nm resolution (see also Figure 4.1). Source: James T. Fulton, 2005, www.sightresearch.net/luminouseffic.htm, updated to neuronresearch.net/vision © 2005, James T. Fulton.

The wavelength form can also be used. It is clear that, if the *same* value of T_{B_ν} is measured *for all* ν, then the object is a true black body. If T_{B_ν} is different for different frequencies, then the object departs from a black body (Example 6.1). The extent to which T_{B_ν} is *not* constant is a measure of how much the object departs from a black body. The brightness temperature of the Sun as a function of wavelength is shown in Figure 6.5 (top curve). If the Sun were a perfect black body, then the brightness temperature curve would be a straight horizontal line in this plot. This is *close* to being true over the wavelength region shown, although there are some minor variations. However, in the radio part of the Solar spectrum (not shown), the departures are much more significant. For example, at the very long wavelength of λ 10 m, $T_{B_\nu} \approx 5 \times 10^5$ K, almost two orders of magnitude higher than shown in Figure 6.5, indicating that processes other than black body emission are occurring (e.g. Section 10.5)[1].

The concept of brightness temperature can be extended to include *any* astrophysical source, even if its spectrum is nowhere near approximating that of a black body, and even if the spectrum has no dependence on temperature at all. For example, T_{B_ν} could be specified for an object whose spectrum is a power law (e.g. Section 10.5). The brightness temperature simply provides a convenient way of expressing the radiation temperature of an object, whether or not a physical (kinetic) temperature is implied. Eq. (6.4), or its wavelength equivalent form, is a straightforward functional relation between specific intensity and brightness temperature. Therefore, to say that an object has a certain brightness

[1] This emission refers to the *radio quiet Sun*. When there are *radio bursts*, the brightness temperature can be up to $10^7 \times$ higher than this!

temperature is essentially equivalent to saying that it has a certain specific intensity and vice versa.

Example 6.1

In a particular kind of telescope used to set up an absolute calibration scale, the signal path is designed to switch back and forth between a target in the sky and a local calibrator. The local calibrator is a rotating vane that must meet the specifications of being a black body at a temperature of 50 K to within ± 2 K over a wavelength range of λ 2 to λ 20 cm. An engineer measures the specific intensity at the two end points of the wavelength range finding, $I_{2\,cm} = 2.57 \times 10^{-5}$ erg s^{-1} cm^{-2} cm^{-1} sr^{-1} and $I_{20\,cm} = 2.24 \times 10^{-9}$ (in the same units). What is the calibrator's brightness temperature at each wavelength? Is it a black body to within the desired tolerance?

The units imply that the λ form of the Planck curve is being used, so we rearrange Eq. (6.2) to solve for the brightness temperature,

$$T_{B_\lambda} = \left(\frac{hc}{\lambda k}\right) \frac{1}{\ln\left(1 + \frac{2hc^2}{\lambda^5 I_\lambda}\right)} \tag{6.5}$$

From this equation, we find that the brightness temperatures at λ 2 and λ 20 cm are, respectively, $T_{B_{2\,cm}} = 50.1$ K and $T_{B_{20\,cm}} = 43.4$ K. The brightness temperature at λ 2 cm is within the desired tolerance but the brightness temperature at λ 20 cm is not. Thus, the calibrator is not yet a black body to the desired specifications.

6.1.2 The Rayleigh–Jeans law and Wien's law

The *Rayleigh–Jeans law* and *Wien's law* are not special 'laws' in a fundamental sense, but are rather approximations to the Planck curve at the low and high frequency (or wavelength) ranges. For example, suppose we observe the Planck curve at a low frequency in comparison with the peak of the curve or, more technically, in the regime, $h\nu \ll kT$ or $\frac{h\nu}{kT} \ll 1$ Then,

$$B_\nu(T) = \frac{2h\nu^3}{c^2}\frac{1}{\left(e^{\frac{h\nu}{kT}} - 1\right)} \approx \frac{2h\nu^3}{c^2}\frac{1}{\left(1 + \frac{h\nu}{kT} + \cdots - 1\right)} = \frac{2\nu^2 kT}{c^2} \tag{6.6}$$

where we have used an exponential expansion (Eq. A.3 of the *online material*). Note that higher orders are not needed in the expansion because the term $\frac{h\nu}{kT}$ is less than 1 and is therefore diminishingly small in higher orders. Eq. (6.6), or its wavelength equivalent (Problem 6.2), is called the *Rayleigh–Jeans law*. The brightness temperature can also be expressed in this limit by replacing $B_\nu(T)$ with I_ν and T with T_{B_ν} in Eq. (6.6).

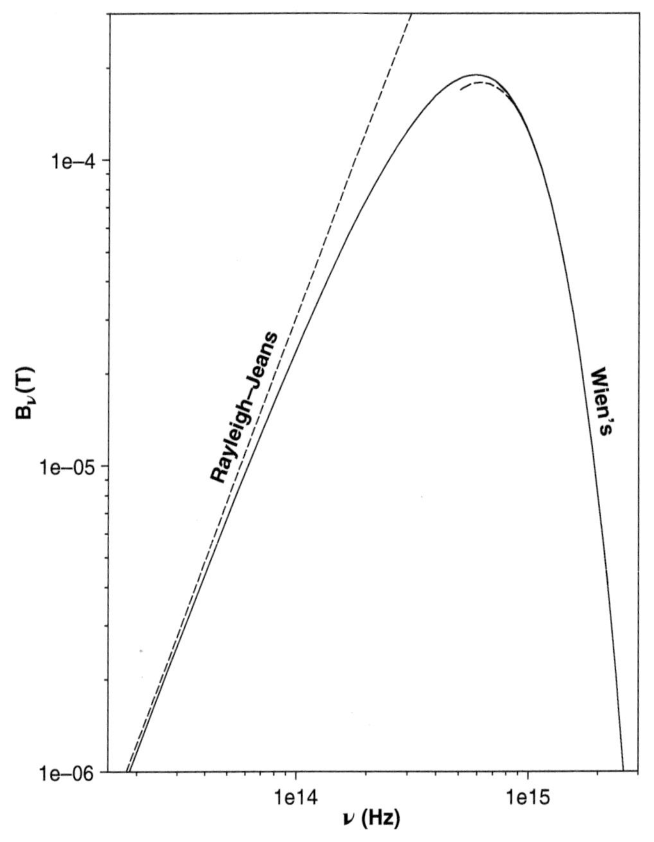

Figure 6.6 A Planck curve, in cgs units, for T = 10 000 K (solid curve) showing the Rayleigh–Jeans and Wien's approximations (both dashed). The dashed Wien's curve is difficult to see at high frequency because it closely matches the Planck curve.

On the other end of the spectrum, where $\frac{h\nu}{kT} \gg 1$, the Planck function can be written,

$$B_\nu(T) = \frac{2h\nu^3}{c^2} \frac{1}{\left(e^{\frac{h\nu}{kT}} - 1\right)} \approx \frac{2h\nu^3}{c^2}\left(e^{-\frac{h\nu}{kT}}\right) \tag{6.7}$$

In this regime, the exponential term dominates over ν^3 so the function diminishes rapidly as the frequency increases. Eq. (6.7), or its wavelength equivalent, is called *Wien's law*.

Examples of the two approximations are shown in Figure 6.6. Example 6.2 indicates when they are valid to within 10%.

Example 6.2

For what values of $\frac{h\nu}{kT}$ do the Rayleigh–Jeans and Wien's approximations depart from the Planck curve by more than 10%?

For the Rayleigh-Jeans Law, this can be determined by letting $x = \frac{h\nu}{kT}$ and setting the ratio of Eq. (6.6) over Eq. (6.1) equal to 1.1, i.e. $(e^x - 1)/x = 1.1$ The solution to this equation is $x = 0.19$. For Wien's law, we set the ratio of Eq. (6.7) over Eq. (6.1) to 0.9 since Wien's approximation gives values lower than the full Planck equation. Then, $e^{-x}(e^x - 1) = (1 - e^{-x}) = 0.9$. In this case, $x = 2.3$. Therefore, the use of the Rayleigh–Jeans law will give an accuracy of better than 10%, provided $\frac{h\nu}{kT} < 0.19$ and Wien's law gives better than 10% accuracy if $\frac{h\nu}{kT} > 2.3$.

6.1.3 Wien's Displacement law and Stellar Colours

To find the wavelength or frequency of the peak of the Planck curve, we need only differentiate Eq. (6.1) or Eq. (6.2), respectively, and set the result to zero, the results being (cgs),

$$\lambda_{max}\,T = 0.290$$

$$\nu_{max} = 5.88 \times 10^{10}\,T \tag{6.8}$$

This is called *Wien's Displacement law.* (Be careful not to confuse this with Wien's law discussed in Section 6.1.2!) Wien's Displacement law indicates how the peak of the Planck curve shifts with wavelength or frequency. Note that $\lambda_{max} \neq c/\nu_{max}$ because the wavelength and frequency forms of the Planck curve are different functions, as noted at the beginning of Section 6.1. Figure 6.1 illustrates that, as the temperature increases, not only does the specific intensity increase, but also the peak of the curve shifts to higher frequencies as Eq. (6.8) specifies.

These simple relationships provide very powerful tools for obtaining the temperature of a black body. While only a single measurement of $B_\nu(T)$, in principle, is required to determine T (Section 6.1), for many objects, stars in particular, the specific intensity cannot be measured because the object is unresolved (see Figure 3.9 and related discussion). In such cases, the flux density, f_ν, is measured instead. Because the solid angle subtended by an unresolved star, Ω, is not necessarily known, the specific intensity cannot be directly determined from $f_\nu = B_\nu(T)\Omega$ (Eq. 3.13 as applied to a black body) and T cannot be found from a single measurement. However, Ω is assumed not to vary with frequency[2]. This means that *the shape of B_ν(T) will be the same as the shape of f_ν*. We show an example in Figure 6.5 in which the flux density of the Sun, rather than its specific intensity, is plotted. For any unresolved star, then, it would be sufficient to plot f_ν, identify the frequency or wavelength of the peak of its spectrum, and using Wien's Displacement law, the temperature can be found. Note that this can be done without knowing the distance to the star!

Wien's Displacement law also explains how the colour of a hot opaque object changes with its temperature. Subtle colour differences of stars are quite visible to the naked eye. Examples can be found throughout the sky, but a good contrast is between the

[2] Small changes in Ω can occur with frequency, but these are generally not sufficient to make significant changes in the shape of the spectrum.

reddish star, Betelgeuse, in Orion's shoulder and the blueish star Rigel in his foot (see Figure 7.8 for locations). The shift in the peak of the Planck curve provides the basis for the colour index used in the observational HR diagram shown in Figure 3.15 and discussed in Section 3.6.3. The ratio of Planck curve values at two different frequencies, or the log of the ratio of two flux densities gives a colour index via Eq. (3.34). The colour index of a star of unknown angular size and distance can be compared to (calibrated against) the colour index based on the ratios of Planck curves of known temperature in order to relate colour index to stellar temperature (see Example 6.3). Such a calibration is shown in Figure T.1. Once a good calibration has been set up, one need only make two measurements of stellar flux density in order to determine the surface temperature. This procedure is even simpler than determining the peak of the spectrum, because to find the peak, a number of measurements would be required. All stars are good approximations to black bodies, whether they are on the main sequence, are red giants, or white dwarfs, so the colour index can yield a good approximation to the temperature.

Many subtleties have been ignored here. For example, stellar spectra do show departures from perfect black bodies, as Figure 6.4 illustrates. Thus, temperatures determined by different techniques are similar but not always exactly the same. The calibrations given in Tables T.5–T.7 also do not include the finer subclasses that are present over a continuum of stellar temperatures. Modern approaches make use of as much information as possible to determine stellar temperatures, including stellar spectral lines, to be discussed in Chapter 8.

Example 6.3

The star, Gomeisa (β Canis Minoris) is measured to have B = 2.79 and V = 2.89. Determine its colour index, B − V, the ratio of its specific intensities at the two observing wavelengths, $B_B(T)/B_V(T)$, estimate its surface temperature, T, determine the wavelength of the peak of its Planck curve, λ_{max}, and provide a best estimate of its spectral type, given that it is a main sequence star. Assume that there is no dust extinction in the line of sight.

From the measurements, B − V = 2.79–2.89 = −0.10. Using Eq. (3.34) for colour index, $-0.10 = -2.5\,\log\left(\frac{f_B}{f_V}\right) - (-0.601 - 0)$, where we have taken the zero point value from Table 3.1 and are using the wavelength form of the Planck curve because we are being asked for λ_{max}. Solving for the ratio of flux densities, $\frac{f_B}{f_V} = 1.9$. This result is the same as the ratio of specific intensities, $\frac{B_B(T)}{B_V(T)}$, because we take the solid angle subtended by the source to be constant with wavelength. Referring to the calibration of Table T.5 for main sequence stars, B − V = −0.1 corresponds to T = 11 950 K and a spectral type of B8. By Wien's Displacement law (Eq. (6.8)), the spectrum would peak at $\lambda_{max} = 243$ nm. As a check on the ratio of specific intensities, we can use the above temperature in the Planck equation (Eq. (6.2)) and re-compute the ratio. Using the central wavelengths for B and V bands (Table 3.1), we find, $\frac{B_B(11\,950)}{B_V(11\,950)} = 1.7$ which is within 10% of the value of 1.9 above and within 5% of it when allowing for a range of temperatures between spectral types B5 and A0 in Table T.5.

6.1.4 The Stefan–Boltzmann law, Stellar Luminosity and the HR Diagram

The total intensity of a black body can be found by integrating under the Planck curve over all frequencies, or all wavelengths,

$$B(\mathrm{T}) = \int_0^\infty B_\nu(\mathrm{T})\, d\nu = \int_0^\infty B_\lambda(\mathrm{T})\, d\lambda = \mathcal{K}\mathrm{T}^4 \ \mathrm{erg\ s^{-1}\ cm^{-2}\ sr^{-1}} \tag{6.9}$$

where,

$$\mathcal{K} = \frac{2(k\pi)^4}{15\, c^2\, h^3} \tag{6.10}$$

If we now use the relationship between intensity and flux given in Eq. (3.14) for emission escaping in the outwards direction from a surface, then the resulting astrophysical flux through the surface of such a black body is,

$$F_{\mathrm{BB}} = \sigma \mathrm{T}^4 \tag{6.11}$$

where $\sigma = \pi\mathcal{K} = 5.67 \times 10^{-5}$ erg s^{-1} cm^{-2} K^{-4} is the *Stefan–Boltzmann constant.* Eq. (6.11) is usually written in the more familiar form,

$$F = \sigma \mathrm{T}_{\mathrm{eff}}^4 \ \mathrm{erg\ s^{-1}\ cm^{-2}} \tag{6.12}$$

Equation (6.12) defines the *effective temperature*, $\mathrm{T_{eff}}$, which is *whatever temperature results in the flux observed from the object.* Like the brightness temperature, the effective temperature allows for departures from a perfect black body. In the former case, there could be a different brightness temperature at each wavelength, but in the latter case, there is only one effective temperature for any object. Also, although a brightness temperature may be quoted for any object, whether approximating a black body or not, the effective temperature is usually only quoted for objects whose emission resembles a Planck curve, especially stars. Eq. (6.11) or (6.12) is referred to as the *Stefan–Boltzmann Law.* The quantity, F, is equivalent to the astrophysical flux, defined in Eq. 3.10.

Using this flux, it is straightforward to calculate the total (i.e. bolometric) luminosity of the object since, by Eq. (6.9), an integration over all frequencies (or wavelengths) has already taken place. For a sphere of radius, R, the result is (see Eq. 3.10),

$$L = 4\pi R^2 \sigma \mathrm{T}_{\mathrm{eff}}^4 \tag{6.13}$$

It is now clear that the luminosity of a star or any black body is a strong function of its temperature. This is why, for any mixture of stars, the luminosity tends to be dominated by hot, high-mass stars (Problem 6.7) even though they are outnumbered by smaller, cooler stars.

This equation also helps to clarify the placement of the stars on the colour-magnitude diagram (the observational HR diagram) which is shown in Figure 3.15. (Refer to this figure for the following discussion.) Taking the logarithm of both sides of Eq. (6.13)

$$\log(L) = 4\log(T_{eff}) + 2\log(R) + \text{constant} \tag{6.14}$$

The log of the luminosity is related to the absolute visual magnitude (see Eq. 3.33) which is the ordinate of the diagram. The temperature, T_{eff} is related to the colour index, B – V, as discussed in Section 6.1.3. Different colour indices could be used, such as $G_{BP} - G_{RP}$, as in Figure 3.15, but the colour index still yields the temperature, with appropriate calibration. The HR diagram, therefore, can be understood as a relationship between the *luminosities* and *temperatures* of stars. This concept was introduced in Section 3.6.4, but now we can see a more direct link between stellar observational and physical properties.

If all stars were the same size, then a plot of $\log(L)$ against $\log(T_{eff})$ would yield a straight line of slope, 4. (Recall that temperature increases to the *left* on the diagram which makes the slope negative.) Let us consider the main sequence. Here, we see that cooler stars on the right are indeed much dimmer than the brilliant hotter stars on the left as Eq. (6.13) predicts. In fact, a plot of $\log(L)$ versus $\log(T_{eff})$ yields a main sequence slope that is not perfectly straight, and slightly steeper than 4. This tells us that *stars are not all the same size.* From Eq. (6.14), we conclude that *along the main sequence, hotter stars are also larger.* This is shown more quantitatively in Table T.5. Some of the curvature is due to second order effects, such as the nature of the star's opacity (Sect. 7.4.1).

There are many stars in the upper right region of the HR diagram as well. For the same temperature, these stars are much more luminous than their main sequence counterparts. By Eq. (6.13), this means that they must be much larger, hence the designation, 'giants'. A G2III star, for example (Table T.6) has about the same surface temperature as the Sun, yet is 10 times larger. The internal structure of such stars is different from that of those on the main sequence, as discussed in Section 5.3.2. What about the white dwarfs whose sequence can be seen at the bottom left of the HR diagram? These are still black bodies, so given that they are so hot, they must be very small to make their luminosities so low. Notice that some very strong conclusions can be drawn about stars by simply plotting their location on the HR diagram. It is from these and other observational starting points that we can actually piece together the evolution of stars.

6.1.5 Energy Density and Pressure in Stars

Because stars are close to being black bodies, this implies that the radiation field is close to (but not exactly) isotropic at any interior location. This means that the intensity, I, is approximately equivalent to the mean intensity, J. That being the case, we can use Eq. (3.18) to write the energy density (erg cm^{-3}) as,

$$u = \frac{4\pi}{c}I = \frac{4\pi}{c}B = \frac{4\pi}{c}\mathcal{K}T^4 = aT^4 \tag{6.15}$$

where we have used Eq. (6.9) and the constant, $a = 7.57 \times 10^{-15}$ erg cm^{-3} K^{-4}, is called the *radiation constant* (Table T.1). Thus, the energy density is also a strong function of temperature.

We have already seen how the radiation pressure, P_{rad} (dyn cm^{-2}), is related to the energy density in an isotropic radiation field (see Eq. 3.24), so now we can relate it to the temperature as well.

$$P_{rad} = \frac{1}{3}u = \frac{1}{3}aT^4 \tag{6.16}$$

This radiation pressure must be added to the particle pressure (Eq. 5.11), as indicated in Eq. (5.16), to determine the total pressure at any location within a star. For low mass stars like our Sun, the radiation pressure is negligible with respect to the total. However, for hotter more massive stars, the radiation pressure, being a strong function of temperature, can become dominant. For very high-mass stars, for example, radiation pressure can become so important that it affects the stability of stars and contributes to mass loss of the outer layers (e.g. see the discussion of Wolf–Rayet stars in Section 7.1.2.2). The importance of radiation pressure for stellar dynamics will be discussed further in Section 7.4.2.

6.2 GREY BODIES AND PLANETARY TEMPERATURES

Opaque bodies at a single surface temperature are black bodies if they absorb all incident radiation. However, in most cases, some fraction of the incident radiation will be reflected away. The fraction of light that is reflected is called the *albedo, A*, from the Latin, 'albus', meaning white. Values for the albedos of the planets are given in Table T.3. A surface that absorbs all incident radiation (a true black body) would have an albedo of 0 and a surface that reflects all incident radiation (such as a 'perfect' mirror) has an albedo of 1. We implicitly considered 'perfect absorption' and 'perfect reflection' previously when dealing with radiation pressure in Section 3.5. The introduction of the albedo now permits a quantification of intermediate cases. Objects that do not absorb all incident radiation are called *grey bodies.*

The term, 'grey', is normally meant to apply to a quantity that is independent of wavelength (i.e. 'grey' or 'colourless'). This is technically only true for opaque objects that reflect the same fraction of light at all wavelengths, i.e. A is independent of λ, but the terminology also tends to be used for many objects. For example, the Earth's albedo is not constant with wavelength, but it does not vary by orders of magnitude either. As Figure 6.7 shows, the albedo in the optical band varies by less than about 30%. The optical band is most important because this is where the Sun's Planck curve peaks and therefore is the band over which most of the energy is absorbed.

Planets, asteroids, comets, moons (see Iapetus, Figure 6.8), and dust grains are all *grey bodies*, as are normal opaque objects in everyday life, like the walls of a room, the surface of a human being, or the hound dog shown in Figure 6.9. *It is the reflected light that allows us to see the object.* Our eyes detect the portion of light that is reflected (i.e. the albedo, A) from walls, dogs, or planets that falls into the visual band. If the reflection is like a mirror with an albedo that is constant with wavelength, then the spectrum of the reflected light

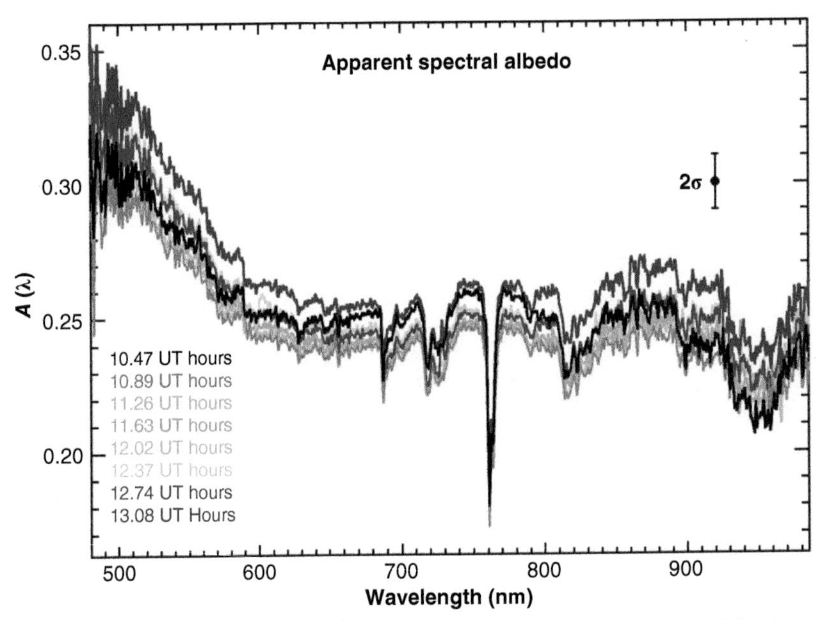

Figure 6.7 Plot of the Earth's albedo as a function of wavelength in the optical band, measured on 19 November 1993 at the times indicated. The albedo increases towards the blue (left) due to Rayleigh scattering (see Section 7.1.1.3) and a number of spectral features can be seen, including one at λ 760 nm due to O_2. These wavelength-dependent values are comparable to the Bond Albedo of $A = 0.306$ (Table T.3) which is the value applicable over all wavelengths. Source: Montañés-Rodriguez et al. [17]. © 2020 The American Astronomical Society.

is just the spectrum of the infalling light. If that light is sunlight, then the reflected light will have the spectrum of the Sun. If the incoming light is ceiling lights, then the reflected light will have the spectrum of those lights. More likely, however, the albedo will not be exactly constant with wavelength and so the reflected spectrum will be a modification of the spectrum of the incoming light.

If light falls on an opaque object, the portion that is not reflected *must be absorbed.* The result of the *absorbed portion* (1-A) is that the object is *heated.* It is possible that absorbed light is not the only source of heating and there could also be an internal contribution. Whatever the source of heating, the object will have a temperature and it will then *radiate* with a Planck spectrum dictated by that temperature. Therefore, the full spectrum of a grey body will look like the spectrum of black body whose temperature is determined by the absorbed portion of infalling light plus any internal sources (if present) – *plus* – the spectrum of the reflected portion of the infalling light. It is unlikely that these two contributions to the spectrum will be in the same spectral band. For example, reflected Sunlight will occur mostly in the optical part of the spectrum because that is where the Solar spectrum peaks, whereas typical temperatures of Solar system objects result in Planck curves that peak in the *infrared.* We will explore this further in Section 6.2.1.

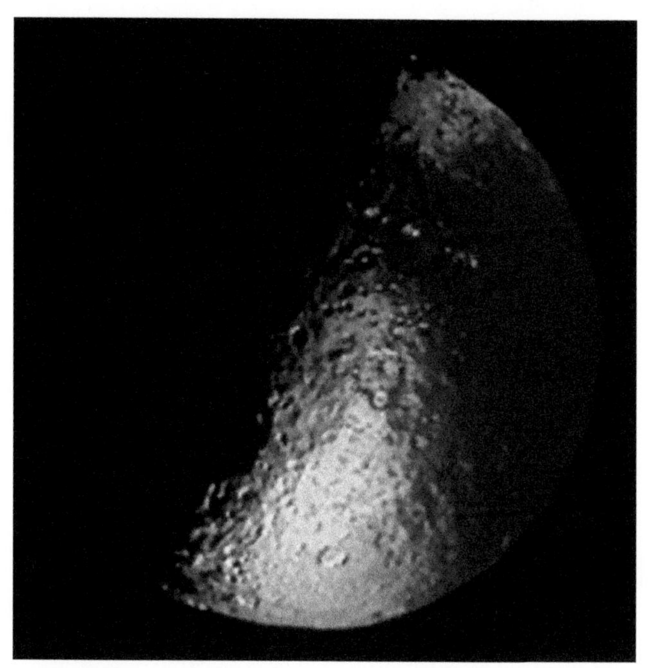

Figure 6.8 Image of Iapetus, a moon of Saturn (equatorial diameter = 1436 km), taken at 'half-moon' phase from the Cassini spacecraft on October, 2004, with a resolution of 7 km. Iapetus is unique in the Solar System because of the strong contrast in albedo between its two hemispheres. This contrast can be seen here as dark and bright sections on the illuminated side of the moon. (The narrow white streak in the dark region is a chain of icy mountains.) The dark section is as dark as coal ($A = 0.03 \rightarrow 0.05$, almost a black body) while the bright section is more typical of Saturn's other icy moons ($A = 0.5 \rightarrow 0.6$). The dark material is a thin (approx. 10 cm) layer on top of the icy surface. Although the reasons for this dichotomy aren't entirely clear, it could be a runaway effect from some initially dark material that came from an external source. The dark material will be warmer, causing the volatile icy (light) component within it to sublimate out, thereby keeping the dark component dark. The slow ($\approx$ 80 days) rotation of Iapetus helps to ensure that these effects are not smeared out over time (see [20]). Source: Reproduced courtesy of NASA/JPL/Space Science Institute.

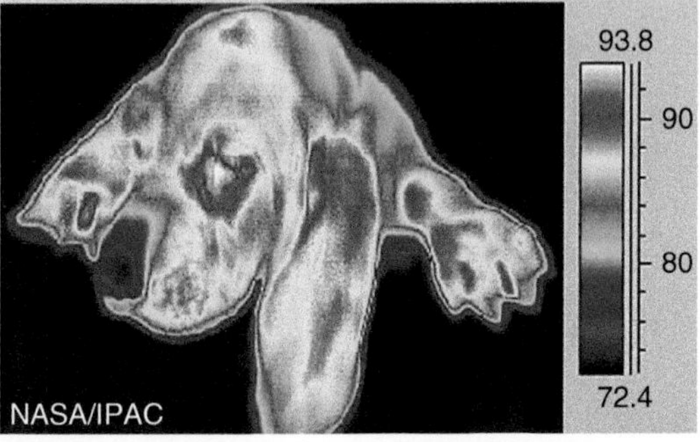

Figure 6.9 This hound dog is a 'grey body'. A fraction of the light that falls on the dog is absorbed and the remaining fraction is reflected. It is the reflected light in the optical part of the spectrum that allows us to see the dog normally as shown on the left. The absorbed light and the dog's internal heat generation must be balanced by its emission to maintain a constant temperature at any given position. The emitted spectrum will be a Planck curve which peaks in the IR for the range of surface temperatures shown in the false colour image on the right. The scale is in Fahrenheit degrees. Source: Reproduced by permission of SIRTF/IPAC.

6.2.1 The Equilibrium Temperature of a Grey Body

For a grey body in the Solar System, in most cases, the dominant heating source is the Sun. If a grey body is to maintain an *equilibrium temperature* that remains constant with time, then the rate at which the grey body radiates must balance the rate at which it absorbs energy. Because a fraction of the incident energy is reflected away, the temperature of a grey body will not come into equilibrium with the temperature of the incident radiation field but rather with a temperature that corresponds to the absorbed fraction of the incident radiation. The grey body then emits a Planck spectrum at its equilibrium temperature. Example 6.4 shows how the equilibrium temperature of an externally heated grey body can be determined. Grey bodies in our Solar System have Planck curves that peak in the infrared (Problem 6.6). Dust that is detected in our Milky Way, such as in the Lagoon Nebula shown in Figure 5.13, as well as the dust that is observed in other galaxies such as in Arp 220 (Figure 6.10) also tend to be at temperatures resulting in λ_{max} in the IR or sub-mm.

There are two cases, however, in which a grey body may have a temperature that departs from the equilibrium temperature expected from absorbed external radiation.

The *first* is if the object has its own internal source of heat. There are a variety of possible internal heat sources. For living, warm-blooded creatures, for example, chemical reactions turn food into an internal heat source. For the outer gas giant planets, Jupiter, Saturn, and Neptune, there are significant internal heat sources[3], possibly remnant heat from their formation (Jupiter) or heat released by the precipitation and slow sinking of

[3] These outer planets radiate significantly more energy (of order a factor of 2) than they are absorbing from the Sun. By contrast, the temperature of the planet, Uranus can be explained solely by external Solar heating.

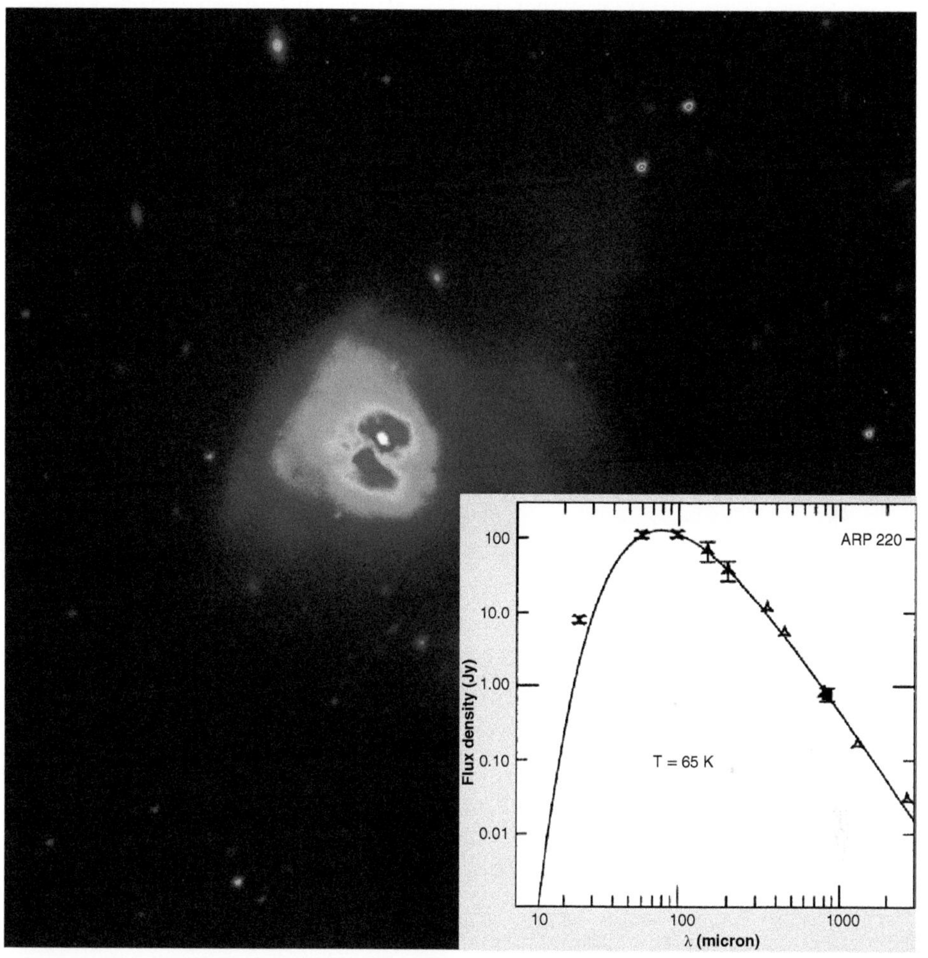

Figure 6.10 Image of the ultra-luminous infrared galaxy, Arp 220. This galaxy is undergoing an intense 'starburst', meaning that it is rapidly forming stars, in this case, at a rate of 100 $M_{\odot}$ $year^{-1}$. Thus, there are many hot young stars in this galaxy whose UV photons heat the abundant dust that permeates the system. (An active nucleus is also present.) Most of the galaxy's luminosity is then re-radiated in the IR by dust grains, giving the very high value of $L_{8\rightarrow 1000\,\mu m} = 1.6 \times 10^{12}\,L_{\odot}$. Arp 220 is believed to be the remnant of two galaxies that have merged. The inset shows a Planck spectrum that has been fitted to the dust emission from the galaxy. The spectrum is adjusted slightly from a true Planck spectrum to take into account grain properties. Source: Image reproduced by permission of David Sanders. Inset adapted from Lisenfeld et al. [15].

helium droplets towards the core (Saturn). Heavier particles that sink down in a planet turn gravitational potential energy into heat[4]. Internal heat can also be generated via tidal stresses, such as those that occur within Jupiter's close moon, Io, due to the gravitational forces exerted on it by Jupiter. The *Earth's* internal heat flow amounts to 47 ± 2 TW [14], corresponding to $S_{int} = 0.094 \pm 0.004$ W m^{-2}. An important fraction of this is due to radioactive decay with some residual heat from the time of the Earth's formation, but the exact make-up of the various contributions are not entirely known. Even though the Earth's internal heat sources may be high, they are still insignificant in comparison with heating by the Sun. Compare, for example, S_{int} with the absorbed fraction (70%) of the incident Solar flux of $S_{\odot} = 1367$ W m^{-2} (i.e. the Solar constant, Section 3.2, Figure 3.6).

The *second* is if the heat from the planet that would normally be radiated away freely is somehow trapped. This is what happens from the *Greenhouse Effect.* Planets such as Earth and Venus are shrouded in atmospheres that act like a blanket, retaining heat within it. This occurs because most incident Solar radiation is at optical wavelengths (Figure 6.5) whereas the emitted radiation from the surface is in the infrared. The atmosphere is not as transparent in the infrared as in the optical, as the atmospheric transparency curve for the earth (Figure 4.2) shows. A photon emitted from the surface may be scattered in the atmosphere or re-absorbed and re-emitted, possibly to be absorbed by the surface again. Thus, the atmosphere has the effect of retaining photons for a longer period of time, contributing an extra source of heating. The Greenhouse Effect provides an important and necessary heat source to the Earth, contributing +33 °C to its surface temperature, as Example 6.4 outlines. 'Global warming', which is currently of great concern on our planet today, refers to much smaller changes, of order a degree Celsius, that the Earth has been experiencing over the last century, as Figure 6.12 illustrates.

Example 6.4

Estimate the equilibrium temperature of the Earth. Compare this to the Earth's globally averaged mean temperature of +14 °C (287 K).

The Earth absorbs radiation over only the cross-sectional area, $\pi R_{\oplus}^2$, that faces the Sun (see Figure 6.11 and Example 3.1), where $R_{\oplus}$ is the radius of the Earth. The rate at which energy is absorbed is therefore $L_{abs} = (1 - A)\,\pi R_{\oplus}^2 f$, where f is the flux of the Sun at the distance of the Earth (i.e. the Solar Constant) and A is the Earth's albedo. Expressing the Solar flux in terms of luminosity, $L_{\odot} = 4\pi r^2 f$ (Eq. 3.9), where r is the Earth–Sun distance, the above equation becomes,

$$L_{abs} = \frac{(1 - A) R_{\oplus}^2 L_{\odot}}{4r^2} \qquad (6.17)$$

The Earth can be approximated as a 'fast rotator', which means that the energy absorbed on the daytime side is quickly equalized over the night-time surface. This is a reasonable approximation because day/night variations in surface temperature (of order 10 K) are much smaller than the temperature itself (287 K). The Earth, being a

[4] The mathematical relation between the gravitational potential energy and thermal energy (heat) for a gravitationally bound object is called the *Virial Theorem.*

grey body, then emits a Planck spectrum at a (globally averaged) temperature, T. The rate of energy emission is given by the Stefan–Boltzmann law (Eq. (6.13))[5],

$$L_{em} = 4\pi R_{\oplus}{}^{2}\sigma T^{4} \tag{6.18}$$

For an equilibrium temperature, we require the absorption rate to equal the emission rate, so equating Eq. (6.17) to (6.18) and solving for temperature, we find

$$T = \left[\frac{(1-A)L_{\odot}}{16\pi\sigma r^{2}}\right]^{1/4} \tag{6.19}$$

Using $A = 0.306$ (Table T.3), $L_{\odot} = 3.845 \times 10^{33}$ erg s^{-1}, $r = 1.496 \times 10^{13}$ cm (Table T.2), and $\sigma = 5.67 \times 10^{-5}$ erg cm^{-2} K^{-4} (Table T.1), we find T = 254 K = −19 °C. By Wien's Displacement law (Eq. (6.8)), the peak of the corresponding Planck curve would be in the infrared ($\lambda_{max} = 11\ \mu$m). Note that Eq. (6.19) is independent of the size of the Earth, $R_{\oplus}$. Thus, provided the above assumptions are valid, a planet of *any size* placed at the distance of the Earth would have the same equilibrium temperature if it has the same albedo.

The resulting equilibrium temperature is considerably colder than the Earth's measured globally averaged temperature of $\overline{T} = 287$ K = +14°C [8] (averaged over the years, 1951–1980). The difference can be accounted for by the Greenhouse Effect. It is clear that, without the warming blanket of our atmosphere, the Earth would be a very cold, inhospitable place to live.

Equation (6.19) in Example 6.4 can be used for any grey body in the Solar system whose illumination and radiation geometry are as shown in Figure 6.11. Changes in the geometry of a grey body will not make large changes in the results, however. For example, a hypothetical asteroid shaped like a cube, rather than a sphere, would have an equilibrium

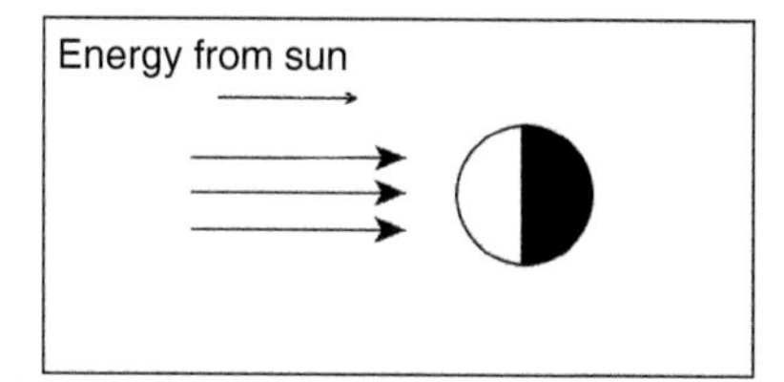

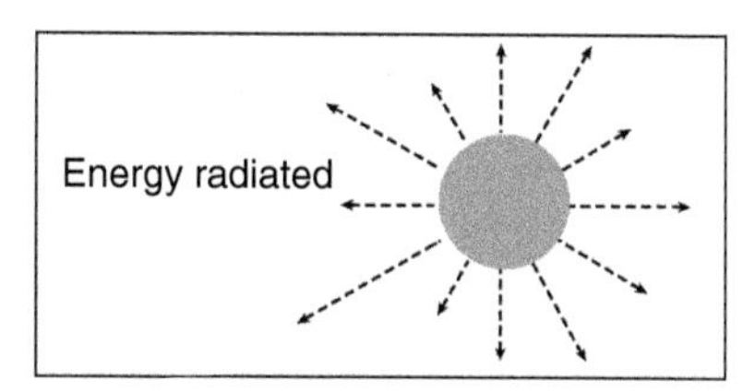

Figure 6.11 A planet receiving light from the Sun intercepts it over its daylight half (left). Being a grey body, a fraction, A (the albedo), of this light is reflected away and a fraction, $1-A$, is absorbed. Assuming that the absorbed energy is equalized quickly over the planet's surface, the planet then radiates the energy over its entire surface (right) at a rate that balances the energy input rate, thus maintaining an equilibrium temperature.

[5] A multiplicative factor is actually required on the right hand side of this equation to account for the emissive properties of the surface material, but the correction is small for the purposes of this example: approximately 0.9 for solid surfaces and 0.65 for gaseous surfaces [9], leading to corrections to T (Eqn. (6.19)) of factors of 0.97 and 0.90, respectively.

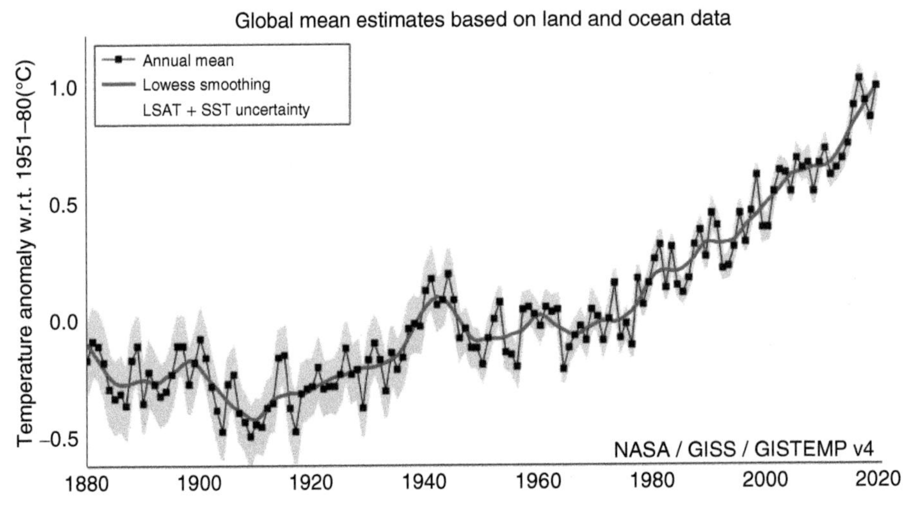

Figure 6.12 Globally averaged temperature anomaly of the Earth from 1880 to 2020, with respect to a zero point of 14 °C that is an average over the period, 1951 to 1980 [8]. The increase in temperature correlates with an increase in carbon dioxide concentration in the atmosphere. GISTEMP Team, 2020: GISS Surface Temperature Analysis (GISTEMP), version 4. Sources: Hansen et al. [8]; Lenssen et al. [13] and NASA Goddard Institute for Space Studies. Accessed 26 March 2020 at https://data.giss.nasa.gov/gistemp, see also [13].

temperature that differs from that given by Eq. (6.19) by a factor of only 1.1. A 'slow rotator' – an object that cannot equalize its temperature between its day/night sides, but rather is hot on one side and cold on the other – will have a hot side temperature that is higher than that of Eq. (6.19) by a factor of 1.2. Therefore, Eq. (6.19) provides a good estimate of the temperature of any grey body in the Solar System whose dominant heating source is the Sun.

What about grey bodies outside of the Solar System? The same kind of arguments apply but now $L_{\odot}$ would have to be replaced with the relevant luminosity, L, of the source that is supplying the incident radiation. Most observations of grey bodies outside our Solar System are of dust which is widely distributed throughout the ISM. With many stars as sources of illumination, Eq. (6.19) would have to be modified to account for that geometry. The smallest dust grains, however, never achieve an equilibrium temperature. These *very small grains* (VSGs, less than 0.01 μm) experience spikes in temperature due to the absorption of individual photons, followed by rapid cooling. Only the larger dust grains, called *classical grains*, reach an equilibrium temperature similar to that described in Example 6.4. Both classical grains and planets are grey bodies so, in a sense, a planet might be thought of as just a very large, compact dust grain! We can easily see the IR emission from dust in the ISM of our Milky Way and in other galaxies as well. But what about planets? In the next section, we will consider direct observations of the emitted black body portion of light from *extrasolar planets*, or *exoplanets*, as they are now more commonly called.

6.2.2 Exoplanets and Their Detection

The rate of discovery of exoplanets in recent years has been quite breathtaking (Figure 6.13). Exoplanet research is now a sophisticated, maturing discipline that is also of keen popular interest. Yet these new discoveries are just the beginning, as ground-based telescopes achieve their diffraction-limited resolutions (Section 4.3.3) and space-based telescopes become larger and more sophisticated (e.g. the James Webb Space Telescope). This subfield of astronomy has moved far beyond simple detections; it is now essentially conducting an exoplanet census. What types of stars have planets, what are the size and mass distributions of these planets, which ones have atmospheres, and could there be any indicators of life?

The earliest searches, which gave tantalizing results [1], were from *radial velocity* measurements of the parent star. (This is accomplished by measuring Doppler shifts, which will be discussed in Section 9.1.1.) Rather than observing the planet directly, the star's motion was seen to be perturbed because of the presence of an unseen companion as both components orbit the centre of mass (Section 9.1.1.2). This approach continues to be a fruitful avenue for discovery (*red* in Figure 6.13), but the planet must be massive enough and close enough to produce a noticeable perturbation of the star.

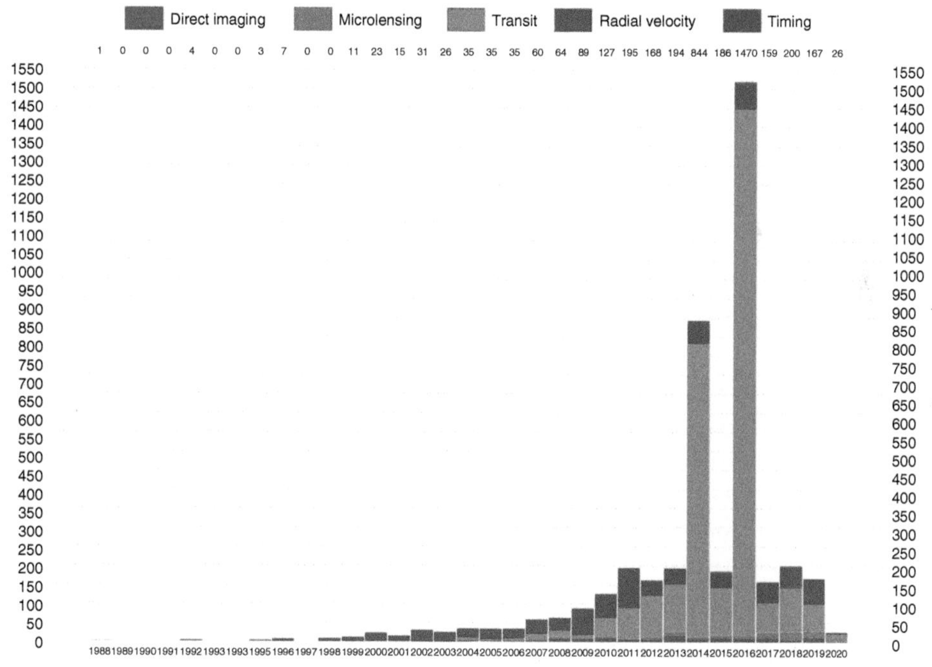

Figure 6.13 Histogram of the number of exoplanets discovered between 1988 and the beginning of 2020. Transits, by far, outnumber the other methods (green). A small number of planets have been found by direct imaging (blue). See text for more information. Source: File created with matplotlib by User:Betseg Retrieved from: https://commons.wikimedia.org/wiki/File:Confirmed_exoplanets_by_methods_EPE.svg licensed under CC BY 4.0.

The first compelling evidence for an exoplanet was in a completely unexpected place. The pulsar, PSR1257 + 12, which should normally display very regular pulses, showed a 'wandering' in the *timing* of its pulses that could only be explained by the effect of two planetary companions [21]. Like the radial velocity method above, the companions are not directly seen, but their dynamical effect on the pulsar causes the pulsar to move periodically in orbit about the common centre of mass. As the pulsar approaches the observer in its orbit, the pulses are closer together in time, and as the pulsar recedes with respect to the observer, the pulses are farther apart in time[6]. Pulsar timing has revealed few planets so far (*violet* in Figure 6.13) since pulsars are rare compared to normal stars and it is a challenge to understand how a planet could have survived the supernova explosion of the parent star en route to becoming a pulsar (Section 5.3.3). Conditions on such exoplanets would be unlike anything that we are familiar with in our Solar System.

Gravitational microlensing (orange in Figure 6.13) will be further explored in Section 9.2.2. This type of detection is relatively rare and requires long-term monitoring of many stars. Again, we do not see the light from the planet itself with such measurements, but only the effect that it has on the light from background stars.

How, then, can we *directly* detect the *emission* from an exoplanet? There are two challenges to overcome. The first is that an entire planetary system is often unresolved, so the flux density of the planet(s) and star are received together in a single beam or pixel. For example, if a relatively close star, say at a distance of 5 pc, had a planet orbiting it at the same distance as Jupiter is to the Sun, then the planet and the star would be separated by only 1 arcsecond. The second challenge is that the star is much hotter than the planet, so starlight swamps the emission from the planet itself. As we pointed out in Section 6.1, a Planck curve at a higher temperature is brighter at *all* wavelengths (Figure 6.1). The perturbation (increase) in flux due to the planet is very small, as Example 6.5 shows. Fortunately, there are two ways in which the emission from a planet *can* be detected. The first is when the star/planet system is unresolved but we measure a small but regular *change* in the flux density; this is what occurs in *exoplanet transits*. The second is if the planet and star can actually be *resolved*. We consider each of these next.

When a planet transits in front of its parent star, it causes the starlight to dim. Since the orbital motion is regular, such dimming can be followed over many orbits, building up signal and helping to offset the dynamic range problem outlined in Example 6.5. Exoplanet transits are essentially *eclipses* so detection requires that the geometry of the orbit be favourable from the observer's point of view. Transits are more likely, then, when the planet is *large* and when it is *close* to the star.

Fortunately, a class of planet has been found, called *hot Jupiters*, that has both of these characteristics. They are as massive as Jupiter, but orbit their stars at a very small radius, of order 0.05 AU, with orbital periods of only a few days. Thus, they are advantageous for a number of reasons. In addition to the higher likelihood of transit, being close to the star means that a hot Jupiter is relatively bright, so a change in flux density over the orbit is more obvious. And because the orbital period is short, it does not take very long to build up a believable signal.

[6] Again, it is *Doppler shifts* which give this effect (Sect. 9.1.1).

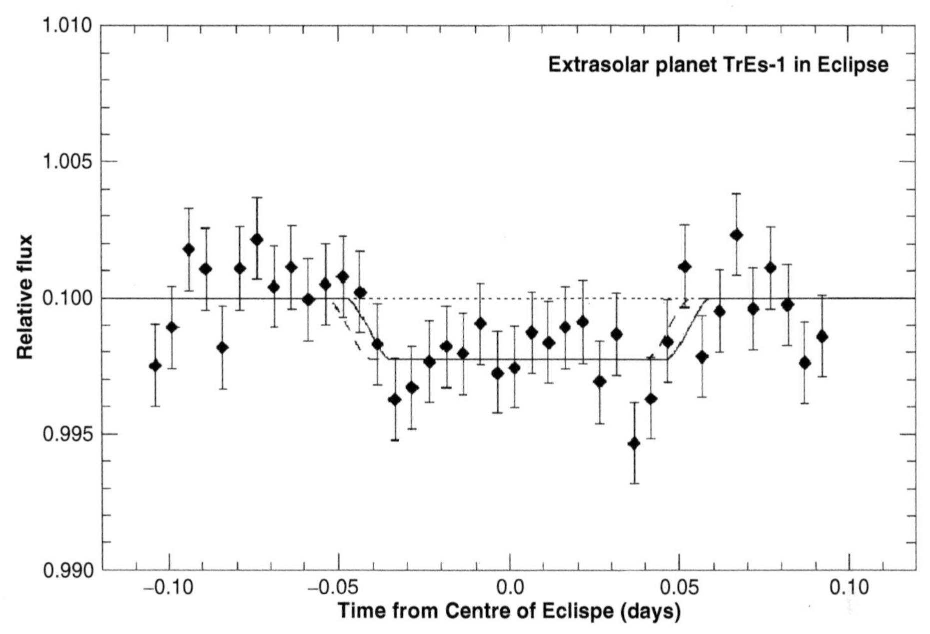

Figure 6.14 Light curve from the extrasolar planet system, TrES-1. At the centre of the plot, the flux is lower because the planet is behind the star. At the right and left, the flux includes both the planet and star. These observations were taken at $\lambda 8\ \mu$m with NASA's Spitzer orbiting telescope [2]. They constitute one of the first two observations, published in 2005, in which photons from a planet have been directly and clearly detected (see also [4]). The planet in this system is a 'hot Jupiter'. The star is of type, K0V, and the planet is a distance, $r = 0.0394$ AU, from it with an albedo of $A = 0.31$ and mass, $M_p = 0.76\ M_J$, where M_J is the mass of Jupiter. Sources: Deming et al. [4] and Charbonneau et al. [2]. © D. Charbonneau.

The first two detections of hot Jupiters in eclipse were made in the year, 2005. The *light curve* (a plot of varying emission with time) of one of these is shown in Figure 6.14. When the star is out of eclipse, the emission is the *sum* of the starlight and planetary light. Thus, the thermal Planck emission of the planet is being directly detected, but in combination with the Planck emission from the star. In the particular case shown in Figure 6.14, it is the *star* that is eclipsing the *planet*!

Detections of exoplanet transits (*green* in Figure 6.13) have gone far beyond hot Jupiters. With this technique, we see an explosion of discoveries, especially since space-based searches have taken place, e.g. 'Convection, Rotation et Transits planétaires' (CoRoT, 2006–2013), Kepler (2009–2018), and the Transiting Exoplanet Survey Satellite (TESS, 2018–). Even Earth-sized planets in a *habitable zone*[7] have been found (e.g. [6], Problem 6.14).

[7] The habitable zone has been defined as the circumstellar region within which liquid water can exist on the surface of an Earth-mass planet which has an atmosphere that contains CO_2, H_2o and N_2, although other modified definitions exist [12].

Example 6.5

A Sun-like star at a distance of d = 5 pc has an Earth-like planet orbiting 1 AU from it. Determine the excess flux density due to the planet at the wavelength of the peak of the planet's Planck curve. Assume that the planet has no atmosphere and that the system is unresolved.

A Sun-like star has a temperature of 5781 K (Table T.2) and, from Eq. (6.19), we know that an Earth-like planet without an atmosphere (no greenhouse effect) will have a temperature of 254 K. By Wien's Displacement law for the planet (Eq. (6.8)), $\lambda_{max} = 1.14 \times 10^{-3}$ cm (11.4 μm). Using Eq. (6.2) at this wavelength, the *star's* specific intensity is, $B_s(5781) = 2.53 \times 10^{10}$ erg s^{-1} cm^{-2} cm^{-1} sr^{-1} and the *planet's* specific intensity is $B_p(254) = 4.32 \times 10^7$ erg s^{-1} cm^{-2} cm^{-1} sr^{-1}.

We can only measure the flux density (not the specific intensity) of an unresolved object (Section 3.3). The flux density of each is $f_\lambda = B_\lambda(T)\Omega$ (Eq. 3.13). The solid angle, $\Omega = \pi(R/d)^2$ (Eqs. I.9, and 4.6), where R is the radius of the object and d is its distance. Using the radius of the Sun (Table T.2) and the Earth (Table T.3), the star's solid angle is $\Omega_s = 6.39 \times 10^{-17}$ sr and the planet's is $\Omega_p = 5.37 \times 10^{-21}$ sr. Then, the flux density of the star and planet are, respectively, $f_s = 1.62 \times 10^{-6}$ erg s^{-1} cm^{-2} cm^{-1} and $f_p = 2.32 \times 10^{-13}$ erg s^{-1} cm^{-2} cm^{-1}.

The total flux density detected is $f_s + f_p \approx f_s$. The planet would therefore make a very tiny perturbation that is only 1.4×10^{-7} times the value of the star's flux density at the peak wavelength (λ_{max}) of the planet. In order to detect such a tiny perturbation of one Planck curve superimposed on another, the dynamic range (see Section 4.5) of the observations would have to be better than 7×10^6 and measurements at a variety of wavelengths are really required to be sure that the spectrum is the sum of two Planck curves.

The second class involves direct imaging (*blue* in Figure 6.13). For planets that are sufficiently far from their parent star, it is now possible to resolve the star/planet system and *image* the planet directly. The planet typically appears as a weak 'pin-prick' of light that is sufficiently separate from the stellar image to be seen as distinct. Such imaging is now greatly helped with the application of adaptive optics (Section 4.3.3). AO was used to find the two earliest examples of direct imaging made in 2004 and 2005 ([3] and [18]). The planetary candidates were each a few Jupiter masses at distances of 55 and 100 AU from their parent stars, respectively.

Because a star is so much brighter than a planet, planet detection in a resolved system is greatly aided by blocking out the light of the parent star. An optical instrument that can do this is called a *coronagraph*. Clearly, a combination of adaptive optics and stellar-light blocking is a better approach. An example is the SPHERE (Spectro-Polarimetric High-contrast Exoplanet Research) instrument on the VLT (Figure 4.17 Right). Figure 6.15 shows an example of direct imaging using this instrument. Here, we see a *protoplanet*, a newly forming planet within a dusty disc around the star, PDS 70. Consecutive images even show the orbital motion of this protoplanet around its parent star [11].

Modern methods include other creative ways to distinguish stars from planets in a kind of no-holds-barred approach to planet-discovery. Even detection of the *reflected*

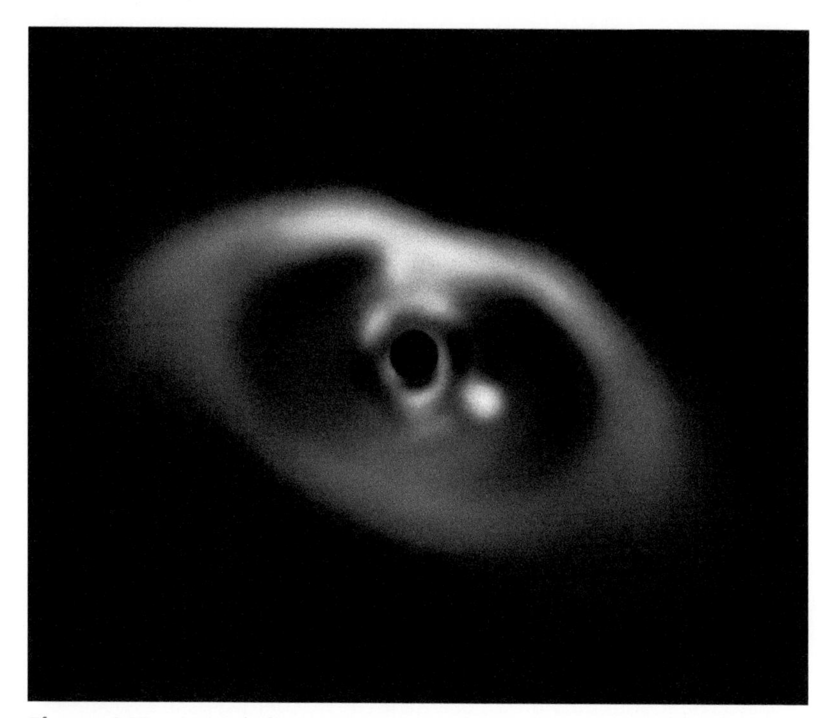

Figure 6.15 A newly forming planet can be seen embedded in the dusty disc around the star, PDS 70. The light from the star at the centre has been blocked out by a coronagraph and the planet is the bright point-like source to the lower right of centre. It is separated from the star by 195 mas corresponding to 22 AU. Adaptive optics have been employed to help resolve the star/planet system. The detector was the SPHERE/IRIDS instrument at the VLT, and the image is in K-band. Source: From Keppler et al. [11]. Credit: ESO and A. Müller et al.

light from planets is now in sight [10]. Because reflected light has polarized properties that are different from emitted light, a separation between star and planet can be aided by examining the polarization characteristics. Spectroscopy, a detailed examination of the spectrum of the star or planet, can provide further differentiation. And radio emission from planets (cf. Figure 10.13 and Section 10.5.1) could also be possible in the future [16]. With new and innovative probes, including *both* the reflected and emitted portions of these grey bodies, exoplanet discovery will move from census-taking to the characterization of new worlds.

PROBLEMS

6.1. (a) Verify that Eq. (6.1) is reproduced when Eq. (6.2) is substituted into Eq. (6.3).

(b) Using Wien's Displacement law, evaluate λ_{max} and ν_{max} for a temperature of 6000 K and show that $\lambda_{max} \neq c/\nu_{max}$.

(c) Evaluate $B_\lambda(6000)$ at $\lambda = \lambda_{max}$.

(d) Evaluate $B_\nu(6000)$ at the frequency corresponding to λ_{max}. Convert the result to the same units as $B_\lambda(6000)$ and verify that the result is the same as in part (c).

6.2. (a) Write the Rayleigh–Jeans law and Wien's law in their wavelength-dependent forms. Will the criteria specified in Example 6.2 for departures from the full Planck curve be the same for the wavelength-dependent functions as for the frequency-dependent functions?

(b) Assuming that the emission can be well-approximated by a Planck curve, determine whether the Rayleigh–Jeans law or Wien's law may be used for the following conditions:

i. an observation of the Sun at λ912 Å;

ii. an observation of a dense molecular cloud of temperature, T = 20 K at $\nu = 230$ GHz;

iii. an observation of the 2.7 K CMB at λ21 cm.

6.3. (a) Determine and plot on the same graph, B_λ(T) for the eye region (white) and the tip of the nose of the hound dog shown in Figure 6.9 (right), using the scale in the figure to estimate the temperature.

(b) Determine λ_{max} (μm) for the same eye and nose.

(c) Assuming that photons leaving both the eye and nose can escape from the surface in all directions outwards, determine the flux density at the surface of the eye and nose.

(d) Is the energy loss rate greater through the eye or through the nose of the hound dog?

6.4. Repeat Problem 6.3 but, instead of the eye and ear of a hound dog, compare a O8 V star to a K7 V star (Table T.5). For part (b), indicate in which part of the spectrum λ_{max} occurs for each.

6.5. In the caption to Figure 6.1, we note that a hot object is brighter than a cool object at *all* frequencies when both emit a Planck spectrum. Yet, observations of the ISM in the infrared show mostly cool dust emission while the hotter stellar emission is negligible in comparison. How can these two statements be reconciled?

To answer this question, consider a cubic region of volume, 1 pc^3 in the Galaxy some distance, *D*, away so that any object in the volume can be considered at the same distance. Assume that there is, at most, one star within this volume and that this star is equivalent to the Sun (same radius, *R*, and effective temperature, T_{eff}). Assume that all dust grains are spherical, evenly distributed, and that no grain 'shadows' (is directly in front of) any other grain. For the dust, let the temperature, T = 100 K, radius, $a = 0.2\ \mu$m, density, $\rho = 2.5$ g cm^{-3}, and the total dust mass in the volume is $2.65 \times 10^{-4}\ M_\odot$. Find the ratio of the stellar flux to the dust flux at IR (100 μm) and visible (550 nm) wavelengths, assuming both emit perfect Planck spectra. Comment on the results.

6.6. Compute the temperature range over which an object would have a Planck curve peak, λ_{max}, in the infrared. Comment on how many Solar System objects you expect would emit in the infrared.

6.7. Assume, for simplicity, that a region of space contains one O8V star, one A0Ib star, 100 G5V stars, and 1000 M2V stars. Refer to Tables T.5, T.6, and T.7 and,

(a) determine the total stellar mass (units of $M_\odot$) in the region and indicate which type of star dominates the mass;

(b) determine the total stellar luminosity ($L_\odot$) in the region and indicate which type of star dominates the luminosity.

6.8. (a) Write an expression for the ratio of the flux of a star to the flux of the Sun, $f_*/f_\odot$, in terms of the radii of these objects and their effective temperatures. Assume that the two are located at the same distance and radiate as perfect black bodies.

(b) Using the results of (a), find the flux ratio for the neutron star, J0538 + 2817, compared to the Sun. The neutron star's radius is $R = 1.68$ km, and temperature is $T = 2.12 \times 10^6$ K. Determine λ_{max} for this neutron star and indicate in what part of the spectrum this occurs.

(c) Presently, thermal black body radiation has been measured from the surfaces of only a 'handful' of neutron stars (e.g. [19])[8]. Based on the results of part (b), explain why this might be the case.

6.9. Compute the equilibrium temperature of Mars. Compare it to the average measured temperature[9] of $-63\ ^\circ$C and comment on whether Mars might have a significant internal heat source or experiences a Greenhouse Effect.

6.10. (a) Rewrite Eq. (6.19) for a grey body that is a *slow rotator*, i.e. determine T for the Sun-facing side of a spherical body when the temperature is negligible on the hemisphere opposite to the Sun.

(b) On the image of Iapetus (see Figure 6.8), the bright and dark sections are sometimes separated by no more than a single pixel of width, 7 km. If an astronaut were to take a stroll over such a distance at the bright/dark boundary (both sides of the boundary facing the Sun), by how much would the surface temperature change? Which is the warm side? Assume that Iapetus is a slow rotator and that its heat is supplied only by the Sun.

6.11. From the information given in Figure 6.14 for the extrasolar planet system, TrES-1, as well as Table T.5, find the following:

(a) the temperature, T_s, radius, R_s, and luminosity, L_s, of the star;

(b) the temperature, T_p, of the planet, assuming that it is a fast rotator;

(c) the ratio of flux densities of the planet to the star, $f_{\lambda p}/f_{\lambda s}$ at the observing wavelength of $\lambda 8\ \mu$m;

(d) the ratio of specific intensities, $B_\lambda\ (T_p) = B_\lambda\ (T_s)$ at $\lambda 8\ \mu$m;

(e) the ratio of solid angles, Ω_p/Ω_s;

(f) the radius of the planet, R_p. Express this result in terms of Jupiter radii, R_J.

---

[8] Neutron stars are usually detected as pulsars and it is the stronger *non-thermal radiation* (see Section 10.1), rather than their black body radiation, that is usually detected.

[9] The temperature on Mars varies significantly. See https://mars.nasa.gov/insight/weather for a daily weather report at Elysium Planitia.

6.12. (a) Derive an expression for the *reflected* flux, F_{refl}, of an extrasolar planet that intercepts light as shown in Figure 6.11 Left, and reflects it outwards over only the one hemisphere facing the star. The result should be expressed in terms of the albedo, A, the luminosity of the star, L_*, and the distance of the planet to the star, r.

(b) For the *emitted* portion of the light, assuming a fast rotator model, express Eq. (6.19) in terms of the emitted flux at the planet's surface, F_{em}, instead of its temperature.

(c) Show that the fraction of stellar flux that is reflected compared to that emitted from a grey body, for a fast rotator is

$$\frac{F_{refl}}{F_{em}} = \frac{2A}{(1-A)} \tag{6.20}$$

(d) For what value of albedo will the reflected flux equal the emitted thermal flux?

(e) Suppose the planet were a slow rotator, as in Problem 6.10. Find F_{refl}/F_{em} for this case.

6.13. Do a rough hand-sketch of the spectrum of the asteroid, Ceres, ranging from the infrared through to the optical for a geometry in which all of the reflected light can be directly seen. Ceres' rotation period is nine hours and its distance from the Sun is 2.8 AU. $A = 0.034$.

6.14. The exoplanet, TOI-700 *d*, has been found to orbit the M2 V star, TOI-700, with a semi-major axis of $a = 0.163$ AU. The effective temperature and radius of the star are $T_{eff} = 3480$ K and $R_* = 0.420\ R_\odot$, respectively. The planet has a radius of $1.19\ R_\oplus$.

(a) Rewrite Eq. (6.19) in terms of the stellar radius and temperature, rather than its luminosity.

(b) What is the equilibrium temperature of the planet? Assume that $A = 0.3$, there is no Greenhouse Effect, and it is a fast rotator. How does this result compare to the Earth?

(c) At what wavelength does the stellar spectrum peak? In which part of the spectrum does it fall? Repeat this for the planet.

(d) What is the astrophysical flux of the star? What is the flux of the star at the distance of the planet and how does this compare to the Solar Constant?

(e) At λ_{max} of the *planet*, what is $B_\lambda(T)$ for both the planet and star? How much brighter is the star than the planet?

(f) What is the ratio of Ω_p/Ω_*, where Ω_p is the solid angle subtended by the planet and Ω_* is the solid angle subtended by the star?

(g) The planet/star system is unresolved. If the system is observed at a wavelength of λ_{max} of the planet, what is the ratio of the flux density of the planet over the flux density of the star? What dynamic range is required to measure a transit?

JUST FOR FUN

6.15. Suppose you were standing on the exoplanet of Problem 6.14 whose mass is $M_p = 2.19 \times 10^{27}$ g and whose orbital period is 37.426 days. How much would you weigh (lbs) on this planet? How much larger would the alien sun appear in the sky compared to our Sun? From night to night, over what angle (degrees) would the constellations shift? (Assume that an alien day equals an Earth day.) How old are you in alien years?

6.16. A Borg Cube, which is slowly losing power, is near an F0V star. A recently assimilated alien race has given them knowledge of a heat-induced power regeneration sequence. But it requires that they increase the hull temperature to 1000 °C. They move the Cube, which can be approximated as a black body, to a distance of 0.15 AU from the star. Is this close enough for power regeneration?

PART IV

The EM Signal Perturbed

An astronomical signal does not simply emerge from an object and travel through a vacuum unperturbed. Rather, it may travel partially through the object that emitted it or through interstellar clouds that could alter or extinguish it. It travels through space which is itself expanding, and could pass near massive objects that bend its path. Thus, we cannot hope to understand an astronomical object unless we first understand how its signal may have become corrupted en route.

There is another way of looking at this, however. Every time the signal is changed, it may be possible to learn something about the object or medium that produces the change. Thus, an alteration in the signal is not always an unwanted nuisance, as was the case for telescopes and (for astronomers at least) the atmosphere. Instead, the astronomical signal itself might be a useful background source that probes the space and matter through which it travels, like a flashlight illuminating the intervening material for us. If the intervening matter neither emitted its own radiation nor had any affect on a background signal (or nearby objects), it would be impossible to learn anything about it.

Thus, we now focus more intently on astrophysical processes that can alter a signal and what insights they might provide either about the signal itself or the altering medium. We will first consider interactions in which the signal has contact with matter (the interaction of light with matter) and then those interactions that are 'non-touching' (the interaction of light with space).

Chapter 7
The Interaction of Light with Matter

> You see the motes all dancing, as the sun Streams through the shutters into a dark room. ... From this you can deduce that on a scale Oh, infinitely smaller, beyond your sight, Similar turbulence whirls.
>
> – the Roman poet, Lucretius, first Century B.C. [12]

The interaction of light with matter spans the boundaries of classical and quantum physics and of microscopic and macroscopic physics. One approach is to consider how an individual photon or a wavefront might interact with an individual particle. Another is to consider many particles together as representing a smooth medium that collectively alters the intensity and/or the trajectory of an incoming wavefront. Both approaches (and their variants) are valid and, if both exist, the resulting mathematical description of the physical phenomenon must agree. In this chapter, we will consider the interaction of light with matter in some detail, visiting each method along the way. There are other

Astrophysics: Decoding the Cosmos, Second Edition. Judith A. Irwin.

Companion website: www.wiley.com/go/irwin/astrophysics2e

examples of scattering – for example, electrons can scatter off of magnetic field lines, a process that we will consider in Chapter 10. In this chapter, however, we only consider light interacting with *matter.*

There are many types of matter that a photon could encounter in its flight across the cosmos – hot ionized hydrogen gas, a cold cloud containing complex molecules, a dust grain or even a planet. Such interactions with matter may produce a change in the photon's path and/or its energy. Although there are many types of matter, interactions with individual particles fall into two categories, *scattering and absorption.* In the case of absorption, a photon will disappear completely, its energy going into some other form such as the thermal energy of the gas. For scattering, on the other hand, a photon emerges from the interaction, though not necessarily at the same energy. The combined effects of scattering and absorption, both of which remove photons from the line of sight, are referred to as *opacity* (usually used for gases) or *extinction* (for dust).

The importance of an interaction with a particle depends on the cross-section presented to the photon by the particle (e.g. Example 5.5) as well as the number density of particles that the incoming light beam encounters as it travels through a cloud. Therefore, in the next few sections, we focus on the *scattering cross-section*, σ_s, and the *absorption cross-section*, σ_a (both in units of cm^2) for the interaction of interest. The dependence on particle density is considered later in Section 7.4.1. The special case of dust grains is presented in Section 7.5.

7.1 THE PHOTON REDIRECTED – SCATTERING

Scattering is a process by which the direction of an incoming photon changes as a result of an interaction. As we will see, the interpretation of an interaction as 'scattering' rather than 'absorption' is sometimes unclear. This is because, at an atomic or subatomic level, scattering involves the absorption and then re-emission of a photon. However, scattering implies that the re-emission occurs 'immediately', meaning on a short enough timescale that no other processes (e.g. collisions between particles) could interfere and remove the energy supplied by the incident photon. Scattering can be represented by the process: *photon → matter → photon.*

With a sufficient number of interactions, scattering tends to randomize the direction of the photons. Thus, when an observer looks back along a line of sight, the signal is attenuated (weaker) because many photons will be scattered into directions other than the line of sight, as shown in Figure 7.1a. However, it is also possible for scattering to be preferentially into a specific direction (symmetrically forwards and rearwards, for example) depending on the properties of the scatterers and the wavelength of the radiation. *Reflection*, for example, is just a special case of backwards-directed scattering, or 'back-scattering'.

Depending on the geometry, scattering can also redirect light *into* the line of sight. One example is when a dusty cloud surrounds a star. Photons that are emitted in directions away from the line of sight will be scattered by the dust and a fraction of them will scatter in a direction towards the observer (Figure 7.2a). A scattering dusty cloud around a star is called a *reflection nebula*, an example of which is shown in Figure 7.3.

A more spectacular example of scattering into the line of sight can occur when there is a sudden increase in the brightness of an object, such as when a supernova (Section 5.3.3) or gamma-ray burst (Section 7.1.2.2) goes off or when a star experiences a bright outburst as can occur in certain types of variable stars. The light, travelling at c, scatters

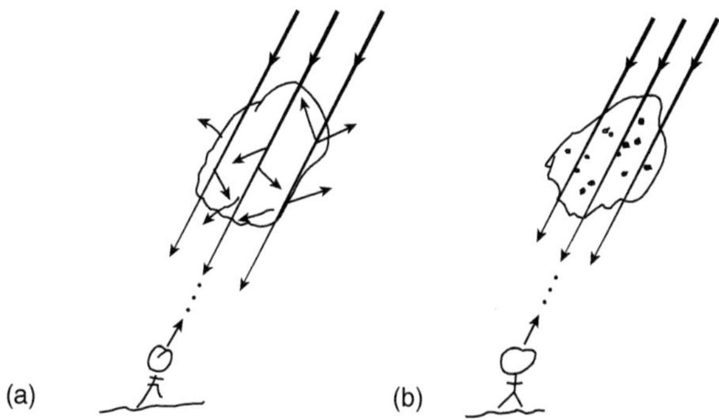

Figure 7.1 Illustration of (a) random scattering and (b) absorption along a line of sight. The observer is meant to be far from the source as well as the scattering medium in this diagram. Darker arrows convey a more intense signal.

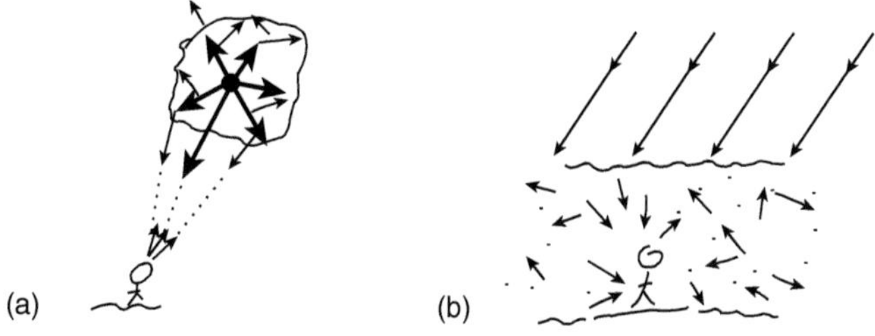

Figure 7.2 Two examples of scattering *into* the line of sight. (a) Dust particles surrounding a star scatter its light randomly, but some of these photons will be directed towards the observer. Thus, the observer sees a faint glow around the star, much like the glow around a streetlight on a misty night. (b) When the observer is *within* a randomly scattering medium, light can enter his eye from all directions even though the original source of light is in one direction only.

Figure 7.3 The reflection nebula around the bright star, Merope, one of the stars in the Pleiades star cluster. The cross on the star is an artefact of the telescope, and the other stars in the image are in the background or foreground. Source: Reproduced courtesy of Yuugi Kitahara, Astronomy Picture of the Day, 1 March 1999.

from surrounding dust clouds at specific times after the event dependent on the distance to the cloud. The result is called a *light echo*, an example of which is shown in Figure 7.4. A more familiar example of light scattering *into* the line of sight occurs when the observer is within the scattering medium itself, as is the case for the Earth's atmosphere. Our sky looks bright even in directions away from the direction of the Sun in the sky, because of such scattering (Figure 7.2b).

The probability that a photon will undergo a scattering interaction is proportional to the scattering cross-section, σ_s, a quantity that must be derived or measured for each type of scattering process. Since a particle may present a different effective scattering cross-section to photons of different wavelengths, this quantity is often wavelength-dependent. Several scattering cross-sections and further details of some of the scattering processes that are important in astrophysics are given in Appendix D of the *online material*. We provide a summary of them in the following subsections.

An important observational consequence of scattering is that the scattered emission, depending on viewing geometry, is *polarized*. The degree of polarization, D_p (introduced in Section 3.7), is the fraction of the light that is polarized. If emission is observed at some position, (x, y), on the sky and some fraction, D_p, is polarized, then this fraction of the

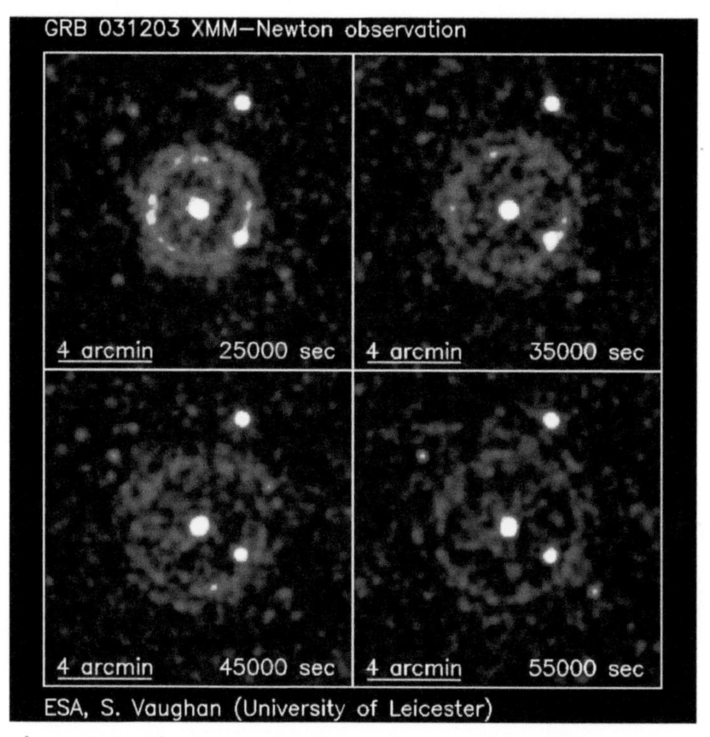

Figure 7.4 The X-ray *afterglow* of the gamma-ray burst (GRB) source, GRB 031203, is seen at the centre of these images. The GRB was detected in December 2003 and is in a distant galaxy. The four X-ray images, taken from the XMM-Newton satellite, span the interval from 25 000 seconds (6.9 hours) to 55 000 seconds (15.3 hours) after the GRB was first detected. The rings are caused by X-rays scattering from different dust screens in our own Milky Way that are illuminated sequentially. The larger rings are seen at later times because it takes longer for light to travel from the GRB over the large angle and then to the observer, in comparison with light travelling closer to the GRB's direct line of sight. The images give the appearance of an expanding ring, but are only produced by dust reflections. This is the first observation of a light echo seen at X-ray wavelengths. Source: Reproduced by permission of Simon Vaughan, University Leicester [Source ID: GRB 031203].

total intensity *may* have been generated at a different location and then scattered by particles at (x, y) into the line of sight. As indicated above, reflection is a kind of scattering, and the light from a reflection nebula, as shown in Figure 7.3, will show polarization. We have also introduced this concept as a way of distinguishing a planet from a star, provided the polarized reflected planetary light is observed (Section 6.2.2). Some care must be exercised in this interpretation, however, because, as indicated in Section 3.7, polarized emission can also result from certain processes intrinsic to the energy generation mechanism. Generally, other available information about the source is also considered in order to decide whether or not the polarization is due to scattering[1]. Figures D.1 and D.6 of the *online material* provide example visual representations of polarization.

7.1.1 Elastic Scattering

If there is no change in the energy of the photon as a result of the interaction, then the scattering is said to be *elastic*, analogous to a small particle bouncing off a wall without losing kinetic energy. If the light is treated as a wave, then such interactions are called *coherent*, meaning that the outgoing energy, and therefore wavelength (recall $E = hc/\lambda$) has not changed from the incoming one. Some examples follow.

7.1.1.1 Thomson Scattering from Free Electrons

Thomson scattering is the elastic scattering of light from free electrons in an ionized or partially ionized gas. The scattering will be elastic provided the energy of the incoming photon is much less than the rest mass energy of the electron, $E_{ph} \ll m_e c^2$. Otherwise, some energy would be imparted to the electron (see Section 7.1.2.2 for that case). Thomson scattering is a very common process in astrophysics and occurs wherever ionized gas is found[2], from regions as diverse as stellar interiors (Example 5.5) to the diffuse intergalactic medium.

A characteristic of Thomson scattering is that the scattering cross-section is *independent of wavelength* (i.e. it is 'grey'). This means that a radio photon is equally as likely to be scattered as an optical or UV photon from a free electron. If the observed emission is the same at all wavelengths, then this is a clue that Thomson scattering could be taking place.

A nice example is the nuclear region of the galaxy, NGC 1068 (Figure 7.5). Vectors in part (c) of the figure show the *polarization* of the emission. Polarization can result from scattering from free electrons and also from scattering by dust or molecules, so without more information, we would not know which. The degree of polarization, D_p, that was introduced in Section 3.7, depends on the number of scatterers, so it is naturally going to be different at different locations in the image. The additional information that we need, however, is the observation that D_p is *constant* with wavelength at any given location. This is a clear sign of Thomson scattering, since scattering by dust or molecules would result in different D_p at different wavelengths.

---

[1] Examples are the nature and waveband of the spectrum. For example, if polarization is seen at *radio* wavelengths, some or all of the polarized emission might be due to synchrotron radiation (Section 10.5).

[2] Wherever there is ionized gas there will also be free protons. However, the proton scattering cross-section is negligible in comparison to the electron scattering cross-section (see Appendix D.1.1 and Eq. D.2 of the *online material*).

This galaxy's *active galactic nucleus* (AGN) is obscured from direct view, but the free electrons in the ionized gas around the nucleus scatter some AGN photons into the line of sight. Moreover, since the scattering is independent of λ, the scattered spectrum looks like the spectrum of the AGN. Thus, the ionized gas acts like a mirror that allows us to see into the AGN itself! This is a good example of how a process that might be thought of as 'corrupting' the signal has actually turned into an indispensable tool for studying one of the most powerful types of objects in the Universe. Appendix D.1.1 of the *online material* provides more information.

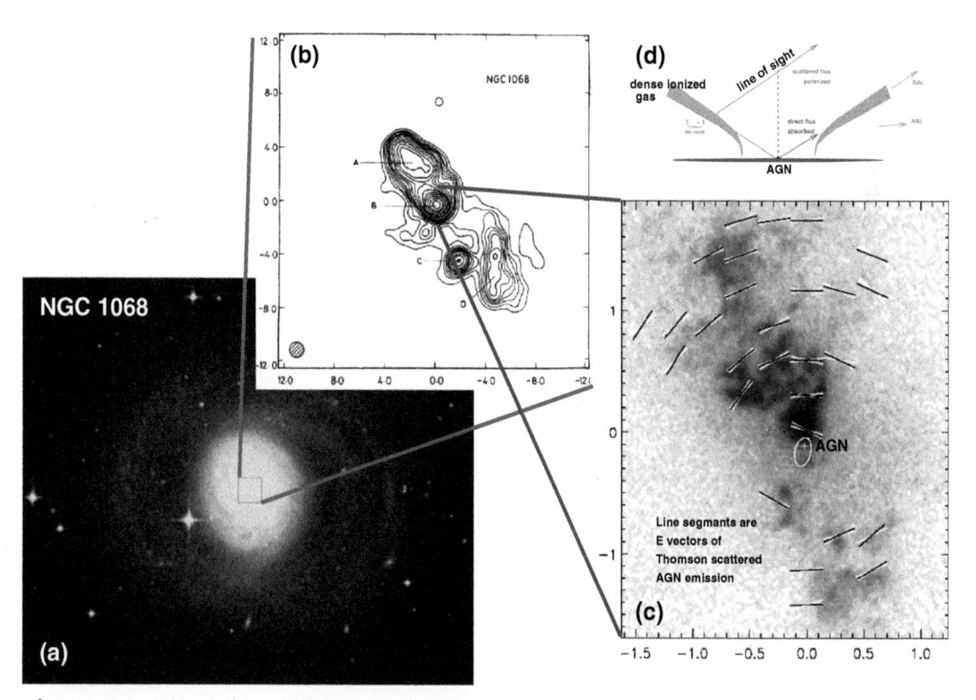

Figure 7.5 The galaxy, NGC 1068 showing different size scales. NGC 1068 belongs to a class of spiral galaxy that has an *active galactic nucleus* (AGN). (a) An optical image showing the entire galaxy. Source: Reproduced courtesy of Pal. Obs. DSS. (b) Blow-up of the boxed region in the previous image showing a collimated jet at the radio wavelength of λ73 cm [18]. Source: Courtesy of the NASA/IPAC Extragalactic Database. (c) Blow-up of the central region of the previous image showing a UV image in colour [10]. The AGN is marked, and the vectors show the polarized emission (in this case, the orientation of the electric field) of *Thomson scattered UV emission from the AGN*. The measured degree of polarization, D_p, ranges from 10% to 44% (Eq. 3.37). Source: Reproduced by permission of M. Kishimoto. (d) A possible model for Thomson scattering of AGN emission [5]. The AGN itself is obscured from direct view but electrons in the ionized region scatter the emission into the line of sight. Source: Elvis [5]. © 2000 American Astronomical Society.

7.1.1.2 Resonance (Line or Bound–Bound) Scattering

When the frequency of incoming light matches or is close to the frequency of an energy transition of a bound electron in an atom or molecule, the electron responds strongly and in *resonance* with the forcing of the wave's oscillating electromagnetic field. This is called *resonance scattering* and it is coherent when the bound electron is excited to an upper state by absorbing the photon and then de-excites to the initial level again, re-emitting a photon at an equivalent energy[3] but in a different direction. Resonance scattering results in a polarized signal at the frequency of the line [23].

An example of a resonance scattering line is the Ly α line (a transition between quantum number n = 1 and n = 2) of hydrogen. This is a UV transition at a rest wavelength of λ122 nm with an upper state lifetime of only 10^{-9} s. This is a much shorter timescale than the time between collisions under most interstellar conditions (Example 7.1), so incident UV radiation near the Lyα wavelength is strongly scattered in this line.

Example 7.1

A Ly α photon travels through a pure hydrogen gas of density, $n = 10^3$ cm^{-3} and temperature, $T = 10^4$ K. If the gas is mostly ionized and the photon excites an upwards Ly α transition in one of the small fraction of particles that are neutral and in the ground state, will the outcome be scattering or absorption? Consider the permitted $2p \rightarrow 1s$ transition only.

Once the Ly α photon has excited the atom, the electron will spend a time, $\tau_{2,1} = \frac{1}{A_{2,1}} = 1.6 \times 10^{-9}$ s in the energy level, n = 2, before spontaneously de-exciting to the ground state, n = 1, again. Here, $A_{2,1}$ is the Einstein *A* coefficient for the $2p \rightarrow 1s$ transition (Table C.1 of the *online material*).

In the meantime, there are collisions between particles and there is a possibility that de-excitation will occur via a collision with a free electron[4] or that collisional excitation will occur to higher levels. Considering only downwards permitted transitions from Table 5.2, the relevant collision rate coefficient for de-excitation of this transition at the given temperature is $\gamma_{Ly\alpha(2p \rightarrow 1s)} = 6.8 \times 10^{-9}$ cm^3 s^{-1} so the timescale between de-exciting collisions is (Eq. 5.19) $\bar{t} = 1/[(10^3)(6.8 \times 10^{-9})] = 1.5 \times 10^5$ s.

Since $\tau_{2,1} \ll \bar{t}$, the de-excitation will occur spontaneously. Similarly, it can be shown that the downwards collisional timescale via the $2s \rightarrow 1s$ transition, as well as upwards collisional timescales are longer than the spontaneous de-excitation timescale under the conditions described (see also Section 10.4). Once spontaneous de-excitation occurs, the re-emitted Ly α photon (after having travelled a mean free path) can excite another particle, followed by spontaneous de-excitation again, and so on. Thus, the photon will be *scattered*[5].

---

[3] The emitted photon may have a slightly different energy since the emission probability is distributed in frequency according to the Lorentz profile (Figure D.2 of the *online material*). Other effects will broaden the line even more (Section 11.3) so coherence is not perfect but the shift in energy is small in comparison to the energy of the line itself.

[4] Collisions with protons are highly unlikely, since protons are massive and slow-moving in comparison to electrons.

[5] Note that we have neglected the possibility of photo-absorption from the n = 2 level to a higher level, but this will only be important in a very strong radiation field.

Table 7.1 Sample photon interaction cross-sections[a].

Type	Description	Wavelength or energy	Cross-sesction (cm^2)
σ_{T^b}	Thomson scattering	≪0.51 MeV	6.65×10^{-25}
σ_{K-N^c}	Compton scattering	0.51 MeV	2.86×10^{-25}
		5.1 MeV	8.16×10^{-26}
σ_{R^d}	Rayleigh scattering (N_2)	532 nm	5.10×10^{-27}
	(CO)	532 nm	6.19×10^{-27}
	(CO_2)	532 nm	12.4×10^{-27}
	(CH_4)	532 nm	12.47×10^{-27}
σ_{b-b^e}	Ly α (natural)[f]	121.567 nm	7.1×10^{-11}
	Ly α (10^4 K)[g]	121.567 nm	5.0×10^{-14}
$\sigma_{HI \to HII^h}$	H ionization	13.6 eV	6.3×10^{-18}
σ_{f-f^i}	free-free absorption	21 cm	2.8×10^{-27}

[a] Cross-sections apply to a single scattering event from a single particle.
[b] Thomson cross-section (Eq. D.2 of the *online material*).
[c] Klein–Nishina cross-section for Compton scattering (see Figure D.4 of the *online material*).
[d] Rayleigh scattering cross-section for a temperature of 15 °C and pressure of 101 325 Pa [24].
[e] Resonance, bound–bound scattering from the line centre.
[f] From the natural line shape using Eq. (D.12) with data from Table C.1 of the *online material.* Note that only the permitted transitions have been included.
[g] As in the previous row but assuming that the line is *Doppler broadened* (Section 11.3) at the temperature indicated [20].
[h] Photoionization cross-section from the ground state (Eq. C.9 of the *online material*).
[i] Free–free absorption cross-section for the conditions: $n_e = 0.1$ cm^{-3} and T = 10^4 K. The cross-section will vary with these quantities and also decreases with increasing frequency [25].

The Ly α photon scattering cross-section is given for the *natural linewidth* in Table 7.1. The natural linewidth would be applicable to isolated atoms at rest. However, it is important to note that thermal motions of many atoms in a gas (Section 11.3) will result in a line that is much wider in frequency and has a lower peak scattering cross-section than a line from the unphysical situation of atoms at rest. Therefore, a sample cross-section for 10^4 K gas is also given in Table 7.1. Even when one considers this more realistic linewidth, it is clear that the resonant scattering cross-section is quite high in comparison to other values in the table.

HII regions (e.g. Figures 5.13, 10.4) are good examples of objects in which Ly α resonance scattering is important. In HII regions, photons at the wavelength of Ly α are numerous, both from Ly α recombination line emission as well as the continuum of the central ionizing star. As explained in Section 5.4.7, an HII region is almost entirely ionized but, at any time, recombinations are occurring and a small fraction of atoms will have electrons in the ground state. Therefore, once a Ly α photon is emitted, it will be reabsorbed by other neutral atoms, exciting the electron into the state, n = 2, followed by re-emission to the ground state again, and so on. There will be many such scatterings within the nebula because, not only is the resonance scattering cross-section large (Table 7.1), but also the densities of HII regions tend to be low, minimizing the possibility that a collision could interfere with the scattering process. Even though an HII region has a very low density in comparison to the interior of a star (Table 5.1), Ly α photons in an HII region take a random walk out of the nebula much like we saw for the interior of the Sun (Figure 5.15). Because Ly α photons are scattered so strongly, the nebula is actually opaque (Figure 6.2) at this wavelength[6], as explained in Example 7.2. Other Lyman lines are similarly scattered, though to a lesser extent than Ly α.

Example 7.2

What is the mean free path for scattering of a Ly α photon in an HII region of density, $n = 100\ cm^{-3}$?

HII regions are typically at a temperature of $T = 10^4$ K (Table 5.1). The scattering cross-section is $\sigma_{b-b} = 5.0 \times 10^{-14}\ cm^2$ at this temperature (Table 7.1) so the mean free path of a photon is,

$$\bar{l} = \frac{1}{n\sigma_{b-b}} = \frac{1}{(100)(5.0 \times 10^{-14})} = 2 \times 10^{11} cm \tag{7.1}$$

This is less than three times the radius of the Sun and is a very short distance in comparison to a typical HII region which can be hundreds of parsecs in size. Therefore, the Ly α photon will take a random walk in an HII region in order to escape.

Another example of Ly α resonance scattering can be seen in the *Ly α forest*, an example of which is shown in Figure 7.6. The top figure shows the spectrum of a *quasar* (Section 3.6.5) that is relatively nearby, and the bottom figure shows a distant quasar. The peaks of the two images represent the Ly α emission line from the quasars, themselves. However, the distant quasar (bottom) shows many apparent absorption lines on the shorter wavelength (blueward, or left in the image) side of the quasar. These lines must be from clouds that are intermediate in distance between us and the quasar. They result from absorption and re-emission (resonance scattering) from the Ly α line in many individual intergalactic HI clouds, each at a different distance along the line of sight towards the quasar. The differing distances correspond to different redshifts and therefore frequencies along the line of sight, as will be described in Section 9.1.

The Ly α forest lines are commonly referred to as 'absorption lines' because what the observer actually sees is removal of the background quasar light over these many different frequencies. However, the process is one of resonance scattering because, at the densities of these clouds (Table 5.1), an atom that has absorbed a photon will reemit a photon (typically out of the line of sight) before it will collide with another particle. The HI clouds that produce most of the observed Ly α forest lines presumably permeate space but are too faint to be observed directly in emission with current telescopes. They are only

[6] The situation in which downward transitions to the $n = 1$ level are immediately re-absorbed in HII regions is called *Case B*. *Case A* assumes that all such emission escapes from the nebula without any interaction. However, the density of a Case A nebula would have to be so low that it would typically be too faint to be detectable. Case B is considered to be the most realistic situation [16]. See also Section 11.4.2.

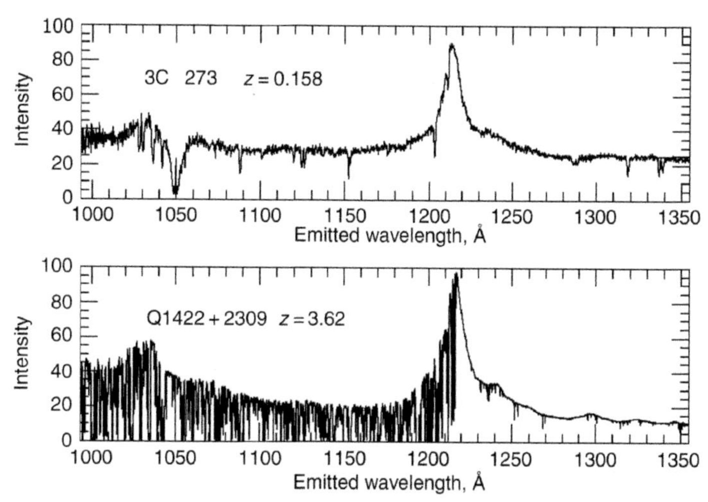

Figure 7.6 Spectra of the quasars, 3C273 (top) and Q1422 + 2309 (bottom). The quasar, 3C273 is relatively nearby at a redshift of $z = 0.158$ (see Section 9.1.2 for a discussion of the expansion redshift) whereas Q1422 + 2309 is much farther away. The spectra have been aligned so that the Ly α line emitted from the quasars, themselves, is shown at its rest wavelength. In between us and the more distant quasar, Q1422 + 2309, there are many HI clouds at a variety of distances, creating the many *apparent* absorption lines seen to the left of the quasar emission. This is called the *Lyman alpha forest*. Source: William Keel.

seen in absorption in front of bright background sources. However, numerical simulations suggest that they may form a *cosmic web* similar to that shown in Figure 7.7. The nature of this web is a subject that is actively being studied. It is likely that much of the hydrogen gas is ionized, rather than neutral, but the age of the Universe when it transitioned from mostly neutral (the 'dark ages') to mostly ionized is currently uncertain (e.g. [15] and Section 7.2.2).

Although hydrogen is the most abundant atom, other elements are present in astrophysical sources and many of these atoms also have resonance lines (see [23] for examples). They can also play an important role in the transfer of radiation through a source.

Appendix D.1.2–D.1.5 of the *online material* provides more information about scattering from bound elections.

7.1.1.3 Rayleigh Scattering and the Blue Sky

The most famous example of Rayleigh scattering is the scattering in the Earth's atmosphere leading to the blue colour of the sky. It was Leonardo da Vinci who, around 1500, first suspected the reason for the blue sky, in particular by his observation that wood

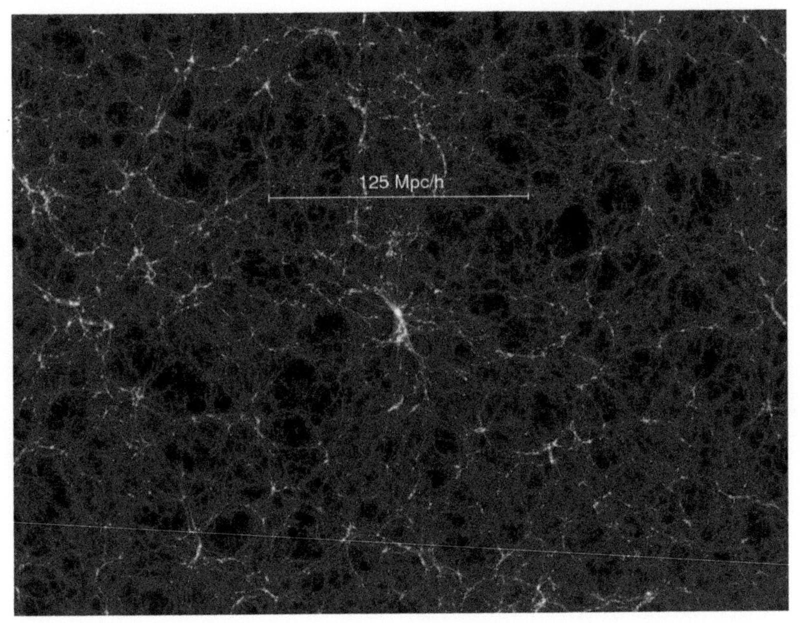

Figure 7.7 One view of the *cosmic web* from the 'Millennium Simulation' which used 10^{10} particles in a computer model to simulate the formation of structure in the Universe [26]. The image represents the expected structure of dark matter (Section 5.2), but the structure of the Ly α forest may be similar. This image is for a redshift of $z = 1.4$ (Section 9.1.2), corresponding to a time, 4.7 Gyr after the Big Bang. The unitless quantity, h, is defined as $H_0/(100\ \mathrm{km\ s^{-1}\ Mpc^{-1}})$ (Appendix F of the *online material*). Source: Reproduced by permission of V. Springel and the Virgo Consortium.

smoke looked blue when observed against a dark background [9]. The effect was finally described quantitatively in the year 1899 by Lord Rayleigh whose name is now associated with the phenomenon.

Rayleigh scattering occurs when the incoming signal has a wavelength, $\lambda_{\mathrm{signal}}$, that is much larger than the resonance wavelengths of a bound electron in an atom or molecule. By resonance wavelengths, we mean the wavelengths, λ, corresponding to any of the electronic energy transitions, ΔE, in the atom or molecule (recall that $\Delta E = hc/\lambda$). In our atmosphere, the most abundant molecules are nitrogen and oxygen which are only about 0.3 nm in size whereas optical light ranges from about 400 to 700 nm. This means that $\lambda_{\mathrm{signal}} \gg$ the size of the scattering particle, and consequently $\lambda_{\mathrm{signal}} \gg$ the wavelengths of any transitions within those molecules as well. Therefore, atmospheric scattering is firmly in the Rayleigh regime.

Because of the strong dependence of the Rayleigh scattering *cross-section* on wavelength, $\sigma_s \propto 1/\lambda^4$, shorter wavelength blue light[7] ($\lambda \approx 470$ nm, Table T.4) will scatter

[7] Violet light is scattered even more but our eyes are less sensitive to violet than blue.

Figure 7.8 A time exposure (ISO 200, f/2.8, 69 s) reveals the blue colour of the night sky in this un-retouched image. Source: Reproduced by permission of © Joseph A. Shaw, Montana State University Bozeman, Montana.

more efficiently than longer-wavelength red light ($\lambda \approx 660$ nm). For a geometry as shown in Figure 7.2b, blue light is more likely to scatter *into* the line of sight of the observer than red light. Thus, the *yellow* sun produces a *blue* sky to an observer on the Earth. Small dust particles also contribute, but the dominant scattering is due to molecules. A consequence is that the sky would still look blue even if it were dust-free. Though less obvious, the *nighttime sky* is also blue. Although the feebleness of the light at night makes it impossible to discern by eye, a time exposure reveals the colour, as Figure 7.8 shows.

If there were no atmosphere, the daytime sky would be black except at the position of the Sun itself. The fact that the clear sky atmosphere is transparent (Example 7.3) means that most photons travel through it unimpeded with only a small fraction scattered. This is why, on a clear day, the brightness of the Sun is much greater than the brightness of the blue sky.

Example 7.3

What is the mean free path of an optical photon to Rayleigh scattering in the Earth's atmosphere?

The mass density of the atmosphere at sea level (the densest air) is $\rho \approx 1.2 \times 10^{-3}$ g cm^{-3}. The mean molecular weight of air is $\mu = 28.97$ so the number density of air at sea level is (Eq. 5.11), $n = \rho/(\mu m_H) \approx 2.5 \times 10^{19}$ cm^{-3}, where m_H is the mass of a hydrogen atom. The most likely interaction is with the most abundant molecule, N_2, so its Rayleigh scattering cross-section at 532 nm (Table 7.1) together with Eq. (5.18) yields, $\bar{l} = 1/(n\sigma_R) \approx 80 \times 10^5$cm. At higher altitude, the density decreases, so the mean free path increases. Thus, $\bar{l} > 80$ km. Recall that this lower limit is well into the ionosphere (Section 4.3), confirming what we already know – that the clear sky is optically transparent.

For a geometry as shown in Figure 7.1a, blue light is more likely to be scattered *out* of the line of sight than red light. Thus, any light-emitting object above the Earth's atmosphere will be *reddened* (first introduced in Section 4.3.5) and also *dimmer*, due to Rayleigh scattering. The Sun is redder than its true colour even when it is high overhead. If the line of sight through the atmosphere is longer, such as when viewing the Sun at sunrise or sunset (Figure 4.14), then the redder colour is more enhanced and more obvious to the eye[8]. The same effect can be observed for other objects such as the Moon, planets or stars.

It is important not to confuse the wavelength-dependence of the scattering cross-section, with the fact that Rayleigh scattering is *elastic*. Individual scattered photons have the *same* wavelength as the incident photons, so Rayleigh scattering represents elastic, or coherent, scattering. However, the scattering cross-section is strongly wavelength-dependent, so the probability that blue light will be elastically scattered is much greater than the probability that red light will be elastically scattered.

Rayleigh scattering produces polarized light in the same way as Thomson scattering. Even though the Sun emits optical light that is intrinsically unpolarized, its *scattered* light will be polarized if viewed at an angle of 90°. This can be very nicely verified by looking near the horizon with a polarizing filter when the Sun is overhead. Also, like Thomson scattering, Rayleigh scattering provides a way of looking at a source via its 'reflection', although with its spectrum weighted by the $1/\lambda^4$ scattering cross-section dependence. It is thus possible to see the Sun's (weighted) spectrum by pointing a spectrometer at some position in the sky away from the Sun itself. The Sun's Fraunhofer absorption lines[9], for example, can easily be seen in such a fashion.

On a final note, the blue sky of the atmosphere can be contrasted with the greyer colour associated with water droplets in clouds. Water droplets are *not* small with respect to the wavelength of light, so scattering from these particles is not in the Rayleigh scattering regime. The wavelength dependence of large particle scattering is flatter than for Rayleigh scattering, hence the grayish colour of clouds. Details on Rayleigh scattering can be found in Appendix D.1.5 of the *online material*.

7.1.2 Inelastic Scattering

If the scattered photon is of lower energy (longer λ) than the incident photon, then the scattering is said to be *inelastic* or *incoherent*. Several examples follow.

[8] Scattering (and some absorption) from dust, water vapour and large molecules can also contribute to reddening (see also Problem 8.1).

[9] The Sun's Fraunhofer lines are the absorption lines formed in the Sun's *photosphere*, as described in Section 8.4.2. See also Figure 8.4.

7.1.2.1 Bound–Bound Inelastic Scattering and Fluorescence

Inelastic scattering can occur from the bound–bound excitation of an atom to a high-energy state followed by cascading to other energy levels between the high state and the initial one. Thus, a high-energy photon is degraded into multiple lower-energy photons. The emission of light in this cascading process is called *fluorescence*. There are various possible outcomes for bound–bound scattering, depending on the incident photon energy and the final state of the atom.

If the final energy state is the same as the initial one, then the sum of the energies of the emitted photons equals the energy of the incident photon, and the situation is one of *pure inelastic* scattering (Problem 7.3).

If the final state of the atom is an excited state higher than the initial state, then some energy has been lost to the excitation of the atom. What might have happened to that energy? Eventually, an atom will always decay spontaneously from any excited state to its ground state, but if an energy-exchanging collision occurs in the meantime, then energy is lost in the collision. In that case, after a bound–bound excitation, a fraction of the energy goes into lower energy photons as the electron cascades downwards and then a fraction of the energy goes into heating the gas by the collision. Processes in which energy is 'lost' (or rather exchanged with particles) are called *absorption* and will be considered in Section 7.2. In this case, we would say that there is inelastic scattering followed by absorption.

7.1.2.2 Compton Scattering at High Energies

Compton scattering is the scattering of a photon from a free electron in the high-energy limit, i.e. when the energy of the incident photon is of order or greater than the rest mass energy of the electron, $E_{ph} \gtrsim m_e c^2 = 0.51$ MeV. In Compton scattering, some of the photon energy is imparted to the electron giving it a velocity. The energy of the scattered photon, $E'_{ph} = hc/\lambda'$, is less than the energy of the incident photon, the difference going into the kinetic energy of the electron. Because the scattered photons have less energy (longer wavelengths) than the incident photons, the result is the *cooling* of the radiation field. In practical terms, the brightness temperature, T_B, of the radiation field, which we know can be used for *any* type of radiation (Eq. 6.4), decreases.

Since 0.51 MeV is in the gamma-ray part of the spectrum (Table I.1), Compton scattering is associated with highly energetic objects, such as active galactic nuclei (AGN), binary stars in which at least one member is a compact object such as a neutron star, in regions around black holes and in supernovae. Compton scattering will occur in any source in which there are both gamma-rays and free electrons. Included in this group is an intriguing highly energetic source of transient gamma-rays, that is, *gamma-ray bursts* (GRBs). These were first introduced in the light echo example of Figure 7.4. Let us look at them in a bit more detail now.

GRBs are bursts of gamma-ray emission of very short duration, from milliseconds to 2 seconds for the *short duration bursts*, and from 2 seconds to several minutes or even hours for the *long duration bursts*. Discovered by spy satellites in the 1960s, GRBs were

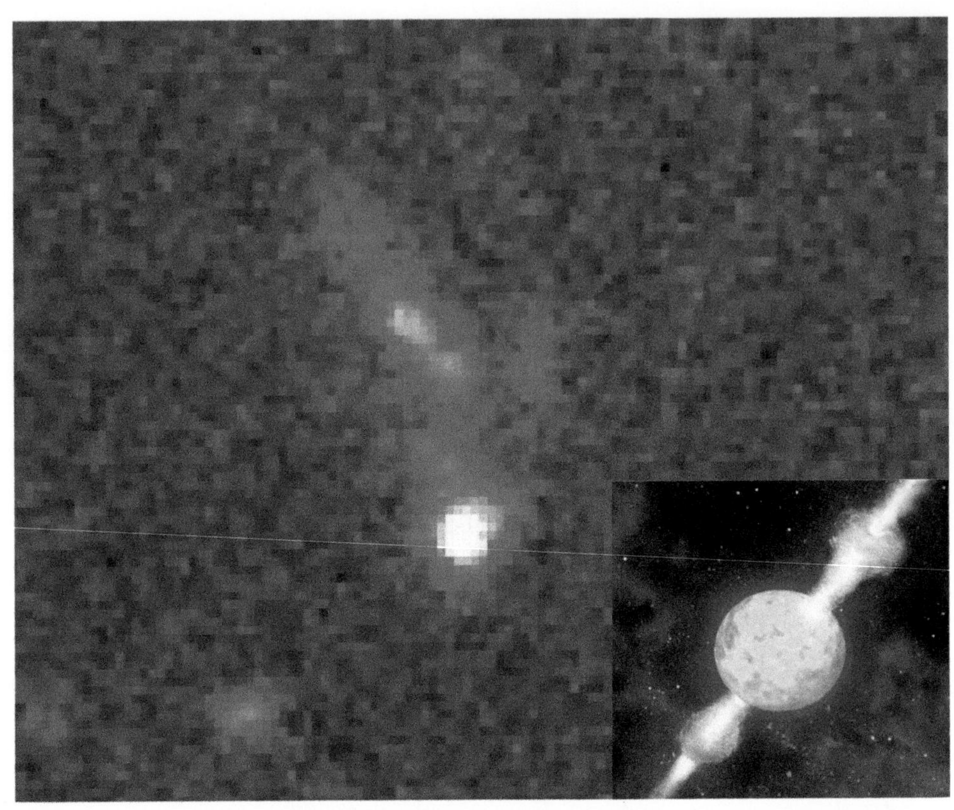

Figure 7.9 The optical afterglow of the powerful gamma-ray burst, GRB 990123, seen here as a bright point source, was caught by the Hubble Space Telescope in this image. The fact that this GRB was found to be associated with a galaxy (faint extended emission) provides powerful evidence that GRBs are located in distant galaxies. Source: GRB 990123, reproduced by permission of the HST/GRB collaboration/NASA. The inset shows an artist's concept of an extremely massive star beginning its gravitational collapse with light at gamma-ray energies being beamed into two jets. Compton scattering, among other processes, takes place in the initial phases of the GRB. Source: Dana Berry, SkyWorks Digital, NRAO.

found to be associated with distant faint galaxies in the 1990s (see Figure 7.9). GRBs represent the most energetic known sources in the Universe and can be 100 times more powerful than supernovae. For either type, the end-result is a black hole.

Short duration GRBs are likely associated with colliding neutron stars or neutron star – black hole collisions [4]. These are the most promising sources for gravitational wave detection and indeed, as we found in Section 2.9, the GW source, GW170817, was found to be a short duration GRB source. This source was also observed in *afterglows*, which are emissions at other wavelengths that are observed with some time delay (seconds to days) after the initial GRB itself.

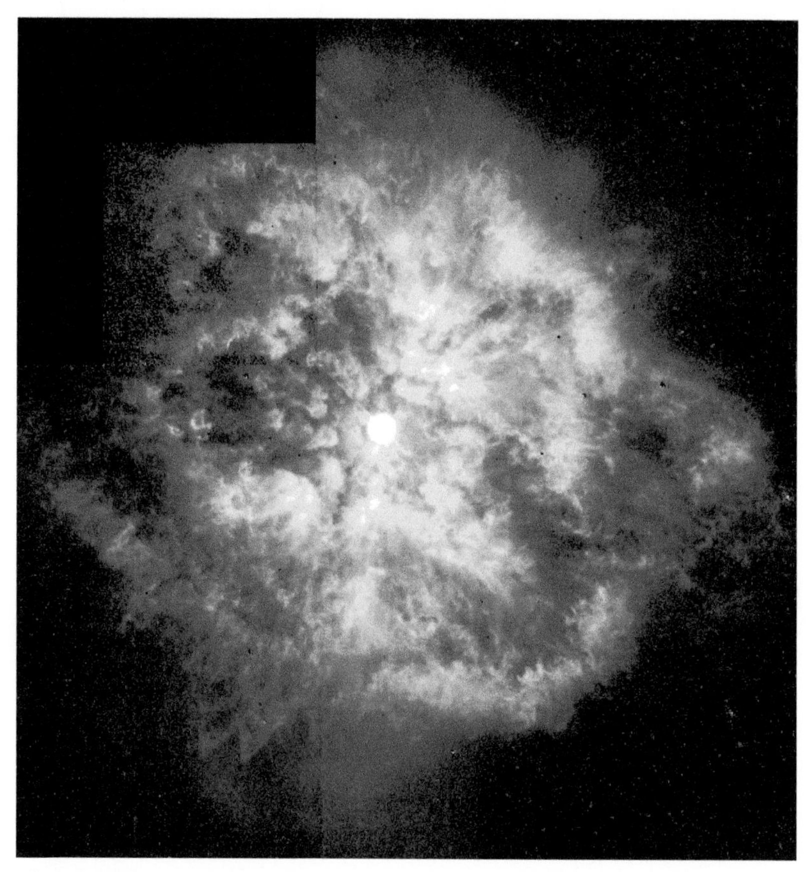

Figure 7.10 The 2.5 pc diameter nebula, M1–67, shown in this H α image from the Hubble Space Telescope, is caused by a massive stellar wind from the Wolf–Rayet star at the centre. Wolf–Rayet stars are examples of the youngest, hottest, and most massive stars known and are losing mass via stellar winds at rates of 10^{-5}–10^{-4} $M_{\odot}$ year^{-1}. This nebula has a mass of 1.7 $M_{\odot}$ and an age of no more than 10^4 year, consistent with a very high mass-loss rate [7]. Source: Reproduced by permission of Y. Grosdidier et al. [7] and NASA.

The long duration GRBs are related to *hypernovae* – very powerful supernovae that originate from precursor stars more than 20 times the mass of our Sun. A possible precursor is a *Wolf–Rayet star* which is known to have high mass loss via stellar winds (see Figure 7.10). The interaction of the hypernova with wind material may also play an important role in forming *afterglows*. It is difficult to explain the high energies of GRBs without requiring that the detected emission be *beamed* in our direction, as the inset to Figure 7.9 suggests (see also Problem 3.4). Thus, long-duration GRBs appear to originate from ultra-relativistic jets during the collapse of the cores of these massive stars, a view that is substantiated by numerous observations of these powerful sources and their afterglows over a variety of wavelengths (e.g. [13]).

Compton scattering is the dominant scattering process for gamma-ray photons in the range, 0.51 to a few MeV, which is the low energy part of the gamma-ray spectrum. In this regime, the cross-section, called the Klein–Nishina cross-section, decreases strongly with increasing photon energy and is smaller than the Thomson cross-section (Figure D.4 of the *online material*), ensuring that the probability of interaction is low. Thus, gamma-rays are quite penetrating and strong effects are only observed in regions of high density, such as near the source itself (Problem 7.8). At higher energies, other processes also become important, such as the conversion of a gamma-ray into an electron/positron pair, or the interaction of the gamma-ray with atomic nuclei or even with other photons. Consequently, modelling a highly energetic source, like a GRB, requires that a variety of processes be considered (e.g. Figure 12.1).

Appendix D.2 of the *online material* provides more information about Compton scattering as well as an expression for the energy of the scattered photon when the electron is initially at rest. However, a more general development takes into account the initial kinetic energy of the electron and reveals that the *inverse* energy exchange can also occur. That is, if the electron is more energetic than the photon (Eq. D.29), then the electron can impart an energy to the photon, a process called *inverse Compton* (*IC*) *scattering*. In the case of IC scattering, the *photons* shift to *higher* energies. The inverse process results in radiation that *looks like* an emission process, so we defer discussion of IC radiation to Section 10.6.

7.2 THE PHOTON LOST – ABSORPTION

Absorption occurs when a photon loses all of its energy (disappears) because of an interaction. In Section 7.1.2.1, we gave an example of how a fraction of the energy of an incident photon could be absorbed if a complex series of interactions occur; for example, an incident photon results in a bound–bound excitation, followed by a bound–bound de-excitation to a higher energy state than the original, followed by energy loss due to a collision. True absorption, however, requires that there is no scattered photon at all; the entire incoming photon energy is lost. Following the symbolism of Section 7.1, absorption can be represented as: *photon* → *matter*. Like the scattering cross-section, the *absorption cross-section* refers to the effective cross-sectional area that a particle presents to an incoming photon for an absorptive process. If a photon is absorbed, energy must still be conserved, so the incoming photon's energy must be converted into some other form. There are primarily two possibilities for the absorption of light by matter: *kinetic energy gain* (which includes heating), or a *change of state*.

7.2.1 Particle Kinetic Energy – Heating

If a photon is absorbed by matter, imparting a motion to it, then the photon energy ($E = h\nu$) is converted into kinetic energy ($E = \frac{1}{2}mv^2$). This occurred for Compton scattering (Section 7.1.2.2) but, in that case, a lower energy scattered photon remained so the process was categorized as inelastic scattering. When we considered *radiation pressure* for the case of perfect absorption (Section 3.5), the absorbed photons imparted some kinetic energy to that surface. With a sufficiently powerful beam, the surface could be pushed along like a sail. Some of the *absorbed* photons, however, will contribute to *heating* the sail. For matter in any phase, heat involves the random internal motions of its constituents and therefore represents the collective kinetic energies of the particles.

For *solids*, such as the Solar sail, a dust grain, or a planet, these internal motions refer to vibrations of the particles within their internal crystalline structure. As was described in Section 6.2.1, a dust grain, asteroid or planet that does not move from its position with respect to the light source will eventually achieve and maintain an equilibrium temperature as a result. For dust, it is only the larger grains that actually maintain an equilibrium temperature, since *very small grains* (VSGs, Section 5.5.2) cannot retain internal heat and tend to lose it quickly before another photon can be absorbed. The interaction of light with dust is a more complex process and will be considered separately in Section 7.5.

If photons impinge on a *gas*, then they can contribute to the kinetic energy of individual particles in the gas. The randomly moving gas particles share energy via collisions (Section 5.4.1), so any absorption process that is immediately followed by an energy-exchanging collision will result in the loss of the photon. An example is if a particle experiences a bound–bound excitation followed immediately by a collision (the opposite conclusion to Example 7.1).

If a photon *ionizes* the gas and if there is excess energy after ionization, then that excess energy goes into the electron kinetic energy (see Section 7.2.2) which is then shared via collisions.

Another important absorption process in ionized gases occurs when a free electron happens to be in the vicinity of a positive nucleus so that it feels the electrostatic force of the nucleus as it travels past. In such a case, a background photon can be absorbed by the electron, potentially giving the electron a higher kinetic energy and different trajectory than it originally had. This is called *free–free absorption* [10]. The free–free cross-section depends on frequency, temperature and density. A sample value is given in Table 7.1.

7.2.2 Change of State – Ionization and the Strömgren Sphere

Absorbed photons can also produce a change of state, such as the transition from a solid to a gas (*sublimation*), from a neutral gas to an ionized gas (*ionization*), or from a molecular gas to a neutral atomic gas (*dissociation*). These processes could also occur via particle collisions of sufficient energy, so terms such as *photoionization* are sometimes used to distinguish the photo-absorptive process from *collisional ionization*.

Photoionization, in particular, is very common in astrophysics and will occur anywhere that photons are energetic enough to ionize neutral material that may be nearby. It is useful to remember that there are many constituent elements in the Universe and each has its own ionization energy. These elements can be very important in terms of the

[10] The inverse of this effect, free–free emission, will be discussed in Section 10.2.

heating and cooling balance in various objects (neutral and singly ionized *carbon* in the ISM is a good example). For the most abundant element, hydrogen, the required energy for ionization from the ground state (its *ionization potential*) is 13.6 eV which is in the UV part of the spectrum (Table I.1). Any photon of equivalent or greater energy, if absorbed by the atom, would ionize it and any excess energy would go into the kinetic energy of the ejected electron, contributing to heating of the ionized gas. The photoionization cross-section for hydrogen in its ground state is given in Table 7.1.

Photoionization occurs in a variety of astronomical objects. Some examples are planetary nebulae (Figure 5.7) in which the source of ionizing photons is the central white dwarf, the outer gaseous 'edges' of the discs of galaxies for which the source of ionizing photons is the weak extra-galactic radiation field, the regions around active galactic nuclei (e.g. Figure 7.5) ionized by the AGN itself, and HII regions (Figures 5.13, 10.4) which are ionized by a hot central star or stars.

Even the Universe itself had a period of reionization. At the earliest times, the Universe was an ionized plasma which we now see as the cosmic microwave background (CMB, Section 5.1). As the Universe expanded, electrons recombined with protons to form neutral hydrogen. As indicated in Section 7.1.1.2, the epoch during which hydrogen was mostly neutral is called the 'dark ages' (introduced in Section 3.4). Once the first stars and proto-galaxies formed, some of which would harbour AGNs, the hydrogen was ionized again. It is not yet clear whether the reionization was caused by the first stars or the first AGNs (Problem 7.11). The quest to understand which sources first 'turned on' in the Universe is currently of much interest and the time period over which this occurred is referred to as the *Epoch of Reionization* (EOR).

If an ionized region is not expanding or shrinking and the degree of ionization (the fraction of particles that are in an ionized state) is constant, then the ionization rate must equal the recombination rate and we say that the region is in *ionization equilibrium*. Such an argument is similar to the one used for temperature equilibrium in Section 6.2.1. This fact can then be used to determine some physical parameters of the region.

In Example 7.4, we do this for the simple case of an HII region around a single star. The star must be a hot, massive, O or B star since only these stars have a sufficient number of UV photons to form HII regions (Problem 7.10). In an HII region, the gas is almost completely ionized but a very small fraction of particles will be neutral in accordance with ionization equilibrium. The assumption is that every photon of sufficient energy leaving the star will eventually encounter a neutral atom that it then ionizes. Such a development was first presented by B. Strömgren, and so an HII region around a single star is sometimes called a *Strömgren sphere* with a radius called the *Strömgren radius*.

Example 7.4

Find the Strömgren radius of an HII region consisting of pure hydrogen of uniform electron density, $n_e = 1000\ \mathrm{cm}^{-3}$ and temperature, $T = 10^4$ K around a single central O8 V star.

As described in the text, $N_i = N_r$, where N_i is the number of ionizing photons per second leaving the star and N_r is the number of recombinations per second within the nebula.

Beginning with the recombination rate, a recombination can occur whenever an electron 'collides with' a proton. We use the total collision rate per unit volume, ν_{tot}, from Eq. (5.21) multiplied by the total volume, V, of the HII region and use the collision rate coefficient appropriate to recombination, called the total *recombination coefficient* α_r (Table 5.2),

$$N_r = \nu_{tot}\, V = n_e n_p \alpha_r V = n_e^2 \alpha_r \frac{4}{3}\pi {R_S}^3 \tag{7.2}$$

where n_p is the proton density, taken to be equivalent to n_e for pure ionized hydrogen and R_S is the Strömgren radius.

We now need to equate N_r to N_i and rearrange the result to find,

$$R_S {n_e}^{2/3} = \left(\frac{3}{4\pi}\frac{N_i}{\alpha_r}\right)^{1/3} \tag{7.3}$$

(all cgs units). The RHS of this equation contains information about the exciting star (plus constants)[11] and the LHS contains information about the HII region, often combined into a single parameter called the *excitation parameter*, $\mathcal{U}$,

$$\mathcal{U} \equiv R_S {n_e}^{2/3} \tag{7.4}$$

The excitation parameter has cgs units of cm cm^{-2} but is often expressed in units of pc cm^{-2} (e.g. Section 10.2.2). If the exciting star is visible and can be identified (implying that N_i is known, see below), then Eq. (7.3) can be used to find the excitation parameter which is the quantity that contains the desired information about the HII region itself. Then, if either the density or size of the HII region is known, the other can be calculated.

The missing quantity is still the number of ionizing photons per second, N_i. This can be computed from the ionizing luminosity, L_i (erg s^{-1}), of the star divided by the photon energy, $h\nu$ (erg). The photon energy, however, can have a variety of values as long as it is greater than the ionization potential of hydrogen, that is, $\nu \geq \nu_i$, where ν_i is the threshold frequency required to ionize hydrogen (13.6 eV corresponding to $\nu_i = 3.28 \times 10^{15}$ Hz). L_i can be determined from an integration of the Planck curve, $B_\nu(T)$ (Eq. 6.1), using Eqs. (3.10) and (3.14),

$$N_i = \frac{L_i}{h\nu_{(\geq \nu_i)}} = 4\pi^2 R_*^2 \int_{\nu_i}^{\infty} \frac{B_\nu(T)}{h\nu} d\nu \tag{7.5}$$

where R_* is the radius of the star.

For an O8 V star, T = 37 000 K and $R_* = 10\, R_\odot = 6.96 \times 10^{11}$ cm (Table T.5). Numerical integration with $\nu_i = 3.28 \times 10^{15}$ Hz yields $N_i = 7.95 \times 10^{48}$ photons s^{-1}. Using this value in Eq. (7.3) along with $n_e = 1000$ and $\alpha_r = 2.59 \times 10^{-3}$ cm^3 s^{-1} (Table 5.2), gives $R_S = 1.93 \times 10^{18}$ cm or 0.63 pc.

Early computations [17] simply tabulated values of $\mathcal{U}$ for different types of stars so that computations like Eq. (7.5) need not be continually carried out. More accurate derivations would consider non-uniform densities and also take into account

[11] **The recombination coefficient is not exactly constant but varies only slowly with temperature.**

some absorption of photons by other elements such as helium, or by dust particles. Modern codes are very sophisticated and include realistic dust models, formation of molecular hydrogen on grains, photodissociation, non-equilibrium cooling, full chemical composition, and many other effects [19][12]. Nevertheless, Eq. (7.3) provides a first approximation to $\mathcal{U}$ and gives the basic link between the ionizing star and the HII region around it.

As indicated in Example 7.4, this simple development links the stellar type to the size and density of the HII region. In principle, if the stellar type is known and the density determined by other means (e.g. Section 10.2.2) the angular size can be compared to R_S to find the distance. HII regions have sharp boundaries (Example 7.5) so an angular size measurement of a resolved HII region is straightforward. If the distance is known, the size can be used to determine the kind of star that may be ionizing the nebula or whether more than one star is required. In general, the modern approach is to model the HII region in the sophisticated fashion as the last paragraph of Example 7.4 has outlined, and apply any other knowledge of the region that might be available. We will see other ways, for example, in which quantities like the density of an HII region can be obtained (see Chapters 10 and 11).

The arguments outline above apply to *any* ionized region, not just an HII region around a single hot star. If the source of ionizing radiation is an AGN at the centre of a spherical region in a galaxy, then one need only insert the spectrum of an AGN (a power law) instead of the Planck function in Eq. (7.5). Other geometries could also be imagined.

Example 7.5

Why do HII regions and planetary nebulae have sharp boundaries?

If we assume a density of $n = 100\ \mathrm{cm}^{-3}$ and use the cross-section for photoionization in Table 7.1, the mean free path (Eq. 5.18) of a photon to the ionization of hydrogen is $\bar{l} = 1/(n\sigma_{\mathrm{HI}\rightarrow\mathrm{HII}}) = 1/[(100)(6.3\times10^{-18})] = 1.6\times10^{15}\mathrm{cm} \approx 100$ AU. Thus, the transition region between an ionized and a neutral gas of this density is about 100 AU.

A close HII region is the Orion Nebula at a distance of 460 pc, and a close planetary nebula is the Helix Nebula at a distance of about 140 pc. The angular sizes of transition regions for these two objects would be 0.23 and 0.76 arcseconds, respectively (Eq. I.9). HII regions and planetary nebulae are normally much farther away so the boundaries are typically unresolved using standard telescopes without adaptive optics.

7.3 THE WAVEFRONT REDIRECTED – REFRACTION

Refraction is another example of light interacting with matter. As light passes from one kind of transparent medium to another that has a different index of refraction, n, the wavefront changes direction according to *Snell's law* (Table I.2). Figure 7.11 shows how the

[12] The photoionization code, 'CLOUDY', is often used, see www.nublado.org.

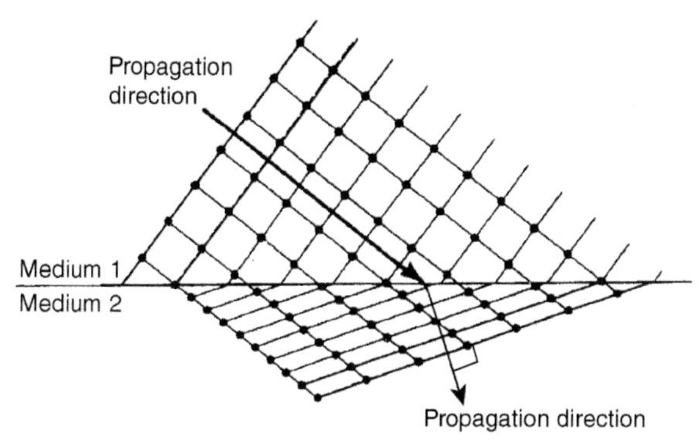

Figure 7.11 This simple diagram is a nice illustration of how refraction can be thought of in terms of both photons (light 'particles') and waves. The direction of propagation is from Medium 1 to the denser Medium 2 which has a higher index of refraction. The arrows show the direction of propagation of the wavefront and the dots represent the motion of individual photons. When the photons hit Medium 2, they slow down because of the interaction with matter particles. In this illustration, the photons themselves do not change direction when they interact with particles (consistent with scattering that is preferentially along the line of sight) but the wavefront bends. This is a convenient illustration to understand what is happening at the boundary but, collectively, photons will move perpendicular to the wavefront in both media. Source: Shu [27]. © 1982 University Science Books.

bending of a wavefront can be thought of as the result of interactions that slow the speed of light, v, through the medium that has the higher index of refraction ($n \equiv c/v$)[13]. Although other subtleties are involved, this illustration is useful in providing a visualization of the refraction phenomenon from both a wave and particle viewpoint. An example that we have already seen is the systematic bending of light as it enters the Earth's atmosphere (Section 4.3.1).

Aside from the Earth's atmosphere, refraction can also occur under other astrophysical conditions. For example, two media of the same composition but a different density should have different indices of refraction. Thus, in principle, any gas within which density variations exist should also experience some refractive effects. For neutral gas, the

[13] A similar decrease in the speed of light in a medium was considered for the diffusion of photons out of the Sun (see Problem 5.7), though in that case, the medium was opaque and scattering was in random directions.

degree of bending turns out to be negligible. However, refraction can become important when a signal propagates through an ionized gas (a *plasma*) or through a neutral gas that has a significant fractional ionization.

For a plasma, we can express the index of refraction using its definition (Table I.2)[14] together with Eq. (E.7) of the *online material*, and write the result in terms of the frequency, ν, of the incoming wave,

$$\mathrm{n} = \left[1 - \left(\frac{\nu_\mathrm{p}}{\nu}\right)^2\right]^{1/2}, \qquad \nu_\mathrm{p} \equiv 8.91 \times 10^3\, n_\mathrm{e}^{1/2}\ \mathrm{Hz} \tag{7.6}$$

where ν_p is the *plasma frequency* and n_e is the electron density. Note that we require $\nu > \nu_\mathrm{p}$ for n to be real, implying that a signal with a frequency less than ν_p cannot propagate through the plasma. A signal with $\nu < \nu_\mathrm{p}$, therefore, is reflected, rather than transmitted. On the other hand, if $\nu \gg \nu_\mathrm{p}$, then $\mathrm{n} \approx 1$ in which case there is no bending. In order to see the dependence of n on the electron density, we can use Eqs. (7.6) and E.6 of the *online material* with a binomial expansion for small ν_p/ν in order to find,

$$\mathrm{n} = 1 - n_\mathrm{e}\left[\frac{0.0063}{\left(\frac{\nu}{\mathrm{MHz}}\right)}\right]^2. \tag{7.7}$$

Refraction will be greatest for media in which n differs the most from 1. Therefore, from Eq. (7.7), we can see that refraction is most important at *low frequencies* and/or *high* n_e. The lowest radio frequencies observed are seldom less than several hundred MHz, so this expression shows that departures of the index of refraction from unity and therefore changes in the angle of propagation of the signal are very small unless electron densities exceed $\sim 10^7\ \mathrm{cm}^{-3}$ (for example, the coronas and atmospheres of stars).

However, both a sufficient electron density as well as some specific geometry, such as sheets or filaments, are required for refraction from ionized gas to be important in any *systematic* sense. Note that Eq. (7.6) implies that $\mathrm{n} \leq 1$, because $\nu > \nu_\mathrm{p}$ is required for propagation, so an intervening plasma along a line of sight can act like a *diverging lens*[15] in front of a background signal. If we repeated the atmospheric bending example given in Section 4.1 of the *online material*, but for the ionized ionosphere instead, the bending of light would be *away from the normal* in Figure 4.1 of that section.

From the point of view of astrophysical gases, refraction is mostly observed in terms of its random effects. Even in the very low-density ISM ($n_\mathrm{e} \approx 10^{-3}\ \mathrm{cm}^{-3}$), for example, refraction can be important for background point sources such as pulsars or distant point-like radio-emitting quasars. As the signal travels through regions of varying density in the ISM en route to the Earth, even a very small change in angle due to refraction can move the

[14] As indicated in the table, the index of refraction can be written as a complex number for cases in which some fraction of the incoming light is absorbed. For the astrophysical plasmas discussed here, a complex index of refraction is not needed.
[15] From the definition of n, this implies that $v > c$, but here v is the phase velocity (see footnotes to Table I.2) which can be greater than c without violating Special Relativity since it is the group velocity that carries information.

signal in and out of a direct path to the detector on Earth. This, along with other effects such as time variable scattering, causes the brightness of the signal to fluctuate with time, an effect called *interstellar scintillation*. This effect, a kind of *space weather* beyond the Solar System, is similar to the scintillation (Section 4.3.4) that we have already seen for the Earth's atmosphere, so the ISM acts like a *galactic atmosphere*. The refractive index is frequency-dependent, so the ISM is also *dispersive*, like a prism. That is, if a broad spectrum of wavelengths is present in the incident beam (which is the case for pulsars and AGN) then the lower frequency light will be bent more than the higher frequency light.

Pulsars, in particular, tend to have high velocities so, statistically, the component of their velocity that is transverse (i.e. perpendicular to the line of sight) will also be rapid (of order 100 km s^{-1}). Therefore, a background pulsar will 'illuminate' different parts of the ISM with time. Studying the properties of the resulting signal as a function of both time and frequency provides important information on the properties of the ISM, such as ISM turbulence. Scintillation due to refraction, for example, is variable on timescales of days for a pulsar at a distance of 1 kpc moving at 100 km s^{-1} and an ISM electron density fluctuation of order a factor of 2. The implied size scale for the perturbing cloud is then $\sim 1.5 \times 10^{12}$ cm [21], or one-tenth of an astronomical unit. If dispersive properties of the ISM are considered, size scales of 10^{10} cm or smaller can be studied. This is smaller than the radius of the Sun! Estimates for the size scales of interstellar density fluctuations that have been studied in this way range from about 10^7 to 10^{18} cm and the number density of clouds follows the power law, $N \propto D^{-11/3}$, where D is the cloud diameter.

Thus, analysing the perturbation in the signal from a background source is a powerful tool for studying the richness of the structure of the ISM on size scales that would otherwise be inaccessible to direct imaging (Problem 7.14). While one might see the ISM as just 'in the way' of understanding a background signal, instead, the background signal is a probe of the nature of the ISM itself. More information about a plasma and its index of refraction is provided in Appendix E of the *online material*.

7.4 QUANTIFYING OPACITY AND TRANSPARENCY

7.4.1 Total Opacity and the Optical Depth

So far, we have discussed gases in terms of whether they are 'transparent' (mean free path is greater than the size of the gas, such as the Earth's clear sky) or 'opaque' (mean free path is less than the size of the gas, such as in the Sun). An illustration of these two cases can be seen in Figure 6.2. We now wish to *quantify* these terms. This will allow the characterization of cases that are intermediate between being fully opaque and fully transparent, such as a cirrus cloud in front of the Sun, a mist around a street lamp, or a partially transparent HI cloud in the ISM. That is, *how transparent*, or *how opaque* is the gas?

As we have seen from the previous sections, there are many possible ways in which a photon can interact with matter. A key parameter that helped to determine whether an interaction might occur was the mean free path, $\bar{l}$ (Eq. 5.18), already discussed in a variety of contexts for both particle–particle collisions as well as photon–particle collisions (e.g. Examples 5.4, 5.5, 7.2, 7.3). Although the mean free path is a 'representative' path length, not every photon (or particle) will travel exactly this distance before interacting. It would therefore be a rather challenging job to characterize all processes that might be interfering with an incoming signal. Is there a more straightforward approach?

Since we want to understand what fraction of the incoming beam is lost as it travels through an intervening cloud, a simpler approach is to *fold together* the processes that, from the point of view of the observer, eliminate photons from the line of sight, that is, *all* scatterings and *all* absorptions. This is why the term 'absorption' is often used even if scattering is actually occurring. From the point of view of an observer who is trying to understand the incoming signal, it *might* not matter. The important issue is to know what fraction of the light is lost and what fraction is transmitted through the gas. However, at some level, these details need to be grappled with, as we illustrate below. The loss of a signal from scatterings and absorptions, when folded together, is referred to as total *opacity* (or *extinction* which is used for dust, see Sections 5.5.1, 7.5).

We now introduce a unitless parameter called the *optical depth*, τ_ν such that an infinitesimal change in optical depth, $d\tau_\nu$, along an infinitesimal path length, dr, is given by,

$$d\tau_\nu = -\sigma_\nu \, n \, dr = -\kappa_\nu \, \rho \, dr = -\alpha_\nu \, dr \tag{7.8}$$

where σ_ν is the *effective cross-section* for the interaction (cm^2), κ_ν (cm^2 g^{-1}) is the *mass absorption coefficient* and α_ν (cm^{-1}) is the *absorption coefficient*, each at a given frequency, ν. The quantities, n, and ρ are the particle density (cm^{-3}) and mass density (g cm^{-3}), respectively. The negative signs in Eq. (7.8) indicate that, while the coordinate, r, increases in the direction that the incoming beam is travelling, the optical depth is measured *into* the cloud as measured from the observer (Figure 8.2).

Even though 'absorption' is used for these coefficients, this term is often a catch-all to include both absorptions and scatterings as mentioned above and illustrated in Figure 7.1a and b[16]. More accurately, however, these terms should be separated into a true absorption and a scattering portion (e.g. Eq. (7.12), below).

The total optical depth is then determined from an integration over the total line of sight through the source. Using the version with the mass absorption coefficient,

$$\tau_\nu = -\int_l^0 \kappa_\nu \rho \, dr \approx \kappa_\nu \rho \, l \tag{7.9}$$

where we integrate from the near to the far side of the cloud and l is the line of sight thickness of the cloud. The latter approximation of Eq. (7.9) is often used for situations in which the density and mass absorption coefficient can be considered constant, or approximately constant, through the cloud.

[16] The coefficients could also include corrections for scattering *into* the line of sight.

The optical depth provides a quantitative description of how transparent or opaque a gas is. If $\tau_\nu < 1$, then the probability that the photon will be scattered or absorbed is <1 and the cloud is *optically thin* or transparent (Figure 6.2 top). If $\tau_\nu > 1$, then the probability that a photon will be scattered or absorbed is high and the cloud is *optically thick* (Figure 6.2 bottom). If $\tau_\nu = 1$, then the cloud is 'just' optically thick and the mean free path is equal to the line of sight thickness of the cloud. The optical depth is the number of mean free paths through an object along a line of sight.

The shape of the function, τ_ν, when plotted against frequency, shows how the effective absorption cross-section varies with frequency for a cloud of given density. From Eq. (7.8) it is clear that the optical depth depends on (i) the type of material, (ii) the frequency of the radiation, (iii) the density of material and (iv) the line of sight distance. It is possible, for example, for a small dense cloud to have the same optical depth as a large diffuse cloud of the same material. It is also possible for two clouds of the same size and density, but different material, to have very different optical depths. Example 7.6 further elaborates.

Example 7.6

Consider two ionized clouds of pure hydrogen, each with the same fractional ionization of 99.9%, density of 10^{-2} cm^{-3} and temperature of $T = 10^4$ K. Cloud 1 has a line of sight thickness of $l = 1$ pc and Cloud 2 has $l = 100$ pc. Compare the optical depth of Cloud 1 at a wavelength of $\lambda = 121.567$ nm to the optical depth of Cloud 2 at a wavelength of $\lambda = 200$ nm. Assume that only scattering processes are important (see Table 7.1) and ignore scattering into the line of sight.

Cloud 1: Most particles are ionized, so electron scattering must be considered. In addition, however, the incoming photons are exactly at the wavelength of the resonance scattering of the Lyα line so we also need to consider whether Lyα scattering will be important for those few particles that are neutral (all of which we assume are in their ground states, Section 5.4.6). Then, from Eqs. (7.8) and (7.9),

$$\begin{aligned}\tau_\nu &= \sigma_\nu n\, l = \sigma_T(0.999)\, n\, l + \sigma_{Ly\,\alpha}(0.001)\, n\, l \\ &= (6.65 \times 10^{-25})\,(0.999)\,(10^{-2})(3.09 \times 10^{18}) \\ &\quad + (5.0 \times 10^{-14})\,(0.001)\,(10^{-2})\,(3.09 \times 10^{18}) \\ &= 2 \times 10^{-8} + 1.5 = 1.5\end{aligned} \tag{7.10}$$

where we have used the Thomson scattering and Lyα cross-sections from Table 7.1. This cloud is optically thick, with an opacity dominated by Lyα scattering from a small fraction of neutral particles.

Cloud 2: In this case, the incoming photons are not at the frequencies of any resonance lines so we need only consider Thomson scattering,

$$\tau_\nu = (6.65 \times 10^{-25})\,(0.999)\,(10^{-2})\,(100)\,(3.09 \times 10^{18}) = 2 \times 10^{-6} \tag{7.11}$$

This cloud, although 100 × larger than the previous one, is highly optically thin.

As indicated above, the total mass absorption coefficient of Eqs. (7.8) and (7.9) must include all processes that may be occurring. We can therefore write a general expression for κ_ν that takes this into account,

$$\kappa_\nu = \kappa_{\mathrm{bb}_\nu} + \kappa_{\mathrm{bf}_\nu} + \kappa_{\mathrm{es}} + \kappa_{\mathrm{ff}_\nu} + \dots \tag{7.12}$$

where the subscripts indicate,

bb: all bound–bound processes such as line absorptions and scatterings
bf: bound–free processes, i.e. ionizations
ff: free–free processes, i.e. free–free absorptions of photons by free electrons near nuclei, as introduced in Section 7.2.1.
es: electron scattering (Thomson scattering)
+ ... : whatever other processes might be occurring[17].

Notice that electron scattering has been written without a subscript, ν. Although the other terms are frequency-dependent, Thomson scattering is not, as we pointed out in Section 7.1.1.1. Fortunately, not all processes are occurring in any given environment in which case Eq. (7.12) can be simplified accordingly. In fact, in many situations, a single process may dominate all others, as shown in Example 7.6 for the two clouds.

In the case of *stars*, all of the processes listed in Eq. (7.12) *do* have to be considered. This is not an easy or straightforward process since some of the absorption coefficients, themselves, depend on density and temperature and can also be highly frequency dependent. Bound–bound opacities, κ_{bb}, for example, will be significant at frequencies corresponding to the quantum transitions in atoms (of which there are many) and zero otherwise. For the bound–free and free–free absorption coefficients, it has been found [22] that they can be approximated by functional forms that depend on mass density, ρ, and temperature, T,

$$\overline{\kappa_{\mathrm{bf}}} = 4.34 \times 10^{25}\, f_{\mathrm{bf}}\,(Z)\,(1+X)\,\frac{\rho}{T^{3.5}} \quad \mathrm{cm^2\,g^{-1}} \tag{7.13}$$

$$\overline{\kappa_{\mathrm{ff}}} = 3.68 \times 10^{22} g_{\mathrm{ff}}\,(1-Z)(1+X)\,\frac{\rho}{T^{3.5}} \quad \mathrm{cm^2\,g^{-1}} \tag{7.14}$$

here, X, Y and Z represent the stellar composition (Section 5.3.1) and note that if X and Z are known, then Y is determined (Eq. 5.2). The correction factors, f_{bf} and g_{ff}, are of order $\approx 0.01 \rightarrow 1$ and ≈ 1, respectively[18]. The 'overline' indicates that the quantity is a *Rosseland mean opacity* for the process being considered, which is a statement that a weighted average over frequency has already been carried out. Any opacity that has the functional form $\propto \rho/T^{3.5}$ is referred to as a *Kramers' law*.

[17] There could be other sources of opacity in special environments. For example, some very high energy processes such as Compton scattering or the absorption of a high energy photon in an atomic nucleus are possibilities, or absorption processes specific to extremely dense environments such as those that are degenerate.
[18] The factor, $f_{\mathrm{bf}} = g_{\mathrm{bf}}/t$, where g_{bf} is the mean bound–free Gaunt factor for the various types of particles (of order 1, see also Eq. C.9 of the *online material*), and t is called the *guillotine factor* which is of order 1–100. The factor, g_{ff} is the free–free Gaunt factor (also of order 1) averaged over frequency. See Figure 10.2 for frequency-dependent values (Section 10.2).

Although these functions are approximations, they do show how the opacities depend on density and temperature. The bound–free opacity is dominated by metals in the interior of the Sun since hydrogen and helium are mostly ionized. However, the free–free opacity is dominated by free electrons from hydrogen and helium because these elements are far more numerous than the metals [22]. Note that, as the temperature increases, the bf opacity decreases because more atoms are already ionized and therefore there are fewer bound electrons to release from the atoms. The ff opacity also decreases because, at higher temperature, the free electrons are moving faster on average, and have a lower probability of absorbing an incident photon, given the shorter time spent near nuclei.

For electron (Thomson) scattering in regions of complete ionization, the function is

$$\kappa_{es} = 0.200\ (1 + X) \qquad \mathrm{cm^2\ g^{-1}} \tag{7.15}$$

(Problem 7.17), where a mean over frequency is not necessary since Thomson scattering is frequency-independent.

In stars in which there are few heavy metals, the most important processes are free–free absorption and electron scattering. In stars with significant metallicity and in the cooler outer layers of stars in which there are more bound electrons, κ_{bf} will dominate. The most important source of opacity in the outer layers of the Sun, for example, is bound–free absorption from the H^- ion. It is clear that the opacity of stars is a very sensitive function of density, temperature and composition, and can vary strongly between the deep interior and the surface. See [8] for opacity values that are consistent with the Solar interior model of Table T.8.

Determining the opacities in stars and elsewhere is an extremely important pursuit in astrophysics and many astronomers have spent their careers calculating and improving upon previously published values. Without a knowledge of opacities, we could not understand how the signal is affected as it travels through matter, be it stellar, interstellar, atmospheric or any other material. Without a knowledge of opacities in stellar interiors, moreover, we could not relate the luminosities that are observed at the surfaces of stars to the sources of energy generation at their cores, nor could other stellar parameters be accurately determined as a function of radius, such as densities, temperatures, ionization states, etc. In fact, the very *structure* of a star is intimately connected with its opacity and the variation of opacity with radius. As we will see in the next section, the opacity can even have important consequences for stellar *dynamics*.

7.4.2 Dynamics of Opacity – Pulsation and Stellar Winds

If an incident beam passes through a gas with little probability of interaction ($\tau_\nu \ll 1$) then that beam will have little or no dynamical effect on the gas. However, if the probability of absorption is very high ($\tau_\nu \gg 1$) then, as indicated in Section 3.5, the photons exert a pressure against the gas, as if hitting a wall. Thus, anywhere that the opacity is sufficiently high, dynamical effects can become significant. We now look at two important examples of the dynamics of opacity.

7.4.2.1 Stellar Pulsation

The first occurs in certain types of *variable stars* (stars that vary in brightness) that do so because they are radially expanding and contracting. *Cepheid variables* are examples of this, named after the proto-type star, δ Cephei. These stars are essential tools in obtaining distances in the Universe because they are bright enough to be seen over large distances and also because their pulsation periods (which are easily measured and vary from about 1 to 100 days) are related in a known way to their intrinsic luminosities. If the apparent magnitude of a Cepheid is measured and the intrinsic luminosity is found from the pulsation period, then the distance can be found using Eq. (3.33). Objects like this, whose intrinsic luminosities are known, are called *standard candles.* Cepheids are used to measure distances out to about 25 Mpc, which includes many galaxies other than the Milky Way.

It is common for small perturbations to exist in stars. Examples are perturbations from convection, rotating and distorted magnetic fields, the diffusion of heavier elements downwards, etc. If a small perturbation, say a compression, occurs in a layer within a star, both the density and temperature will increase slightly. From Kramers' laws of Eqs. (7.13) and (7.14), the temperature has a greater effect than the density so the mass absorption coefficient, κ_{ν}, normally decreases. Because $\tau_{\nu} = \kappa_{\nu}\, \rho l$ (Eq. (7.9)), a decrease in κ_{ν} in the perturbed layer offsets the increase in ρ so that there is little change in the optical depth. This means that the perturbation does not increase the opacity sufficiently to produce significant dynamical effects and stars tend to be dynamically stable to such perturbations.

However, there are certain regions within stars that are called *partial ionization zones.* These are at depths corresponding to temperatures at which hydrogen and helium, the latter corresponds to a deeper layer, become ionized. In these layers, if there is a small compression, instead of the temperature increasing as it did before, energy goes into ionization instead, which is a change of phase. The effect is similar to one in which energy is continually added to ice, raising its temperature to the melting point at which time the energy goes into the ice-to-water phase change rather than an increase in temperature.

The temperature rise in the stellar ionization zones is now very small, so the opacity, described by Kramers' Laws, *increases* with the increasing density of the perturbed layer. The optical depth then also increases. Photons from the deeper interior then exert a greater pressure against the perturbed, higher opacity layer, causing an expansion. The opposite occurs during expansion as electrons recombine, releasing energy, increasing the temperature and therefore decreasing the optical depth again. Thus, during compression, the partial ionization zone absorbs heat, and during expansion, the layer releases heat, acting like a *heat engine.* During expansion, the layer may overshoot, that is, the velocity given to the layer by radiation pressure perturbs it from gravitational (or *hydrostatic*) equilibrium and therefore it will fall back again gravitationally, causing the cycle to repeat. This process is called the *kappa mechanism* after the mass absorption coefficient, which acts like a valve, within the star.

Not every star experiences such pulsation. Stars that are too hot have ionization zones too close to the surface in regions in which the density is too low for the kappa mechanism to be effective. Stars that are too cool have deep convection zones which disrupt the

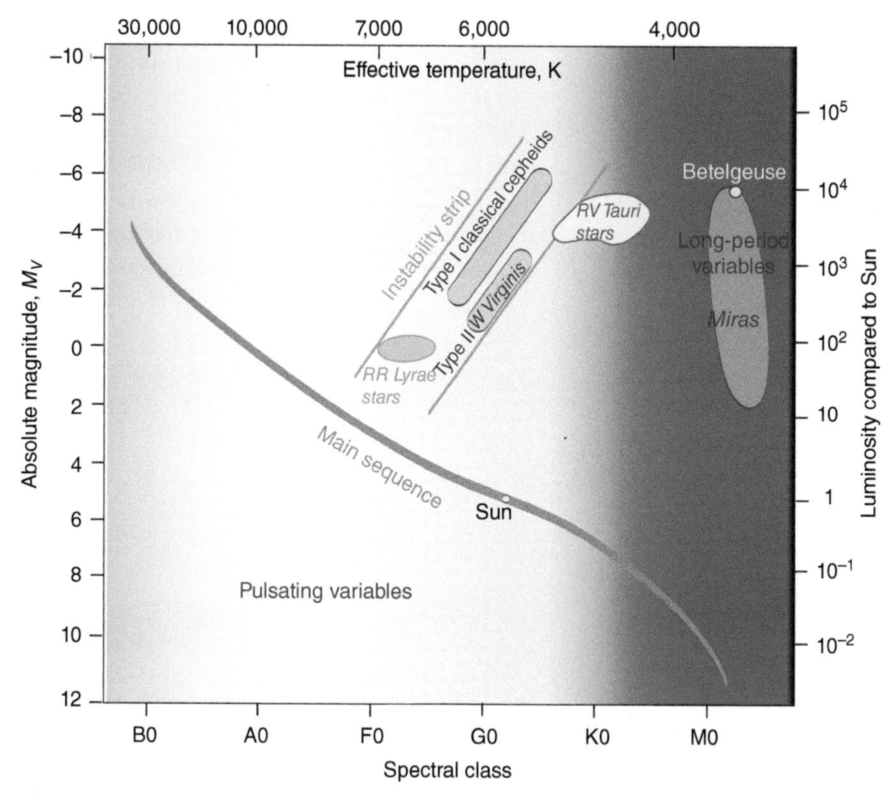

Figure 7.12 A stylized version of the HR Diagram showing a number of different kinds of pulsating variable stars. The instability strip contains stars whose pulsation is believed to be driven by the kappa mechanism. Source: CSIRO.

kappa mechanism. Thus, only certain regions of the HR diagram are amenable to this particular mechanism. An *instability strip*, in which HeII ↔ HeIII appears to be most effective, is shown in Figure 7.12. Note that there are many different kinds of variable stars and a variety of mechanisms, from minor brightness variations related to magnetic fields such as in the Sun (Figure 3.6), to variations resulting from geometric effects such as eclipsing variables, to more cataclysmic events such as irregular flaring. Stellar pulsation can result in mass loss and, if the outer layers of the pulsating star are extended and of low density, such as in a red giant star, then mass loss is more likely. Opacity-driven dynamical effects, such as the kappa mechanism, therefore play a crucial role in the evolution of stars.

7.4.2.2 The Eddington Limit

The second important example involves stars or other objects of very high luminosity. Most stars, like our Sun, are stable because their internal pressure provides support to balance their inwardly pulling gravity. The internal pressure is the sum of all pressures that might be significant (Eq. 5.16). For low mass stars like our Sun, it is particle pressure that dominates, as given by the perfect gas law (Eq. 5.11). However, in stars that are more massive, hotter, and more luminous, the radiation pressure (Eq. 6.16) dominates. In very high mass stars, the radiation pressure is so strong that the star's gravity is insufficient to retain the outer gaseous layers and they are blown off in *stellar winds* (Figure 7.10). This, in fact, puts a theoretical upper limit on the masses of stars.

In Section 5.3.2, we noted that a lower limit to the mass of a star is set by the inability of low mass stars to ignite nuclear burning in the core. Now we have an argument for an upper limit to the mass of a star. The highest luminosity that any object can have before radiation pressure starts to 'blow the object apart' is called the *Eddington Luminosity or Eddington Limit*. Our Sun's luminosity is 'sub-Eddington' because its luminosity is well below this limit[19]. However, we should not see any *stable stars* that are 'super-Eddington'. This theoretical upper limit is explored in Example 7.7. Although other processes can also contribute to stellar mass loss, such as magnetic fields, the simple argument of radiation pressure leads to values that are in rough agreement with observations.

Example 7.7

Estimate the highest stellar mass possible before radiation pressure starts to cause mass loss.

Consider the outer layer of a star with a thickness equivalent to a photon's mean free path, $\bar{l}$ (taken to be small in comparison to the star's radius). Then a typical photon that passes through this region will 'just' be absorbed and $\tau = \kappa_\nu \rho \bar{l} = 1$. The net pressure in such a layer is directed outwards and we can then use Eq. (3.25) with $\theta = 0$, to find the radiation pressure on this outer layer,

$$P_{rad} = \frac{f}{c} = \frac{L}{4\pi r^2 c} \quad (7.16)$$

where f is the star's flux, L is its luminosity, r is its radius, and c is the speed of light. The force per unit area due to gravity acting inwards on this layer is,

$$P_{grav} = \frac{F}{4\pi r^2} = \frac{1}{4\pi r^2}\frac{G M_* m}{r^2} = \frac{G M_* (\rho V)}{4\pi r^2 r^2} = \frac{G M_* \rho 4\pi r^2 \bar{l}}{4\pi r^2 r^2} \quad (7.17)$$

where m is the mass of the layer, ρ is its density and $V = 4\pi r^2 \bar{l}$ is its volume. For balance, we equate Eq. (7.16) to Eq. (7.17) and replace $\rho\bar{l}$ with $1/\kappa_\nu$ to give,

$$L_{Ed} = \frac{4\pi c G M_*}{\kappa_\nu} \quad (7.18)$$

where L_{Ed} is the Eddington Luminosity. That is, if $L > L_{Ed}$, then the star will lose mass through radiation pressure. It is therefore unlikely that stars with luminosities significantly greater than L_{Ed} exist. For the very high temperature, low-density outer layers of the most luminous, massive stars, the dominant source of opacity should be electron scattering. Therefore, using Eq. (7.15) with $X = 0.715$ (Eq. 5.4), expressing the mass in units of $M_\odot$ and the luminosity in $L_\odot$, then

$$L_{Ed} = 3.8 \times 10^4 \frac{M_*}{M_\odot} L_\odot \quad (7.19)$$

[19] The Sun does emit a Solar Wind of high energy particles, but these are accelerated by magnetic fields and are unrelated to stellar stability as discussed in this section. The mass-loss rate from the Sun is ≈ 9 orders of magnitude *less than* the mass-loss rate from stars near the Eddington limit.

It is now possible to compare this luminosity to luminosities that are actually observed for stars.

It is quite clear from Eq. (7.19) that the Sun, with a mass of 1 $M_\odot$, has an Eddington luminosity that is more than 10^4 times higher than its actual luminosity! The Sun is clearly not at risk of tearing itself apart. But what about the most massive stars? These are actually difficult to find because there are so few in comparison to low mass stars and their lifetimes are so short (Section 5.3.4) but a good place to look is in young stellar clusters. Here, we find that the most massive stars have $M \approx 136\ M_\odot$ [14], for which $L_{Ed} = 5.2 \times 10^6\ L_\odot$ (Eq. (7.19)). These stars are observed to have $M_{bol} \approx -11.9$ which, using Eq. (3.33), corresponds to an observed luminosity of $L_{obs} = 4.5 \times 10^6\ L_\odot$. Thus, the observed upper limits are in good agreement with the Eddington value (see also [6]). The 'limit', however, starts to become fuzzy, rather than sharp because of those strong winds. A few stars do show higher masses (~200–300 $M_\odot$) but mass loss is significant for these objects and it is also possible that they have resulted from mergers [3].

The Eddington limit can refer to objects other than stars as well. For example, the Eddington luminosity argument is often applied to active galactic nuclei (AGN) that exist at the centres of some galaxies (e.g. Figures 3.4, Section 7.5). AGN activity is believed to be powered by a *supermassive black hole*, that is, black holes in the range, 10^5–$10^{10}\ M_\odot$. Although no light can escape from the black hole itself, the material that is surrounding it, either accreting onto the black hole or rapidly orbiting it, is in a region in which the gravitational energy is extremely high. This material is heated to very high temperatures, producing radiation and its associated pressure. Just as for stars, the radiation pressure must not exceed that produced by the gravity of the black hole itself or else the material will be blown away. Thus, Eq. (7.19) provides a way of relating an *observable* AGN luminosity to the mass of an *unobservable* black hole (BH)!

Of course, this process will only work if the luminosity is actually at the Eddington value. It has indeed been argued that most AGN are radiating at or close to their Eddington limits, i.e. within a factor of ≈ 3 ([11], see also [2]). Typical values of luminosity for powerful AGNs are in the range, $10^{45} \rightarrow 10^{47}$ erg s^{-1}, nicely corresponding (Eq. (7.19)) to black hole masses of $\approx 10^7 \rightarrow 10^9\ M_\odot$ at the centres of galaxies. As can be imagined, these estimates are challenging. Accretion discs have a different geometry and are far more dynamic than the outer layers of stars. Considerable flux is also carried off in jets. The accretion discs are also subject to instabilities. Still, *super*-Eddington accretion is thought to be a relatively rare event, perhaps present in the early universe when supermassive BHs are first forming [1].

7.5 THE OPACITY OF DUST – EXTINCTION

Much as already been said about dust in Section 5.5. Here, we wish to connect some of the observational aspects of dust outlined in that section with the scattering and absorption properties of dust. Recall that scattering and absorption both take light out of the line of sight and is collectively referred to as *extinction*. This primarily occurs in the optical and UV parts of the spectrum which is the focus of this section. If we shifted our observations to the infrared, we would see the dust emit, rather than obscure light, as illustrated in Figure 5.21.

The interaction of light with dust grains is complex since dust grains have irregular shapes, come in a variety of compositions with individual grains also being of non-uniform composition, span a range of sizes, and some fraction of them may be aligned with the magnetic field. As a result, details of the interaction may involve scattering, absorption, diffraction and polarization. A convenient starting point is to assume that the grains are spherical, span a range of sizes in comparison to the incident light, and have uniform composition (though the composition may differ for different grains). An extensively used theory for dust is called *Mie Theory* which folds the absorption and scattering characteristics of a spherical grain into a *complex* index of refraction. We describe the basics of this theory in Appendix D.3 of the *online material*.

The main result is a determination of how efficiently light interacts with the grain as a function of wavelength and grain size. This is measured by the *extinction efficiency factor*, $Q_{ext\,\lambda}$, of a dust grain. $Q_{ext\,\lambda}$ is the ratio of the effective to the geometric extinction cross-section of a grain, and it is a function of $x = 2\pi a/\lambda$, where a is the grain radius and λ is the wavelength of the incident light. A plot of $Q_{ext\,\lambda}$, including its absorption and scattering parts, is shown as an inset in Figure 5.20 for one specific grain composition. The strongest interaction occurs when the wavelength of the light is of order the grain size.

Using Eqs. (7.8), (7.9), and D.30 of the *online material*, the optical depth of a set of grains, assuming that there is no change in properties as a function of line-of-sight distance, is,

$$\tau_\lambda = n_d\, \sigma_\lambda\, l = n_d\, Q_{ext\,\lambda}\, \pi\, a^2\, l \tag{7.20}$$

where n_d is the number of grains per unit volume, σ_λ is the effective grain cross-section (equivalent to $C_{ext\,\lambda}$ in Appendix D.3) and l is the line-of-sight depth of the region.

Once $Q_{ext\,\lambda}$ is determined for a variety of different grains, it is then possible to calculate a theoretical extinction curve for various admixtures of grain sizes and compositions. The result can then be compared with the observed extinction curve shown in Figure 5.20. The admixture of grain sizes and types which produces the closest fit is more likely to represent the dust that is actually present. Example 7.8 shows how one might approach this problem for a the simple case of a single grain size and composition (also Problem 7.19).

Example 7.8

Assume that all grains are identical and possess the properties illustrated by the inset to Figure 5.20 (a larger version is shown in Figure D.5 of the *online material*). Describe how the observed extinction curve shown in Figure 5.20 can be related to the inset image.

From Eqs. (3.13) and (5.29), the absorption in some waveband, λ, can be written,

$$A_\lambda = -2.5\,\log\left(\frac{I_\lambda}{I_{\lambda_0}}\right) \tag{7.21}$$

where I_λ is the specific intensity of the object undergoing extinction and I_{λ_0} is its specific intensity in the absence of extinction. Anticipating a result from Section 8.2 for the transmission of light through a medium that absorbs, but does not emit optical light (Eq. 8.9),

$$\frac{I_\lambda}{I_{\lambda_0}} = e^{-\tau_\lambda} \tag{7.22}$$

Combining the above two equations,

$$A_\lambda = 1.086\tau_\lambda \tag{7.23}$$

This result shows that an optical depth of *one* is approximately equivalent to an extinction of one magnitude[20]. Using Eq. (7.20),

$$A_\lambda = 1.086\, n_d\, Q_{ext\lambda}\, \pi\, a^2\, l \tag{7.24}$$

From this equation, Eq. (5.33), and our assumption that there is no change in dust properties along the line of sight,

$$\frac{E_{\lambda-V}}{E_{B-V}} = \frac{Q_{ext\,\lambda} - Q_{ext\,V}}{Q_{ext\,B} - Q_{ext\,V}} \tag{7.25}$$

The only wavelength-dependent quantity on the right-hand side of this equation is $Q_{ext\,\lambda}$ in the numerator.

Because we have taken the grain size to be constant, the abscissas of the $Q_{ext\,\lambda}$ plot (Figure 5.20 inset) and the extinction curve (Figure 5.20 main image) *both* are functions of $1/\lambda$. Therefore, if all of our assumptions are correct, then the *shape* of the curve of $Q_{ext\,\lambda}$ (except for some scaling) should be the same as the shape of the extinction curve. A visual comparison indeed shows that there is some similarity. However, there are also significant differences, leading to the conclusion that all dust grains are not identical.

PROBLEMS

7.1. [*Online*] What would be the scattering cross-section of a photon from a free proton (see Eq. D.2)? In an ionized gas, what fraction of scatterings will be due to free protons in comparison to free electrons?

7.2. [*Online*]

(a) From the relationships in Appendix D.1, show that the Einstein $A_{j,i}$ coefficient between energy levels, j and i can be written,

$$A_{j,i} = \frac{0.667}{\lambda_{i,j}^2}\frac{g_i}{g_j} f_{i,j} \tag{7.26}$$

where $\lambda_{i,j}$ is the wavelength of the transition, g_j, g_i, are the statistical weights of the upper and lower states, respectively, and $f_{i,j}$ is the absorption oscillator strength. Verify that this relationship gives the Einstein $A_{j,i}$ coefficients of the Ly γ and H γ lines given in Table C.1.

[20] This would be true for any medium (dust or gas) that does not, itself, produce emission at the same wavelength as the incident light.

(b) Show that the cross-section given by Eq. (D.12) can be written,

$$\sigma_S = \frac{\lambda_{1,2}^2}{2\pi}\left(\frac{g_2}{g_1}\right) \tag{7.27}$$

for the centre of the Ly α line in the absence of any other line broadening mechanisms, where $\lambda_{1,2}$ is the wavelength of the line centre. Verify that this result gives the natural bound–bound cross-section given in Table 7.1 of this text.

7.3. [*Online*] For a scattering process in which a Ly γ photon is absorbed by hydrogen and immediately followed by the emission of an H β and then Ly α photon, verify that the absorbed energy is equal to the total emitted energy.

7.4. [*Online*] Consider resonance scattering from bound electrons in the hydrogen atom (Appendices D.1.2 to D.1.4) in which the linewidths are given by the natural linewidth[21].
   (a) Which of the spectral lines in Table C.1 has the widest frequency response to photon scattering?
   (b) Under typical interstellar conditions, almost all neutral hydrogen atoms are in their ground states (Section 5.4.5, 5.4.6). Assuming that there are an equal number of photons at all frequencies impinging upon a hydrogen atom in the ISM, which transition is the most probable and why?

7.5. Consider Rayleigh scattering from bound electrons (Section 7.1.1.3).
   (a) In the Rayleigh scattering limit for any atom or molecule, how much greater (i.e. by what factor) will the scattering cross-section be for blue light (λ470 nm) in comparison to red light (λ660 nm)?
   (b) [*Online*] Calculate the total Rayleigh scattering cross-section, σ_R, for the hydrogen atom at the two wavelengths adopted in part (a) using data from Table C.1 and Eq. (D.24). Consider only the lines for which the Rayleigh limit is applicable. Determine the ratio of cross-sections for the two wavelengths and comment as to why there might be a difference with the results of part (a).

7.6. In Example 7.3, the mean free path of a photon in the Earth's atmosphere was found to be greater than 80 km. Above this altitude, the atmosphere is largely ionized (the *ionosphere*). From the information given at the beginning of Section 4.3, calculate the mean free path of a photon to Thomson scattering in the ionosphere. Is the ionosphere *more* or *less* transparent to optical photons than the atmosphere at sea level?

7.7. [*Online*]
   (a) What is the mean value of the increase in wavelength, $\overline{\Delta\lambda}$, for Compton scattering in an isotropic radiation field?

---

[21] Note that the natural line width is much narrower than observed in nature (see Section 11.3).

(b) Show that, after N Compton scatterings of equal angle, θ, each time, the final photon energy, E_N, can be written in terms of the initial photon energy, E_i, via,

$$E_N = \frac{E_i}{1 + \frac{N E_i}{m_e c^2}(1 - \cos\theta)} \tag{7.28}$$

(c) If a photon of energy, 5.1 MeV is repeatedly Compton scattered at an angle of 60°, how many scatterings are required to reduce its energy to 0.51 MeV?

7.8. What is the mean free path of a 0.51 MeV photon to Compton scattering through (a) the intergalactic medium of mean density, $n_e = 10^{-4}$ cm^{-3} and (b) through a Wolf–Rayet star of mass, 20 $M_\odot$, and radius, 20 $R_\odot$? Comment on how easily this gamma-ray photon can escape from the star and how easily it can travel through intergalactic space in the absence of other interactions.

7.9. For conditions at the centre of the Sun,

(a) Calculate a typical photon wavelength and the corresponding energy in eV. In what part of the electromagnetic spectrum does this fall?

(b) Would we expect Compton scattering or Thomson scattering at this location?

7.10. Determine the Strömgren radius of the resulting HII region if the Sun were to enter an HI cloud of the same density as the HII region described in Example 7.4, i.e. let n_e after ionization = n_H before. Assume that the same recombination coefficient can be used. Based on the result, qualify the statement: 'only O and B stars create HII regions'. [Hint. Either $B_\lambda(T)$ or $B_\nu(T)$ may be used. Consider whether either the Rayleigh–Jeans Law or Wien's Law might be used to simplify the integration.]

7.11. Suppose a spherical, pure HI primordial galaxy of uniform density is present in the dark ages (Section 7.2.2) before any stars or AGN exist. The galaxy has a mass and radius of $M_G = 10^{10}$ $M_\odot$ and $R_G = 10$ kpc, respectively. If an AGN of ionizing luminosity, $L_i = 10^{46}$ erg s^{-1} suddenly 'turns on' at the centre of this galaxy, would the entire galaxy become ionized and, if so, would photons leak out into the intergalactic medium? If not, what fraction of the galaxy would be ionized? For simplicity, assume that all ionizing photons from the AGN are exactly at the energy required to ionize hydrogen.

7.12. [*Online*] Consider the development given in Appendix E and write the plasma frequency equation (Eq. E.6) for ions of charge, *Ze*, instead of electrons. Determine the ratio of the plasma frequency oscillation period of ions compared to electrons for a pure hydrogen gas, i.e. how many times will an electron oscillate for each ion oscillation?

7.13. (a) Compute the electron plasma frequency, ν_p, for a typical maximum density of $n_e = 10^6$ cm^{-3} in the Earth's ionosphere.

(b) Look up the FM, AM and shortwave radio band ranges from some convenient reference. Indicate which frequencies in these bands, emitted by a ground-based transmitter, could reflect off of the ionosphere and therefore be detected by a radio receiver that is over the physical horizon. Repeat for the amateur ('Ham') radio frequencies in your country.

(c) [*Online*] Refer to Figure E.2 which shows a Type III radio burst. Estimate the maximum and minimum electron densities in the plasma between the Sun and the Earth encountered by this Solar flare.

(d) [*Online*] Indicate over what range of frequencies the radio burst of part (c) could have been detected from the Earth's surface. How much of the emission shown in Figure E.2 could have been detected from the Earth's surface?

7.14. What would be the angular size (arcsec) subtended by the smallest interstellar cloud described in Section 7.3 if it were at the fairly nearby distance of 50 pc? Determine how large a radio telescope (let $\lambda = 20$ cm) would be required to image such a cloud directly. Comment on the technique of using interstellar scintillation for probing these size scales in comparison to direct imaging.

7.15. Compute optical depths for the following cases and indicate whether the cloud or medium is optically thick or optically thin (see Table 7.1):

(a) 5.1 MeV photons travelling 100 Mpc through intergalactic space of density of $n_e = 10^{-4}$ cm^{-3}.

(b) An incident beam containing 13.6 eV photons impinging on an ISM HI cloud of density, $n_H = 1$ cm^{-3} and thickness, $l = 1$ pc.

(c) Incident photons at λ500 nm in an upper layer of a star that is 100 km thick with a density, $\rho = 10^{-8}$ g cm^{-3} and a total mass absorption coefficient of $\kappa_{500\,\mathrm{nm}} = 0.83$ cm^2 g^{-1}.

7.16. Consult Table T.9 and explain why the Sun's limb appears sharp to the unaided eye (assuming no glare as would be the case if the Sun is viewed behind a thin, partially transparent cloud).

7.17. Using the Thomson scattering cross-section, $\sigma_T = \sigma_{es}$, derive Eq. (7.15). Equations (7.8) and (5.14) will be useful.

7.18. From the information given in Table T.8 or Figure T.2, estimate κ_{bf}, κ_{ff} and κ_{es} at positions, (i) $R = 0.1\ R_\odot$, (ii) $R = 0.5\ R_\odot$ and (iii) $R = 0.9\ R_\odot$, within the Sun. Adopt $f_{bf} = 0.1$ and $g_{ff} = 1$. Comment on the relative importance of these three processes (bound–free absorption, free–free absorption, and electron scattering) as a function of depth.

7.19. [*Online*] Estimate $\frac{E_{\lambda-V}}{E_{B-V}}$ for a wavelength, λ210 nm, assuming that all dust grains are identical with radii of $a = 0.1$ μm and have the properties shown in Figure D.5, which is reproduced as an inset in Figure 5.20 (see Example 7.8). Compare the result to the observed value shown in the main image of Figure 5.20. Estimate how much this would change if all grains had radii a factor of 2 larger and comment on the results.

JUST FOR FUN

7.20. It can be disconcerting to stand on ice and look through to the water below, and possibly to the bottom of the pond. How thick does the ice have to be before it becomes opaque? The extinction coefficient for ice varies depending on surface texture and impurities, but a reasonable value is $\kappa_{ext} = 0.009$ cm^2 g^{-1} at λ500 nm.

7.21. You have landed on an alien planet (near a Sun-like star) that has a pure methane atmosphere. Is the sky blue? Is the sky blue on Titan? Why or why not?

Chapter 8
The Signal Transferred

> Astronomers ... speak of million degree temperatures, of densities smaller than our lowest vacuum; they study light that left its source two hundred million years ago. From a fleeting glimpse, they reconstruct a whole history.
>
> – Cecilia Payne-Gaposchkin [6]

8.1 TYPES OF ENERGY TRANSFER

The fact that we see light in the Universe means that the energy contained in that light – a tiny fraction of what was actually generated – has been transported from its point of origin to our eyes or our telescopes. With no intervening matter, the energy transferred is just that contained in photons diminished by distance (Eq. 3.9). With intervening matter, as we have seen, photon–particle, particle–particle, or other interactions can transfer or 'share' energy. We have so far considered how such interactions might divert a photon out of (or possibly into) the line of sight. We now wish to consider the net transfer of energy *along a path* as a beam of light moves through matter. This will allow us to relate the interactions that are occurring at a microscopic level to the light that we actually observe. In the process, we can obtain information about the material within which the energy transport is occurring.

The transport of energy along a path through matter can occur in four ways: (i) *thermal conduction,* (ii) *convection,* (iii) *radiation, and* (iv) *electrical conduction.* The first three are routinely cited as heat transport mechanisms in non-astronomical objects. *Thermal conduction* involves collisions between particles and therefore the sharing of particle kinetic energy, with no net bulk motion of the material. If one side of a material is hot and

Astrophysics: Decoding the Cosmos, Second Edition. Judith A. Irwin.

Companion website: www.wiley.com/go/irwin/astrophysics2e

the other cold, eventually collisions between adjacent particles will cool the hot side and warm the cool side. This process is important in solids. *Convection* involves the physical motion of a fluid (liquid or gas) and only exists in the presence of a gravitational field. In a gas, hotter pockets ('cells') move upwards into cooler regions via buoyancy and start a circulation or 'boiling' motion[1]. Energy is thus transported by the physical motion of hot gas upwards. *Radiation* is taken to mean the transport of energy via emission and eventual absorption again of photons, as we considered at length in Chapter 7. *Electrical conduction* refers to the motion of electrons through a medium in which the nuclei are fixed. It is an important mechanism for transporting energy in white dwarfs. Electrical currents can also exist in some regions in which there are strong magnetic fields, such as regions near sunspots.

In our own Sun (Figure 5.15), thermal conduction is negligible as an energy transport mechanism and electrical conduction will not be important until the Sun evolves off the main sequence and the Solar core becomes dense enough to be considered a white dwarf precursor (Section 5.3.2). It is convection and radiative transfer (ii and iii in the above list) that are the important mechanisms for transporting energy from the core to the surface – radiation out to 70% of the Solar radius and convection in the remaining outer 30%.

Convective energy transport dominates when the temperature *gradient* is high[2], and this occurs in the outer 30% of the Solar radius. Figure 8.1 shows a beautiful illustration of convective cells at the surface of the Sun, with the Earth superimposed for scale. The convective pattern is referred to as *granulation*. In other stars, the same energy transfer mechanisms are occurring although the interior locations and relative importance of radiation and convection may differ. Stars of lower mass than the Sun have deeper convective zones for example, whereas high-mass stars may have no convection in the outer layers, but instead have convective cores.

Radiative energy transport, as illustrated in Figure 5.15, is a diffusive process since the mean free path for a photon is very small. Although this process dominates out to 70% of the Solar radius, that 70% actually corresponds to 97% of the Solar mass, so radiative transfer is the dominant energy transport mechanism in the Sun.

Indeed, radiative transfer is arguably the most important energy transport mechanism in astrophysics. It involves the kinds of absorptions, scatterings, and re-emissions of photons that have been described in some detail in Chapter 7. Beyond its importance in stars, this is also how radiation from any background source transfers through intervening material that may lie along its path. This includes the signal from a quasar in the distant Universe, light from an interstellar cloud, or starlight passing through

[1] Convection is a complex process that does not lend itself well to a detailed mathematical description. An approach that considers how far a particular cell will rise before its temperature matches its surroundings and that calculates the resulting transfer of energy is called *Mixing Length Theory*. See [1] for more details.

[2] More specifically, convection occurs when $d\ln(P)/d\ln(T) < \gamma/(\gamma - 1)$, where P is the pressure, T is the temperature, and $\gamma \equiv C_P/C_T$ is the ratio of specific heats, where C_P and C_T are the amount of heat required to raise the temperature of 1 g of material 1 K held at constant pressure and constant temperature, respectively.

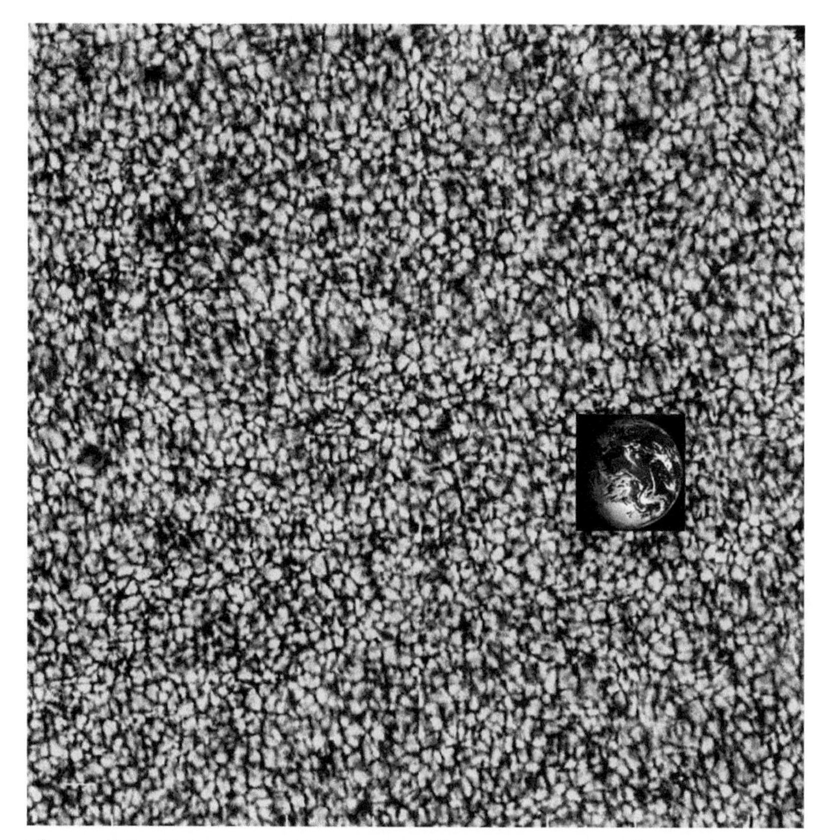

Figure 8.1 This granulation pattern illustrates convection in the outer layers of the Sun. Brighter regions are hotter and darker regions are cooler. This section is 91 000 km x 91 000 km in size, and the image was taken at a wavelength of λ403.6 nm using the Vacuum Tower Telescope on Tenerife in the Canary Islands. Source: Reproduced by permission of W. Schmidt, Kiepenheuer-Institut für Sonnenphysik. *Inset*: The Earth, *to scale*, as taken from the Galileo Spacecraft. Source: Reproduced by permission of NASA/Goddard Space Flight Centre.

the Earth's atmosphere. Moreover, the intervening material itself might not only be a source of opacity, but also a source of radiation. In the next section, then, we focus on radiative transfer which takes into account the propagation of a signal through absorbing, scattering, and emitting matter.

8.2 THE EQUATION OF TRANSFER

Let us consider an incoming signal of specific intensity, I_ν, passing through a cloud. By 'cloud', we mean any gaseous region containing matter, including sections of larger clouds, layers in the interiors of stars or their atmospheres, or whatever other gaseous

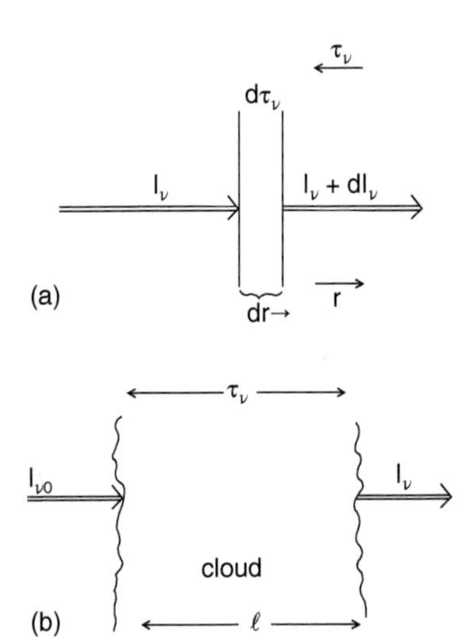

Figure 8.2 (a) Illustration of the change in intensity of a beam of light as it passes through material of infinitesimal thickness, *dr*, or optical thickness, $d\tau_\nu$. The outgoing intensity may be higher or lower than the incident intensity. (b) As in (a), but showing the background incident beam and the total thickness, *l*, or total optical thickness, τ_ν, of the cloud. The development of Section 8.2 does not require the beam to be perpendicular to the cloud, as shown here.

region, is of interest. As shown in Figure 8.2, the cloud has some total thickness, *l*, and the observer would be to the far right. The direction, *r*, is in the same direction as the incoming beam, and the optical depth, τ_ν, is measured in the opposite direction, *into* the cloud as seen from the observer.

As the beam passes through a small thickness of width, *dr*, its specific intensity will change from I_ν to $I_\nu + dI_\nu$, where

$$dI_\nu = dI_{\nu\,\mathrm{loss}} + dI_{\nu\,\mathrm{gain}} \tag{8.1}$$

The first term on the right includes all scattering and absorption processes that remove photons from the line of sight at the frequency being observed (from now on, referred to collectively as 'absorptions'). The second term contains all processes that add photons to the beam, such as all emission processes from the matter itself at the frequency being observed[3]. Note that dI_ν could be positive or negative. Using the absorption coefficient, α_ν ($\mathrm{cm^{-1}}$) introduced in Eq. (7.8), and introducing the *emission coefficient*, j_ν ($\mathrm{erg\ s^{-1}\ cm^{-3}\ Hz^{-1}\ sr^{-1}}$), Eq. (8.1) can be written,

$$dI_\nu = -\alpha_\nu\, I_\nu\, dr + j_\nu\, dr \tag{8.2}$$

Recall that the absorption coefficient represents the mean number of absorptions per centimetre that would be experienced by photons as they travel through a material (the inverse, $1/\alpha_\nu$, being a photon's mean free path). Therefore, if there are more incident photons, there will be more absorptions, which is why the loss term is a function of I_ν. The emission coefficient, on the other hand, is a quantity that represents all radiation-generating mechanisms per unit volume of material, and it does not depend on the strength of the incident beam[4].

[3] We ignore scatterings into the line of sight from other angles in this development.
[4] An exception is *stimulated emission* which is a process by which an incoming photon causes an atom to de-excite without the absorption of the incident photon. This is an emission process. However, since it depends on the strength of the background radiation, if it is occurring, it can be incorporated into the loss term as a correction factor. It is typically important only in very strong radiation fields.

Rearranging Eq. (8.2), using Eq. (7.8) and recalling that r increases in the direction of the incident beam while τ_ν increases into the cloud from the point of the view of the observer, we arrive at two forms of the *Equation of Radiative Transfer* (or simply, the *Equation of Transfer*),

$$\frac{dI_\nu}{dr} = -\alpha_\nu I_\nu + j_\nu \qquad \text{(Form 1)} \tag{8.3}$$

$$\frac{dI_\nu}{d\tau_\nu} = I_\nu - \frac{j_\nu}{\alpha_\nu} = I_\nu - S_\nu \quad \text{(Form 2)} \tag{8.4}$$

where

$$S_\nu \equiv j_\nu / \alpha_\nu \tag{8.5}$$

is called the *Source Function* and has units of specific intensity (erg s^{-1} cm^{-2} Hz^{-1} sr^{-1}). The Source Function is a convenient way of folding in all the absorptive and emissive properties of the cloud into one term.

The solution of the Equation of Transfer links the observed intensity, I_ν, to the properties of the matter, because τ_ν contains information on the type and density of the intervening material through Eq. (7.8). This link between the observed light and the physical properties of matter is at the heart and soul of astrophysics and provides the key to deciphering the emission that is seen. It is helpful to remember that, for a single physical situation, the Equation of Transfer could be written many times, once for every frequency for which data are available. For example, matter that 'absorbs only' at one frequency could 'emit only' at a different frequency (e.g. dust at optical and infrared wavelengths, respectively, see Figure 5.19 or 5.21).

8.3 SOLUTIONS TO THE EQUATION OF TRANSFER

We will now show the solution to the Equation of Transfer for some specific cases. In a sense, this section is somewhat repetitious. For example, a general solution could be found first, followed by simplification to specific situations. However, it is useful to see a progression from simple to more complex cases so that we can pause and consider the implications of each. It is worth taking this slower approach, given the importance of this equation to astrophysics. Eqs. (8.3) and (8.4) are equivalent, so we will adopt whichever form is most convenient in the process.

8.3.1 Case A: No Cloud

If there is no intervening cloud, then there is no absorption or emission, i.e. $\alpha_\nu = j_\nu = 0$. Using Eq. (8.3),

$$\frac{dI_\nu}{dr} = 0 \Rightarrow I_\nu = \text{constant} \tag{8.6}$$

Thus, we come to the same conclusion as first presented in Section 3.3, that is *the specific intensity is constant and independent of distance in the absence of intervening matter.*

8.3.2 Case B: Absorbing, but Not Emitting Cloud

If the cloud absorbs background radiation at some frequency, but does *not* emit at the same frequency, then the emission coefficient, $j_\nu = 0$ in which case (Eq. (8.5)) $S_\nu = 0$ and (Eq. (8.4)) becomes

$$\frac{dI_\nu}{d\tau_\nu} = I_\nu \tag{8.7}$$

Integrating from the front to the rear of the cloud,

$$\int_{I_\nu}^{I_{\nu 0}} \frac{dI_\nu}{I_\nu} = \int_0^{\tau_\nu} d\tau_\nu \tag{8.8}$$

where $I_{\nu 0}$ is the initial specific intensity when the beam first enters the cloud (the background intensity), I_ν is the emergent intensity on the near side of the cloud, and τ_ν is the total optical depth of the cloud along the line of sight being considered. The solution is

$$I_\nu = I_{\nu 0}\, e^{-\tau_\nu} \tag{8.9}$$

We see that the intensity diminishes exponentially with optical depth.

This equation was used to describe the extinction of optical light when travelling through a dusty medium (Example 7.8). It is also relevant to the opacity of the Earth's atmosphere which is an absorbing, but not emitting medium at optical wavelengths[5]. (Recall that we are using 'absorption' to include absorption and scattering in this section.) Thus, Eq. (8.9) allows us to quantify the atmospheric reddening that was discussed in Sections 4.3.5 and 7.1.1.3 (see Problem 8.1).

8.3.3 Case C: Emitting, but Not Absorbing Cloud

Using Eq. (8.3) with $\alpha_\nu = 0$, we find,

$$I_\nu = I_{\nu 0} + \int_0^l j_\nu\, dr \tag{8.10}$$

Unless we are modelling the cloud in some detail, in many cases an assumption that the emissive properties of the cloud do not change along a line of sight (i.e. j_ν independent of r) is adequate for estimating the needed physical parameters of the cloud. For such cases, Eq. (8.10) can be written,

$$I_\nu = I_{\nu 0} + j_\nu\, l \tag{8.11}$$

In either case, an expression for j_ν is required in order to relate the emission to the source parameters. Expressions for j_ν for several emission processes are provided in Chapter 10.

[5] Since the atmosphere does show some optical emission lines, this development applies to frequencies that are off the line frequencies.

8.3.4 Case D: Cloud in Thermodynamic Equilibrium (TE)

In thermodynamic equilibrium (Section 5.4.4), the radiation temperature is constant and equal to the kinetic temperature, T, everywhere in the cloud. The specific intensity, I_ν, is therefore given by the Planck function, $B_\nu(T)$ (Section 6.1), at every point in the cloud. Thus, there can be no intensity gradient ($dI_\nu/dr = 0$) so I_ν = constant and (Eq. (8.3)) becomes,

$$0 = -\alpha_\nu \, B_\nu(T) + j_\nu \quad \Rightarrow \quad B_\nu(T) = \frac{j_\nu}{\alpha_\nu} \tag{8.12}$$

Equation (8.12) is known as *Kirchhoff's Law* and is another way of stating that emissions and absorptions are in balance in such a cloud. By the definition of the Source Function (Eq. (8.5)), for a cloud in TE, we therefore have

$$I_\nu = S_\nu = B_\nu(T) \tag{8.13}$$

If such a cloud existed in reality, a photon that impinged on it would simply be absorbed and 'thermalized' to the temperature of the cloud. However, as indicated in Section 5.4.4, a cloud in true thermodynamic equilibrium is difficult to find. The best example is the cosmic background radiation whose spectrum is showed in Figure 6.3.

8.3.5 Case E: Emitting and Absorbing Cloud

When the cloud is both emitting and absorbing, the general solution for a case in which the emission and absorption properties do not change along a line of sight[6] is (Eq. (8.4)),

$$I_\nu = I_{\nu_0} \, e^{-\tau_\nu} + S_\nu(1 - e^{-\tau_\nu}) \tag{8.14}$$

This expression allows us to see the various contributions to the observed specific intensity. The first term on the RHS is identical to Eq. (8.9) and indicates that a background signal will be attenuated exponentially by a foreground cloud of optical depth, τ_ν. The second term includes two contributions. S_ν is the added specific intensity from the cloud itself, and $-S_\nu \, e^{-\tau_\nu}$ accounts for the cloud's absorption of its own emission.

Let us now consider the two extremes of optical depth: opaque ($\tau_\nu \gg 1$) and transparent ($\tau_\nu \ll 1$). Then, Eq. (8.14) reduces to,

$$I_\nu = S_\nu \qquad (\tau_\nu \gg 1) \tag{8.15}$$

$$I_\nu = I_{\nu_0} \, (1 - \tau_\nu) + S_\nu \tau_\nu \qquad (\tau_\nu \ll 1) \tag{8.16}$$

$$= I_{\nu_0} \, (1 - \tau_\nu) + j_\nu \, l \tag{8.17}$$

where we have used an exponential expansion (Eq. A.3 of the *online material*) for the case, $\tau_\nu \ll 1$. In Eq. (8.17), l is the line of sight distance through the cloud and we have used the

[6] See [8] for the derivation of this result and also for the case in which the properties change with line-of-sight distance.

definition of the Source Function (Eq. (8.5)) and the equations relating optical depth to the absorption coefficient (Eqs. 7.8, 7.9), assuming there is no variation in these quantities through the cloud. These equations will be very useful to us when we look at expressions for S_ν and j_ν in Chapter 10.

8.3.6 Case F: Emitting and Absorbing Cloud in LTE

In LTE, as described in Section 5.4.4, the temperature of a pocket of gas is approximately constant over a mean free photon path (see Example 5.5 for the interior of the Sun) and therefore, over this scale, the emissions and absorptions must be in balance. We therefore associate the Source Function with the Planck function,

$$S_\nu \equiv \frac{j_\nu}{\alpha_\nu} = B_\nu\,(T) \tag{8.18}$$

There is a net flux of radiation through the cloud, however, so I_ν is not constant and there is a nonzero intensity gradient. Then, following Eq. (8.14), we can write

$$\boxed{I_\nu = I_{\nu 0}\; e^{-\tau_\nu} + B_\nu\,(T)\,(1 - e^{-\tau_\nu})} \tag{8.19}$$

Equation (8.19) is, arguably, one of the most important equations in radiation astrophysics and can be applied to any *thermal cloud* (i.e. one whose emission depends on its temperature), provided the gas is in LTE. As indicated in Section 5.4.4, moreover, even if a cloud is not in LTE, sometimes an assumption of LTE is a useful starting point and can provide initial, approximate parameters for the cloud. We therefore explore some of the implications of the LTE solution in the next section.

8.4 IMPLICATIONS OF THE LTE SOLUTION

8.4.1 Implications for Temperature

Equation (8.19) leads to some important conclusions in specific circumstances. Following the development in Section 8.3.5, we can write down the result for the optically thick and optically thin extremes,

$$I_\nu = B_\nu\,(T) \qquad (\tau_\nu \gg 1) \tag{8.20}$$

$$I_\nu = I_{\nu 0}\,(1 - \tau_\nu) + B_\nu(T)\tau_\nu \qquad (\tau_\nu \ll 1) \tag{8.21}$$

$$= I_{\nu 0}\,(1 - \tau_\nu) + j_\nu\, l \tag{8.22}$$

Equation (8.20) is a statement that an optically thick cloud in LTE emits as a black body. A background source, if present, does not contribute to the observed specific intensity (although it might contribute to heating the cloud) since the foreground cloud completely blocks its light. Stars are good examples of this case (Section 6.1). *The only information obtainable from an optically thick source in LTE is the temperature* because the Planck curve is a function of temperature only. We can see into such a source only to a depth of the mean free path which is less than the depth of the source itself (Figure 6.2 bottom). We have no way of obtaining any information from the far side of an optically thick cloud and therefore no way of obtaining other information about it, such as its density.

On the other hand, Eqs. (8.21) and its equivalent (8.22) show that, for an optically thin source, the observed intensity depends on the intensity of the background source, if present, because the background source can be seen through the cloud. Moreover, Eq. (8.21) indicates that, *for an optically thin source, the observed intensity depends on both the temperature and density of the foreground cloud* because $B_\nu(T)$ is a function of temperature and τ_ν is a function of density via Eqs. (7.8) or (7.9). The signal that we detect has probed every depth through the cloud in such a case (Figure 6.2 top). The emission coefficient of Eq. (8.22) will also be a function of both density and temperature (see, for example Section 10.2).

Let us now consider a cloud *without a background source* ($I_{\nu_0} = 0$). If the cloud is *optically thick*, Eq. (8.20) does not change, and if the cloud is *optically thin* ($\tau_\nu < 1$), Eq. (8.21) becomes $I_\nu = B_\nu(T)\tau_\nu$. That is, for an optically thin cloud, the observed specific intensity is the Planck intensity *diminished by* the optical depth. Therefore, regardless of the value of τ_ν, it will always be true that,

$$I_\nu \leq B_\nu(T) \qquad \text{(no background)} \qquad (8.23)$$

In Section 6.1.1 Eq. 6.4, we found that the specific intensity of *any* signal, I_ν, can be represented by a brightness temperature, T_{B_ν}. Similarly, a black body is described by the Planck function which can be represented by the kinetic temperature, T. Therefore, Eq. (8.19) could be written with I_ν and I_{ν_0} expressed in terms of T_{B_ν} and $T_{B\nu_0}$, respectively, and $B_\nu(T)$ could be expressed in terms of T. As temperature increases, so does the respective specific intensity. Then, following Eq. (8.23) for the case with no background source, we find,

$$T_B \leq T \qquad (8.24)$$

Equation (8.24) is a simple, yet important result because it means that, for sources in LTE without background emission, *the observed brightness temperature places a lower limit on the temperature of the source.* The brightness temperature can be determined for any source by simply inserting the observed value of I_ν into Eq. (6.4). Once T_B is known, the true kinetic temperature must be the same or higher.

8.4.2 Observability of Emission and Absorption Lines

A straightforward observation is whether a spectral line is observed in absorption or emission in the presence of a background source. This observation provides an important clue as to whether the region in which the line formed is hotter or colder than the background, as Example 8.1 outlines.

Example 8.1

Under what conditions will a spectral line be seen in absorption or emission?

We will take a simple case in which a foreground cloud is capable of interacting with the background signal and therefore forming a spectral line at only one specific frequency (with a small line width) as Figure 8.3 illustrates. 'Off the line' here will mean any frequency that is outside of the frequency of the spectral line.

The background signal emits at a wide range of frequencies both on and off the line, but only interacts with the cloud over the frequencies of the line. So, off the line, there is essentially 'no cloud' as in Case A of Section 8.3.1. This means that $I_\nu = I_{\nu_0}$ off the line as Figure 8.3 illustrates.

What happens *on* the line? The condition for an *absorption line* is if the specific intensity at the line centre is less than the value of the background, i.e. $I_\nu < I_{\nu_0}$, as in Figure 8.3a and the condition for an *emission line* is if $I_\nu > I_{\nu_0}$, as in Figure 8.3b. Let us consider the *absorption case* and use Eq. (8.19). Then,

$$I_\nu = I_{\nu_0} e^{-\tau_\nu} + B_\nu(T)(1 - e^{-\tau_\nu}) < I_{\nu_0} \tag{8.25}$$

where I_{ν_0} is the specific intensity of the background continuum, $B_\nu(T)$ is the Planck function at the temperature of the cloud, and τ_ν is the optical depth of the cloud at the frequency of the line centre[7]. Rearranging Eq. (8.25) and repeating the process for an emission line lead to,

$$B_\nu(T) < I_{\nu_0} \quad \text{(absorption line)} \tag{8.26}$$

$$B_\nu(T) > I_{\nu_0} \quad \text{(emission line)} \tag{8.27}$$

From the Planck function and the definition of brightness temperature, this becomes,

$$T < T_{B_{\nu_0}} \quad \text{(absorption line)} \tag{8.28}$$

$$T > T_{B_{\nu_0}} \quad \text{(emission line)} \tag{8.29}$$

Therefore, if the temperature of the foreground cloud is less than the brightness temperature of the background source, then the line will be seen in absorption (Figure 8.3a) and if the temperature of the foreground cloud is greater than the brightness temperature of the background source, then the line will be seen in emission (Figure 8.3b). Keep in mind that the background source need not be in LTE nor does it have to be a source whose emission is related to a temperature at all, because a brightness temperature can be defined for any source of emission.

If there is *no* background source, then, $T_{B_{\nu_0}} \approx 0$ and the line is always seen in emission *even if the cloud is cold*[8]. Also, if $T = T_{B_{\nu_0}}$, then *no line will be seen.*

[7] Note that, since a spectral line has some intrinsic width (see Section 11.3), the optical depth varies over the line. It is highest at the line centre.

[8] However, the minimum background brightness temperature is 2.7 K which is set by the CMB (Figure 6.3).

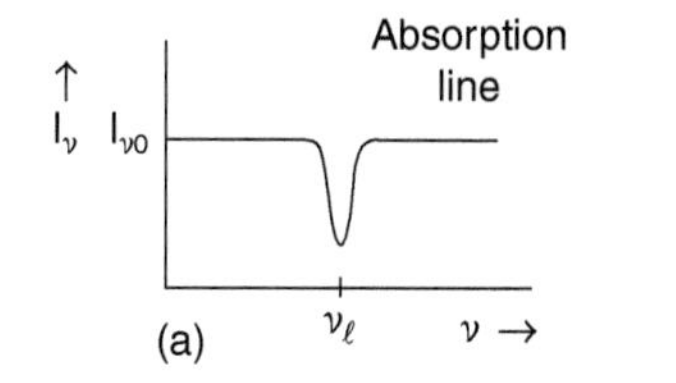

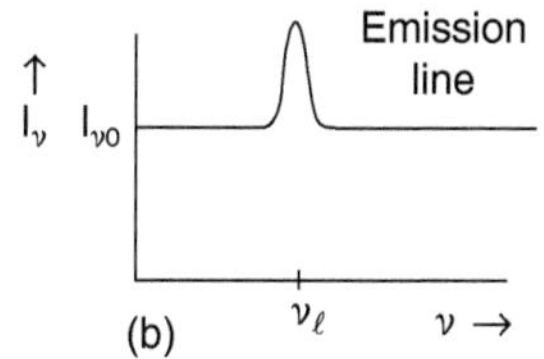

Figure 8.3 In this example, a cloud, such as is sketched in Figure 8.2, is capable of forming a single spectral line at a frequency, ν_ℓ. If the line is lower than the background level, I_{ν_0}, it is an absorption line (a) and if it is higher than the background, it is an emission line (b). Eqs. (8.26) and (8.27) give the conditions for the two cases. Note that the x-axis could be plotted as a function of frequency, wavelength, or velocity (for the latter, see Eqs. 9.3 and 9.4).

Some interesting conclusions can be drawn from Example 8.1. Suppose, hypothetically, that there were some very cold distant primordial HI clouds in the Universe (similar to the Lyα absorbers discussed in Section 7.1.1.2) and these clouds were at the same temperature as the 2.7 K cosmic microwave background. Then, we would have no way of detecting the λ21 cm line such a cloud, either in absorption or in emission. Similar arguments could be made for other lines from cold molecular clouds, should they exist.

8.4.2.1 Absorption and Emission Lines in the Sun

The Sun's lower atmosphere is an environment within which a rich array of absorption lines is formed, as can be seen in the Solar spectrum of Figure 8.4. These lines are called *Fraunhofer lines* and the fact that they are seen in absorption, rather than emission, indicates that they are formed in a region that is cooler than the background[9]. The structure and properties of the Solar atmosphere are shown in Figure 8.5, and details of the Solar *photosphere* are given in Table T.9. The photosphere is a region from just below the surface (the surface is taken to be where the optical depth is unity)[10] to a height of about 500 km at which the temperature is lowest.

Above the photosphere is the *chromosphere*, or 'sphere of colour', after its first identification as a thin reddish ring during Solar eclipses (Figure 6.6). The chromosphere is the region in which the temperature rises again, but hydrogen still remains predominantly neutral. The chromosphere also contains much activity such as *prominences* which are denser features, often loop-like, that are seen on the limb of the Sun and are related to

[9] A better comparison would take into account departures from LTE that occur in these regions.
[10] More accurately, a stellar 'surface' occurs at the depth from which most photons originate, on average. This actually corresponds to an optical depth of $\tau = 2/3$ (see [1] for details). For the Sun, however, the difference is only 12 km and will be neglected. Sometimes, the word 'photosphere' is used to denote just this surface, rather than the region shown in Figure 8.5.

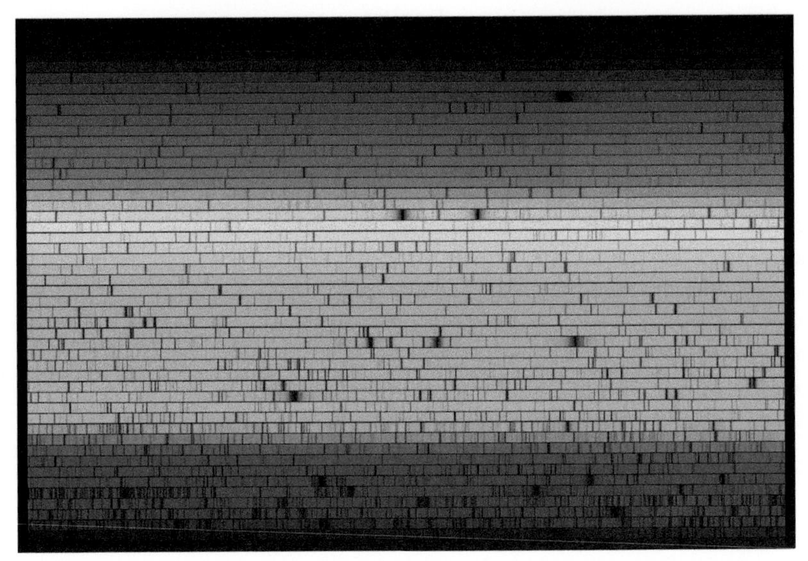

Figure 8.4 The Sun's Fraunhofer lines are seen as dark vertical bands superimposed on background colour. The wavelength increases from left to right along each strip which covers 60 Å. There are 50 horizontal strips with increasing wavelength from bottom to top, for a total spectral coverage from 4000 to 7000 Å. An intensity trace across these strips would result in a spectrum similar to that shown in Figure 6.5 centre. Image from a digital atlas based on observations taken with the McMath–Pierce Solar Facility on Kitt Peak National Observatory in Arizona. Source: Reproduced by permission of NOAO/AURA/NSF.

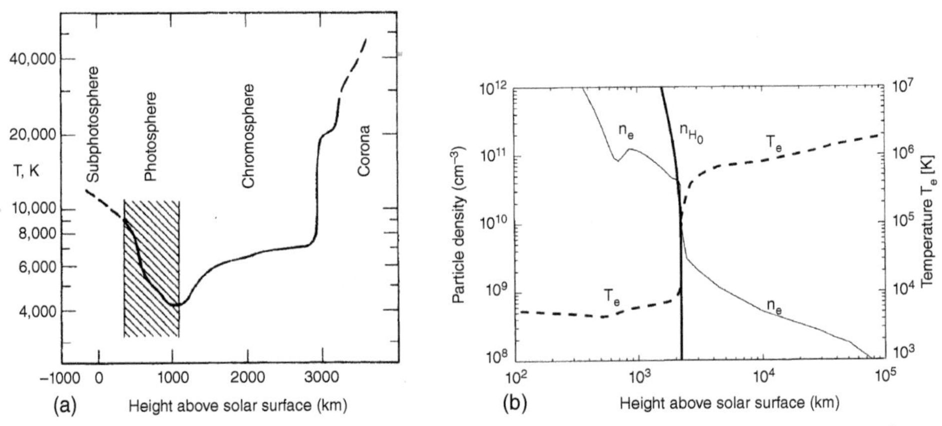

Figure 8.5 Plots of physical conditions in the Sun's atmosphere as a function of height. A height of 0 corresponds to the Solar surface which we take to be at $\tau = 1$.
(a) Temperature as a function of height, with labelling showing the parts of the Sun from below the photosphere to the lower corona. Source: Adapted from [3].
(b) Density and temperature as a function of height over a larger region from the upper photosphere to the corona. Note the change in hydrogen ionization state at a height of about 2000 km called the *transition region*. This marks the boundary between the chromosphere (hydrogen mostly neutral) and the corona (hydrogen mostly ionized). Source: M. J. Aschwanden.

the magnetic field structure. As Figure 8.5a suggests, it is in the upper photosphere and chromosphere that absorption lines will form, the hotter background radiation being supplied by the Sun's surface whose spectrum is given by the Planck curve.

The Sun also provides us with a good example of *emission lines*. As the temperature of the Solar atmosphere continues to rise with height, hydrogen becomes more fully ionized, marking the boundary between the chromosphere and Solar *corona*. The corona has a typical temperature of $\approx 10^6$ K but can be an order of magnitude higher or more during *Solar flares*, which are extremely high energy (up to 10^{32} erg) outbursts lasting minutes to hours. The corona extends well beyond the photosphere, as Figure 8.6 shows.

With temperatures higher than the Solar surface, any spectral lines in the corona would be seen in emission above the black-body spectrum from the surface. The density is low, however, so these emission lines are weak. Optical emission lines, for example, do not stand out sufficiently, in comparison to the background photospheric Planck curve, to be seen. Instead, they are more easily observed at positions in the corona outside of the Solar

Figure 8.6 This Solar eclipse image shows us glimpses of the reddish narrow chromosphere as well as the extended hot corona with its delicate filaments. The discrete features at the Solar limb are prominences and are reddish in this image because of H α emission. The optical corona is too faint to be seen in normal daylight and is only visible by eye during an eclipse. Source: Reproduced by permission of L. Viatour and the GNU Free Documentation Licence.

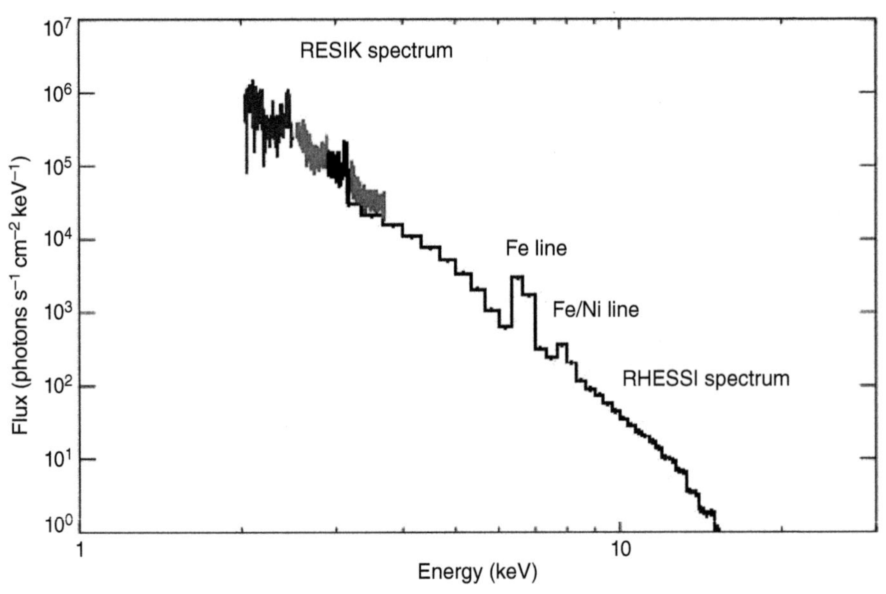

Figure 8.7 Two X-ray emission features at 6.7 and 8.0 keV are seen in this X-ray spectrum of a Solar flare. Each feature consists of a number of spectral lines that are unresolved, mainly from FeXXIV, FeXXV, FeXXVI, and some ionized Ni. The underlying continuum spectrum is not from Solar black-body radiation which is very weak at X-ray wavelengths, but rather from optically thin Bremsstrahlung and free-bound recombination radiation (see Sections 10.2, 10.3). This spectrum was obtained by the Reuven Ramaty High Energy Solar Spectroscopic Imager (RHESSI) telescope (NASA). To the left are data from the RESIK spectrometer aboard the Russian CORONAS-F Solar telescope launched in 2001. Source: K. J. H. Phillips and the RHESSI team.

disc where there is no background. At such high temperatures, emission lines of highly ionized species are present and these can occur at UV or X-ray wavelengths. Figure 10.11, for example, shows a UV emission line image of a large *coronal loop*. X-ray emission lines are also observed and can be seen against the Sun's disc, since the Sun's black-body photospheric emission is negligible, in comparison, at X-ray wavelengths. Figure 8.7 shows an emission line from a Solar flare superimposed on other continuum-emitting processes that are also present in the spectrum.

8.4.2.2 Absorption and Emission Lines in Other Stars

As in the Sun, the spectra of other stars also provide us with important information about the physical parameters of stars. Absorption lines, in particular, are observed in the spectra of other stars, as Figure 8.8 shows. Just like the Sun, these lines must be formed in regions that are cooler than the background photospheres of the stars. Moreover, the change in absorption line strength with stellar spectral type can be understood in terms

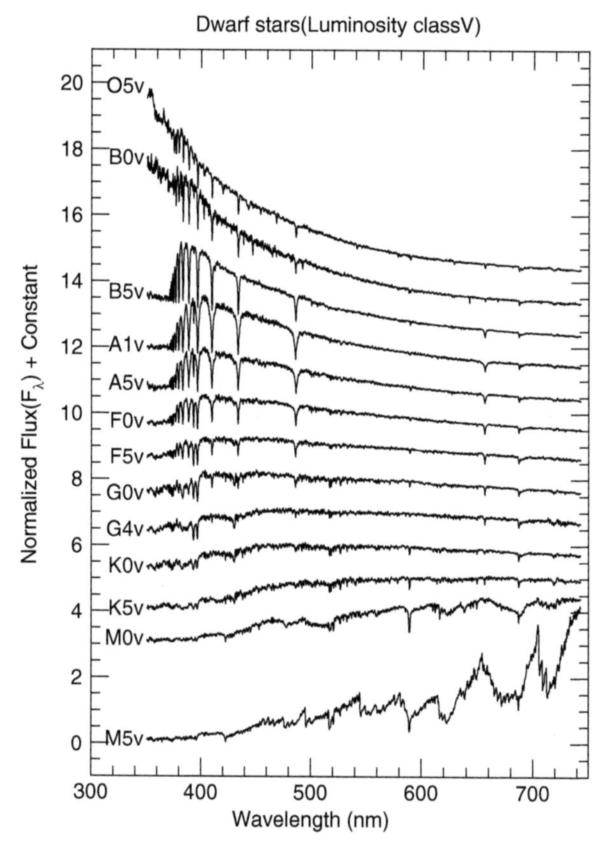

Figure 8.8 This plot shows a comparison of spectra for various types of stars of luminosity class, V (see Table T.5 for related stellar data). The shape of the background emission in each case roughly follows the shape of the Planck curve corresponding to the temperature of the star's photosphere. The locations and depths of the absorption lines provide a unique 'fingerprint' for the types and amounts of elements that are present. The Balmer lines, Hα (λ656.3 nm), Hβ (λ486.1 nm), and Hγ (λ434.0 nm), can be easily seen and are strongest (in this plot) at spectral type, A1V. The Sun's spectrum would lie between G0V and G4V. Source: Richard Pogge, with data from [4].

of a *temperature sequence*, as quantified in Tables T.5–T.7 for different stellar luminosity classes. An analysis of the strengths of various lines and their ratios further leads to tighter constraints on the temperature and density of stellar atmospheres (Section 11.4.2). Thus, absorption lines provide a wealth of information about stars, including the elemental abundances that were discussed in Section 5.3.4.

Because stars are generally unresolved, it is the absorption lines that are normally seen. It is not possible, for example, to isolate the coronas of these stars to look for weak emission lines as is possible for the Sun. There are, however, some stars that do show emission lines. Such stars usually have 'e' included in their spectral type, for example B[e] stars. Historically, such stars have shown Balmer emission lines, often variable, which are in the optical part of the spectrum and therefore should not normally be seen in *emission* in front of the star's photospheric Planck curve. Although there are various types of emission stars, it is believed that most of the emission is related to circumstellar material such as shells or discs (e.g. [7]). Wolf–Rayet stars, for example, are one type of emission line star (Figure 7.10).

8.4.3 Determining Temperature and Optical Depth of HI Clouds

As indicated in Section 5.4.5, any interstellar HI cloud has its hydrogen atoms with their electrons in the ground ($n = 1$) state. Therefore, the only detectable spectral line from hydrogen in a neutral cloud is from the spin-flip transition of the ground state, resulting in the λ 21 cm HI line. More information about this line can be found in Appendix C.4 of the *online material.* The lifetime for a spontaneous transition from the upper to the lower state of this hyperfine transition is very long, of order 10^7 years, so transitions are more likely to result from collisions whose timescales are shorter (Problem 8.5a). The fact that collisions are driving the line excitation means that the line is thermalized ($T_{ex} = T$ in the Boltzmann equation, Section 5.4.5) and the LTE solution to the Equation of Transfer may be used for this line. A nice simplification is that the λ21 cm line is always observed in the Rayleigh–Jeans limit ($h\nu \ll kT$, Problem 8.5b). Thus, the specific intensity is directly proportional to the brightness temperature (Eq. 6.6) and we can rewrite Eq. (8.19) as,

$$T_{B_\nu} = T_{B_{\nu 0}}\, e^{-\tau_\nu} + T\,(1 - e^{-\tau_\nu}) \quad \text{(with background source)} \tag{8.30}$$

$$T_{B_\nu} = T\,(1 - e^{-\tau_\nu}) \quad \text{(no background source)} \tag{8.31}$$

where T_{B_ν} is the observed brightness temperature, $T_{B_{\nu 0}}$ is the brightness temperature of the background source, and T and τ_ν are the kinetic temperature (taken to be equal to the temperature of the gas[11], see Section 5.4.5) and optical depth of the HI cloud, at a frequency, ν, in the line, respectively.

For the sake of completeness and also for future reference, in an optically thick case, both Eqs. (8.30) and (8.31) become

$$T_{B_\nu} = T \quad (\tau_\nu \gg 1) \tag{8.32}$$

and in an optically thin case with *no* background

$$T_{B_\nu} = T\tau_\nu \quad (\tau_\nu \ll 1) \tag{8.33}$$

[11] The 'spin temperature', T_S, is often used in these equations instead of T, just in case the assumption of LTE breaks down and T_S differs from T for some reason (see Appendix C.4 of the online material).

Equations (8.32) and (8.33) show more explicitly that the brightness temperature will always be less than the kinetic temperature (cf. Eq. (8.24) for a gas in LTE.

Suppose we wish to know the physical parameters of an HI cloud, specifically its temperature and density. If the HI cloud is optically thick, then we can find only its temperature, as indicated by Eq. (8.32), and the density is inaccessible to us.

Most HI clouds in the ISM, however, are *optically thin*. If there is no background source, then Eq. (8.31) can be used, and if there is a background source, then Eq. (8.30) can be used. In either case, it is clear that we have only one equation in two unknowns, which is insufficient. To find *both* the temperature and optical depth of an HI cloud, we need to find a cloud that both has – and does not have – a background source. Then, Eqs. (8.30) and (8.31) can both be used, giving us two equations in two unknowns: T and τ_ν. The optical depth is related to the cloud density (Eq. 7.9), so such a method would make both the temperature and density of the cloud accessible.

Fortunately, nature has provided many such examples, because most HI clouds that we wish to understand are relatively nearby in the ISM of the Milky Way and therefore have large angular sizes. By contrast, the background sources, which are typically radio-emitting quasars in the distant Universe, are small in angular size. Figure 8.9 shows the relevant geometry for a simple case.

When the telescope is pointed at the cloud directly in front of a background quasar, we call this 'ON'. The quasar typically has a brightness temperature higher than the temperature of the cloud in which case an absorption line is seen, as in Figure 8.3a. The emergent brightness temperature in this case can be called, T_{B_ν} (ON). When the telescope is pointing at the cloud but immediately *adjacent* to the background source, which we call 'OFF', then an emission line from the cloud is seen and the emergent brightness temperature is

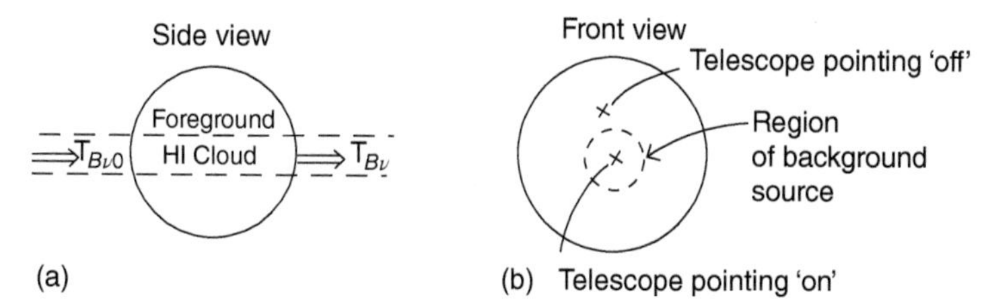

Figure 8.9 Geometry showing how a background radio-emitting source can be used to probe a foreground HI cloud. (a) The emission from a background source of brightness temperature, $T_{B_{\nu_0}}$, impinges on a foreground HI cloud and emerges at a brightness, T_{B_ν}, on the near side of the cloud. (b) The angular size of the background source (size exaggerated) is shown with a dashed circle. When the telescope points at the cloud directly along the line of sight to the background source, it measures a brightness temperature, T_{B_ν}(ON). When it points on the cloud but away from the direction of the background source, it measures T_{B_ν}(OFF).

T_{B_ν} (OFF) meaning with no background. Subtracting Eq. (8.31) from Eq. (8.30) for these two cases and rearranging yields the optical depth,

$$\tau_\nu = -\ln\left[\frac{T_{B_\nu}(\text{ON}) - T_{B_\nu}(\text{OFF})}{T_{B_{\nu 0}}}\right] \tag{8.34}$$

Notice that all of the quantities on the RHS are measurable. The numerator is the difference between the ON and OFF measured brightness temperatures, and the denominator, $T_{B_{\nu 0}}$, can be obtained by measuring the continuum of the background source just *off the line frequency* (analogous to $I_{\nu 0}$ in Figure 8.3). We take the optical depth here to refer to the centre frequency in the line, although it does not have to be. (See Chapter 11 for more information on line shapes and widths.) Once τ_ν is known, either Eq. (8.30) or (8.31) can then be used to obtain T.

Figure 8.10 shows a realistic situation in which a region of sky contains many sources, both in the ISM and outside of the Milky Way. The left image shows the broadband continuum emission. Some of the emission may not be truly 'background', but the marked quasar certainly is. The right image shows one frequency in the HI λ21 cm line. Here, the background emission (i.e. the left figure) has already been subtracted so that only emission and absorption in the λ21 cm line are shown. There is now a black dot where the quasar was. A measurement ON and OFF the quasar would yield the temperature and optical depth of the HI cloud at this position (Problem 8.6). Numerical values are given in the figure caption.

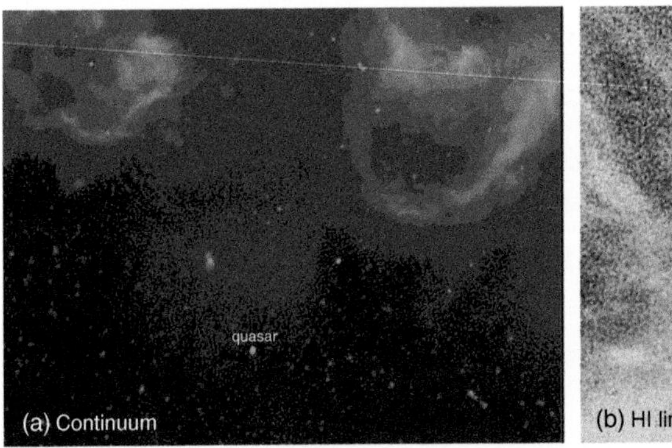

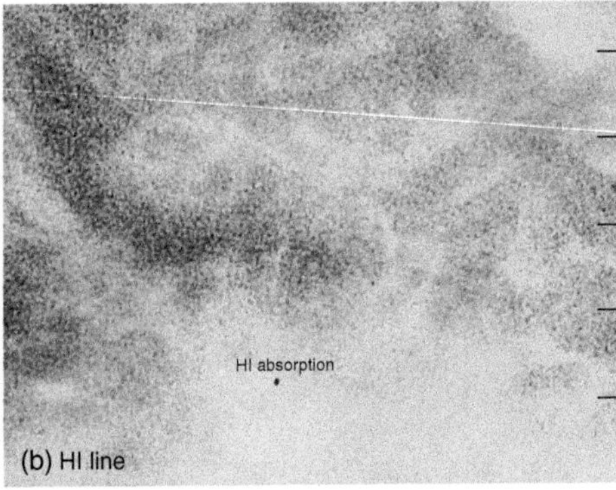

Figure 8.10 Two images of the same region near the plane of the Milky Way taken at λ21 cm. Tick marks separate 0.5° (about the diameter of the full Moon) on the sky, showing what a large region of the sky these images cover. (a) Radio continuum emission at a frequency just off the HI line. The two large structures at the top are HII regions in our Milky Way, whereas the myriad point or point-like sources are quasars in the distant Universe, one of which is labelled and has a brightness temperature of $T_{B_{\nu 0}} = 471$ K. (b) HI emission and some absorption from neutral HI clouds in our Milky Way. This image, which is at one particular frequency, ν, in the line, has *already had the continuum subtracted* (i.e. it represents $T_{B_\nu} - T_{B_{\nu 0}}$ at every position). The value at the location of the absorption feature is −126 K and the average value immediately adjacent to it is 55 K. Source: Reproduced by permission of The Canadian Galactic Plane Survey (CGPS) team [9].

Although nature has provided many examples of background sources behind HI clouds, one also needs to be aware of some subtleties. For example, this process will work only if the temperature and optical depth of the cloud are the same in the ON and OFF positions. Cloud temperatures should not change dramatically from place to place, but τ_ν depends on both the density and line of sight distance through the cloud (Eq. 7.9). Both of these quantities may vary with position, and therefore, a number of OFF measurements should be made in different directions as close as possible to the ON position and the results averaged. This development also assumes the presence of only a single HI cloud along a line of sight whereas there may be several. The fact that clouds are moving at different velocities, however, ameliorates this problem somewhat. The frequency of the HI line will be Doppler-shifted (Section 9.1.1) by different amounts for each cloud, so this helps in treating them as discrete objects along a given line of sight. Another issue is the possibility of variations within the spatial resolution of the telescope. For example, the telescope's beam in the ON position may have a larger angular size than the background quasar, in which case some fraction of the beam may be 'filled' with a signal that has no background.

A more fundamental problem is that the ISM is known to contain not only HI that is cool ($T \approx 100$ K), fairly dense ($n_{HI} \approx 1$ cm^{-3}) gas that is absorbing, called the *cold neutral medium* (*CNM*), but also hot (of order 8000 K), diffuse (≈ 0.1 cm^{-3}) HI, called the *warm neutral medium* (*WNM*). Contributions from the hot HI to the beam affect the results although, because the hot component is less dense, the cold component has greater effect. For example, if half of the beam is occupied by the CNM at $T = 100$ K and half by the WNM at $T = 8000$ K, this method will result in a temperature of $T = 200$ K, an error of a factor of two [5].

With these caveats in mind, this method is still a very powerful tool for probing the physical conditions in HI clouds. More sophisticated analyses that take the hot contribution into account find that the cold component has temperatures that range from about $20 \rightarrow 125$ K with a median temperature of 65 K, with warmer clouds in this range more common than colder ones [2]. Optical depths take on a range of values depending on the cloud size and density.

PROBLEMS

8.1. [*Online*]

(a) Let the Earth's atmosphere be approximated by a uniform density plane parallel slab (a layer as shown in Figure 4.1 of the *online material*) and modify Eq. (8.9) so that the specific intensity, I_ν, is written as a function of the specific intensity above the atmosphere, $I_{\nu 0}$, the vertical optical depth, $\tau_{\nu\,(z=0)}$ and the zenith angle, z.

(b) Use the result of part (a) to write an expression for the ratio of observed specific intensities at two different frequencies, I_{ν_1}/I_{ν_2}, for a source whose intrinsic spectrum is flat (i.e. the true specific intensity is the same at all frequencies).

(c) For clear sky optical depths of $\tau_{\mathrm{blue}(z=0)} = 0.085$ and $\tau_{\mathrm{red}(z=0)} = 0.060$, plot a graph of $I_{\mathrm{red}}/I_{\mathrm{blue}}$ as a function of zenith angle over the range, $0° \leq z \leq 75°$ and comment on the result.

8.2. An H α emission line of specific intensity, $I_\nu = 2.7 \times 10^{-4}$ erg s^{-1} cm^{-2} Hz^{-1} sr^{-1} at the line centre is observed from an ionized gas that is in LTE. Find a lower limit to the temperature of this gas.

8.3. Indicate whether, in the following cases, a spectral line would be seen in emission, in absorption, or not at all, and briefly explain why. In each case, the vantage point is the surface of the Earth.

(a) A region of the Sun's photosphere of temperature, $T = 5000$ K, measured in a direction towards the centre of the Sun's disc.

(b) A solar prominence of temperature, $T = 5000$ K, seen in the chromosphere during a solar eclipse.

(c) The Earth's atmosphere at night.

(d) A gas cloud at a temperature of 2.7 K with no discrete background source.

(e) An HII region in front of a background quasar. The quasar has a brightness temperature of $T_B = 100$ K.

(f) As in (e) but the quasar has a brightness temperature of $T_B = 80\,000$ K.

8.4. A B0V star (unresolved) shows some absorption lines like those shown in Figure 8.8, as well as Balmer emission lines. Argue that there is circumstellar material (e.g. a disc) around this star.

8.5. [*Online*]

(a) Compare the collisional timescale for de-excitation of the λ21 cm line in a typical interstellar HI cloud to the mean time for spontaneous de-excitation of this line. Table C.1 may be of help, as well as Tables 5.1 and 5.2 in this text. Confirm that the λ21 cm line is more likely to be collisionally, rather than spontaneously de-excited.

(b) Show that the λ21 cm line is always observed in the Rayleigh–Jeans limit.

8.6. (a) From the information given in Figure 8.10 (see also Section 8.4.3), find the temperature, T, and optical depth, τ_ν, of the HI cloud at the position of the labelled quasar. Is the cloud optically thick or optically thin?

(b) Assuming that the depth of the cloud along the line of sight is 5 pc, find the number of absorptions per cm, α_ν, in this cloud at the frequency of the line.

(c) Assuming that the cloud density is constant, how deep (pc) would the line of sight distance have to be for the cloud to become optically thick?

(d) What cloud temperature would be required for an HI emission line to be seen at the position of the quasar?

JUST FOR FUN

8.7. The Romulans have developed a new cloaking device to render their spaceships (which are effectively black bodies) invisible to their enemies. They call the cloaked ship, the 'Chameleon warbird', and can rapidly change the effective temperature of the hull to match whatever background is present. As Romulan chief engineer, what temperature limits would you recommend for Chameleon excursions into the globular cluster, NGC 6397, that is discussed in Section 12.2.3?

Chapter 9
The Interaction of Light with Space

There's no such thing as the unknown, only things temporarily hidden, temporarily not understood.

–Captain James T Kirk, Classic Series, Episode *The Corbomite Maneuver*

We now wish to consider how light interacts with space. Matter is not ignored in this process since it is matter that emits light and also affects the light's path. However, the interaction is not one of direct contact, such as we saw for scattering and absorption previously. If the signal did not intercept a single intervening particle or another photon, and even if we had perfect measuring instruments that did not alter it, the signal could still be perturbed from the state it was in, when emitted. In Sections 2.1–2.3, we introduced the concepts of *space–time*, the bending of space near a mass, and the path of light through curved space. Now, let us consider some of the more practical consequences of this

Astrophysics: Decoding the Cosmos, Second Edition. Judith A. Irwin.

Companion website: www.wiley.com/go/irwin/astrophysics2e

paradigm. How does a signal become altered as it travels through a vacuum? In the next sections, we consider some possibilities.

9.1 REDSHIFTS AND BLUESHIFTS

If a signal is emitted at a frequency, ν_0, it may be observed at a frequency, ν, that is shifted from the emitted frequency. Since $\lambda = c/\nu$, the observed wavelength will also be altered from the emitted value, i.e. the *rest wavelength*, λ_0. If λ has increased in comparison to λ_0, the light is said to be *redshifted* and if it has decreased, the light is *blueshifted*. For a photon, $E = h\,c/\lambda$, so redshifted and blueshifted photons have lower and higher energies, respectively.

Because, on cosmological scales, observed wavelengths are redshifted (Section 9.1.2), changes of wavelength are often described by a parameter called the *redshift parameter* or just *redshift*, z, defined by,

$$z \equiv \frac{\Delta\lambda}{\lambda_0} = \frac{\lambda - \lambda_0}{\lambda_0} => z + 1 = \frac{\lambda}{\lambda_0} \tag{9.1}$$

Thus, z describes the fractional change in the wavelength. An observed increase in wavelength corresponds to positive z and a decrease in wavelength to negative z. There are three possible origins for such a shift, as will now be described.

9.1.1 The Doppler Shift – Deciphering Dynamics

9.1.1.1 Kinematics – The Source Velocity

As a source moves through space, its velocity, v (the *space velocity*), can be expressed in terms of a component that is directed towards or away from the observer, called the *radial velocity*, v_r, and a component that is in the plane of the sky, called the *transverse velocity*, v_t (Figure 9.1a). Then, by Pythagoras' theorem, $v = \sqrt{v_r^2 + v_t^2}$.

The *radial velocity* produces a *Doppler shift*[1] as the source moves through space. A motion away from the observer results in a redshift, and a motion towards the observer gives a blueshift. Note that v_r is the radial velocity of the source *with respect to the observer*. In practice, though, the redshift is usually corrected to a *rest frame* that is common to all observers. This is because, as the Earth rotates and also orbits the Sun, a radial velocity measured from any location on the Earth's surface will vary with time. Radial velocities are often referenced to the Sun, in which case they are referred to as *heliocentric velocities*, $v_\odot$. Other options are to measure a radial velocity with respect to a point at the location of the Sun that is moving in a perfect circle about the centre of the Galaxy. Such an imaginary point is called the *Local Standard of Rest* (LSR). A radial velocity corrected to the LSR, v_{LSR}, differs from $v_\odot$ because the Sun's orbit about the centre of the Galaxy is not

[1] We do not consider the Doppler shift for *transverse* motion in this section, which is only important when speeds are relativistic.

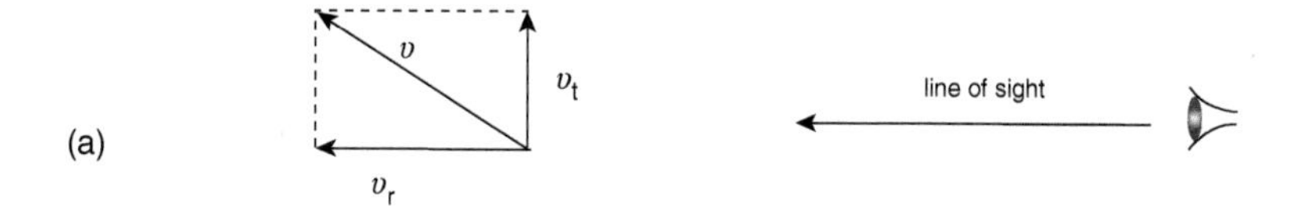

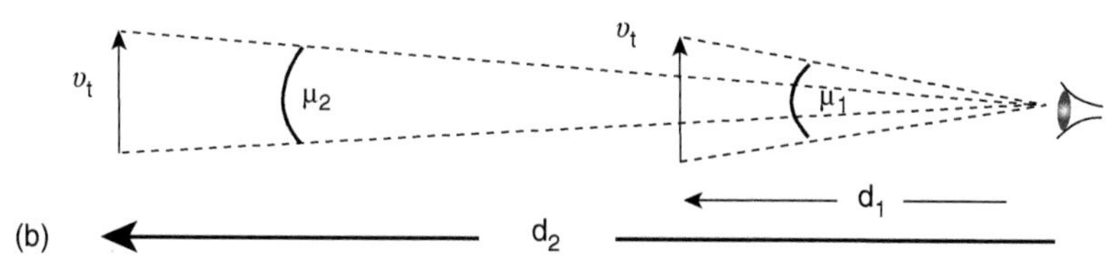

Figure 9.1 (a) An object's space velocity, v, can be represented as the vector sum of the transverse velocity in the plane of the sky, v_t, and the radial velocity in the line of sight, v_r. (b) The transverse velocity can be determined from the proper motion if the distance is known, i.e. $v_t = \mu d$. A nearby object (at d_1 in the figure) will have a higher proper motion than a distant object (at d_2) for the same v_t.

a perfect circle. Currently, the difference between these two measurements is less than about 20 km s^{-1}.

From the Doppler shift formulae for radial velocities of Table I.2 and Eq. (9.1),

$$z + 1 = \left(\frac{1 + \frac{v_r}{c}}{1 - \frac{v_r}{c}}\right)^{1/2}, \qquad z = \frac{v_r}{c} \quad (v_r \ll c) \tag{9.2}$$

The second expression, which is the approximation for radial velocities much less than the speed of light, is all that is needed for most cases (Problem 9.1). This simple relation is very powerful because it directly relates an observable quantity, z, to a physical parameter of the source, v_r, independent of any other knowledge of the source, including its distance. All that is necessary is to ensure that some spectral feature, usually a spectral line (Problem 9.2) can be identified, as Example 9.1 illustrates.

Example 9.1

The Hα line, emitted from a location near the centre of the Andromeda galaxy, M 31, is observed at a wavelength of $\lambda(\mathrm{H}\alpha) = 655.624$ nm. Write an expression for the radial velocity of a galaxy, assuming $v_r \ll c$, and determine the radial velocity of the Andromeda galaxy.

From Eqs. (9.1) and (9.2), the radial velocity is given by,

$$v_r = \left[\frac{\lambda - \lambda_0}{\lambda_0}\right] c = \frac{\Delta\lambda}{\lambda_0} c \tag{9.3}$$

Equation (9.3) can also be written in terms of frequency,

$$v_r = \left[\frac{\frac{c}{\nu} - \frac{c}{\nu_0}}{\frac{c}{\nu_0}}\right] c = \left[\frac{\nu_0 - \nu}{\nu}\right] c = \frac{-\Delta\nu}{\nu} c \tag{9.4}$$

We need only one of these equations and, given $\lambda_0(\text{H}\alpha) = 656.280$ nm from Table C.1 of the *online material*, as well as the above value for $\lambda(\text{H}\alpha)$, Eq. (9.3) leads to $v_r = -300$ km s^{-1}. Thus, the Andromeda galaxy is approaching us. Because the Andromeda galaxy is at a distance of $d = 779$ kpc, a straightforward estimate of the time at which it will start to merge with the Milky Way is $t = d/v_r = 2.5$ Gyr, assuming constant velocity and that the motion is radial only.

The problem of obtaining the transverse velocity is less easily solved. If the object is relatively nearby, then it may be possible to measure its angular motion in the plane of the sky over time (Figure 9.1b). This is called the *proper motion*, μ,

$$v_t = \mu d \tag{9.5}$$

and is simply the small-angle formula (Eq. I.9) expressed as a rate. Equation (9.5) will give v_t in km s^{-1} for μ in rad s^{-1} and d in km. The proper motion is typically quite small, often measured in units of arcseconds per year. Note that, unlike radial velocities, finding a tangential velocity requires that the distance to the object be known. Since μ is small, this kind of measurement has historically only been possible for the nearby stars (Problem 9.3).

New, more accurate measurements, however, are now pushing the proper motion limits. For example, the Gaia satellite that was introduced in Section 3.6.4 obtained proper motions of groups of stars in the Andromeda galaxy [26] as shown in Figure 9.2. The yellow arrows show the average proper motions of groups of stars in this galaxy. These arrows only show motions of stars in the plane of the *sky*, so they do not show the true space velocities which actually follow the rotation of the galaxy. Additional measurements of radial velocity, if combined as in Figure 9.1a, would reveal that rotation. The Gaia measurements have revealed proper motions of order tens of microarcseconds per year (μas yr^{-1})! Combined with previous data, it was found that the Andromeda galaxy has a global net transverse motion of $v_t = 57$ km s^{-1}. This means that its space velocity, although dominated by the radial velocity, is not perfectly radial, i.e. $v = \sqrt{v_r^2 + v_t^2} = \sqrt{(-300)^2 + (57)^2} = 305$ km s^{-1}. As a result, the time estimate for merging with the Milky Way (Example 9.1) has been adjusted upwards to 4.5 Gyr.

The velocity of the centre of mass of a system is referred to, logically, as the *systemic velocity*, v_{sys}. A cluster of stars has a systemic velocity, as does a galaxy, as does a cluster of galaxies. If proper motion measurements are also available and the distance is known so that Eq. (9.5) can be used to obtain v_t for the system, then the space velocity is also known, as demonstrated above for the Andromeda galaxy[2].

[2] This does *not* apply to motions for which v_{sys} is dominated by the expansion of the Universe (cf. Section 9.1.2).

Figure 9.2 The Andromeda galaxy's UV light is shown in greyscale in this image. The blue dots show the locations of stars for which stellar motions were measured using the Gaia satellite (Section 3.6.4) and yellow arrows show the average proper motion of stars at various positions, from [26]. Copyright: ESA/Gaia (star motions): NASA/Galex (background image): R. van der Marel, M. Fardal, J. Sahlmann (STScI). Reproduced with permission.

The space velocity can sometimes be inferred, however, based on geometry. For example, suppose that the proper motions of stars in a stellar cluster can be measured but the distance is not known so that Eq. (9.5) cannot be used to find v_t. On the sky, the proper motion vectors of individual cluster stars will appear to point towards a common 'vertex', or 'convergent point' on the sky, a consequence of geometrical perspective. Then straightforward measurements of radial velocities and a measurement of the angle from the cluster to the convergent point can be used to determine the space velocity and distance of the cluster. This is called the 'moving cluster method' and is most often used to find distances to stellar clusters.

For more distant objects for which proper motions are too small to be measured, statistical arguments can sometimes be used. For example, in a stellar cluster, we might expect some average measurement of v_r (measured with respect to v_{sys}) to be the same as some average value of v_t with respect to the galaxy's centre, i.e. $v_r \approx v_t$. In other words, measurements of v_r are sufficient to imply v_t, provided some symmetry is assumed.

In a worst-case scenario, v_t is simply unknown and a measurement of a Doppler velocity of v_r places a lower limit on its space velocity.

9.1.1.2 Dynamics – The Central Mass

As early as 1619, Johannes Kepler realized that the orbital motions of the planets were governed, not by the mass of the planet itself, but by the mass of the Sun. In modern form, Kepler's third law[3] states (cgs units),

$$\mathcal{T}^2 = \frac{4\pi^2 a^3}{G(M + m)} \tag{9.6}$$

where $\mathcal{T}$ is the period of the orbit, G is the universal gravitational constant, and a is the semi-major axis of the elliptical orbit (see Appendix A.5 of the *online material* for properties of the ellipse). Eq. (9.6) indicates that, for $m \ll M$, the period depends only on the mass of the object being orbited, M. The period, in turn, depends on the *velocity* of the orbiting body and its separation from M, so it is sufficient to measure the velocity, the separation of the orbiting object and the period, to find the interior mass. This is a very important and powerful concept in astronomy because it is relatively straightforward to measure radial velocities via Doppler shifts (Example 9.1). Thus, the orbiting body or bodies act like test particles that probe the central mass about which they move. Moreover, this method can be generalized to include extended mass distributions, rather than the simple point-like mass of the Sun at the centre of the Solar System, as we show below.

It is frequently of interest, then, to extend our concepts of finding a velocity with respect to the Sun or the LSR, and determine a velocity with respect to some other 'rest frame'. For example, one could adopt the rest frame of the centre of our Milky Way galaxy and consider the Sun's orbital velocity about this centre. Similarly, we might want to consider the orbital motion of extra-solar planets about their parent star or the motions of stars about the centre of another galaxy. As indicated earlier, if the entire system has some velocity, this motion is called the systemic velocity, v_{sys}. It is common, then, to measure a velocity with respect to v_{sys}, or otherwise to specify the reference standard that is being used. This will help us get to the heart of the *dynamics* of the system which is our goal here.

Let us now confine our discussion to the simple case of *circular motion.*

Rotating *thin discs* are examples of extended mass distributions for which the orbits of individual particles can be closely approximated by *circles.* They are frequently observed in many contexts in astronomy and have a common geometry. By *thin*, we mean that the disc

[3] Kepler's first law states that the orbits of the planets are ellipses with the Sun at one focus, and the second law indicates that the planets 'sweep out' equal areas of the ellipse in equal units of time. The 2nd law refers to the fact that the planets move fastest when they are near perihelion and slowest near aphelion.

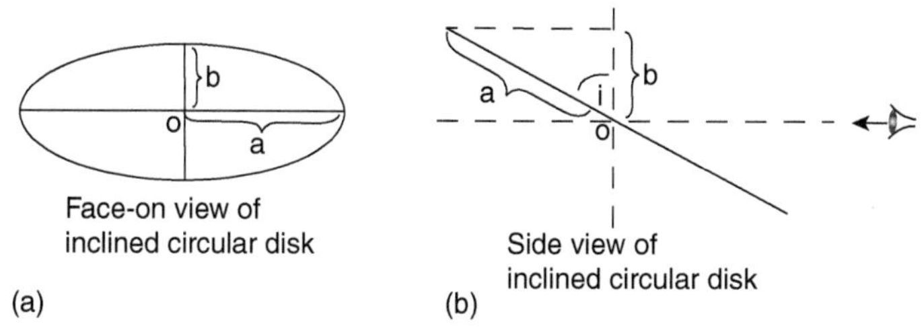

Figure 9.3 (a) A circular galaxy that is inclined to the line of sight will look elliptical in projection. The galaxy's true radius is *a*. When inclined, its apparent semi-major axis is *a* but its semi-minor axis is *b*, as shown. (b) This side view shows the minor axis of the inclined galaxy of (a) and its inclination.

thickness is much less than the disc diameter. A thin disc would consist of individual particles, each with its own rotational speed, υ_c (R), measured with respect to υ_{sys}, at different radii, R, from the centre of the system.

Examples of thin discs are the Solar System-sized dusty discs around stars, the galaxy-sized dusty discs in elliptical galaxies (Figure 5.21), and hot, gaseous discs (called *accretion discs* because their material accretes onto the central object) around compact objects like neutron stars and black holes. *Spiral galaxy* discs can also be modelled as thin, the individual particles being all those objects making up the disc such as stars, gas, dust, and nebulae. These include the nearly face-on galaxies like M 51 (Figure 5.8), galaxies of intermediate inclination like NGC 2903 (Figure 4.11), and galaxies that are edge-on to the line of sight like NGC 3079 (Figure 5.16) or the Milky Way from our vantage point (Figure 5.3). The *inclination* of a galaxy is measured with respect to the plane of the sky, so that a face-on galaxy has $i = 0°$ and a galaxy that is edge-on to the line of sight has $i = 90°$.

A thin circular disc looks elliptical in projection and the inclination can therefore be determined by the geometry of the ellipse, as shown in Figure 9.3,

$$\cos(i) = \frac{b}{a} \tag{9.7}$$

where b is the semi-minor axis as projected in the plane of the sky and a is the galaxy's semi-major axis. The quantities, a and b, can be expressed in linear units (e.g. cm, kpc) or angular units (e.g. radians, arcminutes) as long as the same units are used for a and b in Eq. (9.7).

The geometry of a rotating disc and how it relates to the observed Doppler shifts of spectral lines is shown for a galaxy disc in Figure 9.4. For any point at radius, R, *along the major axis*[4]

$$\upsilon_r = \upsilon_{sys} + \upsilon_c(R)\sin(i) \tag{9.8}$$

[4] At a point *off* the major axis, the equation is $\upsilon_r = \upsilon_{sys} + \upsilon_c(R)\cos(\theta)\sin(i)$. The angle, θ, is measured in the plane of the galaxy with a vertex at the galaxy's centre. It is the angle between the major axis and the line between the galaxy centre and the point off the major axis that is being considered. Notice that *on* the major axis, $\theta = 0$ so the equation reduces to Eq. (9.8). Along the minor axis, $\theta = 90°$ so $\upsilon_r = \upsilon_{sys}$.

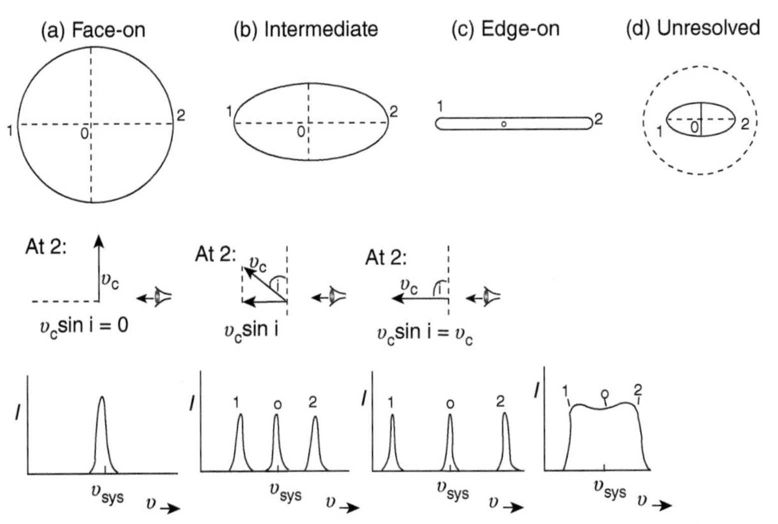

Figure 9.4 (a) The top left image shows a face-on galaxy ($i = 0°$) that is rotating in the plane of the sky. Points 1 and 2 at the ends of the disc are marked, as is point o at the centre. Below this, the middle left diagram shows a side view, indicating that there is no component of circular velocity, v_c that projects onto the line of sight. The bottom view indicates the spectrum from such a galaxy. Since there is no radial motion with respect to the centre, all points lie at the same radial velocity (with a small line width) which is just the systemic velocity, v_{sys}. (b) As in (a) but for a galaxy of intermediate inclination. In this case, there is a radial velocity component with respect to v_{sys} so points 1 and 2 are separated from point o in the bottom spectrum. (c) As in (a) but for an edge-on disc ($i = 90°$). Points 1 and 2 are separated by the maximum amount in the spectrum. (d) Example of a galaxy of intermediate inclination that is *unresolved* spatially. The dashed circle in the top image shows the size of the resolution element. In this case, emission from all points in the galaxy is received at once so the spectrum shows a continuous distribution of velocities centred on v_{sys}. The maximum values of v_r on either side correspond to points 1 and 2.

At the galaxy's centre, the circular velocity is zero, so the radial velocity measured at the galaxy's centre is just v_{sys}. Since i is known from Eq. (9.7), $v_c(R)$ can be found for any position, R, along the major axis, provided a spectral line can be identified and v_r measured for that position. The function, $v_c(R)$, is called the galaxy's *rotation curve*, as will be further described below. It is even possible to determine $v_c(R)$ for our own Galaxy, although the geometry is somewhat more complicated because of our unique position within it (Figure 5.3), and will not be described here.

The development described above for obtaining $\upsilon_c(R)$ for a rotating disc is very powerful because it allows us to determine the *dynamics* of galaxies or any other object for which circular velocities can be derived. For a stable rotational configuration, the gravitational force, F_G, can be equated to the centripetal force, F_c (see Table I.2) which, re-arranging and solving for central mass becomes,

$$M(R) = f\frac{R[\upsilon_c(R)]^2}{G} \tag{9.9}$$

Here, $M(R)$ is the mass *interior to* the point, R, and G is the Universal gravitational constant. The unitless quantity, f, is a geometrical correction factor to account for the fact that the distribution of interior mass may be flattened rather than spherical. For a spherical mass distribution (like a spiral galaxy with a dark matter halo, see below) and for a situation in which virtually all mass is contained in a central point (like the Solar System), $f = 1$. If we want to obtain the total mass of an object, such a calculation should be done at the outermost measurable point so as to include as much interior mass as possible. For values expressed in common astronomical units, the above equation becomes,

$$\frac{M}{M_\odot} = 2.32 \times 10^5\ f\left[\frac{R}{\text{kpc}}\right]\left[\frac{\upsilon_c}{\text{km s}^{-1}}\right]^2 \tag{9.10}$$

where the dependence of mass and velocity on R is taken to be understood.

Equation (9.10) can be used for any circularly rotating disc consisting of individual freely orbiting particles, or any single circularly orbiting body *whether the interior mass is observed or not.* All that is required is that a spectral line be identified and measured to obtain υ_r, the inclination must be known (Eq. (9.7), and the distance to the object be known in order to convert an angular distance along the major axis to R (Eq. I.9). The total mass of a galaxy (both light and dark) within the outermost measurable points, can be found as Example 9.2 describes.

Example 9.2

A spiral galaxy at a distance of 20 Mpc and $\upsilon_{sys} = 525$ km s^{-1} has semi-major and semi-minor axes of $a = 5.0'$ and $b = 2.3'$, respectively. The $\lambda 21$ cm line of HI at the farthest measurable point along the major axis has an observed wavelength of $\lambda = 21.155190$ cm. Find the mass of the galaxy interior to the measured point.

From Eq. (9.7), the inclination is $i = \cos^{-1}(b/a) = \cos^{-1}(2.3/5.0) = 62.6°$. The outermost measurable point along the major axis (5′) is at a distance (Eq. I.9) of $R = (20 \times 10^3)(1.45 \times 10^{-3}) = 29.1$ kpc. The radial velocity of this point, from Eq. (9.3) and using $\lambda_0 = 21.106114$ (Table C.1 of the *online material*), is 697.1 km s^{-1}. Therefore, this side of the galaxy is *receding with respect to its centre*. From Eq. (9.8) and known values of inclination and systemic velocity, we find a circular velocity of $\upsilon_c = 193.8$ km s^{-1}. Finally, we adopt a spherical mass distribution ($f = 1$) to account for the more spherical distribution of dark matter that we believe to be present in the galaxy (see text). Then from Eq. (9.10), we find $M = 2.5 \times 10^{11}\ M_\odot$.

Our dynamical study can go even farther. The *mass distribution* of a galaxy can be found from the rotation curve, $v_c(R)$. For spiral galaxies observationally, it is found that plots of $v_c(R)$ rise in the inner parts (within the first few kpc) and then flatten off over the rest of the disc. This means that, throughout most of the disc and certainly at larger R, $v_c \approx$ constant, in which case (Eq. (9.10)) $M \propto R$. The consequence is that, even though the *starlight* is seen to fall off exponentially with R, the *mass* continues to increase. This is one of the strongest arguments for *dark matter* in and around galaxies. The exact distribution of the dark matter alone is not yet known, but it is known that this matter tends to be in a much broader more spherical distribution than the thin stellar disc. This means that the mass distribution can be described by $f \approx 1$ in Eq. (9.10) even though the light of a spiral galaxy is mostly in a thin disc.

In the other extreme, if virtually all mass is contained in the central point, all test particles 'feel' the central mass regardless of their distance. This implies that $v_c \propto 1/\sqrt{R}$ (since $M \approx$ constant in Eq. (9.10)). This is called a 'Keplerian fall-off'. An example is the Solar System, for which $M \sim M_\odot$. Substituting $a = R$ and $v_c = 2\pi R/\mathcal{T}$ into Eq. (9.6) leads to the same Kelperian fall-off for v_c.

For situations in which the entire rotating disc is spatially unresolved, the emission from all parts of the disc is received in a single resolution element, as shown in Figure 9.4d. All radial velocities represented in the galaxy are then blended into a single broad feature as indicated in the spectrum, although the shape may vary depending on the distribution of material in the disc and its inclination. If it is known from some other measurement that the object is indeed a rotating disc, then it is still possible to obtain the interior mass by noting that the central velocity of the spectrum should correspond to v_{sys} and the extreme points of the spectrum should correspond to the radial velocity at the maximum value of R as the figure indicates (Problem 9.6).

The geometry of a rotating disc is very common in astronomy and exists over a large range of size scales, from dusty systems around individual stars to entire galaxies. In principle, however, this kind of dynamical study can be applied to any gravitationally bound system. We have already seen how the presence of an exoplanet could be inferred from the perturbed motion of the parent star as it orbits the centre of mass of the system (Section 6.2.2).

The stars in elliptical galaxies (Figure 5.21) are not rotating uniformly the way they are in the disc of a spiral galaxy. These stars are moving in individual orbits that are dictated by the gravitational potential of the elliptical galaxy as a whole. A statistical analysis of such stellar motions, specifically the velocity dispersion, σ_v, then leads to a determination of the mass – both light and dark – of the galaxy. Going to a significantly larger scale, by studying the motions of many individual galaxies that are in gravitationally bound galaxy clusters (e.g. Figure 10.8), the mass of the cluster can also be found.

Thus, from a tiny Doppler shift in the wavelength of light (e.g. Example 9.2), a picture of the dynamical nature of matter can be built up, from planets to the largest scale structures that are known to exist. And that little Doppler shift has revealed the *dark* nature (Figure 5.4) of this strange and unusual Universe in which we live.

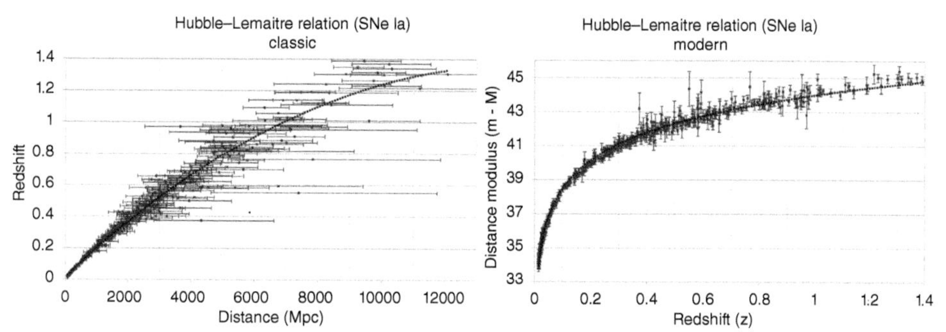

Figure 9.5 Two versions of the Hubble–Lemaître relation using the *same* Type Ia Supernovae (SNe Ia) data from [24]. The observer (us) is at the origin. The redshift, z, can be converted to units of km s^{-1} through a multiplication by c. *(Left)* This plot is shown in its classical form with distance on the x-axis and redshift on the y-axis. The Hubble constant ($H_0 = 71$ km s^{-1} Mpc^{-1} in this text) is the slope of this curve near the origin. The dashed curve is a simple logarithmic fit to the data. *(Right)* A modern version of the same data, but now with redshift on the x-axis and distance modulus on the y-axis. Recall that the distance modulus is related to the logarithm of the distance through $(m - M) = -5 + 5\log(d/\text{pc})$ (Eq. 3.32). The dashed curve is a simple polynomial fit to the data. Source: Data from Suzuki et al. [24].

9.1.2 The Expansion Redshift

As early as 1914, it was realized through the work of the American astronomer, Vesto Slipher, that galaxies preferentially showed *redshifts*, rather than blueshifts. The relatively nearby Andromeda galaxy of Example 9.1 bucks this trend since it shows a blueshift as it moves towards us locally. However, as more data were acquired, it became clear that almost all galaxies showed redshifts, that is, galaxies showed *recessional velocities.* The Belgian priest, Georges Lemaître, recognized that the recession velocities of the galaxies could be explained by an *expansion* of the Universe [16, 17] and the American astronomer, Edwin Hubble, combined his own measurements with those of Slipher to form the first *Hubble Relation*: a plot showing that galaxies at larger distances showed larger redshifts [12]. Today, this relation is called the Hubble–Lemaître law[5].

In Figure 9.5, we present the Hubble–Lemaître relation out to a redshift of $z \sim 1$ (see explanation below). It is interesting to consider the leaps-and-bounds improvement in this plot since Hubble's first efforts in 1929. Hubble's plot extended out to a distance modulus of only 26.5 (a distance of 2 Mpc, Eq. 3.32) – too small to even show up on Figure 9.5! In Sections 2.1 and 2.2, we discussed the concept of space having properties such that it can stretch or distort around a mass. We can now look at the stretching of space on the largest cosmological scales. Although the large-scale geometry of our Universe appears to be flat (Section 2.2), it can still expand and, as it does so, it *carries the galaxies with it.*

[5] Renamed from the Hubble Law by the International Astronomical Union in August, 2018; see https://www.iau.org/news/pressreleases/detail/iau1812/?lang

It is hard to imagine an explanation, other than expansion, for the Hubble–Lemaître plot. Like raisins in an expanding raisin cake (Figure F1 of the *online material*), more distant raisins will appear to be travelling at a higher velocity away from you, no matter what raisin you are standing on. Similarly, in an expanding universe, more distant galaxies have higher redshifts, regardless of the origin. A photon that leaves a distant galaxy must travel through this expanding space and so the observed photon will be 'stretched out' by this expansion in comparison to its wavelength in the rest frame in which it is emitted. Galaxies can still travel *through* space due to local gravitational effects. The blueshifted motion of the Andromeda galaxy (Example 9.1) is a good example. However, these motions tend to be quite small compared with the expansion velocity, as Example 9.3 illustrates. Much of the *scatter* in Figure 9.5 is due to such peculiar motions[6].

Before proceeding, let us consider what distance is actually being measured since there are different ways of defining 'distance' in an expanding universe. Figure 9.5 plots *luminosity distances.* This quantity is *defined* by how the flux falls off with distance, $f \propto \frac{L}{(d)^2} \Rightarrow d \propto \sqrt{\frac{L}{f}}$ (cf. Eq. 3.9 or Eq. F.11 of the *online material*). In Section 9.2.1, we will see another definition of distance.

Example 9.3

Estimate a typical distance beyond which a galaxy's redshift is dominated by the expansion of the Universe rather than its peculiar motion through space.

From Eqs. (9.1) and (9.3), we know that $v_r = cz$. Since H_0 (km s^{-1} Mpc^{-1}) represents the slope of the Hubble–Lemaître relation close to the origin (Figure 9.5 left), we can write (for small z)

$$cz = H_0 d \tag{9.11}$$

where d is the distance. We now wish to know the distance, d, at which,

$$cz(\text{Expansion}) = H_0 d > v_r(\text{Doppler}) \tag{9.12}$$

We will take the cluster of galaxies, Abell 2218 (Figure 9.7), as a typical example of the largest, most massive structures in the Universe and therefore the ones capable of producing the highest galaxy velocities *through* space. A typical velocity of a galaxy in this cluster is $v_r = 1370$ km s^{-1}. Using $H_0 = 71$ km s^{-1} Mpc^{-1} (Table T.2), we find $d > 19.3$ Mpc.

In terms of extragalactic sources (those outsides of the Milky Way), this is not very distant. It is about the same distance as the *Virgo Cluster*, which is the closest substantial cluster of galaxies to us. Therefore, beyond approximately 20 Mpc, the redshift is dominated by the expansion of the Universe and peculiar motions through space become less important as z increases.

Although redshift, z, can be converted into a velocity through a multiplication by the speed of light, c, it is important to remember that the 'recessional velocity of a galaxy' is

[6] Gravitational lensing (next section) also introduces uncertainty in the distances [2].

really a description of the expansion of space itself. Indeed, this redshift is better understood as how much the Universe has expanded between the time the light has been emitted, t_{em}, and the time that it is observed, t_{obs}. The fractional increase in the wavelength of light due to expansion, z, is then equal to the fractional increase in the size of the Universe between these two times,

$$z = \frac{a(t_{obs}) - a(t_{em})}{a(t_{em})} => z + 1 = \frac{a(t_{obs})}{a(t_{em})} = \frac{\lambda}{\lambda_0} \tag{9.13}$$

where $a(t)$ is a dimensionless *scale factor* that describes how the scale ('size') of the universe changes with time. A galaxy with a measured redshift of $z = 6$, for example, emitted its light when the Universe was only 1/7th of its current size. The right-hand side of Eq. (9.13) is a repetition of Eq. (9.1).

As can be imagined, hiding in Figure 9.5 is a great deal of information about the nature and history of our Universe. *We* are at the origin and any measurement near the origin describes the nature of the universe *now*. As distance and redshift increase, we are looking at the nature of the Universe at earlier times. A key concern is the rate of expansion which is described by the *Hubble parameter*, $H(t)$. The value of the Hubble parameter *now* is called the *Hubble constant*, H_0 (units of km s^{-1} Mpc^{-1} or cgs units of s^{-1}) and it can be measured from the slope of the Hubble–Lemaître relation close to the origin.

The classic version of the plot (Figure 9.5 left) shows redshift on the y-axis and distance on the x-axis. Thus, the slope of this plot close to the origin directly gives H_0 when the redshift (y-axis) is multiplied by the speed of light, c. A great deal of effort has been expended in trying to measure this quantity, starting with Hubble's initial, grossly overestimated measurement of 500 km s^{-1} Mpc^{-1}. The current value is much lower, though it still continues to be refined and debated (e.g. [18]). For example, [21] give $H_0 = (67.4 \pm 0.5)$ km s^{-1} Mpc^{-1}, [9] report $H_0 = (76.8 \pm 2.6)$ km s^{-1} Mpc^{-1}, and other variants exist within approximately 10% of these values with a range from about 68 to 75 km s^{-1} Mpc^{-1} [23]. In this text, we will use $H_0 = 71$ km s^{-1} Mpc^{-1}, as listed in Table T.2.

Figure 9.5 Right shows *exactly the same data and error bars* as in the left panel, but with the x and y axes reversed and the distance modulus (Eq. 3.32) plotted instead of the distance. More modern versions of the Hubble–Lemaître relation tend to be presented in this form; the distance modulus, for example, is closer to what is actually measured in practice, in comparison to the distance. What is quite obvious from both panels is the *curvature* of the relation. With the axes reversed, one would think that the curvature would also switch, but this is not the case because of the different ways in which distance is represented.

Conceptually, it is easier to understand $H(t)$, i.e. how the expansion changes with epoch, from the curvature of the left (classic) panel. As can be seen, at higher redshifts (earlier times in our Universe), the slope is *flatter*, meaning the Universe was expanding at a slower rate at earlier times. The conclusion is that the Universe is currently *accelerating*.

The reason for this acceleration is not known, but the energy that is involved is called *dark energy* and it accounts for 70% of the energy density of the Universe, as depicted in Figure 5.4. The Universe has not always been accelerating. Some estimates suggest that the acceleration began at a time corresponding to a redshift of $z_t \approx 0.67$ [10]. The race is on to understand the Universe on the largest scales and the Hubble–Lemaître relation plays a key role.

Since redshifts are easily obtained but distances are not, the calibrated Hubble–Lemaître relation is used to *find* the distances to galaxies and quasars as long as the redshift is dominated by expansion (Example 9.3). Calibrating the relation in the first place, however, requires a set of objects for which we know the distance ('standard candles'). Excellent calibrators are Type Ia supernovae (introduced in Section 5.3.3) which are bright enough to be seen over cosmological distances. These are the sources that are plotted in Figure 9.5.

From the calibrated Hubble–Lemaître relation, a measurement of redshift leads to a distance (with some uncertainty, as the plot suggests). For small redshifts ($\lesssim$0.1), Eq. (9.11) can be used to find d. Expressions for intermediate ($1 > z \gtrsim 0.1$) and higher ($z > 1$) redshifts can be found in Appendix F of the *online material* (Eqs. F.16 and F.32, respectively). This appendix also shows a Hubble–Lemaître plot for redshifts out to $z \approx 7$ [18], as well as further information about some kinematics and dynamics of our Universe.

9.1.3 The Gravitational Redshift

The gravitational redshift was first introduced in Section 2.3 and involves local space–time distortions, rather than global cosmological behaviour as was just discussed. Any mass will distort space–time around it, so a photon that leaves the surface of a mass must emerge from a region of curved space–time. Such a photon must climb out of a potential well, as illustrated in Figure 2.1, and will be redshifted as a result. This gravitational redshift is a result of time dilation near the gravitating mass. Imagine a signal, such as repeated short pulses of light, emitted at regular time intervals from a region near the mass. A distant observer will measure these time intervals to be longer than the intervals he would measure in his own rest frame for an identical signal. Thus, the observed frequency of these pulses ($\nu = 1/t$), as measured by the distant observer, will be lower. If the signal is now a wave of some frequency, rather than individual pulses, then the observed wavelength will be longer, or redshifted, in comparison with the rest wavelength.

It is sometimes useful to look at this in a more classical fashion such as would be done for a particle with a rest mass, rather than a photon. That is, it takes energy for a photon to climb out of a gravitational potential well. The emergent photon's energy decreases during this process, so its wavelength must increase. A gravitational blueshift is also possible for an incoming photon as it gains energy approaching a mass.

An expression for the gravitational redshift is derivable from the Schwarzschild metric (Section 2.1), the result being,

$$z + 1 = \frac{1}{\sqrt{\left(1 - \frac{r_S}{r}\right)}} \qquad r_S < r \tag{9.14}$$

where z, measured by a distant observer, is the redshift of a photon that originates at a distance, r, from the centre of the mass[7] and r_S is the Schwarzschild Radius, defined by,

$$r_S \equiv \frac{2GM}{c^2} \tag{9.15}$$

The Schwarzschild Radius is the radius of a black hole, i.e. the radius of a region of space within which the gravity is so strong that not even light can escape[8]. Any mass, M, that is *completely* contained within a radius, r_S, will be a black hole. Section 5.3.3 explained how stellar-mass black holes can be formed and, indeed, are expected from the normal stellar evolution of massive stars.

From Eq. (9.14), it can be seen that the gravitational redshift will not be a strong effect unless the photon leaves an object from a position, r, that is close to its Schwarzschild Radius. Any object that is not a black hole has a radius that is larger than its Schwarzschild Radius (usually *much* larger), so the gravitational redshift is negligible for most objects (Example 9.4). However, it can become important for very compact or collapsed objects like neutron stars and the light-emitting regions surrounding black holes (Problem 9.14).

Example 9.4

Compare the magnitude of the gravitational redshift to the Doppler shift from a Solar-type star that is receding from us at a velocity of $v_r = 20$ km s^{-1}.

From Eq. (9.2), z (Doppler) $= v_r/c = (20/3.00\times10^5) = 6.7\times10^{-5}$. From Eq. (9.15) for a 1 $M_\odot$ star, $r_S = (2G)\,(1.99\times10^{33})/c^2 = 2.95\times10^5$ cm. A photon leaves a Solar-type star at a radius, $r = R_\odot = 6.96\times10^{10}$ cm, so using Eq. (9.14), z (gravitational) $= 2.1\times10^{-6}$, which is more than an order of magnitude smaller than the Doppler shift. It is interesting to compare these extremely low gravitational redshifts with the cosmological redshifts of Figure 9.5!

9.2 GRAVITATIONAL REFRACTION

Just as a photon that leaves a mass must climb out of its potential well, any photon from a background source that passes near a mass en route to us must also pass through curved space–time, as was illustrated in Figure 2.1. If the background light source, the mass and the observer are sufficiently aligned, then the observer may see an image or images of the source displaced from its true position in the sky. The mass is called a *gravitational lens* since it acts like a convex lens, refracting the light rays towards it.

A variety of image geometries may result, depending on the alignment and distances of the source and lens as well as the source light distribution and lens mass distribution. Examples are shown in Figures 9.6 and 9.7.

[7] At $r \le r_S$, the coordinate system of an external observer breaks down and we do not consider this case further.

[8] The Schwarzschild Radius is accurately derived by considering General Relativity. However, a simplified way in which to understand the equation is to set the escape velocity from the object equal to the velocity of light, i.e. $v_{esc} = \sqrt{2GM/r} = c$, from which Eq. (9.15) follows.

Figure 9.6 The foreground galaxy (the lens) is a luminous red galaxy, LRG 3-757, at a redshift of $z = 0.444$. The almost complete blue circle around it is an Einstein ring. The Einstein ring is a distorted image of a background galaxy at a redshift of $z = 2.4$. This Einstein ring has a diameter of 10 arcseconds [7].
Source: ESA/Hubble & NASA. [https://commons.wikimedia.org/wiki/File:Lensshoe_hubble.jpg].

9.2.1 Geometry and Mass of a Gravitational Lens

The geometry of a gravitational lens is shown in Figure 9.8a. The light path is actually curved, but it can be represented by a single *bending angle*, ϕ, defined as the difference between the initial and final directions. It has therefore been drawn as two straight ray paths in Figure 9.8a with a single bend. The bending angle, ϕ, was determined by Einstein for an isolated lens and can be derived from the Schwarzschild metric (Section 2.1). Because the bending angles are small, the result is

$$\phi = \frac{2r_S}{b} = \frac{4\,G\,M_L}{b\,c^2}, \qquad \phi = 1.75'' \frac{\left[\frac{M_L}{M_\odot}\right]}{\left[\frac{b}{R_\odot}\right]} \tag{9.16}$$

where b, the *impact parameter*, is the distance of closest approach to the mass and r_S is the Schwarzschild Radius of a lens of mass, M_L, given by Eq. (9.15). The left equation is in cgs units, and the right equation gives ϕ in arcseconds for inputs in Solar units. This result is exactly twice that expected from Newtonian physics[9] and catapulted Einstein to fame

[9] Space curvature is not expected in Newtonian physics, but the photon has an 'equivalent mass' via $E = h\nu = m\,c^2$ and could be treated as a small particle to find a deflection angle in a Newtonian model.

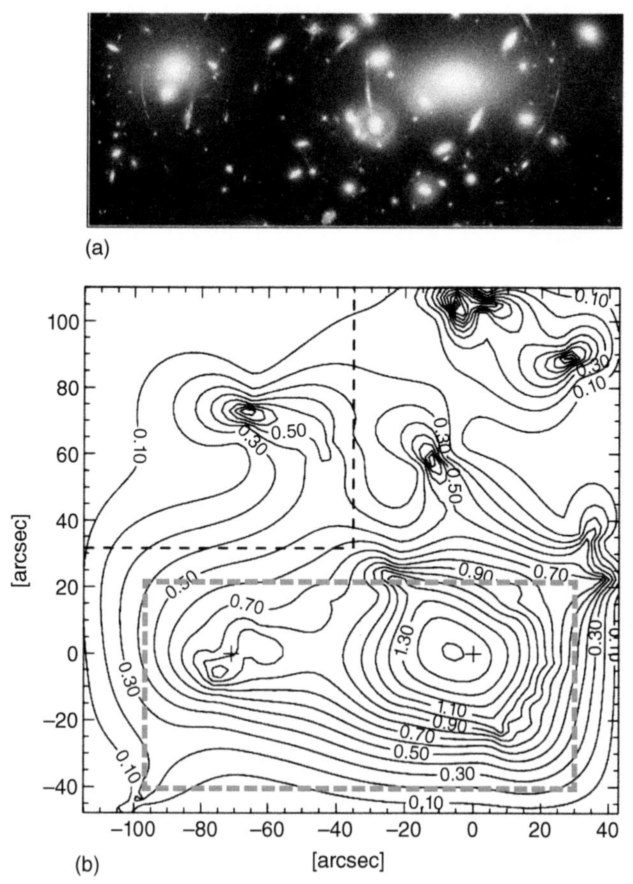

Figure 9.7 (a) An optical view of the galaxy cluster, Abell 2218 ($z = 0.175$), showing over 100 arcs from the gravitational lensing of background sources. The background sources have various redshifts. The velocity dispersion of galaxies in the cluster is 1370 km s^{-1} [27]. The x-axis spans 127″ and the y-axis spans 64″. Source: Reproduced by permission of W. Couch (University of New South Wales), R. Ellis (Cambridge University), and NASA. (b) A reconstructed mass model from [1]. Crosses mark the positions of the two brightest regions in (a) and the dashed green rectangle shows the approximate region displayed in (a). The mass distribution, which includes both light and dark matter, is much more spread out than the visible galaxies. Source: Reproduced by permission of P. Saha and the AAS. AbdelSalam et al. [1]. ©1998 American Astronomical Society.

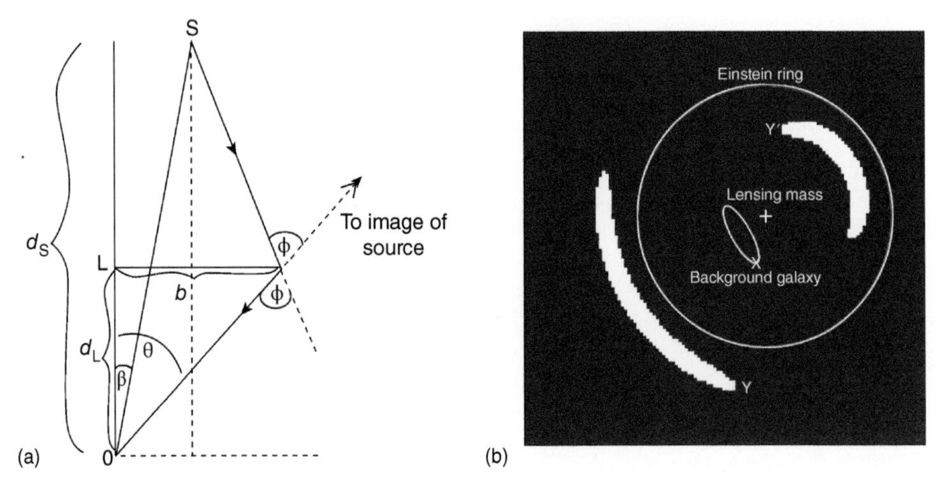

Figure 9.8 (a) Geometry of a gravitational lens (with angles exaggerated) when both the source and lens are points. A point source at S is a distance, d_S, and a true angle, β, from the observer at O. A point lens, L, of mass, M_L, is a distance, d_L, from the observer. The light path, shown here by two straight lines with arrows, is actually a curved path whose bending can be characterized by the gravitational bending angle, ϕ. The impact parameter, b, is the closest distance of the light path to the lens, L. The observer sees an image at an angle, θ. (b) Face-on view of a gravitational lens in which both the source and lens are no longer points. A background galaxy is lensed by a closer spherically symmetric mass distribution centred at the +. The observer sees the two arcs shown. The location of the background galaxy, the lensing mass and its corresponding Einstein Ring are indicated. The point, X, on the source, at an angle, β from the lens, has two images: Y, at an angle θ_1 from the lens, and Y′ at θ_2 from the lens. Source: Adapted from Newbury and Spiteri [19].

when the larger value was confirmed by observations of the displacements of background stars during the solar eclipse of 1919. The potential power of gravitational lensing as an astronomical tool was realized by S. Refsdal in the 1960s [22].

There are two important differences between a gravitational lens as implied by Eq. (9.16) and one made of glass (or other refracting material):

- No focal length: For a glass lens, incoming rays are bent the least near the lens centre (zero at the centre) and greatest near the lens edge, allowing light to intersect at a single point in the focal plane. This is accomplished by machining the surfaces to the desired curvature. For a gravitational lens, on the other hand, the bending angle is inversely proportional to the impact parameter, so light is bent *less* at larger distances from the lens centre. Thus, a gravitational lens has no focal length.
- Wavelength independence: The speed of light in a medium depends on interactions between light and the particles of the medium and such interactions depend on the

wavelength of the light. Thus, for a glass lens, the index of refraction ($n = c/v$, Table I.2) is wavelength-dependent, blue light being bent more than red light. However, the speed of light in a vacuum is constant. Therefore, Eq. (9.16) has no wavelength dependence and a gravitational lens is always *achromatic*. X-rays, IR or radio waves are all bent through the same angle, provided the background source emits in these bands. This is an important clue that gravitational lensing is likely occurring.

From the simplified geometry of Figure 9.8a, which assumes that both the source and lens are *points*, we can calculate the observed displacement of a background object. The bending angle is actually very small, so the impact parameter of the curved path that the light actually follows is about equal to the impact parameter of a path represented by the straight lines shown. From this geometry, Eq. (9.16) and the assumption of small angles, it can be shown (Problem 9.15) that the observed angle of the image, θ, satisfies a quadratic equation, known as the *lens equation*,

$$\theta^2 - \beta\,\theta - \theta_E^2 = 0 \tag{9.17}$$

where,

$$\theta_E^2 = \frac{4GM_L}{c^2}\left(\frac{1}{d_L} - \frac{1}{d_S}\right) \tag{9.18}$$

Equation (9.18) contains information about the lens mass and the source and lens distances. As indicated in Section 9.1.2, a distance can be defined in several ways. For gravitational lensing, the appropriate distance is the *angular size distance*. This distance is defined by the ratio of the object's physical size to its angular size (in radians), i.e. the distance implied by the small-angle formula of Eq. (I.9), $d = s/\theta$. More information is provided in Section F.1 of the *online material* but here, it is sufficient to note that the luminosity distance defined in Section 9.1.2 is about equal to the angular size distance at small redshift, with increasing departures as z increases.

Equation (9.18) can be rewritten in a more simplified and practical form as

$$\theta_E = 3\left(\frac{M_L}{M_\odot}\right)^{1/2}\left(\frac{D_{LS}}{1\ \text{Gpc}}\right)^{-1/2}\ \mu\text{arcsec}, \qquad D_{LS} \equiv \frac{d_L\, d_S}{d_S - d_L} \tag{9.19}$$

D_{LS} has the dimension of 'distance' and is called the *effective lensing distance*[10]. Equation (9.19) allows us to see more easily what size of Einstein ring might be expected. For example, suppose that $d_S = 2.7$ Gpc and $d_L = 1.0$ Gpc, then $D_{LS} = 1.6$ Gpc. We would then expect $\theta_E = 3.4$ arcsec for a 'point' lens galaxy with a mass of $2 \times 10^{12}\ M_\odot$. That is, lensing by galaxies is typically of order a few arcseconds.

The lens equation has two roots corresponding to two image locations,

$$\theta = \frac{\beta}{2} \pm \frac{1}{2}\sqrt{\beta^2 + 4\theta_E^2} \tag{9.20}$$

[10] Here we assume that, on large scales, space is flat so that the distance between source and lens can be written simply as $d_S - d_L$. Many references, however, simply write this distance as d_{LS}.

The value of the square root will always be greater than β, so there will be one positive root (which we call θ_1) corresponding to the image on the same side of the lens as the source, and one negative (θ_2) corresponding to an image on the other side of the lens. Both these angles can be measured, provided the lensing mass location is observed. Then the true position of the source, which is not observable, can be found from the sum,

$$\theta_1 + \theta_2 = \beta \tag{9.21}$$

and θ_E can be determined from the total angular separation between the two images,

$$\theta_1 - \theta_2 = \sqrt{\beta^2 + 4\theta_E^2} \tag{9.22}$$

A special case occurs if the source and lens are perfectly aligned ($\beta = 0$). In this case, $\theta_1 = -\theta_2 = \theta_E$. There is no preferred plane, and instead of two individual images, a ring is formed around the lens called the *Einstein Ring* of radius, θ_E. A nice example is shown in Figure 9.6 in which the ring is almost complete.

Determining θ_E is very important because this quantity contains the desired astronomical information. The distances to both the source and lens can be found from the Hubble–Lemaître relation if their redshifts can be measured. The source redshift should be the same anywhere on the Einstein ring if it is visible, or from any of its images otherwise. Then Eqs. (9.18), (9.21), and (9.22) can be combined to obtain the lens mass,

$$M_L = -\frac{c^2\theta_1\theta_2}{4G}D_{LS}, \qquad M_L = (1.22\times10^8)\theta_1|\theta_2|D_{LS} \quad M_\odot \tag{9.23}$$

The equation on the left is in cgs units with angles in radians. The equation on the right requires θ in arcseconds and distances in Mpc. The significance of M_L is that it includes *all* mass, both light and dark. Thus, *gravitational lenses provide a means of measuring dark matter mass.* We now have an independent method of obtaining masses, other than via the Doppler shifts and Newtonian mechanics described in Section 9.1.1.2.

Neither sources nor lensing masses are 'points'. Figure 9.8b shows an example with an offset background elliptical galaxy of uniform brightness and a spherically symmetric lens centred at the +. The background galaxy can be thought of as consisting of many individual point sources, each of which creates two images as before. For example, point X forms images at Y (corresponding to θ_1) and Y′ (corresponding to θ_2). In general, a gravitational lens distorts the image which, in this case, is clearly no longer elliptical. Because a lens directs light that would normally not be seen towards the observer (consider the case of the Einstein ring, for example), the result is an *amplification* of the signal (see also Section 9.1.2). The flux of a background source will actually be greater as a result of the lensing than it would be in the absence of the lens.

With a variety of possibilities for background light distributions and lensing mass distributions, it is not difficult to imagine that rather complex images can result, as Figure 9.7a illustrates. One must then *model* the system, putting in various mass distributions until the observed images are reproduced (Figure 9.7b). Although some assumptions may be required, such modelling can place strong constraints, not only on the total mass of the lens, but on the spatial distribution of its mass as well.

9.2.2 Microlensing – MACHOs and Planets

There are situations in which the bending angles are so small that the entire system is unresolved. The background source images and lens are collectively seen as a point source, making it virtually impossible to tell that gravitational lensing is occurring in any static orientation. However, the lensing mass will have a proper motion with respect to the background source and observer (Section 9.1.1.1). As the lens passes between the source and observer, there will be an increase and then decrease in the observed flux of the background source due to the amplification of the signal during the lensing phase. This is called *microlensing*. The duration of the microlensing event is approximately the time over which the background source is within the Einstein radius of the foreground lens (Problem 9.19). A characteristic light curve results, such as that shown for one star lensing another star in Figure 9.9.

It is important to be able to identify such a light curve as being due to a microlensing event and not simply to a variable star or a supernova, both of which show light curves that increase, then decrease in brightness. Aside from the shape of the curve, it is possible to separate true microlensing events from other variable phenomena by the frequency dependence of the curve. The light curves of variable stars and supernovae result from processes that involve the transfer of photons through matter which has some opacity. As we have seen earlier (e.g. Section 7.4.2), opacity is highly frequency dependent and, as a result, there tend to be differences, frequency to frequency, in the light curves of variable stars and supernovae. Because microlensing is achromatic, however, light curves should be identical at different frequencies. It is thus possible to identify true microlensing events this way[11].

The detection of microlensing relies on the chance alignment of objects at any time, an event which is intrinsically unlikely. However, modern telescopes, especially robotic ones with customized software, have the ability to monitor millions of objects and routinely extract the lensing candidates. In addition, if fields are chosen for which there are many possible background objects, then the statistics are much better. These approaches have been taken in a variety of experiments. For example, microlensing events have been observed when MACHOs in the halo of the Milky Way (Section 5.2) pass in front of the *Large Magellanic Cloud* which is a nearby companion galaxy to the Milky Way. It is studies like this that have led to the conclusion that there are not enough MACHOs to account for the dark matter that is expected to be in the Milky Way's halo ([3, 4, and 5]).

Figure 9.9 shows another exciting application of microlensing. The main image shows a light curve from a background star (the source) that is being lensed by a foreground star. The background star is in the bulge of the Milky Way and the foreground star (the lens) has a proper motion (Figure 9.1) with respect to the source. As the lens travels past the background star, light from the background star first increases (is amplified), becoming a maximum when the source and lens are closest together in projection, and then decreases again as the lens and source separate. However, a small perturbation on the right of the

---

[11] In practice, the process is somewhat more difficult because each pixel receives the light of many stars. Thus, the problem is approached statistically.

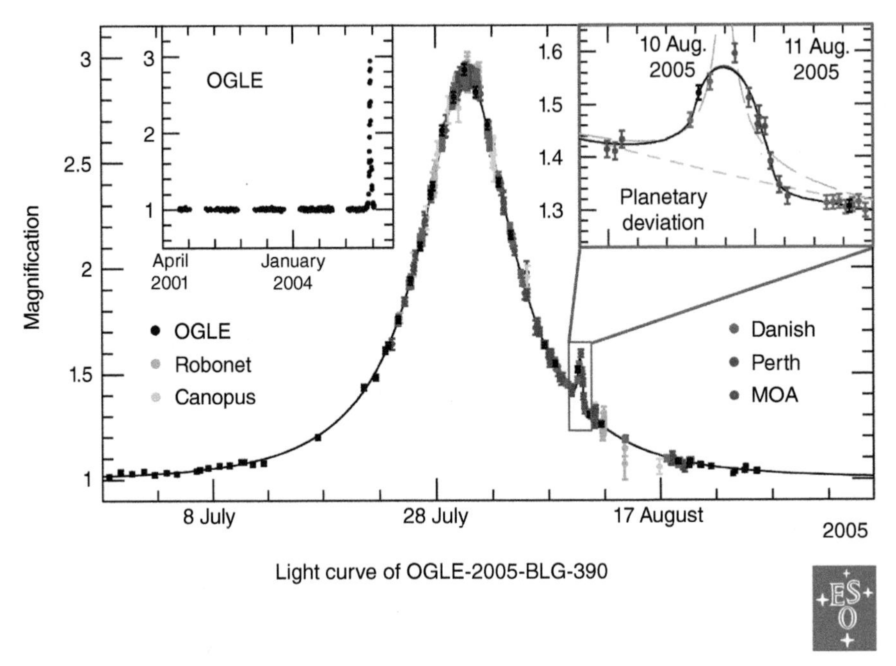

Figure 9.9 Characteristic light curve of a microlensing event. Both the lens and the background source are stars in our Milky Way. The background star is a G4 III giant star at a distance of 8.5 kpc. The lensing star is an M dwarf of mass, $M = 0.22\ M_{\odot}$, at a distance of 6.6 kpc. The left inset shows the light curve, as monitored over four years time. The small bump on the right of the main image, blown up in the right inset, shows the smaller peak due to a planet, called OGLE-2005-BLG-390Lb, orbiting the lensing star. The planet's mass is $M_{pl} = 5.5\ M_{\oplus}$ and is a distance, $r = 2.6$ AU from its parent star. This is the third exoplanet discovered using microlensing techniques. The discovery was made by an international team [6] involved in three microlensing projects: Probing Lensing Anomalies NETwork (PLANET) + RoboNet, Optical Gravitational Lensing Experiment (OGLE) and Microlensing Observations in Astrophysics (MOA). Source: ESO/PLANET/RoboNet, OGLE, and MOA.

light curve, blown up in the right inset, shows another small peak. This peak is due to a 5.5 $M_{\oplus}$ *exoplanet* that orbits the lens star. In other words, the exoplanet is acting like an additional lens that adds to the amplification of the background starlight. It is this observation that led to the exoplanet discovery. Notice how the background star, in this case, is simply being used as a probe of the foreground system of interest.

Exoplanet discovery via microlensing was briefly introduced in Section 6.2.2 as one example in the pantheon of exoplanet detection techniques that are currently available. Only a small fraction of exoplanet discoveries has resulted from gravitational microlensing, as shown by the orange colour in the histogram of Figure 6.13. However, these exoplanets are interesting because their masses tend to be modest, of order a few Earth masses to a few Jupiter masses, and detection is possible in very distant systems. For example, the system, KMT-2019-BLG-1953, is 7.04 kpc from us in a direction towards the Galactic Centre (recall that the distance to the centre of our Galaxy is 8 kpc, Figure 5.3). The lensing star has a mass of 0.31 $M_\odot$ and appears to have two approximately Jupiter-sized planets moving around it [11].

9.2.3 Cosmological Distances with Gravitational Lenses – Time Delays and H_0

Strong gravitational lenses are those for which multiple images can be seen. Such systems can provide additional information regarding *distances* for the special case in which the source brightness is varying with time.

An example is when the background source is a quasar, most of which show time variability. Alternatively, the sudden brightening and slower dimming of a supernova can also be a suitable source (e.g. [14]). An example geometry for the simple case of two images is shown in Figure 9.10. Here, the path length on one side of the lens will be different from the path length on the other side. A change in brightness will therefore be observed in the image that has the shorter path length first (Image i in the diagram), before the longer path length image (Image j)[12].

This variability adds new information to what we already have from 'normal' gravitational lensing. A real lens, however, will not be a point source and so the bending depends on the mass *distribution*, or rather the gravitational potential, Ψ, as a function of angular

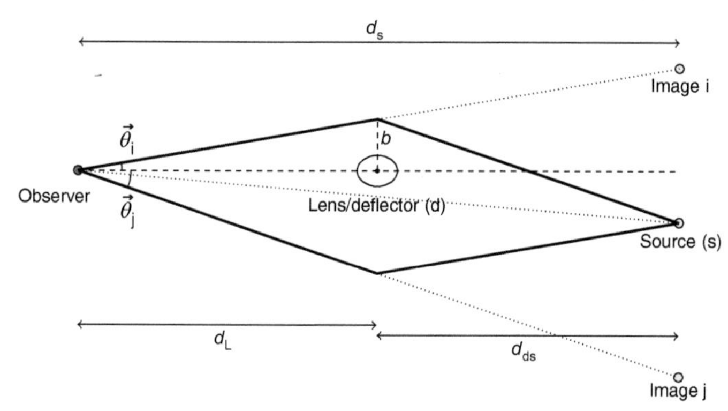

Figure 9.10 Diagram showing the geometry of a gravitational lens that forms two images, i and j. When the source, s, is variable, Image i will show this variation before Image j because it travels a shorter path. Source: Jee et al. [13]. © IOP Publishing Ltd.

[12] There is an additional effect besides the simple geometry shown in Figure 9.10 The upper ray is also travelling through a deeper potential well, as visualized in Figure 2.1, in comparison to the lower ray.

position, θ, from the lens centre, i.e. $\Psi(\theta)$. It can then be shown (e.g. [8, 20, 23]) that the time delay, Δt, is a function of a number of parameters,

$$\Delta t = f(z_L, D_{LS}, \Psi(\theta), \phi_{L,S}) \tag{9.24}$$

Here, z_L is the redshift of the lens, D_{LS} is the effective lensing distance and $\phi_{L,S}$ is a function of the angular positions of the lens and source on the sky. Of these, Δt, $\phi_{L,S}$ and z_L are all measurable. The unknowns are D_{LS} and $\Psi(\theta)$.

Let us now turn around the arguments of Section 9.2.1 and assume that we *know* $\Psi(\theta)$ in which case the only unknown is D_{LS}. This effective lensing distance (Eq. (9.19)) is a combination of the distances to the source and lens. If the distances to the lens and source are not too large, then each individual distance in the equation for D_{LS} can be expressed as $d = cz/H_0$ (Eq. (9.11)). Using this substitution, D_{LS} becomes a function of the redshifts of the lens and source (which are measurable) and is inversely proportional to H_0, i.e. $D_{LS} \propto 1/H_0$. Therefore, if we can solve for this single unknown in Eq. (9.24), then we have a method of *finding* H_0 that is *independent* of any other standard candle that has been used to calibrate the Hubble–Lemaître relation. In other words, we have a new independent measure of distance, leading to an independent measure of H_0. If the redshift is higher, then D_{LS} will also depend on cosmological parameters such as the relative matter and dark energy content of the Universe [20]. However, the strongest dependence is on H_0 [23, 25].

But what about the lens potential that we presumed to be known? Here, we require more information. The important tool at our disposal is a measure of the velocity dispersion, σ_v, of the lens. As we know from Section 9.1.1.2, the dynamics of the stars in a galaxy can tell us about the mass, both light and dark. With some assumptions about the lens mass distribution plus appropriate calibration, one can substitute σ_v for $\Psi(\theta)$ in Eq. (9.24), en route to knowledge of H_0 and other cosmological parameters.

How well is this approach working? There have been very few gravitational lenses measured in this way in comparison with SNe Ia [9, 13]. However, the geometric nature of this technique and the fact that a different measure of distance is used (angular size distance rather than luminosity distance) has permitted tackling the H_0 problem from a fresh perspective. One recent result from this approach, which includes three strong lenses, is $H_0 = (76.8 \pm 2.6)$ km s^{-1} Mpc^{-1} [9]. Currently, the statistical error on H_0, however, is believed to be about 7–10% [23]. The main uncertainties are in how to model the lens potential (including *all* matter along the line of sight from the source to the observer) and in how accurately Δt can be measured [25]. Time delays, which are typically months, require well-sampled light curves as well as an understanding of the possible contribution of additional microlensing that might be occurring from stars in the lens for best results [23].

9.3 TIME VARIABILITY AND SOURCE SIZE

A simple, but powerful argument related to fluctuations in the brightness of a source has proven to be extremely useful in constraining source sizes which otherwise would be too small to measure. Consider a source as in Figure 9.11. The size of the region that is variable must be,

$$d \leq c\,\Delta t \tag{9.25}$$

where Δt is the timescale of the variability. This condition is a requirement of causality. If it were *not* the case, then the light that is emitted from a varying source at point A (see figure) would not be able to travel across the source to point B in a time, Δt. Point B, on the other side of the source, would therefore not 'know about' the variation at A so could not be in synchronicity with it. If points on a source are varying independently, then no global variation would be detected. This is similar to arguments presented earlier as to why planets shine and stars 'twinkle' (Section 4.3.4). If variability *is* observed, then Eq. (9.25) places a limit on the size of the varying source.

Some sources are variable over a variety of timescales. It can be seen from Eq. (9.25) that the shortest measured timescale puts the tightest restrictions on source size. X-ray variability is particularly important because the fact that X-rays are observed indicates that the source is associated with high energy phenomena. X-rays tend to be emitted near very compact objects that have high gravitational fields, such as white dwarfs and neutron stars, or active galactic nuclei that are thought to be powered by supermassive black holes (e.g. Section 7.4.2.2). The shortest variability timescale then provides a limit on the size of the emitting object (Problem 9.20). If information about the mass of an object is available from dynamics (Section 9.1.1.2) and information about the size of the object is available from time variability, then strong constraints can be placed on the type of object that is present, even if it is spatially unresolved.

9.4 A BRIEF CODA

Before leaving this chapter, it is worth pausing to contemplate the extent to which modern science has *decoded our cosmos.* We are living in an expanding, accelerating universe whose nature is being probed and analysed in every possible ingenious way. The universe, as we understand it, had a beginning called the 'Big Bang' and the fixed speed of light means that we can see back in time as we look to greater and greater distances. Indeed, the distance to the outermost reaches of our universe is so vast that our ancestors could not even have imagined such immensity. Space itself has properties that can produce such astonishing distortions as arcs and multiple images – at times, it seems like we are living in a fishbowl as we attempt to peek beyond the distortions produced by the water.

Figure 9.11 The size of a variable source must be less than the light travel time across it. At time, t (left), the source is dim and, at a time, $t + \Delta t$ (right), the source has brightened. All parts of the source can vary together, provided, $d \leq c\Delta t$.

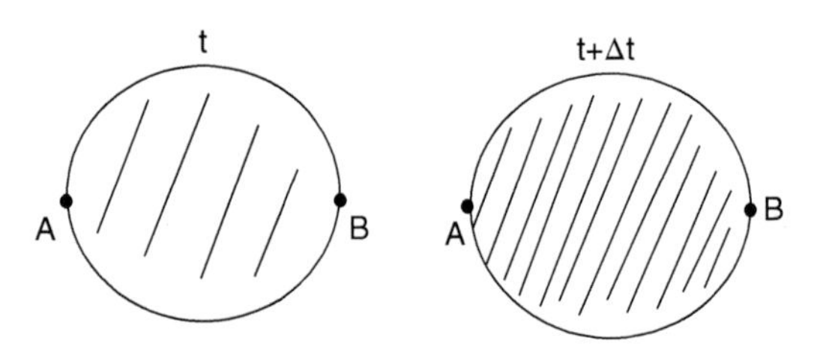

Figure 9.12 Colourized version of a nineteenth-century wood engraving that appeared in Camille Flammarion's book, *L'atmosphère: météorologie populaire*, which depicts a medieval concept of cosmology. Original coloured version by *Houston Physicist*. Source: Reproduced under the Creative Commons Attribution-Share Alike 4.0 International licence.

The contrast between our modern cosmology and that of the medieval concept shown in Figure 9.12 is striking – yet the human sentiment has not really changed.

PROBLEMS

9.1. How high would the radial velocity of an object have to be before z in the low-velocity approximation of the Doppler shift (Eq. (9.2) Right) differs from the accurate expression (Eq. (9.2) Left) by more than 10%? (The solution may be obtained 'by hand' or via computer algebra software.) Suggest an object or objects (if any) for which a Doppler shift of this magnitude might be seen. For most astronomical objects, would you expect to need the exact formula?

9.2. In principle, any spectral feature for which the rest wavelength, λ_0, is known could be used to measure the radial velocity, v_r, of an object. Let us take the peak of the Planck curve of a star as the 'spectral feature'. The star has a radial velocity of $v_r = 50$ km s^{-1} and we assume that its temperature, $T_0 = 5800$ K, has been determined by some other method.

(a) Find the rest wavelength, λ_0, for the peak of the Planck curve of this star and the observed, Doppler-shifted wavelength, λ, of this peak. What is the shift, $\Delta\lambda$?

(b) Find the equivalent Doppler-shifted temperature, T_{sh}, corresponding to the new peak, λ.

(c) Compute the specific intensity of the star at the Doppler-shifted wavelength, λ, if there were no radial velocity (i.e. $B_\lambda[T_0]$). Compute the specific intensity of the moving star at the same wavelength ($B_\lambda[T_{sh}]$). What is the change (in percent) in specific intensity at the wavelength, λ, as a result of the star's motion? If your instrument can detect an amplitude variation of 1%, could you detect this change?

(d) Compute the Doppler shift, $\Delta\lambda$, of the Hα line for the same star. If the width of the Hα line is $\Delta\lambda = 0.4$ Å, can the Hα Doppler shift be easily measured or not?

(e) Comment on the suitability of the Planck curve peak, in comparison with a spectral line, for measuring the radial velocities of stars.

9.3. How close does a star, moving with $v_t = 20$ km s^{-1} have to be, in order for its proper motion to be measurable? Consider the two cases:

(a) A ground-based optical telescope that can measure, at best, $\mu = 1$ arcsec yr^{-1}?

(b) A space-based (e.g. Gaia satellite) measurement of $\mu = 20$ μarcsec yr^{-1}?

(c) Compare each of these results to the size of the Milky Way (Figure 5.3).

9.4. (a) Rewrite Eq. (9.6) with $\mathcal{T}$ in years, a in AU, M in $M_\odot$, and constants evaluated for a case in which $M \gg m$. Simplify your resulting equation so that it applies to the Solar system.

(b) Compute the orbital periods of Neptune and Pluto (Table T.3) and compare them with known values. Discuss what could be producing any variations in the results.

9.5. From information related to the following test particles, find the mass of the central object ($M_\odot$):

(a) A roughly spherical cluster of galaxies for which the outermost galaxies are approximately 1 Mpc from the centre of mass and the maximum measured radial velocities are ≈ 300 km s^{-1} with respect to v_{sys} of the cluster.

(b) A globular cluster of stars for which the average radial velocity with respect to v_{sys} is $v_r = 2.8$ km s^{-1}. The radius corresponding to this average radial velocity is 30 pc. Ignore star–star interactions and assume that the cluster can be represented by a spherical gravitational potential.

(c) A white dwarf orbiting an unseen object. The radius of its orbit is 10 AU, and the circular velocity is 21 km s^{-1}.

9.6. Refer to Example 9.2 and, using the information given in Figures 4.11 and 4.21, determine the mass of NGC 2903.

9.7. A galaxy's nucleus harbours a supermassive black hole of mass, $M = 10^8\ M_\odot$, surrounded by an edge-on accretion disc that is spatially unresolved at X-ray wavelengths. The full width of a spectral line of iron, whose central energy is $E_{line} = 6.4$ keV, is due entirely to disc rotation and is 2.0 keV.
   (a) What is the maximum circular velocity of this disc?
   (b) What is the distance of the accretion disc, at its maximum velocity, from the centre of the black hole (AU)? Should this correspond to the inner edge of the accretion disc or the outer edge?
   (c) Express the radius of part (b) in units of Schwarzschild Radii.

9.8. [*Online*]
   (a) Write an expression for the relative error in d_L that results from using Eq. (9.11) for the Hubble relation instead of Eq. (F.16) when $z < 1$, and evaluate it for the values of q_0 used in the two curves shown in Figure (F.2). The expression will be a function of z.
   (b) Determine the relative error for each of the two values of q_0 for (i) a galaxy with recessional motion of $cz = 5000$ km s^{-1}, (ii) a galaxy at $z = 0.1$ and (iii) a quasar at $z = 0.6$.

9.9. [*Online*] For $z < 1$ and $|q_0| < 1$, verify that the equation for the comoving coordinate, Eq. (F.10), results from Eq. (F.9). Note that, in this regime, any terms including $(t_0 - t)^3$ or higher orders can be neglected.

9.10. (a) Starting with the definition of redshift (Eq. (9.1)), express the quantity, $(z + 1)$, in terms of the frequency of a wave, rather than its wavelength.
   (b) [*Online*] Show that the result of part (a) is equivalent to Eq. (F.14).
   (c) [*Online*] Supernova 1995 K, at a redshift of 0.479, was observed to have a light curve that is time dilated [15]. If nearby supernovae of the same type normally show a light curve width (full width at half maximum) of about 25 days, estimate the width of the light curve of SN1995K.

9.11. [*Online*]
   (a) Starting with Eq. (F.16), derive an expression for the slope of the curve of the Hubble relation shown in Figure F.2 (i.e. $d(cz)/d(d_L)$) as a function of z. Confirm that the slope reduces to H_0 for $z \rightarrow 0$.
   (b) Measure the slope directly from Figure F.2 (dark solid curve) at two values of z and estimate H_0 and q_0 using the expression from part (a).
   (c) Discuss the reasons for any differences between the results of part (b) and the values of q_0 and H_0 that have actually been plotted (see caption).

9.12. Suppose that the Andromeda galaxy had no Doppler velocity with respect to the Milky Way. What would be its expansion redshift?

9.13. [*Online*] For a flat universe in which there is a cosmological constant, there is an early time at which the matter term dominates the behaviour of the expansion and a later time at which the expansion is dominated by Λ. For the epoch at which the energy densities of each component are equal ($\Omega_M = \Omega_\Lambda$), find numerical values for the following (assume $H_0 = 71$ km s^{-1} Mpc^{-1}; $\Omega_{M0} = 0.3$, $\Omega_{\Lambda 0} = 0.7$, and $q_0 = -0.55$):

(i) the density parameters, Ω_M and Ω_Λ, (ii) the deceleration parameter, q, (iii) the cosmological constant, Λ, (iv) the Hubble parameter, H(km s^{-1} Mpc^{-1}), (v) the density, ρ, (vi) the scale factor in comparison to the current value, a/a_0, (vii) the redshift, z, and (viii) the lookback time, $t_0 - t$(Gyr).

9.14. Repeat Example 9.4 for a neutron star of mass, $M = 2\ M_\odot$, and radius, $r_{ns} = 20$ km.

9.15. Derive the lens equation (Eq. (9.17)) from the geometry of Figure 9.8a and Eq. (9.16). (Recall that $\tan\alpha = \alpha$ in radians, for small angle, α.)

9.16. (a) Show that the Einstein ring angle equation (Eq. (9.18)) for a case in which the background source is very distant in comparison to the lens, can be expressed as

$$\theta \approx 0.090\sqrt{\frac{\left[\frac{M_L}{M_\odot}\right]}{\left[\frac{d}{\text{pc}}\right]}}\quad \text{arcsec} \tag{9.26}$$

(b) Using the result of part (a), compute θ_E (arcseconds) for the following cases (approximating each lens as a 'point mass'): (i) a Solar mass star at a distance of 1 pc, (ii) a neutron star of mass, 1.5 $M_\odot$, at a distance of 60 pc, and (iii) a galaxy of mass, $10^{11}\ M_\odot$, at a distance of 10 Mpc.

9.17. (a) Measure the radius of a visible Einstein ring at the location of the bright subcluster on the right-hand side of Figure 9.7a.

(b) Adopt a distance to the background source of $d_S = 1467$ Mpc and calculate D_{LS}.

(c) Estimate the mass of this subcluster.

9.18. (a) Consider four very distant background quasars that are in the true configuration of a square on the sky of length 2″ on a side. One of these quasars is perfectly aligned with the centre of a foreground supermassive black hole of mass $M_{bh} = 4 \times 10^9\ M_\odot$ which is at the core of a galaxy of distance, $d_L = 1$ Mpc. Plot the images of the four stars that result from being gravitationally lensed by the black hole. The result of Problem 9.16a may be useful.

(b) Suppose these four quasars instead delineated the four corners of a uniform brightness background source. Sketch the resulting images on the plot.

9.19. (a) Write an expression for the duration of a microlensing event, t_L, in terms of the transverse velocity of the lens, v_t, the distance to the lens, d_L, and the Einstein radius, θ_E.

(b) Beginning with the result of (a), re-express t_L (in days) in terms of d_L (kpc), d_S (kpc); M_L ($M_\odot$) and v_t (km s^{-1}).

(c) Evaluate t_L (days) for a lensing mass which is a 0.5 $M_\odot$ white dwarf in the Milky Way halo a distance of 5 kpc away with a transverse velocity of 220 km s^{-1}. The background star is in the Large Magellanic Cloud a distance 50 kpc away.

(d) Adopt the information given in the caption of Figure 9.9, evaluate t_L (days) for the lensing star and compare the result to the observed curve duration. Assume that the transverse velocity (net of background star, lens and observer) is 220 km s^{-1}. Comment on the result.

9.20. An X-ray binary system consists of a normal star and either a white dwarf, neutron star, or black hole companion, in which material from the star is accreting onto the collapsed object. In some X-ray binaries, X-ray variability on timescales as short as a millisecond has been observed. If the timescale is this short, what is the maximum size of the source (km)? Is it possible to rule out any of the types of companions for such sources?

JUST FOR FUN

9.21. Suppose you were travelling down the road in a vehicle that could travel at unlimited speeds, almost up to the speed of light. Ahead of you is a red light but you sail through it without stopping. Your excuse to the policeman is that it appeared green. How fast were you going?

9.22. How much closer has the Andromeda galaxy advanced towards you since you started reading this chapter? Express the result in a unit that you can easily visualize.

9.23. A little ant is on a raisin in a raisin cake universe that is expanding at H_0. He can travel at a speed of 0.1 mm s^{-1} through the dough and wants to get to the nearest raisin which is 1 cm away. Will he make it?

9.24. The 'wine glass gravitational lens': Figure 9.13 is a simple way to illustrate the effects of a gravitational lens with nothing more than a flashlight and the stem of a wine glass. While the 'lens' shape departs from a true gravitational lens, many of the effects we expect for astronomical objects can be seen with a little experimentation. Try this at home.

Figure 9.13 The 'wine glass gravitational lens'. *(Left)* A flashlight and wine glass are all that are needed to visualize the effects of a gravitational lens. *(Right)* As the base of the wine glass moves by the light, two arcs can clearly be seen.
Source: Judith A. Irwin.

PART V

The EM Signal Emitted

We have seen how a signal can be altered by our instruments, our atmosphere, and the matter and space that it encounters en route to our detectors. Each of these steps provided important information about the signal, about the intervening matter along and near its path, and even about the large-scale structure of our Universe. It is now time to turn our attention to the source, itself. How is the original signal actually emitted?

This is an important question because, encoded in the signal, is information about the properties of the source. An understanding of the various processes involved in the emission of light will provide us with the keys to unlock these secrets.

Fortunately, there are only a limited number of ways in which charged particles couple to electromagnetic radiation. This means that, although the number of the objects in the Universe is incomprehensibly large, there are actually only a small number of light-emitting mechanisms that are responsible for all of the cosmic radiation that we see. Some of these mechanisms operate only at very high energies and so are relegated to a subset of exotic objects. Others are more common and widespread.

Generally, emission processes fall into two categories: *continuum radiation*, emission that occurs over a broad spectral region, and monochromatic or *line radiation*, emission that occurs at a discrete wavelength with a narrow width in frequency. It is these processes that we now wish to consider.

Chapter 10

Continuum Emission

A scientist must also be absolutely like a child. If he sees a thing, he must say that he sees it, whether it was what he thought he was going to see or not.

– *So Long, and Thanks for all the Fish*, by Douglas Adams

Continuum radiation is any radiation that forms a continuous spectrum and is not restricted to a narrow frequency range, the latter applying to quantum transitions in atoms and molecules. By *emission*, we mean a process that starts with matter and ends with the creation of a photon. This could involve an interaction between a charged particle and another charged particle, a charged particle and an electric or magnetic field, or a charged particle with another photon, the end result being a photon. Following the symbolism of Sections 7.1 and 7.2, the emission is represented as: matter → (matter or photon or field) → photon.

Astrophysics: Decoding the Cosmos, Second Edition. Judith A. Irwin.

Companion website: www.wiley.com/go/irwin/astrophysics2e

Some commonly observed astrophysical continuum processes are included in this chapter, an important exception being black body radiation which was encountered in Chapter 6. The goal, for each type of continuum, is to obtain the intensity and *spectrum* of the signal, a process that involves determining the *emission coefficient,* j_ν (introduced in Eq. 8.2), and/or the *absorption coefficient,* α_ν Eq. (7.8), together with the appropriate solution of the Equation of Radiative Transfer (Section 8.3) to find the specific intensity, I_ν. Other quantities such as flux density, luminosity, etc. follow from I_ν, as described in Chapter 3. To have an expression for I_ν is to have an essential astrophysical tool since it means that properties of the emission, and indeed of the source itself, can be characterized. To this end, it is useful to note that, for a source of fixed solid angle, the flux density, f_ν, and spectral luminosity, L_ν, will have the same spectral shape as I_ν.

10.1 CHARACTERISTICS OF CONTINUUM EMISSION – THERMAL AND NONTHERMAL

For continuum emission, it is helpful to remember that an accelerating charged particle that is not bound to an atom or molecule will radiate. We want to know the spectrum (a function of frequency, ν), so it is also helpful to recall that frequency is inversely proportional to time. For example, an oscillating electron (Appendix D.1.1 of the *online material*) will emit radiation at a frequency $\nu = 1/\tau$, where τ is the period of oscillation; an electron that has a brief 'collision' with an ion of duration, Δt, will emit a photon within some bandwidth $B \approx 1/\Delta t$. Thus, short timescales are related to high frequencies. In any gas in which there are free charged particles, there are many particles, many accelerations, and many timescales involved. Emission over a range of frequencies – a continuous spectrum – will result. As we will see, this range can be quite large, spanning many orders of magnitude in frequency-space. The Planck curve (Section 6.1), for example, takes on all frequencies between zero and infinity, although emission at some frequencies is much more probable than at others. Thus, a characteristic of continuum emission from astronomical sources is that it is *broadband*[1].

Other characteristics of continuum emission depend more specifically on the type of radiation being considered. These fall into two categories – *thermal emission* and *nonthermal emission* – and a few comments can be made for each.

Thermal emission is any process for which the observed signal is associated with a system whose states are populated according to a Maxwell–Boltzmann (M-B) velocity distribution given by Eq. (I.2). Since this distribution defines the kinetic temperature, T, an equation for I_ν may be a function of various parameters but it will certainly be *a function of* T. The emitted radiation must therefore be related to random particle motions in the thermal gas.

[1] See, however, comments in Section 10.5.1.

An M-B distribution can be set up if the particles interact with each other via 'collisions' which includes Coulomb interactions that are sufficiently close to alter the trajectory of a charged particle. For a collisional process, we can apply the LTE solution to the Equation of Radiative Transfer (Eq. 8.19) which, as we found in Section 8.4.1, implies that $T_B \leq T$ (Eq. (8.24)). Recall that T_B applies to *any* emission process, thermal or otherwise, and that it is essentially a *measurable* quantity because it is directly related to the observed specific intensity, I_ν (Eq. 6.4). In astrophysics, we rarely see temperatures above $\approx 10^{7-8}$ K, so an observed brightness temperature *higher than* this range is a clue that the process is likely nonthermal. Therefore, a characteristic of thermal emission is *low brightness temperatures.*

Thermal emission is also intrinsically *unpolarized.* There is no particular directionality to velocities in a thermal gas, so there is no particular directionality to the resulting radiation. If polarization is observed, then either some other process is polarizing the signal after it is emitted (e.g. scattering by electrons or dust, Section 7.1) or else the emission is nonthermal.

Nonthermal emission is everything else. I_ν will be independent of T in any equation describing a nonthermal process. Examples are when the emission depends on the acceleration of a charged particle in a magnetic field or an electric field, or when the emission depends on collisions with other particles whose velocities are *not* Maxwellian. An example is a power-law distribution as we found for cosmic rays (Figure 1.8). Depending on the details of the specific emission process, the brightness temperature could be low or high and there may or may not be polarization of the resulting signal. We now examine these processes in more detail.

10.2 BREMSSTRAHLUNG (FREE–FREE) EMISSION

Bremsstrahlung, which is derived from the German words for 'braking' and 'radiation', occurs in an ionized gas when a free electron travels through the electric field of a positively charged nucleus. The electron is 'braked' by the electrostatic attraction of the nucleus, feeling a Lorentz force, $\vec{F}_e = e\,\vec{E}$ (assuming zero magnetic field, Table I.2) as it passes by (Figure 10.1). The electron is accelerated (its direction is changed) by this 'collision' with the nucleus and will therefore radiate. The process can also be thought of as the scattering of electrons in the electrostatic field of the nucleus. The resulting radiation is equivalently referred to as *free–free emission.* It is the inverse of the *free–free absorption* process discussed in Section 7.4.1 in which an electron absorbs a photon in the vicinity of a nucleus.

Our goal in this and subsequent sections is to find the *spectrum* of the emission, i.e. I_ν, as a function of frequency. Since we are only concerned with emission processes, we will assume initially that there is *no* background source and will focus on the emitting cloud-only.

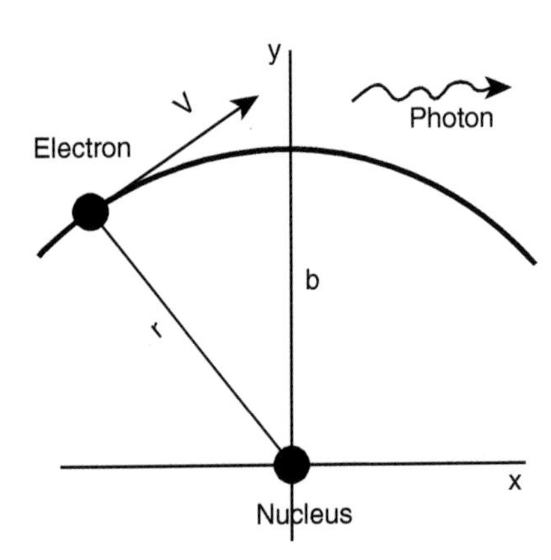

Figure 10.1 Geometry showing thermal Bremsstrahlung (free–free) emission resulting from an electron travelling near a nucleus of charge, Ze, where Z is the atomic number. The heavy curve shows the electron's path. An instantaneous velocity vector is shown for the electron at a position, $\vec{r}$ as well as an x, y coordinate system and the impact parameter, b, which is the distance of closest approach.

10.2.1 The Thermal Bremsstrahlung Spectrum

A derivation of thermal Bremsstrahlung emission can be found in [15, 22] and [34] and will not be repeated here. However, it is useful to offer a qualitative description of this process in order to see the functional dependences on various parameters.

One first considers a single free nonrelativistic (i.e. $v \ll c$) electron in an encounter with a single positive nucleus. Given the mass difference between these particles, the nucleus can be considered at rest. We are dealing with a free–free process, so the electron must not recombine with the nucleus. Rather, the kinetic energy of the electron, $E_k = (1/2)\, m_e v^2$, must be greater than the potential energy at the distance of the closest approach to the nucleus, $E_P = Ze^2/b$, where b is the *impact parameter* and Z is the nuclear charge (atomic number). For a large range of astrophysical conditions, the change in angle is small and the change in electron speed is negligible. The energy of the emitted photon is then much less than E_k of the electron (Problem 10.1), and the interaction can be treated as an elastic collision (Section 5.4.1) of duration, $\Delta t \approx b/v$. The power radiated by an accelerating charged particle is given by the *Larmor formula* (Table I.2) which can be integrated over the duration of the encounter and converted to a dependence on frequency, ν, instead of time[2]. With the aid of some geometry and a knowledge of scattering angles for elastic collisions between charged particles, the result is an expression for the energy radiated by an electron of speed, v, and impact parameter, b.

The next step is to consider the radiation from an ensemble of particles that are all interacting with nuclei in this fashion. For many electrons at a single fixed velocity, a variety of impact parameters are possible so an integration over impact parameter is required. A reasonable minimum value, b_{min}, is one at which the electron would be captured, and a maximum, b_{max}, would be the value at which the perturbation is so small that virtually no change in trajectory results. In a real plasma, corrections to these limits are required, but they are helpful in understanding how the integration over impact parameter might

[2] A proper treatment requires the application of a mathematical function called a *Fourier Transform*. A Fourier analysis allows one to relate processes in the time domain to the observed frequency domain.

proceed. Electrons that travel closest to the nucleus will experience the shortest duration interactions. Therefore, b_{min} determines the upper limit to the emitted frequency. Since the electron cannot come arbitrarily close to the nucleus, there should be a fairly sharp emission *cut-off* at high frequency for electrons of a given velocity.

The final step is to integrate over a distribution of electron velocities. Any velocity distribution that is thought to be present could be introduced at this point (e.g. a power-law distribution or even relativistic velocities). However, we are here considering *thermal* Bremsstrahlung which requires the Maxwell–Boltzmann (nonrelativistic) velocity distribution (Eq. (I.2)). The final result is the *free–free emission coefficient* (cgs units of erg s^{-1} cm^{-3} Hz^{-1} sr^{-1}) that was first introduced in Eq. 8.2,

$$j_\nu = \frac{8}{3}\left(\frac{2\pi}{3}\right)^{1/2} \frac{Z^2 e^6}{m_e{}^2 c^3}\left(\frac{m_e}{kT_e}\right)^{1/2} n_i\, n_e\, g_{ff}(\nu, T_e) e^{-\left(\frac{h\nu}{kT_e}\right)} \tag{10.1}$$

$$= 5.44 \times 10^{-39}\left(\frac{Z^2}{T_e{}^{1/2}}\right) n_i n_e g_{ff}(\nu, T_e)\, e^{-\left(\frac{h\nu}{kT_e}\right)} \tag{10.2}$$

where n_i, n_e are the ion and electron densities, respectively, T_e is the *electron temperature*, and Z is the atomic number. By 'electron temperature', we simply mean the kinetic temperature of the electrons[3].

The function, $g_{ff}(\nu, T_e)$, is a correction factor (for example, it includes corrections to b_{min} for quantum mechanical effects) called the *free–free Gaunt factor*[4] and is a slowly varying function of frequency and temperature. The Gaunt factor takes on different functional forms depending on the temperature and frequency regime (e.g. [9]) and is plotted for a variety of parameters in Figure 10.2. Notice that the plotted frequency range covers the entire EM spectrum from the radio band to gamma-ray (cf. Table I.1). The y-axis is shown as a linear (rather than logarithmic) plot, indicating that there is very little change in this factor within any given waveband. To a first approximation, $g_{ff}(\nu, T_e)$ is of order, one. We first saw the free–free Gaunt factor in Eq. 7.14, when we introduced Kramers' law for the inverse process of free–free absorption.

We showed in Eqs. (8.17) and (8.22) that, in the absence of a background source, the *shape* of a spectrum follows that of the emission coefficient for optically thin radiation ($I_\nu = j_\nu\, l$, where l is the line of sight distance). In fact, most observed radiation in the ISM and intracluster gas is optically thin, so it is worth pausing to examine the behaviour of j_ν.

At very high frequencies (specifically, $h\nu > kT_e$), the exponential term in Eq. (10.1) ensures that the emission coefficient cuts off rapidly. We know that in an MB velocity distribution, the most probable velocity corresponds to an electron kinetic energy of kT_e (Eq. 5.8). Electrons of this energy certainly cannot emit photons whose energies, $h\nu$, are greater than the energies of the electrons. Therefore, the high frequency decline is simply because there are few electrons with energies greater than kT_e.

There is a very useful consequence if the high energy exponential decline can actually be observed. Since the decline depends only on temperature, it is possible to *determine* T_e

[3] For gas that is in true thermal equilibrium, T_e would be the same as the ion temperature.
[4] This quantity is sometimes designated with a bar, $\bar{g}_{ff}$, because it has been averaged over an electron velocity distribution.

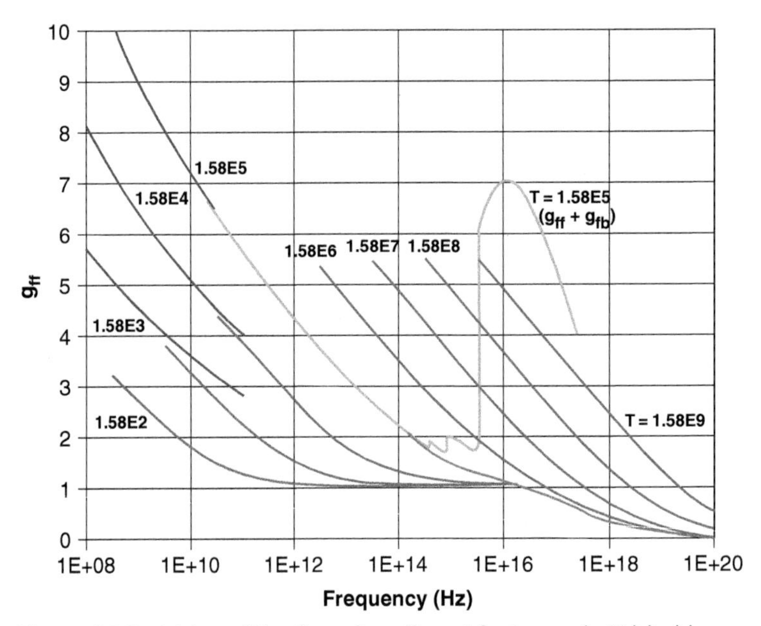

Figure 10.2 Value of the free–free Gaunt factor, $g_{ff}(\nu, T_e)$ (taking $Z = 1$) for a wide variety of conditions. Each curve is labelled with a temperature. Dark solid curves are accurate values taken from [38]. The three shorter curves at the left are the radio wavelength approximations given by Eq. (10.12). The grey curve is the combined free–free plus free–bound Gaunt factor for the temperature, 1.58×10^5 K from [9]. Therefore, $g_{fb}(\nu, T_e)$ for this temperature is represented by the excess above the smooth curve of $g_{ff}(\nu, T_e)$ below it. Sources: Sutherland [38]; Brussaard and van de Hulst [9].

without the requirement of any other measurement (Problem 10.2). For example, if the emission falls to $1/e$ of its maximum value at a measured frequency of ν_e, then $T_e = h\nu_e/k$.

Aside from this high-frequency cut-off, an examination of Eq. (10.1) reveals that the only frequency dependence lies in the functional form of the Gaunt factor. Therefore, the spectral shape of the Gaunt factor gives the spectral shape of the emission coefficient up to the cut-off. The Gaunt factor shows little variation with frequency for an object of a given temperature and so, as we shall confirm below, *the optically thin thermal Bremsstrahlung spectrum is a flat spectrum.* A flat spectrum is therefore a characteristic of this type of emission and is a clue that ionized gas is being observed.

The emission coefficient, j_ν, tells us the emissive behaviour of a pocket of gas without allowance for its internal absorption. In order to derive the full spectrum, I_ν, we also need an expression for the absorption coefficient, α_ν, as required by the Equation of Radiative Transfer (Eqs. (8.3) and (8.4)). In the absence of a background source, the absorption coefficient tells us how much *self-absorption* is occurring in the gas, that is, how effectively the gas absorbs its own radiation.

Because the thermal Bremsstrahlung mechanism arises from a collisional process in a gas with an MB velocity distribution, the LTE equations can be used (e.g. Section 5.4.4). This means that the absorption and emission coefficients are related via the Planck function, $B_\nu(T) = j_\nu/\alpha_\nu$ (Eq. (8.18)). The Planck function and j_ν are known, so we can now find α_ν, which immediately leads to the optical depth [34], τ_ν, (Eqs. 7.8 and 7.9),

$$\tau_\nu = \int \alpha_\nu dl = \int \frac{j_\nu}{B_\nu(T)} dl \tag{10.3}$$

$$= \left(\frac{4e^6}{3\, m_e\, h\, c}\right)\left(\frac{2\pi}{3\, k\, m_e}\right)^{1/2} T_e^{-1/2} Z^2 \nu^{-3}\left(1 - e^{-\frac{h\nu}{kT_e}}\right) g_{ff}(\nu, T_e) \int n_e n_i dl \tag{10.4}$$

$$\approx (3.7 \times 10^8)\, T_e^{-1/2}\, Z^2\, \nu^{-3}\left[1 - e^{\frac{-h\nu}{kT_e}}\right] g_{ff}(\nu, T_e)\, \mathcal{EM} \tag{10.5}$$

In Eq. (10.4), we have assumed that the temperature does not significantly vary along a line of sight. In Eq. (10.5), in addition, we take $n_e \approx n_i$ which is the case for pure ionized hydrogen or for ionized hydrogen with a contribution from singly ionized helium. The quantity, $\mathcal{EM}$, is called the *emission measure*, defined by,

$$\mathcal{EM} \equiv \int n_e^2 dl \approx n_e^2 l \tag{10.6}$$

The emission measure[5] contains information about the number density of particles along the line of sight, l. The approximation of Eq. (10.6) allows for some variation in the electron density, in which case a mean electron density along the line of sight is implied.

It is of interest to examine the *frequency dependence* of τ_ν in Eq. (10.4), because it tells us the frequency at which a cloud is optically thick or optically thin. There is little emission when $h\nu \gg kT_e$ due to the exponential in Eq. (10.1), so we consider the regime, $h\nu \ll kT_e$. Here $\exp[-(h\nu)/(kT_e)] \approx 1 - (h\nu)/(kT_e)$ (Eq. A.3 of the *online material*) so the ν-dependent term in parentheses in Eq. (10.4) equals $(h\nu)/(kT_e) \propto \nu$. The Gaunt factor has only a weak dependence on frequency so it can be considered roughly constant. Therefore, in Eq. (10.4) we are left with $\tau_\nu \propto \nu^{-3}\, \nu \propto \nu^{-2}$. This means that the opacity of the cloud *increases at lower frequencies.*

The final spectrum is given by the LTE solution to the Equation of Transfer (Eq. (8.19)), which we have already seen and which we repeat here for the case of no background source,

$$I_\nu = B_\nu(T_e)(1 - e^{-\tau_\nu}) \tag{10.7}$$

A plot of the thermal Bremsstrahlung spectrum then requires that we insert the Planck curve for $B_\nu(T_e)$ using Eq. (6.1) and we insert the expression for the optical depth, τ_ν, using Eq. (10.4).

Results for two different plasmas, one applicable to 10^4 K gas from an HII region and one applicable to 10^6 K intracluster gas in a cluster of galaxies, are shown in Figure 10.3. The flat nature of the thermal Bremsstrahlung spectrum is quite evident in this figure.

[5] Sometimes Z^2 is included in the definition of emission measure.

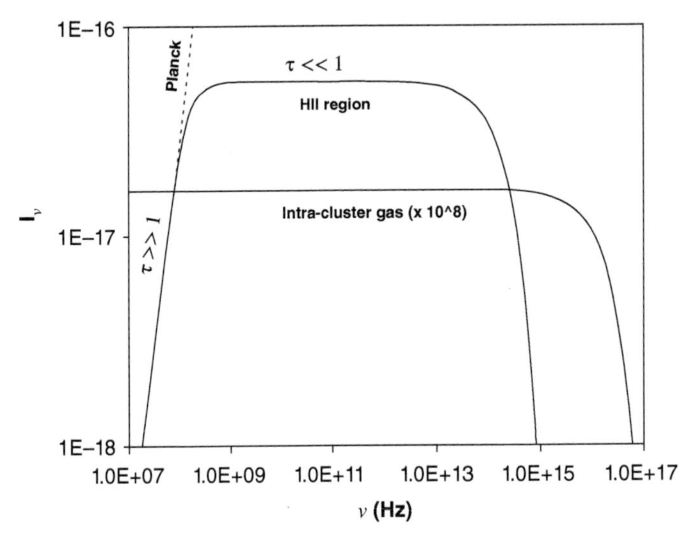

Figure 10.3 Two examples of the thermal Bremsstrahlung spectrum with ordinate (I_ν) in cgs units. For the HII region, the parameters are as follows: $n_i = n_e = 100\ \text{cm}^{-3}$, $T_e = 10^4$ K, $l = 5$ pc, $Z = 1$, $g_{ff} \approx \text{constant} = 6.5$. For the hot intracluster gas: $n_i = n_e = 10^{-4}\ \text{cm}^{-3}$, $T_e = 10^6$ K, $l = 1$ Mpc, $Z = 1$, $g_{ff} \approx \text{constant} = 1.0$, and *the specific intensity has been multiplied by a factor of* 10^8 *to appear on the plot*. For the HII region, the optically thick and optically thin regions are marked as well as the Planck curve at an equivalent temperature. The high-frequency exponential cut-offs are evident for both curves, but the optically thick self-absorbed region for the intracluster gas occurs at a frequency lower than shown on the plot.

As we did in Section 8.4.1, we can look at the optically thick and optically thin parts of the spectrum. In the absence of a background source (cf. Eqs. (8.20) and (8.22))

$$I_\nu = B_\nu(T_e) \quad (\tau_\nu \gg 1) \tag{10.8}$$

$$I_\nu = j_\nu l \quad (\tau_\nu \ll 1) \tag{10.9}$$

In the optically thick limit (Eq. (10.8)), the specific intensity is just the Planck curve. We showed above that the optical depth increases at lower frequencies, so this regime applies to the region in Figure 10.3 where the spectrum turns down at low frequency. Here, the source is just a black body and, as we showed in Section 6.1, the spectrum is a function of temperature only.

As the frequency increases, the optical depth decreases, and the spectrum then departs from the Planck curve. When $\tau_\nu \ll 1$ (Eq. (10.9)), the spectrum has the shape of the emission coefficient (Eq. (10.1)). The frequency response is now flat because it follows the Gaunt factor and then it cuts off exponentially at high frequency. In the flat part of the spectrum, the specific intensity depends on *both* the temperature and density. From Eqs. (10.1) and (10.6)), the specific intensity, $I_\nu \propto \frac{n_e{}^2 l}{T_e{}^{1/2}} \propto \frac{\mathcal{EM}}{T_e{}^{1/2}}$, so the source brightness is much more sensitive to density than temperature.

Now that we know the spectrum of free–free emission and how it depends on the parameters of the source, we can use the *observed* specific intensity of a source to obtain the source parameters. From the Bremsstrahlung spectrum alone, assuming that it is uncorrupted by any other emission processes, this could be accomplished by making a measurement at low frequency in the optically thick regime to obtain T_e. With T_e known, then another measurement could be made at a frequency in the flat part of the spectrum to obtain n_e. Alternatively, we could measure the high-frequency cut-off to obtain T_e and again, make another measurement in the flat part of the spectrum to find n_e.

The line-of-sight distance through the source must also be known, but this is often assumed to be about equal to the source diameter, the latter obtained from the small-angle formula (Eq. (I.9)), assuming that the distance to the source is known.

Once the electron density is found, the hydrogen mass, M_{HII}, or total gas mass, M_g of the ionized region can be found for a completely ionized gas from,

$$M_{HII} = m_H n_{i(HII)} V \approx m_H n_e V = X M_g \tag{10.10}$$

where m_H is the mass of the hydrogen atom, $n_{i(HII)}$ is the number density of hydrogen ions (free protons), V is the volume of the ionized region, and X is the mass fraction of hydrogen (X = 0.7154 for Solar abundance, Eq. 5.4). Note that, although dust may play an important role in intercepting ionizing photons, it contributes negligibly to the total mass (Section 5.4.1). On the other hand, we have assumed throughout that $n_i \approx n_e$. If there is a substantial contribution of free electrons from other constituents, such as HeII, metals, or dust, then $n_e > n_i$ and a correction would have to be made to take this into account.

Free–free emission is most commonly seen in the radio part of the spectrum when the emission is from gas at $\approx 10^4$ K such as HII regions around hot young stars or the diffuse ISM. It is also observed in the X-ray part of the spectrum when the gas is at $\approx 10^{6-7}$ K such as diffuse gas in clusters of galaxies or halos around individual galaxies. As Figure 10.3 illustrates, the spectrum is broader than these wavebands alone and also exists, for example, at optical and UV wavelengths. However, free–free emission is most easily *isolated from other processes* in the radio and X-ray wavebands. We will, therefore, elaborate on these wavebands in the next sections.

10.2.2 Radio Emission from HII and Other Ionized Regions

Regions of ionized gas, such as HII regions (see Figure 10.4), emit by thermal Bremsstrahlung and properties of these regions can be derived through observations of their radio emission. Radio observations are exceedingly important in astrophysics because radio waves are not impeded by interstellar dust. While optical observations

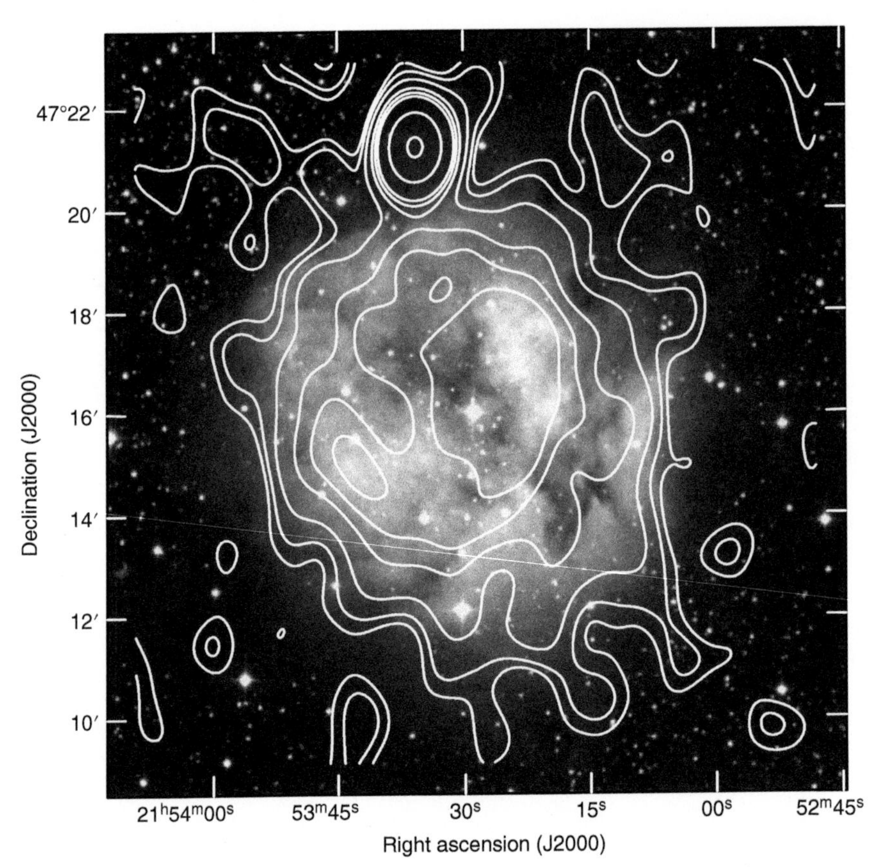

Figure 10.4 The HII region, IC 5146, at a distance of $D = 0.96$ kpc. The greyscale image is an R-band optical image. Source: Reproduced courtesy of the Palomar Observatory – STSc I Digital Sky Survey, Cal Tech. The contour overlays have been made from λ 92 cm data of the Westerbork Northern Sky Survey (WENSS, [31]). Contour levels are at 5, 10, 20, 40, 55, 70, 159, and 300 mJy beam^{-1}, and the map peak is 318 mJy beam^{-1}. The beam has a Gaussian weighting with a FWHM of $\approx 80''$. The total radio flux at this wavelength is $S_{\lambda\,92\,\mathrm{cm}} = 1.44$ Jy, and the electron temperature is $T_e = 9000$ K. The ionizing star, BD + 46° 3474, can be seen at the centre. The strong point source at the top of the image is an unrelated background source.

must contend with extinction (Section 7.5), radio wavelengths are much longer than interstellar grain sizes so can pass through without interaction. Thus, measurements in this waveband can provide information about the source without having to make often uncertain corrections for extinction. In some cases, especially for HII regions that are embedded in dusty molecular clouds or behind many magnitudes of extinction in the plane of the Milky Way, the region cannot be seen at all optically and observations at radio wavelengths may be the *only* way of detecting them (see, for example, Figure 11.6).

For any HII region at radio wavelengths, $h\nu \ll k\mathrm{T_e}$ (Problem 10.4) and certain simplifications can be made to the thermal Bremsstrahlung equations. The Planck curve, for example, can be represented by the Rayleigh–Jeans approximation (Eq. (6.6)), so the *optically thick emission* at low frequency can be represented by (cgs units),

$$I_\nu = B_\nu(\mathrm{T_e}) = \frac{2\nu^2}{c^2} k\mathrm{T_e} \propto \nu^2 \qquad (\tau_\nu \gg 1) \tag{10.11}$$

For *optically thin* emission, we require a value for the Gaunt factor. When $\mathrm{T_e} < 9 \times 10^5$ K, which will be the case for typical HII regions, the Gaunt factor can be represented analytically [19] by

$$g_{\mathrm{ff}}(\nu, \mathrm{T_e}) = 11.962\, \mathrm{T_e}^{0.15}\ \nu^{-0.1} \tag{10.12}$$

Equation (10.12) explicitly shows the flat nature ($g_{\mathrm{ff}} \propto \nu^{-0.1}$) of the Gaunt factor. With these assumptions and expanding the term in brackets in Eq. (10.5), the expression for optical depth can be rewritten in common astronomical units as,

$$\tau_\nu = 8.24 \times 10^{-2} \left[\frac{\mathrm{T_e}}{\mathrm{K}}\right]^{-1.35} \left[\frac{\nu}{\mathrm{GHz}}\right]^{-2.1} \left[\frac{\mathcal{EM}}{\mathrm{pc\ cm^{-6}}}\right] \tag{10.13}$$

The brightness temperature of an optically thin source can now be written by using Eq. (10.13) with Eq. (8.33) to find,

$$\left[\frac{\mathrm{T_{B_\nu}}}{\mathrm{K}}\right] = \mathrm{T_e}\tau_\nu = 8.24 \times 10^{-2} \left[\frac{\mathrm{T_e}}{\mathrm{K}}\right]^{-0.35} \left[\frac{\nu}{\mathrm{GHz}}\right]^{-2.1} \left[\frac{\mathcal{EM}}{\mathrm{pc\ cm^{-6}}}\right] \qquad (\tau_\nu \ll 1) \tag{10.14}$$

Using Eq. (10.14) with the Rayleigh–Jeans relation (Eq. 6.6) again, we can see the flat nature of the spectrum given by the frequency dependence of the specific intensity,

$$I_\nu = \frac{2\nu^2}{c^2} k\,\mathrm{T_{B_\nu}} \propto \nu^{-0.1} \qquad (\tau_\nu \ll 1) \tag{10.15}$$

Note that, in the radio regime, an optically thin HII region is always measured in the flat part of the spectrum as shown in Figure 10.3. It is never measured on the high-frequency exponential tail, because it is always the case that $h\nu \ll k\mathrm{T}$ so that the exponential term in Eq. (10.2) is ≈ 1.

It is sometimes more useful to measure the total flux density, f_ν, of an HII region which is related to the specific intensity via $f_\nu = I_\nu\, \Omega$ (Eq. (3.13)). Together with the Rayleigh–Jeans equation, the brightness temperature can then be expressed in terms of source flux density and Eq. (10.14) can be re-arranged to solve, explicitly, for various parameters of an HII region. These are the electron density, n_e, the hydrogen mass, $\mathrm{M_{HII}}$, the emission measure, $\mathcal{EM}$, and the excitation parameter, $\mathcal{U}$, which was first introduced and defined in Eq. (7.4). For the specific geometry of a spherical HII region of uniform density, the results are [35],

$$\left[\frac{n_\mathrm{e}}{\mathrm{cm^{-3}}}\right] = 175.1 \left[\frac{\nu}{\mathrm{GHz}}\right]^{0.05} \left[\frac{\mathrm{T_e}}{\mathrm{K}}\right]^{0.175} \left[\frac{f_\nu}{\mathrm{Jy}}\right]^{0.5} \left[\frac{D}{\mathrm{kpc}}\right]^{-0.5} \left[\frac{\theta_{\mathrm{sph}}}{\mathrm{arcmin}}\right]^{-1.5} \tag{10.16}$$

$$\left[\frac{M_{HII}}{M_\odot}\right] = 0.05579\left[\frac{\nu}{GHz}\right]^{0.05}\left[\frac{T_e}{K}\right]^{0.175}\left[\frac{f_\nu}{Jy}\right]^{0.5}\left[\frac{D}{kpc}\right]^{2.5}\left[\frac{\theta_{sph}}{arcmin}\right]^{1.5} \tag{10.17}$$

$$\left[\frac{\mathcal{EM}}{pc\ cm^{-6}}\right] = 8920\left[\frac{\nu}{GHz}\right]^{0.1}\left[\frac{T_e}{K}\right]^{0.35}\left[\frac{f_\nu}{Jy}\right]\left[\frac{\theta_{sph}}{arcmin}\right]^{-2} \tag{10.18}$$

$$\left[\frac{\mathcal{U}}{pc\ cm^{-2}}\right] = 4.553\left\{\left[\frac{\nu}{GHz}\right]^{0.1}\left[\frac{T_e}{K}\right]^{0.35}\left[\frac{f_\nu}{Jy}\right]\left[\frac{D}{kpc}\right]^{2}\right\}^{1/3} \tag{10.19}$$

where[6] θ_{sph} is the angular diameter of a spherical HII region as seen on the sky and D is its distance. A measurement of flux density at a radio frequency in the optically thin limit, with a knowledge of the temperature and distance, is sufficient to obtain these quantities (Problem 10.5). Table 5.1 provides some typical results. The excitation parameter can then be related to the properties of the ionizing star or stars, as was described in Example 7.4. Although many HII regions are not perfectly spherical, the line-of-sight distance through the region is not usually significantly different from the diameter, leading to only minor adjustments in the above equations.

Historically, this approach has most often been applied to HII regions but, in principle, the above equations for the radio regime apply to *any* discrete ionized region, provided $T_e \ll 9 \times 10^5$ K.

Planetary nebulae, for example, which are ionized gaseous regions formed by the ejecta of dying stars (Section 5.3.2), have properties that are not very different from HII regions. They are ionized by a central white dwarf and have similar temperatures with (on average) higher densities. Radio observations of free–free emission from these nebulae have also been carried out, revealing their properties (Table 5.1). Planetary nebulae have historically been a more challenging target than HII regions, however, because of their comparatively smaller sizes (less than a pc).

Another example is ionized gas in the Milky Way that is not in discrete regions, but rather spread out throughout the ISM. This is diffuse, low density (e.g. ≈ 0.01 cm^{-3} though it varies) gas, and therefore, the emission from this ISM component is weak. It is called either the *warm ionized medium* (WIM, more often used for the Milky Way) or the diffuse ionized gas (DIG, more often used when observed in other galaxies). The WIM/DIG is likely caused by ionization of interstellar HI by hot stars that are close to the boundaries of their parent clouds. Ionizing photons from these stars then leak out of their surrounding HII regions into the general ISM. The above equations would require a modification for geometry if the line of sight through the gas is significantly different from the diameter, but otherwise are still applicable. The challenge in extracting information about this

[6] The equation for electron density, n_e, includes contributions from singly ionized He, if present, so the hydrogen mass, M_{HII}, computed from Eq. (10.17) should actually be *reduced* by a correction factor, $1/[1 + N(\mathrm{HeII})/N(\mathrm{HII})]$, to account for this, where $N(\mathrm{HeII})/N(\mathrm{HII})$ is the ratio of abundances of singly ionized He to ionized hydrogen. For Solar abundance, this ratio should be less than 8.5%, depending on how many He atoms are ionized, and therefore for our purposes, we have not included the correction factor in Eq. (10.17).

component in the radio regime, however, is that the free–free radio emission tends to be contaminated by synchrotron emission (Section 10.5), requiring that the synchrotron fraction, if known, be subtracted first. Other methods for probing this diffuse component are also available, fortunately, such as observing the signals from background pulsars in the case of the Milky Way (Section 7.3) or measuring optical line emission (Section 11.4.2).

10.2.3 X-ray Emission from Hot Diffuse Gas

The X-ray waveband between approximately 0.1 and 5 keV (the 'soft' X-ray band, corresponding to $\nu \approx 10^{16} \rightarrow 10^{18}$ Hz) is dominated by emission from hot ($\approx 10^{6-8}$ K), low density (10^{-4} to 10^{-2} cm^{-3}) gas that emits via thermal Bremsstrahlung. In this band, the gas is always optically thin (Problem 10.6). Also, because $h\nu$ is *not* $\ll kT_e$, the exponential term in Eqs. (10.1) and (10.2) is not approximately 1 as it was for radio observations of HII regions. At a frequency corresponding to 1 keV, for example (i.e. $\nu = 2.4 \times 10^{17}$ Hz), the gas temperature would have to be higher than 10^8 K for $h\nu \ll kT_e$. Therefore, the exponential term in these equations is important, and the emission is observed on or near the high-frequency exponential tail (Figure 10.3). This also means that the electron temperature can be determined by a fit to the shape of the spectrum in this region, as was described in Section 10.2.1.

A challenge of observing in the soft X-ray band is that these photons are very easily absorbed in the neutral hydrogen clouds that are abundant in the ISM[7]. This is beautifully illustrated by Figure 10.5 which shows the entire sky in the soft X-ray band and in HI. The X-ray map (top picture) is dominated by free–free emission from hot gas, though contributions from other emission processes are also present. Most of the X-ray emission is seen in the direction of the Galactic poles because, along the Galactic plane, X-rays are absorbed in the abundant HI clouds of the disc of the Galaxy. A correction for this absorption must first be applied, therefore, before an accurate X-ray intensity can be found. Because we have a good map of HI for our Galaxy, such a correction is routinely carried out, though it adds to the error bar of the result.

Once a correction for interstellar absorption has been made, it is common to quote an X-ray luminosity (in erg s^{-1}) for the object. This can be obtained by converting the specific intensity (Eq. (10.9)) to a luminosity (Problem 10.7), with knowledge of the distance. Although thermal Bremsstrahlung emission is expected to dominate in the soft X-ray band, some corrections may still be required to account for contributions to the luminosity from emission mechanisms other than thermal Bremsstrahlung alone. Possibilities include free–bound emission (Section 10.3) or, importantly, line emission. Figure 10.6 provides an illustration of some of the challenges involved in isolating the free–free spectrum in the X-ray band. Notice the exponentially declining free–free continuum spectrum superimposed with many spectral lines. The spectral lines would have to be subtracted

[7] The absorption is mainly by bound electrons in the heavy elements (e.g. oxygen) that are within these HI clouds.

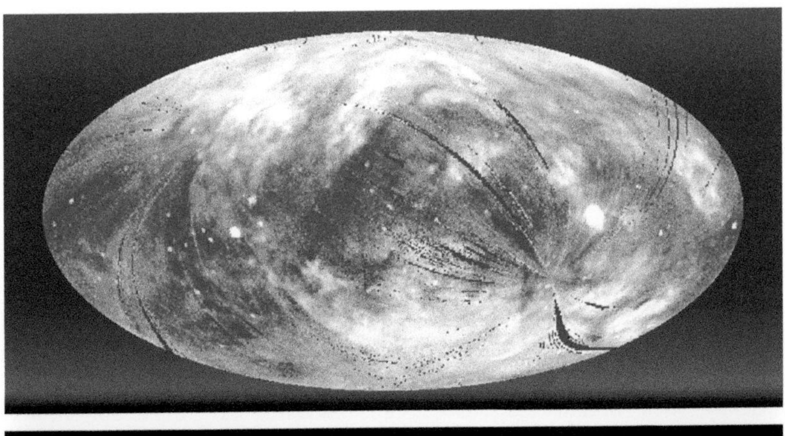

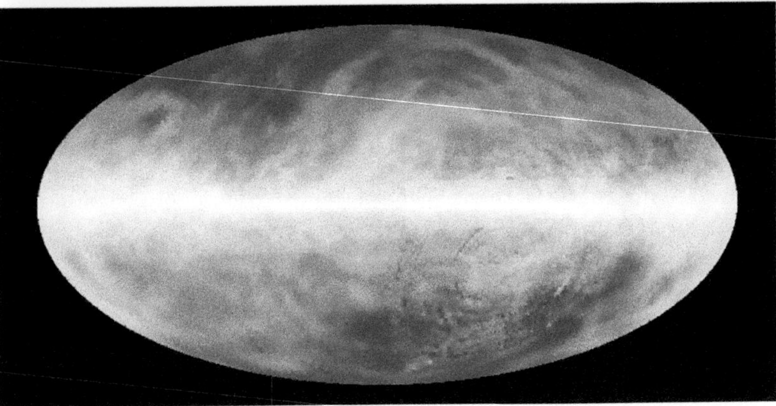

Figure 10.5 Illustration of the *anticorrelation* between soft X-ray emission and total HI distribution in the Galaxy, shown over the whole sky in this Hammer–Aitoff projection. The centre of the map is the direction towards the nucleus of the Milky Way, called the Galactic Centre (GC), and the right and left edges correspond to the *anticentre* direction, 180° from the GC. The top and bottom of the maps correspond to the North Galactic Pole (NGC) and South Galactic Pole (SGP), respectively, and the plane of the Milky Way runs horizontally through the centre. See Figure 5.3 for a map of the Galaxy as it would be seen from the outside. *(Top)* The 0.25 keV soft X-ray band observed with the ROSAT satellite [37]. The long-curved streaks are artefacts. Most of the emission is seen towards the Galactic poles because much of the X-ray emission along the plane of the Galaxy has been absorbed in neutral hydrogen clouds. *(Bottom)* All-sky map of the total (summed over all velocities) neutral hydrogen emission in the Milky Way [13]. Most of the emission is along the Galactic Plane.

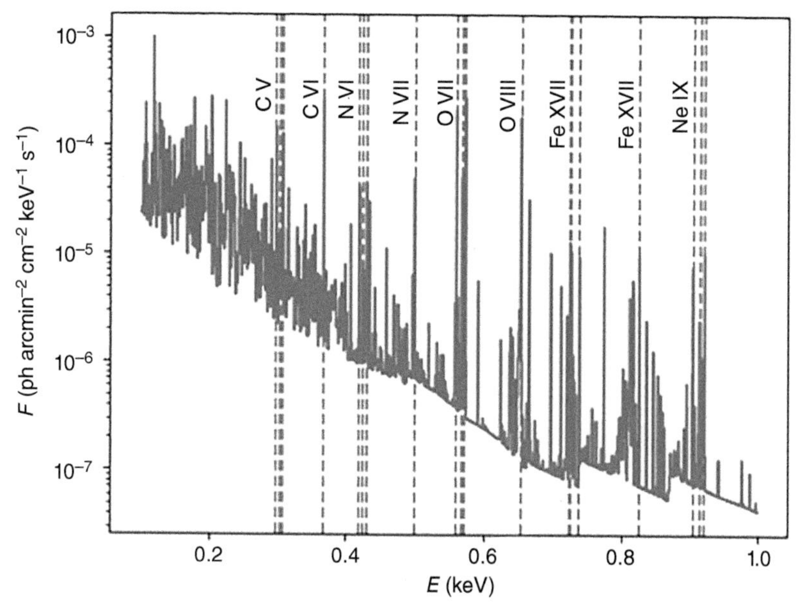

Figure 10.6 A soft X-ray spectrum showing the exponential decline of the thermal Bremsstrahlung continuum superimposed with many spectral lines from highly ionized species, many of which are labelled. The x-axis can be converted to a frequency via $E = h\,\nu$. Notice that the y-axis is a (non-cgs) specific intensity but for number of photons, rather than energy, similar to the cosmic-ray plot of Figure 1.8. The continuum is largely due to hot gas in the Milky Way. Source: Courtesy of Yuning Zhang and Wei Cui of Tsinghua University.

first in order to obtain the underlying continuum and using the equations of this section. This is in contrast to the radio regime where there are very few spectral lines.

Assuming that the observed emission represents only the free–free component, the luminosity of a uniform density, constant temperature plasma with $Z = 1$ and $n_e \approx n_i$ is,

$$L_X = 6.84 \times 10^{-38}\, \frac{n_e^{\,2}}{T_e^{\,0.5}}\, V \int_{\nu_1}^{\nu_2} g_{ff}(\nu, T_e)\, e^{-\frac{h\nu}{kT_e}}\, d\nu \tag{10.20}$$

where, ν_1, ν_2 are the lower and upper frequencies, respectively, of the band in which the emission is observed and V is the volume occupied by the emitting gas.

Taking the Gaunt factor to be a constant ($g_{ff}\,[\nu, T_e] \approx g_X$) and using this equation with Eq. (10.10), the electron density and mass of the ionized hydrogen gas[8] can be found,

$$\left[\frac{n_e}{\mathrm{cm^{-3}}}\right] = (1.55 \times 10^{-19} f_X) \left[\frac{L_X}{\mathrm{erg\ s^{-1}}}\right]^{0.5} \left[\frac{V}{\mathrm{kpc^3}}\right]^{-0.5} \left[\frac{T_e}{\mathrm{K}}\right]^{-0.25} \tag{10.21}$$

[8] Similar comments as given in Footnote 6 of this chapter apply here. However, the error may be larger, depending on metallicity, because at higher temperatures, more electrons may be released from metals.

$$\left[\frac{M_{HII}}{M_{\odot}}\right] = (3.81 \times 10^{-12} f_X)\left[\frac{L_X}{\text{erg s}^{-1}}\right]^{0.5}\left[\frac{V}{\text{kpc}^3}\right]^{0.5}\left[\frac{T_e}{K}\right]^{-0.25} \tag{10.22}$$

where f_X is a unitless function given by,

$$f_X = \left[g_X\left(e^{-\frac{E_1}{kT_e}} - e^{-\frac{E_2}{kT_e}}\right)\right]^{-0.5} \tag{10.23}$$

$E_1 = h\nu_1$ and $E_2 = h\nu_2$ being the lower and upper energies, respectively, of the band over which the Bremsstrahlung spectrum is observed. It is common to express the temperature, not in Kelvins, but rather in energy units (keV) for a more straightforward comparison to the observational energy band. For example, a temperature of 10^7 K corresponds to $kT_e = 1.38 \times 10^{-9}$ erg $= 0.863$ keV. These equations assume that the *filling factor* is unity; that is, the volume is uniformly filled with the emitting gas.

The interstellar absorption illustrated by Figure 10.5 is usually considered to be an undesirable effect that weakens the emission that we wish to observe and introduces error into the result. In fact, all measurements of soft X-ray emission must be corrected for the expected absorption in the intervening clouds before proceeding with the analysis as outlined in the above equations.

Surprisingly, however, absorption due to HI clouds can actually *benefit* our understanding of the X-ray sky. The velocity of any object that emits a spectral line can be determined via the Doppler shift of the line (Section 9.1.1). For HI clouds in the Milky Way, these velocities can be translated into distances by adopting a model for the rotation of the Galaxy (Section 9.1.1.2). HI emits by the λ21 cm line (Section 5.4.5) so it is possible to place constraints on the distances to various HI clouds in the Galaxy. This is not true of continuum emission for which we cannot obtain velocities (e.g. Problem 9.2). When continuum emission is observed, it could be coming from anywhere along a line of sight – from nearby hot gas or from the distant Universe. However, a detailed comparison of maps, like those in Figure 10.5, for which HI cloud distances are known, can help to determine how much X-ray emission is coming from gas in the foreground or background. If an HI cloud is obscuring the X-ray emission, for example, then the X-ray continuum must originate from behind the cloud, placing a limit on the distance to the X-ray-emitting gas. This effect is called *shadowing*, and it has been used to help us understand the hot gas distribution in the Solar neighbourhood.

The results of this kind of analysis are not perfect, given the uncertainties. For example, there is some error in the location of the HI clouds, there are variations in electron density and line-of-sight distance, and sometimes uncertain corrections are required for X-ray emission from our own Sun. However, together with other information and analysis, a picture has emerged that the Sun is located in a *Local Hot Bubble* (LHB) in the Milky Way. Figure 10.7 shows a picture.

The bubble is irregular in shape but appears to be elongated in the vertical direction, roughly perpendicular to the plane. Such vertical low-density regions are seen elsewhere in the Milky Way and also in other galaxies and are called *chimneys*. As the name implies, these are conduits or tunnels in cooler, denser ISM gas through which hot gas may flow

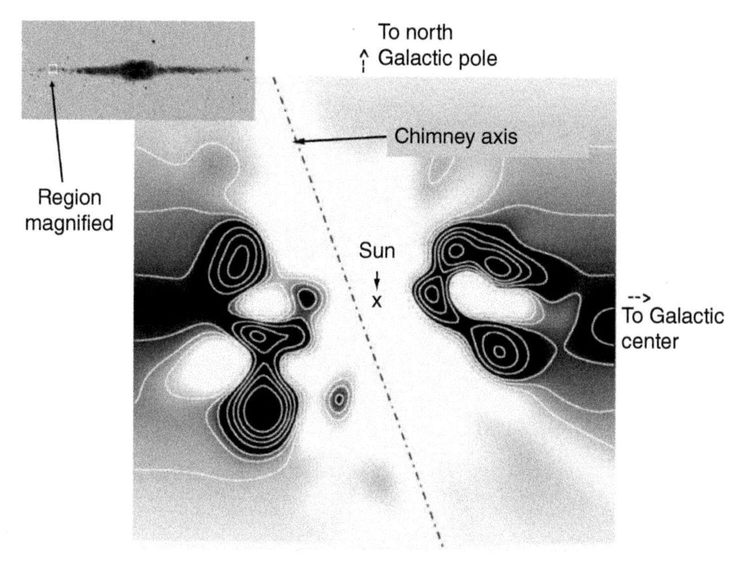

Figure 10.7 Diagram of the *Local Hot Bubble* (LHB) within which the Solar System is located. The top left image shows the region of the Galaxy that has been blown up for the bottom image. The LHB, of temperature, 10^6 K, and density, $\approx 5 \times 10^{-3}$ cm^{-3}, is the irregularly shaped white region around the Sun extending of order 50–200 pc in size. Contours and shading are related to density. Various features are shown, including the chimney axis, which is the approximate direction that hot gas might be venting into the halo [18]. Source: Barry Welsh.

into the Galaxy's halo[9]. They are likely produced by supernovae and/or stellar winds whose collective outflows have merged into larger structures. In the case of the LHB, $n_e = 0.005$ cm^{-3}, and $T_e = 1 \times 10^6$ K [21].

In a number of spiral galaxies, the collective effects of the venting of hot gas have produced an observable soft X-ray halo around the galaxy. These are typically observed in spirals that are edge-on or close to edge-on to the line of sight so that the halo can be clearly seen independent of the disc [20]. They also support a picture in which the disc–halo interface of a galaxy is a dynamic place through which outflows occur and possibly inflows as well if cooler gas descends again like a fountain. There may be other reasons for the presence of X-ray halos around galaxies, for example, leftover gas from the galaxy formation process itself. However, when we see strong X-ray halos from galaxies that have high star formation rates (also called *starburst galaxies*), such as NGC 4631 (Figure 10.8), this argues for outflows. The high mass stars that are produced in starburst galaxies will

[9] It is not always clear whether hot gas is currently flowing into the halo through individual chimneys or whether they are simply structures that result from other factors such as magnetic fields. Therefore, the cooler 'walls' of such features that extend above the plane have also been called *worms, a* name that does not imply hot gas outflow.

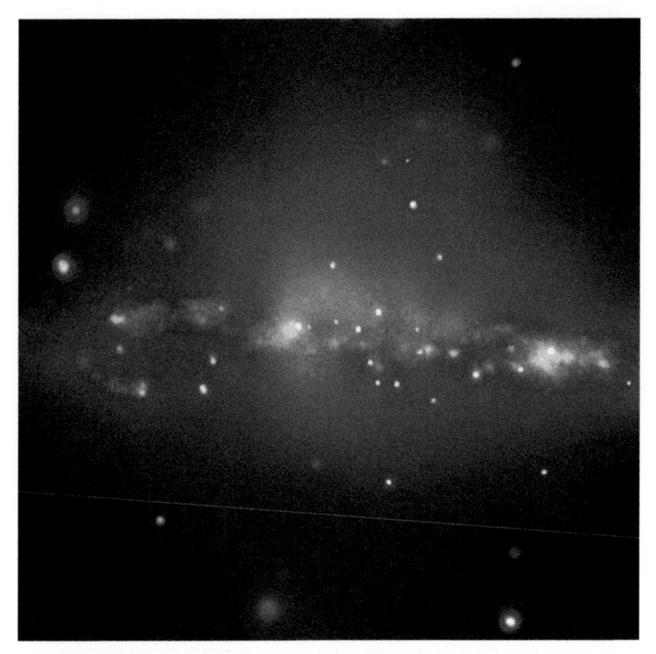

Figure 10.8 The galaxy, NGC 4631 (distance = 7.6 Mpc), is a spiral galaxy that is roughly edge-on to the line of sight. The pink colour shows UV light from hot stars, delineating the disc or plane of the galaxy. Surrounding the galaxy's disc is a halo of X-ray emission shown in blue, which is due to thermal Bremsstrahlung. The broadscale X-ray emission, which extends to about 8 kpc from the plane, can be characterized by a 'temperature' of 0.18 keV. The X-ray luminosity is $L_X \approx 2 \times 10^{39}$ erg s^{-1} in the 0.3–2 keV waveband [41]. Source: X-ray: NASA/CXC/UMass/D. Wang et al., UV: NASA/GSFC/UIT.

produce supernovae (Section 5.3.3) and stellar winds (Figure 7.10) that can initiate these outflows. On kpc scales, interesting regular magnetic field structure is also often seen in galaxies, arguing for magnetic dynamo action as an important, and probably necessary ingredient [43]. Example 10.1 provides some numerical values for NGC 4631.

Example 10.1

Assuming that the soft X-ray halo of NGC 4631 (Figure 10.8) is entirely due to thermal Bremsstrahlung with unity filling factor, estimate the electron density and hydrogen mass of this halo.

Let us first gather together the relevant data that are needed for Eqs. (10.21), (10.22), and (10.23).

From the figure caption, the 'temperature' is 0.18 keV. A conversion to cgs units and using $E = kT_e$, results in $T_e = 2.1 \times 10^6$ K. The radius over which the emission is observed is 8 kpc, leading to a volume of $V \approx 2.1 \times 10^3$ kpc^3, assuming spherical geometry. The luminosity is $L_X \approx 2 \times 10^{39}$ erg s^{-1}.

For the function, f_X, we need the lower and upper bounds of the observing band over which the emission was measured; these are $E_1 = 0.3$ and $E_2 = 2$ keV. To find the Gaunt factor, we will use Figure 10.2 and therefore need the frequency of the band centre. The mid-point of this band ($E = 1.15$ keV) corresponds (via $E = h\nu$) to $\nu = 2.8 \times 10^{17}$ Hz. We can now read the Gaunt factor from the figure, finding $g_X \approx 0.7$. Then from Eq. (10.23), $f_X = 2.75$. Note that, in the exponentials, we can leave E_1, E_2, and kT_e in units of keV, for simplicity.

We now have all the required data. From Eq. (10.21), we find $n_e \approx 0.01$ cm^{-3} and, from Eq. (10.22), $M_{HII} \approx 5.6 \times 10^8$ $M_\odot$.

These results are estimates because this halo is likely to have a varying density with radius and the assumption of pure thermal Bremsstrahlung with unity filling factor must also be considered carefully. Authors who have studied this system [41] have fitted the halo with a model that includes *two* different temperatures, with the hotter gas being closer to the disc. See [41] for further details.

In clusters of galaxies, immense reservoirs of hot gas have been detected between the galaxies, visible via their X-ray emission. An example is the galaxy cluster, Abell 2256, shown in Figure 10.9. This *intracluster* gas is the dominant baryonic component of mass in the cluster. In Abell 2256, for example, the total mass contained in the intracluster gas is 18 times greater than the total mass contained in all the stars of all the galaxies in the cluster [33]! Most of this gas is likely a remnant of the cluster formation process, although a small fraction may have come from supernova outflows from the galaxies. The latter contribution accounts for a small (less than Solar) heavy metal enrichment (Section 5.3.2). The total mass of the cluster is even greater than that of the gas, however, consisting of both light and dark matter components.

The total (light + dark) mass of the cluster may be found via gravitational lensing (Section 9.2.1), via an analysis of the motions of the individual galaxies (Section 9.1.1.2), by the Sunyaev–Zeldovich effect (Section 10.6), or by an assumption of *hydrostatic equilibrium* for the gas. The latter means that the gas, as a whole, is gravitationally held in place and not 'evaporating' away. With this assumption, it is possible to calculate how much mass is required to hold gas of a given temperature in place. The results of such analyses indicate that the gas mass to *total* mass fraction in clusters of galaxies is ≈ 15–20% (Problem 10.8), though a more recent value quotes 12% [23].

These results suggest that a large fraction of baryonic matter exists in low density hot gas on very large scales. With observations of many more galaxy clusters, it has become apparent that the observed baryonic fraction is matching that expected from Big Bang nucleosynthesis. Thus, it is possible that the historical 'missing baryon problem' that was first introduced in Section 5.2 could be resolved by realizing just where the so-called missing baryons may have been hiding. These results also provide a strong argument for pointing telescopes over many different wavebands at the sky, as we argued in Section 4.1.

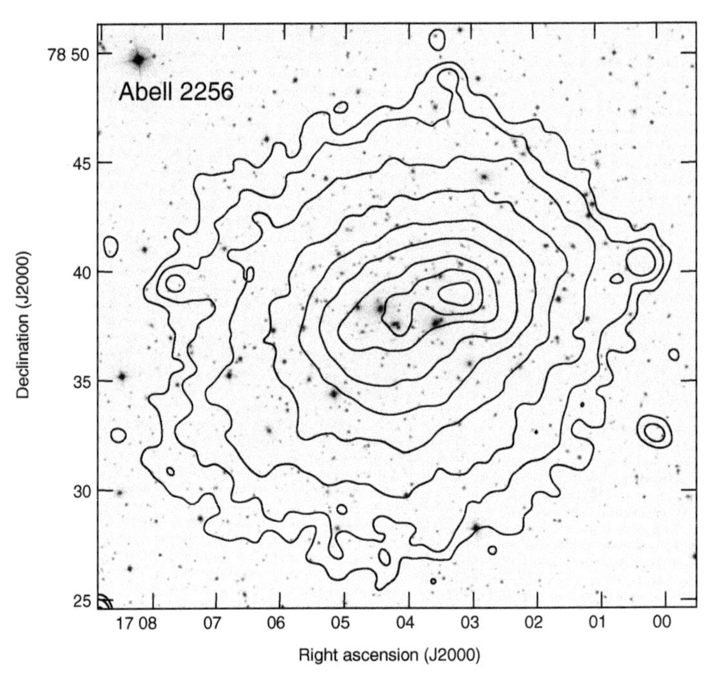

Figure 10.9 The cluster of galaxies, Abell 2256 (redshift, $z = 0.0581$), is shown in the optical (negative greyscale) and in X-rays in the observing band, 0.1–2.4 keV (contours). The gas is emitting at $kT_e = 7.5$ keV and the X-ray luminosity is $L_{0.1\to2.4} = 3.6 \times 10^{44}$ erg s^{-1} [14], of which 78% is due to thermal Bremsstrahlung [7]. The cluster radius is $R = 2.0$ Mpc [33]. The total mass of the cluster, including all light and dark matter, is $M = 1.2 \times 10^{15}$ M$_\odot$ [30]. All values have been adjusted to $H_0 = 71$ km s^{-1} Mpc^{-1}. The optical image is in R band. Sources: Reproduced courtesy of the Palomar Observatory–STSCI Digital Sky Survey, Cal Tech. X-ray data were obtained from the Position Sensitive Proportional Counter (PSPC) of the Röntgen Satellite (ROSAT).

10.3 FREE–BOUND (RECOMBINATION) EMISSION

The free–bound process, or *recombination*, has already been discussed in the context of the ionization equilibrium that is present in HII regions (Section 5.4.7 and Example 7.4) in which the ionization rate must balance the recombination rate. We now wish to consider recombination as an emission mechanism.

Recombination involves the capture of a free electron by a nucleus into a quantized bound state of the atom. Therefore, free–bound radiation must only occur in ionized gases. This now introduces the possibility that *both* free–bound and free–free radiation may occur from the same ionized gas, complicating the analysis of such regions. The free–bound process is essentially the same as described for free–free emission except for the end state, and in fact the free–bound emission coefficient can be written in the same fashion as Eq. (10.2) [19],

$$j_\nu = 5.44 \times 10^{-39} \left(\frac{Z^2}{T_e^{1/2}} \right) n_i n_e g_{fb}(\nu, T_e)\, e^{-\left(\frac{h\nu}{kT_e}\right)} \tag{10.24}$$

Note that this emission coefficient is identical to that of free–free emission except for the use of the *free–bound Gaunt factor*, g_{fb}, so an understanding of the relative importance of free–bound emission compared to free–free emission reduces to an understanding of the relative importance of the two Gaunt factors.

Computation of the free–bound Gaunt factor, $g_{fb}(\nu, T_e)$, requires sums over the various possible final states in atoms and taking into account a Maxwellian distribution of velocities for the initial states. In the limit, if the final state has a very high quantum number that approaches the continuum, then $g_{fb} = g_{ff}$. Gaunt factors have been determined for a range of temperatures, frequencies, and metallicities by various authors (e.g. [9, 24]). In Figure 10.2, we show one result, the sum of $g_{fb} + g_{ff}$, for the temperature, $T_e = 1.58 \times 10^5$ K (grey curve). The free–bound Gaunt factor is therefore represented by the *excess* over the smooth g_{ff} curve below it. Note that there are sharp *edges*, corresponding to transitions into specific quantum levels of energy, E_n (Eq. C.5 of the *online material*), followed by smoother distributions to the right (higher frequencies) of the edges. This behaviour can be understood by considering conservation of energy as electrons are captured, the energy difference going into the emitted photons (Problem 10.9).

At low frequencies (i.e. $h\nu \ll kT_e$), Figure 10.2 shows that the total Gaunt factor is just the free–free value, implying that recombination radiation is negligible in comparison to thermal Bremsstrahlung in such a gas. This is typical of other temperatures as well, validating our assumption (Section 10.2.2) that the radio continuum emission from HII regions and planetary nebulae is due entirely to thermal Bremsstrahlung. In this regime, the electron is perturbed only slightly from its path. However, once the emission frequency is such that $h\nu \gtrsim kT_e$, g_{fb} may become significant. For the temperature of 1.58×10^5 K plotted on the graph, this corresponds to $\nu \gtrsim 3 \times 10^{15}$ Hz which is in the UV part of the spectrum. For HII regions and planetary nebulae with temperatures closer to $\sim 10^4$ K (Table 5.1), the corresponding frequency regime is $\nu \gtrsim 2 \times 10^{14}$ Hz which is in the near-IR and optical part of the spectrum. Thus, depending on temperature and frequency band, the recombination continuum can be a significant or even dominant component to the total continuum.

As the temperature increases, however, the relative contribution of g_{fb} decreases. The free–free Gaunt factor has a weak dependence on T_e (Eq. (10.12) or Figure 10.2), increasing as T_e increases. However, g_{fb} has a stronger dependence on temperature and also *decreases* as T_e increases. At higher temperatures, the electrons will have higher speeds, on average, and are less likely to be captured. Therefore, at very high temperatures, free–free emission dominates the continuum. A final important note is that a proper calculation of Gaunt factors requires all elements to be taken into account. For example, at

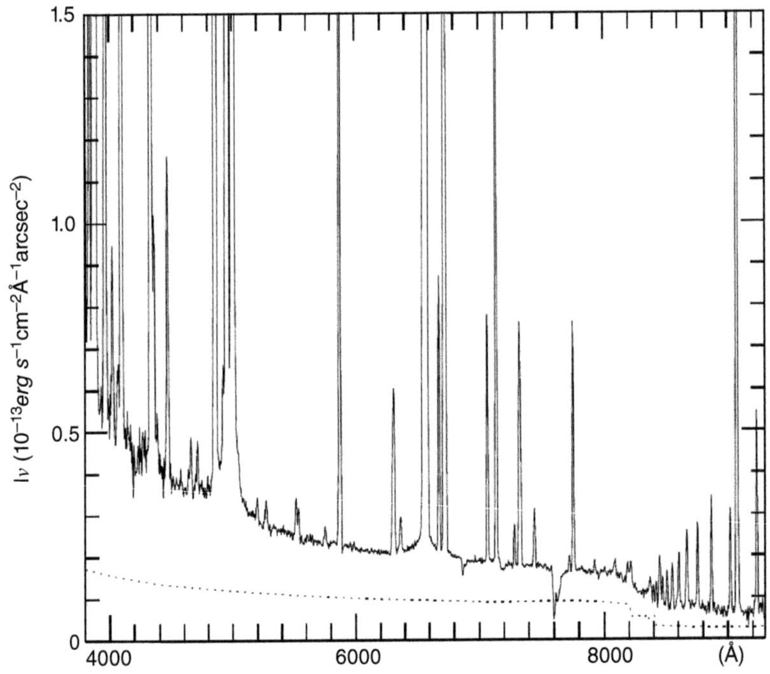

Figure 10.10 A spectrum of an HII region in the optical waveband. Many strong emission lines can be seen superimposed on the underlying continuum. The dashed curve shows the *combined* continuum from both free–bound radiation and two-photon emission (Section 10.4). The remainder of the continuum is due to scattered light from dust. The spectrum is of a region in the 30 Doradus nebula, also called the *Tarantula Nebula*, which is in the *Large Magellanic Cloud*, a companion galaxy to the Milky Way. Source: Darbon et al. [12]. © 1998 Astronomy and Astrophysics.

temperatures below 10^6 K, recombinations to CV, CVI, OVII, OVIII, and NVII are important at high frequencies.

A real spectrum of a combined free–free and free–bound continuum is shown for a Solar flare at X-ray wavelengths in Figure 8.7. Several spectral lines are also visible in that spectrum. Figure 10.10 shows another example, this time for an HII region at optical wavelengths. In this spectrum, the optical continuum consists of free–bound radiation, two-photon emission (Section 10.4) and light that is being scattered from dust. In addition, a great many optical emission lines are superimposed on the continuum.

Notice the increasing level of complexity from the previous sections. In the radio regime, the thermal (free–free) Bremsstrahlung continuum is largely free of spectral lines[10]. The X-ray thermal continuum (Figure 10.6) shows many lines, but the underlying continuum is still often dominated by free–free emission. The optical spectrum

[10] An exception is the 21 cm spectral line of HI. A few other radio spectral lines also exist (e.g. from OH) but they are easy to isolate.

(Figure 10.10), however, not only has many lines, but the underlying continuum results from a combination of processes. This figure illustrates that it is sometimes difficult to isolate emission from a single process alone. It is an example of a *complex spectrum* for which *modelling* is required to understand the various contributions. Complex spectra will be discussed further in Section 12.1.

10.4 TWO-PHOTON EMISSION

Another important contribution to the continuum in low density ionized regions at some frequencies is a process called *two-photon emission*. Two-photon emission occurs between *bound* states in an atom but it produces *continuum* emission instead of line emission. It occurs when an electron finds itself in a quantum level for which *any* downwards transition would violate quantum mechanical selection rules and the transition is therefore highly *forbidden* (e.g. Appendix C.4 of the *online material*). The electron cannot remain indefinitely in an excited state, however, and there is some probability, although low, that de-excitation will occur. The most likely de-excitation may not be the usual emission of a single photon, but rather the simultaneous emission of *two* photons. These two photons can take on a range of energies whose sum is always the total energy difference between the levels. Two-photon emission will occur as long as the electron is not removed from the level via a collisional energy exchange first. Therefore, it occurs in lower density ionized nebulae in which the time between collisions that would depopulate the problematic level is longer than the two-photon lifetime.

The hydrogen atom provides a good example of two-photon emission. We have already seen that the Lyα line (a recombination line, Section 5.4.7), which results from a de-excitation of an electron from the n = 2 to n = 1 levels, has a very high probability. However, the n = 2 quantum level consists of both $2s$ and $2p$ states (Table C.2 of the *online material*). The transition $2p \rightarrow 1s$ is a permitted transition with an Einstein A coefficient of $A_{2p\rightarrow 1s} = 6.27 \times 10^8\ \mathrm{s^{-1}}$, but the $2s \rightarrow 1s$ transition is highly forbidden. The forbidden transition will occur via two-photon emission with a rate coefficient of $A_{2s\rightarrow 1s} = 8.2\ \mathrm{s^{-1}}$, considerably lower than the $2p \rightarrow 1s$ transition rate, but still high enough for this process to be important in nebulae with densities $<10^4\ \mathrm{cm^{-3}}$ [36].

Since the sum of the energies of the two photons must equal the energy difference between the two levels, then so must the frequencies ($E = h\nu$),

$$\nu_1 + \nu_2 = \nu_{\mathrm{Ly}\alpha} \tag{10.25}$$

where ν_1 and ν_2 are the frequencies of the two photons and $\nu_{\mathrm{Ly}\alpha}$ is the frequency of Lyα. The spectrum of photon frequencies can take on any value as long as Eq. (10.25) is satisfied, but the most probable configuration is that the two photons have the *same* frequency. Therefore, the probability distribution of photon frequencies is symmetric and centred on the mid-point frequency which, for Lyα, is $\nu_{0,\mathrm{2ph}} = 1.23 \times 10^{15}$ Hz (λ 243 nm). This is in the ultraviolet part of the spectrum with a width that stretches between $\nu = 0$ and $\nu = \nu_{\mathrm{Ly}\alpha}$. The probability is lower as the photon frequency departs from $\nu_{0,\mathrm{2ph}}$. Note that the *number* of photons with frequencies above $\nu_{0,2ph}$ is equal to the number below, so *the energy* distribution (number of photons times energy per photon), which describes the spectrum, is actually skewed to higher energies.

The strength of this two-photon emission depends on the number of particles in the n = 2 state of hydrogen. This, in turn, depends on the recombination rate to this level which is a function of both density and temperature. Therefore, two-photon emission, although quantum-mechanical in nature, can be thought of as thermal emission and the density dependence is the same as for free–free and free–bound emission, i.e. $j_\nu \propto n_e^2$. The emission coefficient decreases with increasing temperature, described by a somewhat more complex function. Further details can be found in [19].

As noted in Section 10.3, two-photon emission is present in the optical continuum of the HII region shown in Figure 10.10. It tends to be a smaller contributor to the spectrum than free–free or free–bound emission except at frequencies near its peak. For example, at $T_e = 10^4$ K, two-photon emission is about 30% higher than the sum of free–bound and free–free emission at $\nu_{0,2ph} = 1.23 \times 10^{15}$ Hz. However, at $\nu = 3.0 \times 10^{14}$ Hz, it is about six times weaker and, once the frequency departs significantly from $\nu_{0,2ph}$, two-photon emission is negligible [8].

The n = 2 to n = 1 forbidden transition of hydrogen is not the only possibility for two-photon emission. Various others exist over a wide range of principle quantum numbers in the hydrogen atom alone [11]. Two-photon emission is also possible from He as well as heavier species. In hotter gas, two-photon emission from highly ionized species such as N VI, Ne X, Mg XI, Si XIV, SX VI, Fe XXV, and others cannot be ignored if these elements are present. For gas between 10^6 and 10^7 K, two-photon emission can again dominate within certain parts of the observing band between 0.1 and 1 keV [28].

10.5 SYNCHROTRON (AND CYCLOTRON) RADIATION

Synchrotron radiation results from *relativistic electrons* (those with speeds approaching the speed of light) moving in a *magnetic field.* The source of relativistic electrons is the electron component of the cosmic rays that were discussed in Section 1.2. Synchrotron radiation is widespread in the Milky Way and other galaxies and is one of the most readily observed continuum emission processes in astrophysics. There is some similarity between the thermal Bremsstrahlung radiation discussed in Section 10.2 and synchrotron radiation. In the former case, the electron is accelerated (its direction changes) in an electric field, $\vec{E}$, and in the latter case, the electron is accelerated (its direction changes) in a magnetic field, $\vec{B}$. Both represent 'scattering' of a charge in a field. Therefore, synchrotron emission is sometimes referred to as *magnetic Bremsstrahlung* or *magneto-Bremsstrahlung* radiation.

To understand the emission, it is helpful to recall that a charged particle moving at a velocity, $\vec{v}$, in a magnetic field, $\vec{B}$, experiences a Lorentz force which, for an electron of charge, e, and negligible electric field is $\vec{F}_e = e\left(\frac{\vec{v}}{c} \times \vec{B}\right)$ (see Table I.2). Such a force exists whether or not the electron is relativistic. The direction of the force is perpendicular to *both* $\vec{v}$ and $\vec{B}$. Therefore, there is no work done and no energy loss from this process, aside from that from the photon itself[11]. The magnitude of the force is,

$$F_e = \frac{ev}{c} B \sin \phi = \frac{ev}{c} B_\perp = \frac{eB}{c} v_\perp \tag{10.26}$$

where ϕ is the angle between $\vec{v}$ and $\vec{B}$, called the *pitch angle.* As indicated, we can write either $B_\perp = B \sin \phi$ specifying the component of the magnetic field that is perpendicular to the velocity, or $v_\perp = v \sin \phi$ specifying the component of the velocity that is perpendicular to the magnetic field. Figure 10.11 illustrates this geometry. If $\vec{v}$ is perpendicular to $\vec{B}$ ($\sin \phi = 1$), the electron experiences the maximum force and will circle about the field line. If the electron moves only parallel to $\vec{B}$ ($\sin \phi = 0$), then it experiences no force and will not radiate. If the initial velocity is in an arbitrary direction, then the electron will spiral about the field line. Over a *period of gyration* (the time for a loop), the speed of the electron does not change due to this force[12], but its change of direction constitutes an acceleration and it will therefore radiate. Note that *any* charged particle, including protons and other ions, experience this behaviour although, as we will see, the synchrotron emission from ions is negligible in comparison to that of electrons.

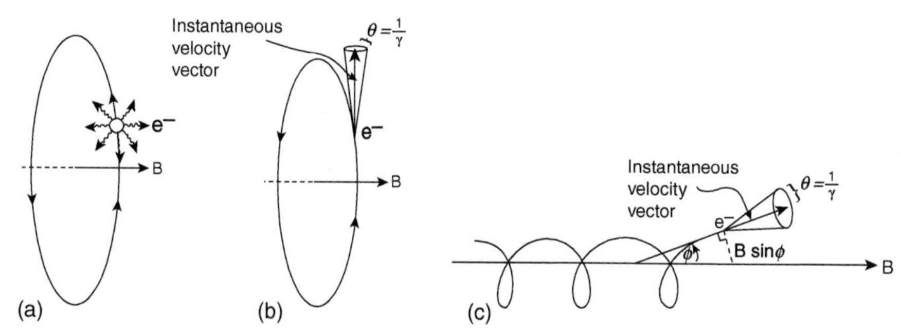

Figure 10.11 Illustration of the radiation given off by an electron (designated e^-) with instantaneous velocity, $\vec{v}$, encircling magnetic field lines of strength, B. (a) Nonrelativistic electron whose velocity is perpendicular to the direction of the magnetic field ($\sin \phi = 1$). The force acting on the electron is perpendicular to both $\vec{v}$ and $\vec{B}$, causing the electron to circle about the field lines. The resulting *cyclotron radiation* is emitted in all directions. (b) Relativistic electron whose motion is perpendicular to the magnetic field. The force on the electron is in the *same* direction as in (a) but the radiation given off (*synchrotron radiation*) is highly beamed in the forward direction into a cone of angular radius, $\theta = 1/\gamma$, where γ is the Lorentz factor (Eq. 1.18). (c) More realistic situation in which an electron has components of its velocity both parallel and perpendicular to the direction of B. The *pitch angle*, ϕ, is the angle between $\vec{v}$ and $\vec{B}$.

---

[11] The emission of a photon, however, will exert a back-reaction on the electron.

[12] Eventually the loss of energy due to radiation will slow the particle, but this occurs over a timescale that is typically very long in comparison to a gyration time.

If the magnetic field within a plasma is strong enough that it dynamically affects charged particles, the plasma is called a *magnetized plasma.* A more specific criterion is if a charged particle completes at least one gyration before it interacts with (collides with) another particle, a condition that may be different between electrons and ions. In a magnetized plasma, the gyrating charged particle is 'coupled' to the field and therefore, in a sense, trapped by it. If the field lines curve, for example, the gyrating particle will follow along this curvature. Any emission that may be given off by such trapped particles can then be used *to map out* the magnetic field configuration. A good example is the prominences seen on the Sun, as shown in Figure 8.6 or the coronal loop of Figure 10.12. Although these two images actually show recombination line emission (Section 11.4) rather than cyclotron or synchrotron emission, they do illustrate how the magnetic field geometry can be traced. Similarly, cyclotron or synchrotron radiation 'illuminates' otherwise invisible cosmic magnetic fields in locations where electrons are dynamically responding to the fields.

Radiation that results specifically from motion about the magnetic field will depend on the strength of the field, B. Therefore, not only do observations of such emission provide a means of determining the presence and orientation of fields, but they also provide a means of obtaining the magnetic field strength (Table 10.1). We will see how this dependence occurs in the next sections. Before considering the details of synchrotron radiation, however, it is helpful to start with its nonrelativistic, and less frequently observed, counterpart – *cyclotron radiation.*

Figure 10.12 A *coronal loop* can be seen at the edge of the Sun. This image, taken with the Transient Region and Coronal Explorer (TRACE) satellite, shows emission at 171 Å from Fe IX and Fe X ions that trace out delicate filamentary and loop-like magnetic field lines in the Solar atmosphere. Source: Reproduced by permission. TRACE is a mission of the Stanford-Lockheed Institute for Space Research, part of the NASA Small Explorer programme.

Table 10.1 Sample magnetic field strengths[a]

Object	B (G)
Interstellar space	10^{-6}
Interplanetary space	$10^{-6} - 10^{-5}$
Solar corona	$10^{-5} - 100$
Planetary nebulae	$10^{-4} - 10^{-3}$
H II region	10^{-6}
Pulsar (surface)	10^{12}
Supernova remnants (SNRs)	$10^{-5} - 10^{-2}$
Earth	0.31
Jupiter	4.28
Saturn	0.22
Uranus	0.23
Neptune	0.14

[a] Ref. [19] except for the planets [3] and SNRs. Planetary fields are surface equatorial values (but note that the value is variable with position and time). Other variations in these values may also be present.
Sources: Lang [19]; Bagenal [3].

10.5.1 Cyclotron Radiation – Planets to Pulsars

When a nonrelativistic electron gyrates about the magnetic field, cyclotron radiation is emitted at the frequency of this gyration, called the *gyrofrequency*, ν_0. For circular motion, the acceleration always points to the centre of the circle and the electric field vector of the emitted radiation follows that of the acceleration vector as seen by a distant observer (see [22], pp. 62–66 for details on the emission of an accelerating charge). If seen from a direction *along* the field line, $\vec{E}$ will rotate with time, giving rise to circular polarization of the emission. Seen from the side, the electric field vector will oscillate linearly (like a dipole), giving rise to linear polarization. From an intermediate angle, the polarization is elliptical. A characteristic of this kind of emission, therefore, is that it is intrinsically *polarized* and high degrees of polarization (D_p, see Section 3.7) have been measured in sources in which cyclotron radiation has been observed. Examples are the planets that have magnetic fields – the Earth, Jupiter, Saturn, Uranus, and Neptune – for which D_p up to 100% has been measured [44].

For spiral motion, the electron has both a velocity component that is parallel to the magnetic field and therefore unaffected by it, $\upsilon_{\parallel}$, as well as a perpendicular component, $\upsilon_{\perp} = \upsilon \sin\phi$. The Lorentz force (Eq. (10.26)) can then be equated to the centripetal force,

$$F_e = \frac{e\,B}{c}\,\upsilon_{\perp} = m_e\,\frac{{\upsilon_{\perp}}^2}{r_0} \tag{10.27}$$

where r_0 is the orbital radius that is perpendicular to the field, called the *radius of gyration* or *gyroradius*. For protons or other charged particles, the appropriate mass and charge must be used (Problem 10.10). Eq. (10.27) can be solved for r_0,

$$r_0 = \frac{m_e v_\perp c}{e\,B} = \frac{v_\perp \mathcal{T}}{2\,\pi} = \frac{v_\perp}{2\,\pi\,\nu_0} \tag{10.28}$$

where $\mathcal{T} = 1/\nu_0$ is the period of gyration and we have used the relation between velocity, distance, and time for motion in a circle ($v_\perp = 2\,\pi r_0/\mathcal{T}$). The gyrofrequency, ν_0, then follows from Eq. (10.28),

$$\nu_0 = \frac{e\,B}{2\,\pi\,m_e\,c} \Rightarrow \left[\frac{\nu_0}{\mathrm{MHz}}\right] = 2.8\left[\frac{B}{\mathrm{Gauss}}\right] \tag{10.29}$$

Notice that the electron gyrofrequency is *independent of velocity* and therefore independent of the kinetic energy of the electron $\left(E_k = \frac{1}{2}\,m_e\,v^2\right)$ when $v \ll c$. Even if there is a range of particle velocities, as is normally the case, all particles will emit at one frequency for a single magnetic field strength, although the radius of gyration will be different.

Since the gyrofrequency is single-valued, in a sense, cyclotron radiation represents monochromatic (or line), rather than continuum, emission for a single-valued magnetic field. We discuss line emission in Chapter 11, but cyclotron radiation is fundamentally different from normal spectral lines which result from quantum transitions within atoms and molecules. Thus, we include cyclotron radiation as a 'continuum process', which we will justify below. Example 10.2 provides some numerical results.

Example 10.2

Determine the radius of gyration, period, and gyrofrequency of a typical electron within the warm ionized medium (WIM) of the Milky Way, in which the magnetic field strength is $B \approx 3\,\mu$G. Can we expect to observe the resulting cyclotron radiation?

From Section 10.2.2, the WIM consists of ionized gas with $T_e \approx 10^4$ K and $n_e \approx 0.01$ cm^{-3}. From Eq. (10.29) with the given magnetic field strength, we find, $\nu_0 = 8.4$ Hz so $\mathcal{T} = 1/\nu_0 = 0.12$ s. A typical electron might be expected to have the mean electron speed in a Maxwellian velocity distribution which, for gas at 10^4 K, is $v = 6.2 \times 10^7$ cm s^{-1} (Eq. 5.9). Therefore, letting $v \approx v_\perp$ and using Eq. (10.28), the radius of gyration is $r_0 = 1.2 \times 10^6$ cm or 12 km. A gyrofrequency of 8.4 Hz is extremely low and this radio emission could not be observed from the Earth's surface because it falls below the plasma frequency of the Earth's ionosphere (about 10 MHz, Figure 4.2) and therefore could not propagate to the ground. Moreover, the plasma frequency in the WIM is (Eq. 7.7) $\nu_p = 0.89$ kHz, more than 100 x higher than the cyclotron frequency, so this radiation would not even escape from the WIM. Therefore, this emission cannot be observed.

The importance of Eq. (10.29) is that the magnetic field strength can be immediately obtained if the cyclotron frequency can be measured. The problem, however, is that this frequency may be very low, as demonstrated by Example 10.2. The cyclotron frequency must be higher than the plasma frequency in which it is generated, in order for the radiation to escape. This favours objects with high magnetic fields (Table 10.1) and/or low electron densities (Problem 10.10). Also, the magnetic field must be higher than about 3.5 G in order for this emission to be detectable from the ground because of the Earth's $\approx$ 10 MHz ionospheric cut-off (Figure 4.2). Thus, the number of objects for which cyclotron emission has so far been observed is limited to those with strong fields, or those for which satellite or space probe data are available, or both. Examples are pulsars, the Sun, and those planets that have magnetic fields (Table 10.1).

For low-velocity electrons ($\upsilon \lesssim 0.03\ c$) in a single-valued magnetic field, only one frequency at ν_0 is observed [5]. In a thermal distribution of particles, this corresponds to $T_e \lesssim 10^6$ K. In hotter gases or in nonthermal distributions with higher velocities, however, not only is the *fundamental frequency*, ν_0, observed, but also weaker *harmonics* at integer multiples of the fundamental. The observation of multiple frequencies at specific intervals helps to confirm that the emission is actually due to the cyclotron process. Such distinct frequencies have been observed in emission in some regions and absorption in others, depending on the presence and strength of the background radiation field. Example 10.3 presents an absorption case for a pulsar. Notice that, for the strong magnetic fields of pulsars, cyclotron lines are in the X-ray part of the spectrum.

Example 10.3

Three absorption lines have been observed in the X-ray spectrum of a pulsar, named 1E1207.4-5209, at energies of 0.7, 1.4, and 2.1 keV [6]. If the gravitational redshift from this neutron star is $z_G = 0.2$, what is the strength of its magnetic field?

These three lines are equally spaced in frequency and are presumed to represent the cyclotron fundamental frequency (at 0.7 keV) and its first and second harmonics. The observed fundamental gyrofrequency is then (via $E = h\nu$) $\nu_0 = 1.69 \times 10^{17}$ Hz. We found in Section 9.1.3 that the gravitational redshift of neutron stars is significant so we need to correct for it. For a gravitational redshift of $z_G = 0.2$, the true fundamental gyrofrequency (Eq. 9.1 with $\lambda\nu = c$) is $\nu_{0true} = \nu_0\ (1 + z_G) = 2.03 \times 10^{17}$ Hz. Then using Eq. (10.29), $B = 7.1 \times 10^{10}$ G.

Other cyclotron spectra may not be quite as straightforward to interpret as is implied by the neutron star of Example 10.3. For example, the magnetic field strength may vary significantly within the region being observed, resulting in a number of fundamental cyclotron frequencies, each corresponding to a different field strength. For a smoothly varying field, the result will be emission that is spread out over a corresponding frequency range, that is, *continuum* emission. This can be seen in Figure 10.13 which shows the low-frequency radio spectra of the planets that have magnetic fields.

Low-frequency planetary radio emission comes from a region around the planets called the *magnetosphere*. The magnetosphere is the region within which charged particles are more strongly affected by the magnetic field of the planet than by other external fields. A diagram of Jupiter's magnetosphere is shown in Figure 10.14. This topology, that of a *magnetic dipole* swept by the Solar wind (Section 1.2.3), is also similar for the other planets that have magnetic fields. Jupiter's field is both larger and stronger than those of the

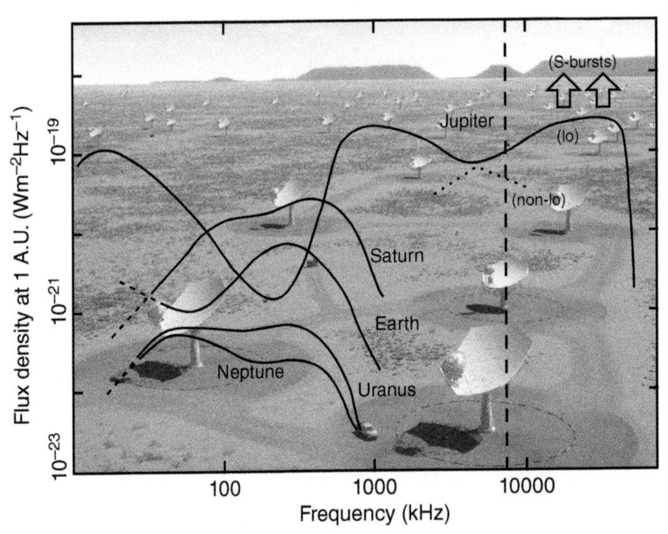

Figure 10.13 Low-frequency radio spectra of the planets that have magnetic fields. The vertical dashed line marks approximately the Earth's ionospheric cut-off. Jupiter has the strongest emission, having a higher magnetic field (Table 10.1) and a larger magnetospheric cross-section for interaction with Solar wind particles. The interaction of Jupiter's magnetosphere with its inner Galilean moon, Io, is responsible for higher frequency emission from this planet as well as the 'S-bursts' which refer to short-duration radio bursts. During radio bursts, peak flux densities can be >10 x higher than shown here and, for Jupiter, >100x higher. The flux density scale assumes that the radio emission has been measured at a distance of 1 AU from each planet. Source: Adapted from Zarka [44]. *Background:* Artist's conception of part of the Square Kilometre Array (SKA) currently under construction. Credit: SKA Organization.

other planets and it also has an additional component related to its Galilean[13] satellite, Io, whose volcanic outbursts supply additional plasma into a torus, called the *Io plasma torus,* around the planet.

The magnetic field strength varies with position within the magnetosphere. The strongest magnetic fields are near the poles of the planet, as indicated by a higher density of field lines in Figure 10.14. From Figure 10.13, Jupiter's maximum gyrofrequency is $\nu_0 \approx 40$ MHz which, by Eq. (10.29), gives a maximum magnetic field strength of $B \approx 14$ G,

[13] The four largest moons of Jupiter, Io, Europa, Ganymede, and Callisto, are called the *Galilean satellites,* after their discoverer.

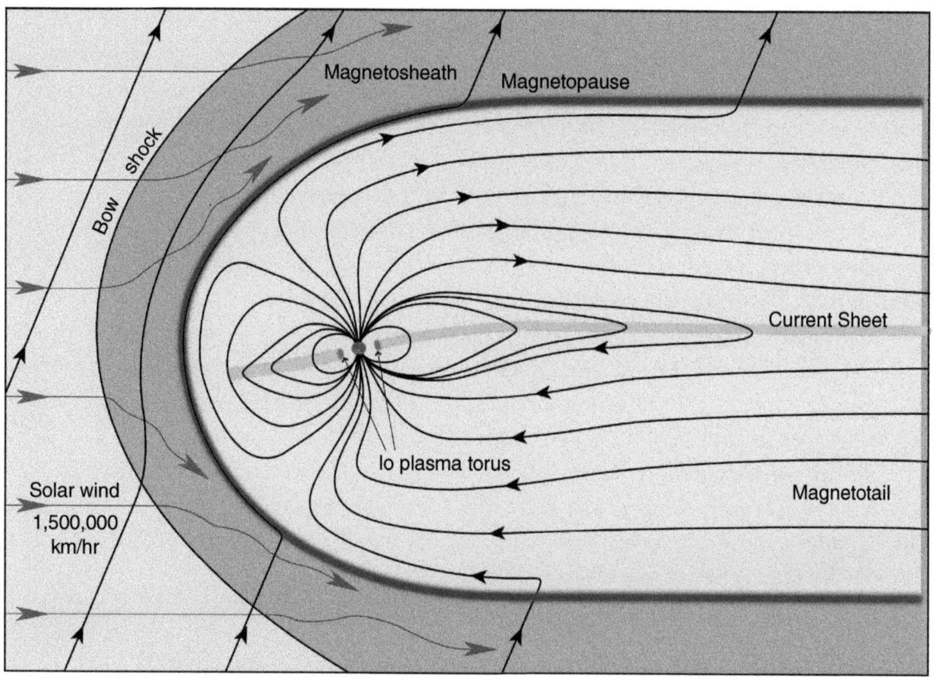

Figure 10.14 Diagram of Jupiter's *magnetosphere*. The planet, itself, is the small brownish dot near the centre. The curves with arrows show the field lines which come closer together (stronger fields) near the poles of the planet. Note the Io plasma torus (red dots on either side of Jupiter) as well as the asymmetric shape of the field due to its interaction with the Solar wind. Source: Originally adapted from http://pluto.jhuapl.edu/science/jupiterScience/magnetosphere.html.

corresponding to a location approximately where the magnetic field lines that pass through the Io plasma torus intersect the planet [44]. However, its equatorial surface field is $B \approx 4$G (Table 10.1). The radio emission also varies with time as the planet rotates, as the various satellites orbit through the magnetosphere, as new material is injected into the magnetosphere, and as the Solar wind varies. The result is strong intermittent *radio bursts*.

An accurate description of the radio spectra shown in Figure 10.13 involves complex modelling that takes a variety of effects into account. For example, for particles that are more energetic or mildly relativistic, the cyclotron frequencies and line widths are functions of electron energy. Therefore, the shape of the spectrum depends on the electron energy/velocity distribution which tends *not* to be Maxwellian (i.e. it is not thermal). At the low frequencies shown for the planets, the electron distribution is actually *inverted*. Rather than the *declining* power-law distribution that we saw for cosmic-ray nucleons (Figure 1.8) and electrons (Eq. 1.22), there are instead more high energy electrons than low energy ones. This is because electrons with high pitch angles (Figure 10.11c), which tend to be those of higher energy, are more easily reflected back along the field lines when they enter the 'bottleneck' region of high magnetic field strength and high plasma densities near planetary poles. Electrons with smaller pitch angles are more likely to suffer collisions in these regions and are not reflected. Therefore, the low energy particles are selectively removed. Other effects, such as resonances between the electron gyrofrequency and ambient radiation frequencies from other emitting electrons, are also important and can result in very intense, low-frequency radio emission[14]. Scattering of electrons from waves in the magnetic field lines (called *Alfvén waves*), plasma waves, and other processes can also affect the shape of the spectrum. Synchrotron radiation (Section 10.5.2) is also produced in the magnetosphere of Jupiter.

Those particles that do leak through to the upper atmosphere in the regions of the magnetic poles can collide with and ionize atmospheric atoms and molecules. When the electrons recombine, the subsequent downwards-cascading bound–bound transitions (i.e. fluorescence, see Section 7.1.2.1) can produce visible light. For the Earth, the result is the *aurora borealis* (northern lights) in the northern hemisphere and the *aurora australis* (southern lights) in the southern hemisphere. These are most often seen after a Solar flare, demonstrating the sensitive interaction between the Earth's magnetosphere and the Solar wind.

If planets in our own Solar System produce cyclotron emission at radio frequencies, what about exoplanets? The prediction is that many exoplanets should indeed produce such radiation, especially 'hot Jupiters' (Section 6.2.2) that orbit close to their parent stars and in which there should be strong star–planet plasma interactions. Radio bursts might be detectable from a variety of exoplanets that have magnetic fields as well (Problem 10.11), providing a possible additional exoplanet probe beyond those discussed in Section 6.2.2. The challenge is that, even though the radio bursts could be very powerful, the distances to these systems means that their emission is very weak and has not yet been detected.

Fortunately, the next generation international radio telescope is under construction and should meet this challenge. The Square Kilometre Array (SKA) will have an increase in total aperture size to one square kilometre which promises to significantly improve sensitivity beyond what is currently available. SKA has both low and high-frequency radio antennae in a system that stretches across two continents: Africa and Australia. The background image in Figure 10.13 shows an artist's conception of the central part of the South African component which operates at higher radio frequencies. Lower radio frequencies, which will be obtainable from the Australian component, are required to detect cyclotron emission from exoplanets. It is estimated that Jupiter-mass exoplanets should be detectable to a distance of 10 pc, a volume within which there are 200 stars [45]. This is just one of the many science goals for the SKA[15].

[14] These effects produce what is called *the cyclotron maser.*
[15] For more information, see www.skatelescope.org.

10.5.2 The Synchrotron Spectrum

There are a number of important ways in which the synchrotron spectrum differs from that of cyclotron emission. As we did for thermal Bremsstrahlung radiation (Section 10.2), we will describe 'semi-qualitatively' the factors that are involved in deriving the spectrum. More detailed descriptions can be found in [5, 32, 34], and elsewhere.

Firstly, the mass of a relativistic particle is greater than its rest mass by a factor, γ, where γ is the Lorentz factor given by Eq. (1.18). The energy of a relativistic electron is, therefore, $E = \gamma\, m_e\, c^2$ Eq. (1.17), where m_e is the rest mass of an electron. The *relativistic gyroradius* and *relativistic gyrofrequency* must take this into account, so using Eqs. (10.28) and (1.17) and setting $v_\perp \approx c$ (note, all in cgs units),

$$r = \gamma r_0 = \frac{\gamma\, m_e\, c^2}{e\, B} = \frac{E}{e\, B} = 2.2 \times 10^9 \frac{E}{B} \qquad (10.30)$$

and using Eq. (10.30) with (1.17),

$$\nu = \frac{\nu_0}{\gamma} = \frac{e\, B}{2\, \pi\, \gamma m_e c} = 2.3 \frac{B}{E} \qquad (10.31)$$

and the period, $\mathcal{T} = 1/\nu$. Eqs. (10.30) and (10.31) reduce to Eqs. (10.28) and (10.29), respectively when the electron is nonrelativistic (i.e. $\gamma \approx 1$). Note that these equations now *do* depend on electron energy, unlike their cyclotron counterparts. Since Lorentz factors can be extremely high (Section 1.2.2), the above two equations indicate that the relativistic gyroradius is much higher than the cyclotron radius and the relativistic gyrofrequency is much lower than the cyclotron gyrofrequency. At first glance, a very low value of ν might suggest that synchrotron-emitting frequencies would be too low to measure. However, this is not the case, as we will see below. Although Lorentz factors up to $\gamma = 10^{11}$ have been measured for nucleons, there are very few particles with such high values (Figure 1.8) and a 'typical' electron energy is lower (Example 10.4).

Example 10.4

Determine the gyroradius, the gyrofrequency, and the period of gyration of an electron with $\gamma = 10^4$ in the ISM with a magnetic field strength of $B = 3\mu G$. What is the speed of the electron compared to the speed of light?

For $\gamma = 10^4$, the electron energy is (Eq. (1.17)) $E = \gamma\, m_e\, c^2 = 8.18 \times 10^{-3}$ erg (5.1 GeV). For a typical ISM magnetic field of $3\mu G$, we find (Eq. (10.30)) $r = 6.0 \times 10^{12}$ cm, or approximately $86 \times$ the radius of the Sun. The gyrofrequency (Eq. (10.31)) is $\nu = 8.4 \times 10^{-4}$ Hz, so its period is $\mathcal{T} = 20$ minutes. By Eq. (1.18), $v/c = 0.999999995$.

The second important difference with cyclotron radiation is that, because of the relativistic motion of the electron, the *transformations* that need to be made between the rest frame of the electron (the frame moving with the velocity of the electron) and the rest frame of the distant observer (the 'lab frame') are significant. These transformations[16] have a number of consequences.

For example, in the rest frame of the electron, the power emitted is given by the Larmor formula (Table I.2) which involves the electron's acceleration. A transformation of this acceleration to the lab frame, however, results in a total power that is boosted by a factor, γ^2. Moreover, the distribution of this power with angle also changes. In the electron's rest frame, the emission is over a very wide angle (see Figure 10.11a). However, a distant observer will see the emission beamed into a narrow cone (Figure 10.11b). The opening angle of the cone depends on the particle's energy such that higher energy particles (higher γ) have narrower cones, the angular radius of the cone (see Figure 10.11c) being,

$$\theta = \frac{1}{\gamma} \tag{10.32}$$

The direction of the beam is the direction of the instantaneous velocity vector of the particle so, as the particle spirals about the field, the cone direction rotates with it and a distant observer will only see emission when (and if) the cone sweeps by his line of sight. The opening angle is very small for a highly relativistic electron (Example 10.5) so the time over which this cone sweeps by the observer is small in comparison to the larger gyration period, resulting in brief, repeated *pulses* of emission.

To obtain the *synchrotron spectrum* of a relativistic gyrating electron requires a conversion from the time to the frequency domain as well as the final important departure from the cyclotron case – the need for a Doppler shift (Section 9.1.1) since the observer is only seeing emission when the particle is moving towards him. As before, the time/frequency conversion requires a more detailed analysis[17]. Just as we saw for thermal Bremsstrahlung radiation (Section 10.2.1), the shortest time period (the pulse duration) sets the maximum frequency as seen by the observer. This maximum, called the *critical frequency*, above which there is negligible emission, is given by,

$$\nu_{\text{crit}} = \frac{3}{2}\gamma^2 \nu_0 \sin\phi = \frac{3\,e}{4\,\pi\,\text{m}_\text{e}\,c}\gamma^2 B_\perp \Rightarrow \left[\frac{\nu_{\text{crit}}}{\text{MHz}}\right] = 4.2\gamma^2 \left[\frac{B_\perp}{\text{Gauss}}\right] \tag{10.33}$$

where we have used Eq. (10.29). Thus, the maximum frequency is, in fact, very high (see Example 10.5) and much higher than the relativistic gyrofrequency. Most of the energy, however, is actually emitted at a frequency somewhat lower than the critical frequency, i.e. at $\nu_{\text{max}} = 0.29\,\nu_{\text{crit}}$. The longest time period, which is related to the gyration period, determines the fundamental frequency, ν_f. The spectrum will contain ν_f *and all its harmonics*, where,

$$\nu_f = \frac{1}{\gamma}\frac{\nu_0}{\sin^2\phi} \Rightarrow \left[\frac{\nu_f}{\text{MHz}}\right] = \frac{2.8}{\gamma \sin^2\phi}\left[\frac{B}{\text{Gauss}}\right] \tag{10.34}$$

using Eq. (10.29) again. The fundamental frequency is very low and its harmonics are so closely spaced (Example 10.5) that the synchrotron spectrum, *unlike* the cyclotron spectrum, is essentially *continuous*. The end result is a very broadband continuous spectrum

[16] The required transformations are called *Lorentz transformations*.

[17] A Fourier Transform is required (cf. Footnote 2 in this chapter).

that peaks at ν_{max} and becomes negligible at $\nu > \nu_{crit}$. Since the spectrum spans frequencies that are typically much greater than the plasma frequency, even small magnetic fields produce observable synchrotron radiation if a supply of relativistic electrons is present.

It can be shown that the power emitted by a single relativistic particle is a strong function of the particle rest mass, i.e. $P \propto 1/m^4$. Therefore, the emission from relativistic protons is negligible.

Example 10.5

For the same conditions as given in Example 10.4, and assuming that the relativistic electron has pitch angle of $\phi = 45°$, determine the total opening angle of the emission cone, the critical frequency, and the spacing of the harmonics for this particle.

For $\gamma = 10^4$ (Eq. (10.32)), $2\theta = 2/\gamma = 2 \times 10^{-4}$ rad or 0.01°. From Eq. (10.33), $B_\perp = B \sin \phi =$ 2.12 μG, so $\nu_{crit} = 890$ MHz. The harmonic spacing (Eq. (10.34)) is every $\nu_f = 1.7 \times 10^{-9}$ MHz. Notice that the critical frequency is well above the lower 10 MHz cut-off of the Earth's ionosphere and is at a radio frequency that is technically easy to measure.

The final step is to consider an ensemble of particles. Unlike thermal Bremsstrahlung emission, the electron velocities do *not* follow a Maxwell–Boltzmann velocity distribution (synchrotron radiation is *nonthermal*). These relativistic particles comprise the electron component of cosmic rays and so form a *power-law distribution* in energy (Eq. 1.22) which can be expressed in the form,

$$N(E) = N_0 E^{-\Gamma} \tag{10.35}$$

where $N(E)$ (cgs units of cm^{-3} erg^{-1}) is the number of cosmic-ray electrons per unit volume per unit interval of electron energy, N_0 (cgs units of erg$^{\Gamma-1}$ cm^{-3}) is a constant [25] and Γ is the energy spectral index of the cosmic-ray electrons (Section 1.2.2). An integral over electron energy will return the total volume density of cosmic-ray electrons, N_{CRe},

$$N_{CRe} = \int_{E_{min}}^{E_{max}} N(E)dE \tag{10.36}$$

where E_{min}, E_{max} are the minimum and maximum electron energies, respectively, that are present in the relativistic gas.

Assuming that the overall distribution of electrons follows a power law as described above and possesses an isotropic velocity distribution, the final emission coefficient, j_ν, and absorption coefficient, α_ν, can be determined. Provided $\Gamma > 1/3$ (which is supported by observations), these are,

$$j_\nu = c_5(\Gamma) N_0 B_\perp^{\frac{\Gamma+1}{2}} \left(\frac{\nu}{2c_1}\right)^{\frac{-(\Gamma-1)}{2}} \tag{10.37}$$

$$\alpha_\nu = c_6(\Gamma) N_0 B_\perp^{\frac{\Gamma+2}{2}} \left(\frac{\nu}{2c_1}\right)^{\frac{-(\Gamma+4)}{2}} \tag{10.38}$$

Table 10.2 Pacholczyk's constants[a]

$c_1 = 6.27 \times 10^{18}$		
Γ	$c_5(\Gamma)$	$c_6(\Gamma)$
0.5	2.66×10^{-22}	1.62×10^{-40}
1.0	4.88×10^{-23}	1.18×10^{-40}
1.5	2.26×10^{-23}	9.69×10^{-41}
2.0	1.37×10^{-23}	8.61×10^{-41}
2.5	9.68×10^{-24}	8.10×10^{-41}
3.0	7.52×10^{-24}	7.97×10^{-41}
3.5	6.29×10^{-24}	8.16×10^{-41}
4.0	5.56×10^{-24}	8.55×10^{-41}
4.5	5.16×10^{-24}	9.24×10^{-41}
5.0	4.98×10^{-24}	1.03×10^{-40}
5.5	4.97×10^{-24}	1.16×10^{-40}
6.0	5.11×10^{-24}	1.34×10^{-40}

[a]From [25]. Note all quantities are in cgs units. Γ is the spectral index of the electron energy distribution given in Eq. (10.35).
Source: Data from Pacholczyk [25]. © 1970 Macmillan Education.

The constant, c_1, and the functions, $c_5(\Gamma)$ and $c_6(\Gamma)$, are collectively (and loosely) called *Pacholczyk's constants* and are provided in Table 10.2. The absorption coefficient is needed to account for the absorption of photons by the relativistic electrons within the synchrotron-emitting cloud, a case called *synchrotron self-absorption.*

What about $B_\perp$? This is the perpendicular component of the magnetic field as introduced in Eq. (10.26) and is the component of the magnetic field that is perpendicular to the velocity vector of the cosmic-ray particles. Because we only see synchrotron emission when a particle's velocity is directed right at us (i.e. in the line of sight, Figure 10.11c), $B_\perp$ is the component of the magnetic field that is in the *plane of the sky.* If the magnetic field is turbulent and isotropic, then $B_\perp \approx B$ and this assumption is often made. In reality, there may be both turbulent and 'regular' components to the field, where the regular field would be one in which the field has a uniform direction (see [4] for details).

Because this is *not* an LTE situation, nor is the emission thermal, we cannot use the Planck function to relate the emission and absorption coefficients as we did for thermal Bremsstrahlung emission in Section 10.2.1. Rather, these coefficients are related by the *Source Function,* defined in Eq. (8.5)

$$S_\nu = \frac{j_\nu}{\alpha_\nu} = \frac{c_5(\Gamma)}{c_6(\Gamma)} B_\perp^{-\frac{1}{2}} \left(\frac{\nu}{2c_1}\right)^{\frac{5}{2}} \tag{10.39}$$

and the optical depth, as usual, is found from the absorption coefficient (Eqs. 7.8, 7.9),

$$\tau_\nu = \alpha_\nu l = c_6(\Gamma) N_0 B_\perp^{\frac{\Gamma+2}{2}} \left(\frac{\nu}{2c_1}\right)^{\frac{-(\Gamma+4)}{2}} l \tag{10.40}$$

where l is the line-of-sight distance through the synchrotron-emitting source and we have assumed no change in properties along the line of sight in Eq. (10.40). Since Γ is *positive* for a typical cosmic-ray energy distribution, Eq. (10.40) indicates that the *optical depth increases with decreasing frequency*. Therefore, a synchrotron-emitting cloud becomes more opaque at lower frequencies.

As we have done before, these equations can be put into the solution to the Equation of Radiative Transfer to determine the *synchrotron spectrum*. Following the equations in Section 8.3.5, when there is no background source,

$$I_\nu = S_\nu\,(1 - e^{-\tau_\nu}) \qquad (\text{any } \tau_\nu) \tag{10.41}$$

$$I_\nu = S_\nu \propto \nu^{\frac{5}{2}} \qquad (\tau_\nu \gg 1) \tag{10.42}$$

$$I_\nu = j_\nu\, l \propto \nu^{-\frac{(\Gamma-1)}{2}} \propto \nu^{\alpha} \quad (\tau_\nu \ll 1) \tag{10.43}$$

where

$$\alpha \equiv -\frac{(\Gamma - 1)}{2} \tag{10.44}$$

is the *frequency spectral index* or just *spectral index* (not to be confused with the absorption coefficient which is distinguished by the subscript, ν). In Eqs. (10.42) and (10.43), we have referred to Eqs. (10.39) and (10.37), respectively, to show how the spectrum varies with frequency. This frequency dependence is important because it can be observed and helps to tell us what emission mechanism is at work.

Synchrotron spectra for several values of Γ determined from Eq. (10.41) are shown in Figure 10.15 along with the observed spectrum of the strongest radio source in the sky (aside from the Sun), the supernova remnant, Cas A. Note that power laws are straight lines in this log–log plot. As can be seen, optically thin emission increases as the frequency becomes lower. As a result, most synchrotron-emitting sources are observed at radio frequencies.

10.5.3 Determining Synchrotron Source Properties

Like any spectrum, that of synchrotron emission allows us to link the received radiation to properties of the source. For simplicity, it is helpful to consider the optically thick and optically thin parts of the spectrum separately.

As indicated by Eq. (10.42), if the emission is *optically thick*, the spectrum *rises* with frequency ($\propto \nu^{5/2}$). This can be seen for the curve with $\Gamma = 3.0$ in Figure 10.15 at the low-frequency end of the spectrum. In this part of the spectrum, the emission is described by Eq. (10.39) and is a function of *only* $B_\perp$ and some constants. This physical situation is not so different from that of an optically thick thermal source for which only the temperature can be determined. Since we cannot 'see' all the way through an optically thick synchrotron source, we cannot detect all the particles that are present in the relativistic gas. Therefore, the brightness of the source is independent of particle density. An important consequence is that *a single measurement of emission from an optically thick synchrotron source leads to a determination of the strength of the perpendicular magnetic field.* For a

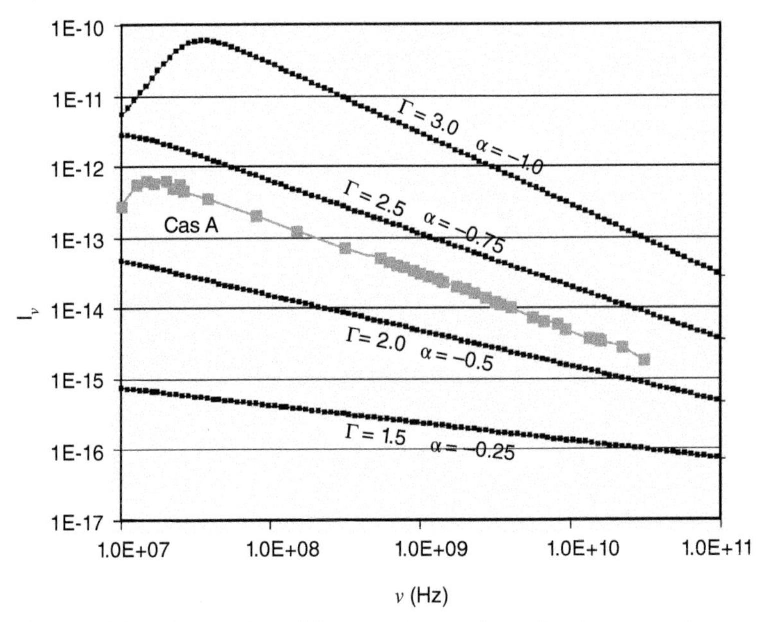

Figure 10.15 Examples of the nonthermal synchrotron spectrum with ordinate (I_ν) in cgs units. All curves with small black boxes are labelled with their electron energy spectral index, Γ, and the corresponding observed frequency spectral index, α, given by Eq. (10.44). These curves have been calculated from Eq. (10.41) using (in Eqs. (10.39) and (10.40)): $N_0 = 10^{-10}$ erg$^{\Gamma-1}$ cm^{-3}, $B = 100\,\mu$G, $l = 10$ pc and values of Pacholczyk's constants from Table 10.2 that correspond to the value of Γ used. The grey curve with larger boxes is the observed spectrum of the supernova remnant, Cas A (see image in Figure 3.2), using data from [2] and an angular diameter of 4.0′. Source: Data from Baars et al. [2].

statistical sample of sources, on average we expect $B_\perp \approx B$ to order of magnitude and it is common to make this assumption for any given source, although modern, more sophisticated approaches are now probing the various components of the field[18].

In reality, it is difficult to find a source that shows a true optically thick synchrotron spectrum. It is the compact sources, such as active galactic nuclei, that are most often optically thick and these tend to consist of a number of CR populations with different power-law energy distributions and optical depths that, together, make a complex spectrum. For other sources, the optically thick turnover may be at too low a frequency to be observed, or else the turnover may be caused by free–free absorption from foreground or

[18] See [4], for example, and also the series of papers from the Continuum Halos in Nearby Galaxies – an EVLA Survey (CHANG-ES) at queensu.ca/changes.

contaminating ionized gas (Section 10.2.1). This is likely what is causing the low-frequency turnover in the supernova remnant, Cas A [16], as described in Example 10.6. If a true synchrotron self-absorption spectrum can be identified, however, the magnetic field strength follows (Problem 10.13).

As indicated by Eq. (10.43), if the emission is *optically thin*, the spectrum *decreases* as a power law ($I_\nu \propto \nu^\alpha$), of which several examples are shown in Figure 10.15. The power-law slope in frequency space, quantified by the spectral index, α, is a characteristic of optically thin synchrotron emission and is a consequence of the power-law slope of the electron energy distribution, Γ, the two being linked by Eq. (10.44). In the optically thin limit, the emission is described by Eq. (10.43) with j_ν given by Eq. (10.37). Thus, the emission depends on *both* the particle density (via the constant, N_0) and the magnetic field strength.

Example 10.6

The supernova remnant (SNR), Cas A (Figure 3.2), has an internal magnetic field of order a few mG [1]. At what frequency will this SNR become optically thick to its own synchrotron radiation?

For the source to be optically thick implies that $\tau_\nu \geq 1$ so we need to solve Eq. (10.40) for ν when $\tau_\nu = 1$. We let $B_\perp = 2 \times 10^{-3}$ G and, to obtain Γ, we must measure the spectral index, α, from the *optically thin* part of Figure 10.15, finding $\alpha = -0.77$. Using Eq. (10.44), this gives $\Gamma = 2.54$. From Table 10.2, Pacholczyk's constants for this value of Γ are, $c_1 = 6.267 \times 10^{18}$, $c_5 = 9.51 \times 10^{-24}$, and $c_6 = 8.09 \times 10^{-41}$ (assuming linear interpolation). For the line of sight distance, we assume that the depth through Cas A is approximately equal to its diameter. From the caption to Figure 3.2, this is $l = 4$ pc. We now have all quantities required in Eq. (10.40) except N_0.

To find N_0, we must again look at the optically thin part of the spectrum. From Figure 10.15, we take an arbitrary point at $\nu = 10^{10}$ Hz and measure a value of $I_\nu \approx 4 \times 10^{-15}$ erg s^{-1} cm^{-2} Hz^{-1} sr^{-1}. Using the applicable equation for the optically thin limit (i.e. Eq. (10.43) together with Eq. (10.37) for j_ν) gives, $N_0 = 2.01 \times 10^{-13}$ erg$^{1.54}$ cm^{-3}.

Finally, putting all required values into Eq. (10.40) and solving for frequency gives, $\nu = 8.3$ MHz at $\tau_\nu = 1$. This is just to the left of the frequency range shown in Figure 10.15 and is significantly to the left of where the turnover actually occurs in Cas A[19]. Therefore, a more likely explanation for the observed turnover is the presence of foreground absorbing ionized thermal gas.

As suggested above, for many sources, observations of pure synchrotron emission in the optically thick limit are not always possible, whereas *optically thin* emission is readily observed. This, then, poses a problem because emission of optically thin radiation depends on both the magnetic field strength and the CR particle density. We cannot obtain one without knowing the other. Moreover, the particle density, N_{CRe} (Eq. (10.36)),

[19] For a pure synchrotron spectrum, the frequency of the peak of the spectrum does not exactly coincide with the frequency at which $\tau = 1$ but, for Cas A, the peak would be about 10^7 MHz which is still to the left of the actual turnover.

does not really give us the most important property of the relativistic gas. There may be few CR electrons present, but the energy associated with those electrons and other CR nuclei can be very high. If we wish to identify the original source that accelerates the CRs, then, some determination of the *energy* contained in the relativistic gas and associated magnetic field is necessary.

The *total energy* represented by all CR particles and magnetic fields together is $U_T = U_{CR} + U_B$, where U_{CR} includes both electrons as well as heavier nuclei and U_B is the energy contained in the magnetic field. Since it is only the CR electron component that emits synchrotron radiation, the heavy particle component must be estimated. It is common to represent the total CR energy as $U_{CR} = (1 + k)\ U_e$, where k is a constant representing the ratio of energy contained in protons and heavy nuclei compared to electrons, and U_e is the energy of the CR electrons only. The constant, k is not always well known and may vary in different environments, but is thought to be in the range, 40–100.

The energy contained in the magnetic field increases with magnetic field strength (see the equation for magnetic energy density in Table I.2). On the other hand, the energy contained in the cosmic-ray particles *decreases* with magnetic field strength. This is a consequence of the fact that particles emit radiation more strongly (Eq. (10.37)), therefore losing energy faster, in higher magnetic fields. Thus, there is a *minimum* in the function, U_T when plotted against B. A conservative assumption is therefore commonly made that the total energy in a synchrotron-emitting source is at or near this minimum. The minimum energy values also turn out to be very close to those of *equipartition of energy*, i.e. the condition in which the energy of the magnetic field is equal to the energy in CR particles. By making the assumption of minimum energy, we introduce one more constraint into the calculations, which allows us to extract information on *both* the magnetic field strength and CR particle energy from observations of synchrotron radiation in the optically thin limit. This method is widely used to obtain magnetic field strengths and energies or energy densities in synchrotron-emitting objects. Further details involving the minimum energy criterion and its development can be found in [25].

There is still one more property of synchrotron radiation that provides important information about the source, and that is its *polarization*. Like cyclotron emission, synchrotron emission is intrinsically *polarized* when there is a component of the magnetic field direction that is regular. In fact, this is usually the case.

For synchrotron emission, it can be shown that the maximum degree of polarization is $D_p \approx 70\%$ for $2 \lesssim \Gamma \lesssim 4$ [40]. In fact, a detection of polarization is the best way of being sure that the emission is indeed synchrotron when measured in the radio regime. As indicated above, a power-law slope is characteristic of optically thin synchrotron emission, but the power-law spectral index, α, depends on the energy injection spectral index, Γ (Eq. (10.44)) which may vary depending on the specific acceleration mechanism and energy losses during propagation. In addition, some compact sources (like active galactic nuclei) may contain a mixture of different values of Γ which result in the superimposition of various values of α. The net result is that, although most optically thin spectra are similar to those shown in Figure 10.15, it is possible for the observed value of α to be flat or even inverted (positive). If a flat slope were to exist, then this would mimic a thermal Bremsstrahlung slope for which $I_\nu \propto \nu^{-0.1}$ (Eq. (10.15)). The solution to the dilemma is a polarization observation. Since thermal Bremsstrahlung emission is never polarized, a clear polarization detection decides the issue.

As is sometimes the case in nature, real situations may not be so clear cut. Most synchrotron sources *do* show polarization. However, as indicated in Section 10.5.2, sources can contain both regular and random components to their magnetic fields. The interstellar medium in the Milky Way, for example, has both uniform and random magnetic field components, each of which is of order, a few μGauss (Table 10.1). In other sources, the magnetic field lines may become 'tangled' and there may be reduced (or no) directionality to the field within the size scale of the telescope's beam. This lowers the degree of polarization. Thus, as indicated in Section 3.7, even strong sources such as radio jets tend to have $D_p \lesssim 15\%$. A clear detection of polarization is therefore indicative of synchrotron radiation but lack of polarization does not rule it out. One needs to pull together as much information as possible to help in such analyses. Very high brightness temperatures, for example, are an additional clue that the emission is nonthermal in nature, as indicated in Section 10.1.

10.5.4 Synchrotron Sources – Spurs, Bubbles, Jets, Lobes, and Relics

Most synchrotron-emitting sources are observed in the optically thin limit at low frequencies where the emission is strongest. Although routine measurements are now made for the stronger sources at optical and X-ray wavelengths, synchrotron radiation is still most readily detected at *radio frequencies.* Since radio waves do not suffer from dust extinction (Appendix D.3 of the online material), this also means that synchrotron radiation is quite easily detected, passing through the ISM of the Milky Way without significant attenuation. Unlike cosmic-ray particles which are easily scattered in the ISM (Section 1.2.3), radio photons come directly from the location at which they are generated, helping us to identify their origin.

Figure 10.16, for example, shows an all-sky image of $\nu = 408$ MHz emission which is predominantly synchrotron radiation. As discussed in Section 1.2.3, supernovae are likely candidates for the acceleration of most Galactic cosmic rays at energies below the 'knee' of Figure 1.8 and the observed spectral indices of supernovae, $-0.8 \lesssim \alpha \lesssim -0.2$, imply $1.4 \lesssim \Gamma \lesssim 2.6$ (Eq. (10.44)) which spans the range, $2.1 \lesssim \Gamma_0 \lesssim 2.5$, believed to describe the initial CR energy spectrum. Two supernova remnants, Cas A and the Crab Nebula, are marked in the figure, and the *North Polar Spur* is part of a local, 250 pc diameter *superbubble* which may also have been formed by one or more supernovae [42]. This superbubble could be related to the Local Hot Bubble discussed in Section 10.2.3.

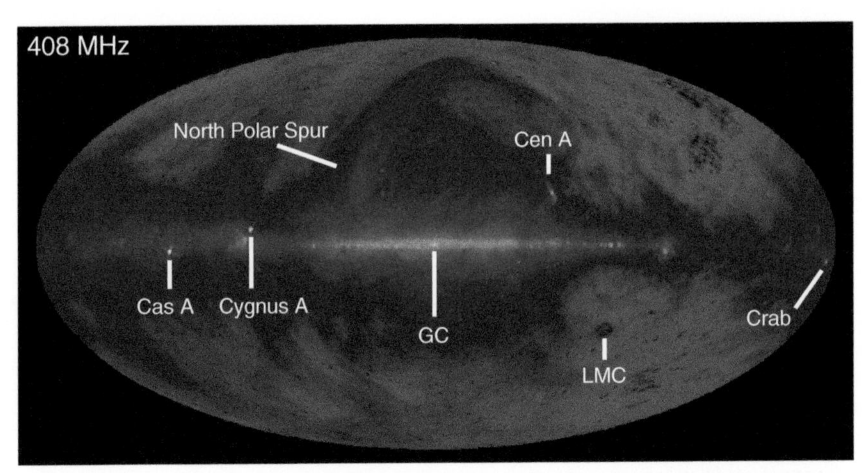

Figure 10.16 All-sky map of synchrotron emission at $\nu = 408$ MHz using the same type of projection as shown in Figure 10.5. (The map is in false colour with brightest to dimmest represented by white/yellow through purple/deep blue and finally red.) Most emission from our own Milky Way is seen along the Galactic plane (horizontal emission in the plot) as well as from spurs (e.g. the North Polar Spur) extending away from the plane. The Galactic Centre (GC) is marked, as well as the supernova remnants, Cas A and the Crab Nebula. Most of the discrete sources (blue dots) away from the plane are extra-galactic, including the Large Magellanic Cloud (LMC, a nearby galaxy), and the radio galaxies, Cygnus A and Centaurus A. The double-lobed shape of the jets in Cen A can be discerned. Source: Reproduced by permission of the Max-Planck-Institut für Radioastronomie.

Although supernova remnants are important sources of synchrotron emission, they are by no means the only sources. In the realm of high energy astrophysics, there are many objects with sufficient energies to accelerate electrons and sufficiently strong magnetic fields to scatter them. In the Milky Way, these include shocks formed by massive stellar winds, stellar jets, some binary stars, pulsars, and others. Synchrotron emission is even observed from Jupiter and the Sun. Cosmic-ray electrons that leak into the ISM from discrete acceleration sites will also emit synchrotron radiation, since magnetic fields are ubiquitous in the disc of the Milky Way. Typical ISM magnetic field strengths of only a few μGauss are sufficient to generate observable emission. Thus, much diffuse emission can be seen along the plane of the Galaxy in Figure 10.16.

Extra-galactic sources include active galactic nuclei (AGN) that are powered by supermassive black holes (e.g. Figure 7.5) and the jets that they generate (e.g. Figure 3.4). The classic *extra-galactic double-lobed radio sources* were first discovered in the radio band and display two radio lobes at the ends of the jets, as can be discerned for the radio galaxy, Centaurus A, in Figure 10.16. The powerful radio galaxy, Cygnus A, is also identified in Figure 10.16 and is shown in more detail in Figure 10.17 Top. The

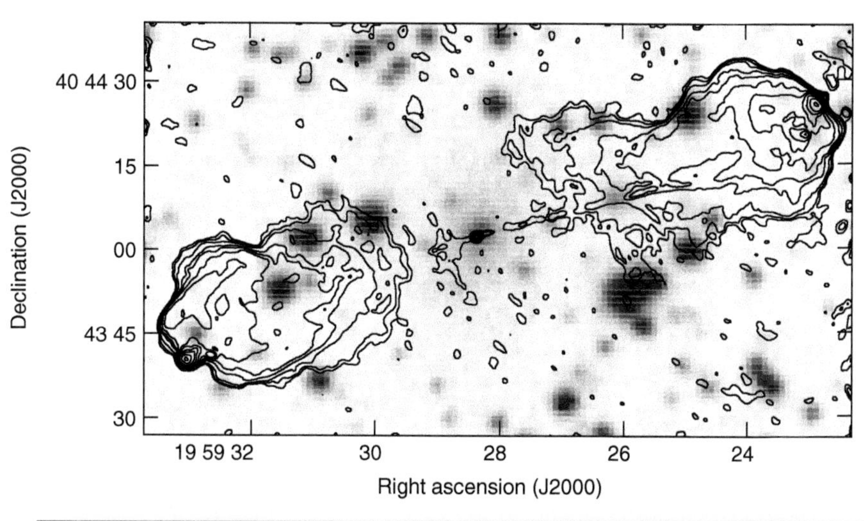

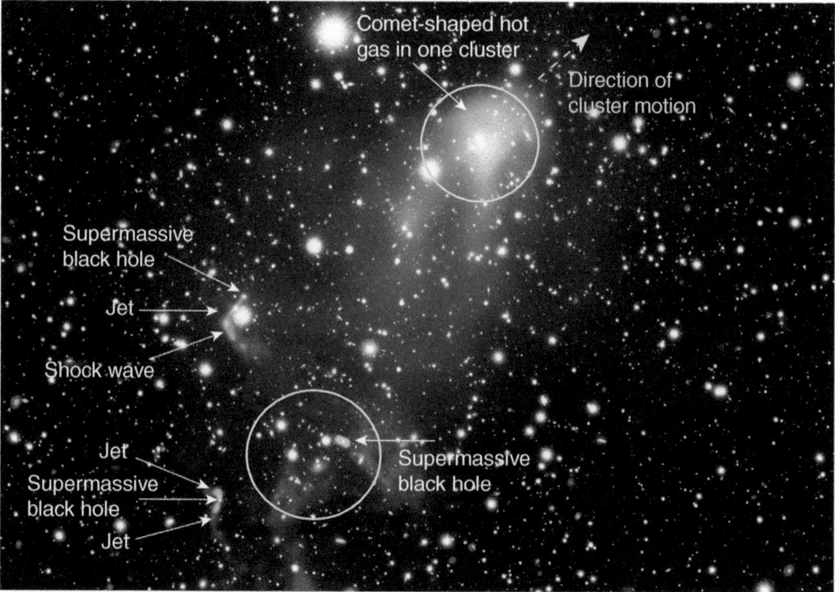

Figure 10.17 *Top:* The double-lobed extra-galactic radio source, Cygnus A (redshift, $z = 0.0562$), shown in contours of $\nu = 4.9$ GHz emission, over the optical image in greyscale (shown with black as highest intensity). The faint grayish region at the centre of the image is the distant elliptical galaxy that harbours a bright AGN at its centre. (Other black spots are stars in our own Milky Way.) Two jets emerge from the AGN in opposite directions, ending in two large bright radio lobes on either side of the galaxy. The total flux density of the source is 213 Jy, and the effective FWHM of a single radio lobe is $\theta_{\text{FWHM}} \approx 10''$. Source: http://skyview.gsfc.nasa.gov/cgi-bin/titlepage.pl. *Bottom:* This image shows two galaxy clusters in collision, Abell 3411 (upper right) and Abell 3412 (lower left) with their centres circled in yellow. X-ray emission from hot gas is shown in blue. Stars are whitish dots with subtle colours. Pink reveals the 610 MHz emission, some associated with active radio galaxies and some separate from the galaxies (the *relics*). A connection between a galaxy and a relic is labelled 'jet' and 'shock wave' in this Figure [39]. Source: X-ray: NASA/CXC/SAO/R. van Weeren et al.; Optical: NAOJ/Subaru; Radio: NCRA/TIFR/GMRT.

inconspicuous elliptical galaxy at the centre, whose optical light comes from stars, belies the true nature of its powerful core. The total energy contained in its magnetic fields and cosmic-ray particles, using the minimum energy assumption (see Section 10.5.3), is a spectacular 10^{60} erg [27]. By comparison, the total kinetic energy of a single supernova is $\sim 10^{51}$ erg. Not only are the particles that emit synchrotron radiation relativistic, but the jets that feed the two lobes also have bulk motions (meaning that the entire synchrotron-emitting plasma is moving in bulk) that are relativistic, with velocities between $0.4c$ and $\sim 1c$. The minimum energy magnetic field strength in the radio lobes is, on average, $50\ \mu$G.

AGN activity is a strong contender for the origin of magnetic fields and cosmic rays in the intergalactic medium of clusters of galaxies from which synchrotron emission has also been detected. In some cases, regions of synchrotron emission are seen in the outskirts of galaxy clusters in regions that are not at the locations of specific galaxies in the cluster. These regions are called *relics*, or sometimes *fossils* or *ghosts*. The origin of this emission has been uncertain because these regions have not appeared to be directly connected to currently active galaxies within the cluster. Rather, the emission may be generated within shocks that formed during the formation of the cluster and/or during later cluster mergers. Alternatively, an older, existing population of cosmic-ray particles that were ejected from radio galaxies in the past could be reaccelerated in such shocks. Evidence for this last idea can be seen in the two clusters of galaxies, Abell 3411 and 3412, that are in the process of colliding, as shown in Figure 10.17 Bottom. Here shocks from the cluster merger can be seen as well as their connection to radio jets in the field, both shown in pink. One shock wave is labelled in the figure and appears to be directly connected to the jet just above it.

10.6 INVERSE COMPTON RADIATION

Inverse Compton (IC) radiation is somewhat different from the other emission discussed in this chapter because it results from the interaction of a particle with a photon. However, since the emission begins with matter and ends with a photon (i.e. matter $\rightarrow$ photon $\rightarrow$ photon) we include it here.

As indicated in Section 7.1.2.2 and further detailed in Appendix D.2 of the *online material*, *Compton scattering* occurs when a high energy photon scatters inelastically from a free electron, boosting the kinetic energy of the electron at the expense of the photon energy.

IC radiation, often referred to as *Inverse Compton scattering*, is the reverse of this process, as its name implies. It occurs when a high energy electron inelastically scatters off of a low energy photon, the result being a loss of kinetic energy for the electron and a gain of energy (a *blueshift*) of the photon. The scattering will go in this direction if the electron energy is greater than the photon energy in the centre-of-momentum rest frame of these two 'particles'. For a highly relativistic electron and low energy photon, this rest frame is that of the moving electron. For such a case, it can be shown that the frequency of the outgoing radiation, as seen by a distant observer, is,

$$\nu_{IC} \approx \gamma^2 \nu \qquad (\gamma h\nu \ll m_e c^2) \tag{10.45}$$

where γ is the Lorentz factor of the electron (Eq. 1.18) and ν is the frequency of the photon before being scattered[20]. For example, for a relativistic electron with $\gamma = 10^4$, a radio photon, at a typical frequency of $\nu = 1$ GHz, would be 'up-scattered' to $\nu_{IC} = 10^{17}$ Hz which is in the *X-ray* part of the spectrum (Table I.1). The result is a redistribution of photon energies with fewer photons in the radio and more photons in the optical or X-ray (depending on γ) than would otherwise be the case. In the presence of a magnetic field, relativistic electrons will also emit synchrotron radiation, so the addition of IC losses implies that the electrons will lose energy faster and the radiated output over all ν will be higher than from synchrotron emission alone.

Any synchrotron-emitting source contains a plentiful supply of high-energy electrons and emits low energy (e.g. radio) photons and these are also the required ingredients for IC emission. Therefore, IC radiation, when observed, will be seen in the same kinds of sources that emit synchrotron radiation. Because the relativistic electrons have a power-law distribution of energies (they are the electron component of cosmic rays), this IC radiation is *nonthermal*. The IC spectrum is, therefore, a power law and so is difficult to distinguish from that of synchrotron radiation which is also a power law as we have just seen (Figure 10.15).

Fortunately (at least in terms of distinguishing synchrotron from IC radiation), not every synchrotron-emitting source is an IC source as well. Synchrotron radiation depends on the strength of the magnetic field, B, whereas IC radiation depends on the number of low energy photons that are available for up-scattering. Thus, for a given number and distribution of relativistic electrons, wherever B is strong and the radiation field is weak, synchrotron radiation will dominate, and wherever the radiation field is strong and B is weak, IC radiation will dominate. To be more precise, it can be shown that the fraction of the total radiative luminosity of a source due to IC radiation, L_{IC}, in comparison with the total radiative luminosity due to synchrotron radiation, L_s, is [36],

$$\frac{L_{IC}}{L_s} = \frac{u_{rad}}{u_B} \approx \left[\frac{T_{Bmax}}{10^{12}}\right]^5 \left[\frac{\nu_{max}}{10^{8.5}}\right] \tag{10.46}$$

where u_{rad} is the energy density of the radiation field (Section 3.4) and u_B is the energy density of the magnetic field (Table I.2). The quantities, T_{Bmax} (K) and ν_{max} (Hz), are the brightness temperature and frequency, respectively, of the source at the peak (the turnover) of the synchrotron spectrum, assuming that only synchrotron self-absorption is producing the turn-over. Example 10.7 shows how the brightness temperature and frequency dependence in this equation results.

[20] Note that an electron moving with a Lorentz factor, γ, will 'see' a photon of energy, $\gamma h\nu$ in its own rest frame.

Example 10.7

Show how the functional dependence on T_B and ν in Eq. (10.46) is obtained.

From Eq. (3.18), the energy density in an isotropic radiation field, $u_{rad} \propto I$, where I is the intensity of the source already integrated over frequency, so $u_{rad} \propto I_\nu \nu$, where I_ν is the specific intensity. Using the definition of brightness temperature (Eq. 6.4) expressed in the Rayleigh–Jeans limit Eq. (6.6) since most of the photons will be in the radio part of the spectrum and $h\nu \ll kT_B$, we have, $I_\nu \propto T_B \nu^2$, which leads to $u_{rad} \propto T_B \nu^3$.

Now, from Table I.2, $u_B \propto B^2$. In the optically thick limit, the magnetic field strength is related to the specific intensity via Eqs. (10.42) and (10.39) and this dependence should still be approximately correct at the peak of the curve where the spectrum turns over. Letting $B \approx B_\perp$, and rearranging Eq. (10.39) to solve for B gives $B \propto \nu^5/I_\nu{}^2$ Thus, $u_B \propto \nu^{10}/I_\nu{}^4$. Converting the specific intensity to a brightness temperature as we did above, then, $u_B \propto \nu^2/T_B{}^4$.

Finally, from the above two results, the ratio, $u_{rad}/u_B \propto (T_B\nu^3)/(\nu^2/T_B{}^4) \propto T_B{}^5\nu$ which is consistent with Eq. (10.46).

The importance of Eq. (10.46) is in the extremely strong dependence of IC losses on the brightness temperature of the source. IC radiation should not be very important until the brightness temperature approaches 10^{12} K at the reference frequency given in the equation (e.g. Problem 10.14). Once brightness temperatures exceed 10^{12} K, however, a situation called the *Compton catastrophe*, the electrons should lose energy so rapidly from IC losses (of order days, [29]) that no sources with such high brightness temperatures should actually exist. Thus Eq. (10.46) suggests a natural upper limit of $T_B \approx 10^{12}$ K for source brightness temperatures[21]. In fact, this limit appears to be approximately correct when compared to observations. There are sources in which higher brightness temperatures are observed, but these are *moving* sources which were not considered when deriving Eq. (10.46). If the entire synchrotron-emitting plasma has a bulk motion that is relativistic in a direction towards the observer, such as in a jet pointed towards the observer, then the emission will be brighter than if the plasma were stationary, a situation called *Doppler beaming*.

The most famous example of inverse Compton scattering is *the Sunyaev–Zeldovich effect* (S–Z effect). This is the IC up-scattering of 2.7 K cosmic microwave background (CMB) photons by very hot (but nonrelativistic) electrons in clusters of galaxies, shifting the CMB photons to higher frequencies.

An abundant supply of ionized gas, and therefore hot electrons, is present in many galaxy clusters, as illustrated in Figure 10.9, and background photons will pass through such cluster gas en route to us. The result is a *distortion* of the 2.7 K CMB spectrum. Rather than the 'perfect' black body that we saw in Figure 6.3 and the dashed curve in Figure 10.18, a distorted spectrum, such as shown by the solid curve in Figure 10.18, is

[21] Other considerations related to equipartition of energy may make the upper limit somewhat smaller, of order $T_B \approx 10^{11}$ K (see [29]).

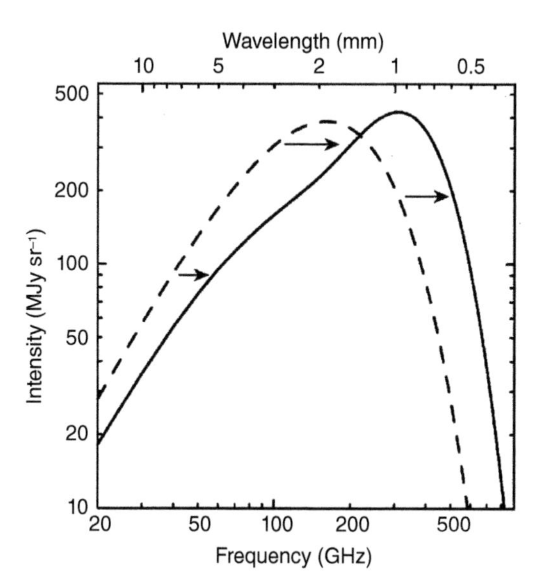

Figure 10.18 Illustration of the Sunyaev–Zeldovich effect, showing the undistorted CMB spectrum (dashed curve) and the spectrum after CMB photons have passed through the ionized gas in a cluster of galaxies (solid curve). The cluster has been made 1000 x more massive than a typical cluster to exaggerate the effect. Source: Carlstrom et al. [10]. © 2002 Annual Reviews Inc.

the result. Since the CMB Planck curve peaks at a frequency of 160.5 GHz (recall Wien's displacement law, Eq. 6.8), frequencies below this value will show a deficit of emission and frequencies above this value will show an excess. The distortion in the spectrum can be measured as a *change* in the black body temperature, $\Delta T_{S\text{-}Z}$. The effect is typically quite small ($\Delta T_{S\text{-}Z} < 1$ mK) and has been exaggerated in the figure. Nevertheless, it has now been measured in many galaxy clusters.

The importance of the S-Z effect is that the observed shift depends on the electron density and temperature (n_e and T_e), as well as the line of sight distance through the cluster, l,

$$\Delta T_{S\text{-}Z} \propto n_e T_e l \tag{10.47}$$

At the same time, the hot ionized gas in the cluster also emits in the X-ray part of the spectrum via thermal Bremsstrahlung radiation. This gas is quite rarefied and is therefore in the optically thin limit. From Eqs. (10.9) and (10.2), we know that the X-ray emitting gas in the cluster has a specific intensity that depends on the same quantities, though the dependences are slightly different,

$$I_\nu \propto n_e^2 T_e l \tag{10.48}$$

If the X-ray emission can be measured, then T_e can be found (Section 10.2.3), and we then have two relations that depend on the quantities, n_e and l. Therefore, both n_e and l can be found. Of these, the quantity that is most of interest is *l because it is related to the cluster distance.* For example, if the line of sight distance through the cluster is the same as the distance across the cluster in the plane of the sky, then an angular measurement of the cluster diameter together with the small-angle formula (Eq. I.9) yields the *angular size distance* (cf. Section 9.2.1). Distances tend to be difficult to find in astronomy, so any method that can determine a distance without using standard candles is extremely useful. This kind of analysis has resulted in a value of the Hubble Constant, H_0, that is independent of other approaches. A recent value is $H_0 = 67 \pm 3$ km s^{-1} Mpc^{-1} [17] which is similar to other values given in Section 9.1.2.

PROBLEMS

10.1. (a) Determine the root-mean-square speed, v_{rms}, and corresponding kinetic energy, E_k, of an electron in an HII region of electron temperature, $T_e = 10^4$ K.
(b) If the electron lost all of its kinetic energy (came to a stop) during an encounter with a nucleus without recombining, what would be the frequency of the resulting emission and in what part of the spectrum does this frequency occur?
(c) Compare the frequency of part (b) with a typical thermal Bremsstrahlung photon at a radio frequency of $\nu = 5 \times 10^9$ Hz from an HII region. Based on this comparison, what fraction of an electron's kinetic energy has actually been lost during an encounter with a nucleus?

10.2. The spectrum of an HII region is observed to have an exponential cut-off such that the emission declines by a factor, $1/e$ (its *e-folding* value), at a wavelength of λ 2.4 mm. What is the temperature of this HII region?

10.3. With the help of a spreadsheet or computer algebra software, determine and plot the spectrum of free–free emission from the Lagoon Nebula shown in Figure 5.13. Ensure that the low frequency optically thick part of the spectrum and the high-frequency cut-off are shown. For simplicity, assume that the ionized gas consists of pure hydrogen and that the Gaunt factor is constant with frequency.

10.4. (a) Show that $h\nu \ll kT_e$ for any HII region observed at radio wavelengths.
(b) Show that Eq. (10.13) results from Eq. (10.5).

10.5. For the HII region, IC 5146 (see Figure 10.4), whose radio emission is in the optically thin limit, assume that the region is spherical of uniform density and find the following:
(a) The frequency at which the HII region becomes optically thick. Do the contours represent optically thick or optically thin emission? Is the optically thick region observable from the ground?
(b) The electron density, n_e, ionized hydrogen mass, M_{HII}, total gas mass, M_g, and excitation parameter, $\mathcal{U}$.
(c) The radius of its Strömgren sphere (assume that the effects of dust are negligible).
(d) The number of ionizing photons per second, N_i, from the ionizing central star, BD $+ 46°$3474.

10.6. Using the parameters of the hot intracluster gas shown by the spectrum in Figure 10.3, determine the frequency at which the emission becomes optically thick. In what part of the spectrum does this occur and is it observable from the ground? Do you ever expect to observe optically thick hot intracluster gas?

10.7. Beginning with Eq. (10.2), show that the luminosity within a frequency band, ν_1 to ν_2, is given by Eq. (10.20) for optically thin thermal Bremsstrahlung radiation from ionized hydrogen gas of constant temperature and density.

10.8. For the cluster of galaxies, Abell 2256, whose soft X-ray emission is shown in Figure 10.9, determine or estimate the following:
   (a) Its distance, D (Mpc).
   (b) The volume, V(kpc^3), of the X-ray emitting gas, assuming that it occupies the entire volume of the cluster.
   (c) The electron temperature, T_e (K).
   (d) The mean electron density, n_e (cm^{-3}).
   (e) The ionized hydrogen mass, M_{HII} ($M_\odot$).
   (f) The fraction, M_{gas}/M_{tot}, where M_{gas} is the total gas mass and M_{tot} is the total light + dark mass of the cluster.

10.9. [*Online*]
   (a) Write an expression for the conservation of energy (kinetic plus potential) for an electron which makes a transition from an initial free state in an ionized hydrogen gas to a final bound state, showing that the difference in energy corresponds to the energy of the emitted photon. See Eq. (C.5).
   (b) Compute the frequency of an emitted photon for a transition to the $n = 1$ *and* $n = 2$ levels of hydrogen when (i) the initial electron velocity is zero and (ii) the initial electron velocity is the most probable velocity in a gas at $T_e = 1.58 \times 10^5$ K.
   (c) Compare the results of part (b) to the curve of g_{fb} shown in Figure 10.2 and explain the shape of this curve.

10.10. Cyclotron emission from electrons moving at $\upsilon \approx 2 \times 10^4$ km s^{-1} in a region of the magnetosphere of Jupiter is observed at the gyrofrequency, $\nu_0 = 540$ kHz.
   (a) Determine the magnetic field in this region.
   (b) Find the gyroradius of the electrons. Is this larger or smaller than the wavelength of the emitted radiation?
   (c) Write an expression for ν_0/ν_p, where ν_p is the plasma frequency (Eq. 7.6), in terms of B and n_e with constants evaluated. What is the maximum possible electron density in order for the cyclotron emission to escape from the region?
   (d) What is the gyrofrequency of a proton, ν_{op}, in the same region? Repeat the first part of step (c) above using ν_{op} and determine the maximum possible electron density that would allow cyclotron emission from a proton to escape from the region.

10.11. Suppose a Jupiter-like planet is orbiting a Sun-like star that is at a distance of 10 pc from the Earth. The system is observed at $\nu = 30$ MHz and is unresolved. Adopt a brightness temperature (Section 6.1.1) of T_B (30 MHz) $= 5 \times 10^5$ K for the star which, like the Sun, departs from a black body at this frequency.

(a) What is the flux density of the star at the observed frequency (cgs units) as measured at the Earth? Assume that the star is *not* undergoing any radio bursts.

(b) From the information given in Figure 10.13, what is the flux density of the planet (cgs) as measured at the Earth? Assume that the planet *is* undergoing a radio burst and has a modulating moon like Io so that its output is $100 \times$ higher than that shown in the figure.

(c) Convert your results from (a) and (b) into Jy and find the ratio of the planet's flux density to the star's flux density. For this frequency where the planet's dominant emission is nonthermal and the star's emission shows departures from a true black body, which object has the higher emission?

(d) If the typical rms noise level of the SKA[22] is 10 μJy beam^{-1}, comment on the prospect of observing radio emission from exoplanets.

10.12. Determine the relativistic gyroradius (kpc) of a CR *proton* that has the highest energy shown in Figure 1.8 in the ISM of the Milky Way. Compare this to the distance of the Large Magellanic Cloud (distance = 50 kpc) and comment on the ability of galaxies, in general, to retain such high energy cosmic rays.

10.13. In 1993, a supernova (SN1993J) was observed in the nearby galaxy, M 81, which is a distance $D = 3.6$ Mpc from us. At a time, 273 days after the observed explosion[23], the following parameters (approximated from [26]) describe the radio emission from this source: (i) a rising spectrum ($I_\nu \propto \nu^{5/2}$) at low frequencies with a flux density of $f_\nu = 72$ mJy at a frequency of $\nu = 1.43$ GHz, (ii) a falling spectrum ($I_\nu \propto \nu^{-1}$) with a flux density of $f_\nu = 25$ mJy at a frequency of $\nu = 23$ GHz, (iii) a radius of $r = 0.0123$ pc, and (iv) a minimum energy cut-off to the electron energy spectrum of $E_{min} = 46$ MeV. Determine the following parameters for this supernova, ensuring that all cgs units are specified:

(a) The electron energy spectral index, Γ.

(b) The solid angle subtended by the source, Ω.

(c) The brightness temperature at $\nu = 1.43$ GHz, T_B.

(d) The strength of the perpendicular magnetic field, $B_\perp$.

(e) The constant of the electron energy spectrum, N_0.

(f) The number density of cosmic-ray electrons, N_{CRe}.

10.14. The total flux density of Cygnus A (see Figure 10.17 Top) is $f_\nu = 2.19 \times 10^4$ Jy at $\nu = 12.6$ MHz, divided roughly equally between the two radio lobes. Assuming that this flux density represents the peak of the spectrum of Cygnus A and that synchrotron self-absorption is the only absorption process that is producing the spectral turnover, determine the ratio of $L_{IC} = L_s$ for one radio lobe. How

---

[22] SKA's frequency range is expected to reach as low as 50 MHz, so not quite as low as adopted in this problem.

[23] The actual explosion occurred 11.7 million years prior to its observation in 1993 since M 81 is 3.6 Mpc = 11.7million light years away!

important is inverse Compton radiation in comparison to synchrotron radiation for this source?

10.15. Verify that the condition in parentheses in Eq. (10.45) is satisfied for the case of CMB photons and hot intracluster gas such as is shown in Figure 10.9.

10.16. Let us summarize continuum emission processes by reviewing some of the images in this chapter. For each image, specify (i) the emission process, (ii) what observing band is shown, and (iii) whether the emission in optically thick or optically thin. Give your reasons.

(a) Figure 10.4 point sources in the greyscale image.
(b) Figure 10.4 contours.
(c) Figure 10.8 pink.
(d) Figure 10.8 blue.
(e) Figure 10.10 dashed curve.
(f) Figure 10.10 difference between the dashed curve and total continuum level.
(g) Figure 10.17 bottom blue.
(h) Figure 10.17 bottom pink.

JUST FOR FUN

10.17. In Section 10.5.4, the energy from magnetic fields and cosmic rays in Cygnus A was specified to be a 'spectacular 10^{60} erg'. How much energy is this, actually? To answer this question, try converting the masses of various objects into energy and find the object or objects that most closely represent 10^{60} erg.

10.18. Along the lines of the previous question, how can we understand the strength of a magnetic field? Look up the magnetic field strengths of the following objects and compare them to the values of Table 10.1. Comment on the results. (i) a refrigerator magnet, (ii) a magnet in a magnetic resonance image (MRI) scanner, (iii) an electromagnet that can pick up a car in a scrap metal yard.

10.19. Suppose you could sit on a magnetic field line of a pulsar. And suppose a cosmic-ray (CR) particle whose energy is at the 'knee' of Figure 1.8 approached you and started to circle around the field. What is the closest you could get to the particle? Could you reach out and grab it, or would it always stay too far away?

Chapter 11
Line Emission

> Have you seen anything so beautiful?
>
> – C. V. Raman, after whom Raman spectroscopy is named, pointing to the evening sky [30]

By the end of the nineteenth century, the German physicist, Max Planck, had explained the continuous black-body spectrum by allowing light to exist as a photon, or 'particle'. His theoretical development introduced the constant, h, now called Planck's constant. The concept of quantized light was further strengthened when Einstein, in 1905, successfully explained the *photoelectric effect* – the release of electrons in some metals and semiconductors via the action of incident light – by treating light as a particle. Still not understood, however, was the series of lines seen in the spectrum of the hydrogen atom.

It was the Danish physicist, Niels Bohr, who brought the concept of quantization into the realm of the atom. He proposed, in 1913, that electrons did not radiate unless they made a transition from one state to another. The difference in energy between these states could be related to the emitted energy of a photon, according to $\Delta E = h\nu$. This profound shift from the classical view of requiring orbiting electrons to radiate, to one in which radiation results only from transitions between states, won Bohr the Nobel Prize in

Astrophysics: Decoding the Cosmos, Second Edition. Judith A. Irwin.

Companion website: www.wiley.com/go/irwin/astrophysics2e

1922. It also set the scene for Heisenberg, Sommerfeld, Schrödinger, Pauli, and others to further develop and refine the new field of quantum mechanics. This chapter is devoted to an understanding of the observed *spectral lines* from atoms and molecules that result from *discrete* internal changes in energy.

While the details of quantum theory are beyond the scope of this book, an introduction to quantum mechanical principles for the hydrogen atom has been provided in Appendix C of the *online material*. Much of the groundwork has been laid out earlier. For example, we have already seen that the population of bound states in an atom under LTE conditions is described by the Boltzmann Equation (e.g. Section 5.4.5) and recombination lines were introduced in Section 5.4.7. The conditions for forming absorption and emission lines were described in Section 8.4.2, and the information that can be obtained from detecting the λ21 cm line of hydrogen in both emission and absorption was presented in Section 8.4.3. In Section 9.1, we also discussed how the relatively simple observation of a redshift of a spectral line can lead to some very powerful conclusions, from information about black hole masses to an understanding of the nature of our expanding Universe.

We focus now on what spectral lines can be formed in atoms and molecules and how these spectral lines provide us with information about the sources that emit them. Although a 'line spectrum' can form from pure cyclotron emission (Section 10.5.1), we will restrict the discussion to spectral lines that are formed from discrete quantum transitions in atoms and molecules as well as their ions and isotopes, that is, *bound–bound transitions*. We also focus here on the line *emission* process and therefore will deal with 'downwards' (higher to lower energy) transitions, ignoring the possible presence of a background source[1]. The latter can be handled by considering the appropriate solution to the Equation of Transfer, as described in Chapter 8.

11.1 THE RICHNESS OF THE SPECTRUM – RADIO WAVES TO GAMMA RAYS

What kinds of bound–bound transitions are possible? These depend on the quantizations that occur within atoms and molecules. The possibilities are as follows: *electronic transitions, vibrational transitions, rotational transitions, and nuclear transitions*, all of which are quantized[2].

11.1.1 Electronic Transitions – Optical and UV Lines

Electronic transitions occur when an electron changes energy states in atoms and molecules, or their ions and isotopes. Because we are focussing on emission lines, the

[1] The 2.7 K CMB background will always be present, but we will assume that its brightness is negligible, in comparison to the line.

[2] More complex molecules, such as the PAHs introduced in Section 5.5.2, can also experience *stretching* or *bending* modes which result in unique spectral features, but we do not pursue those details here.

transitions would be from an upper principal quantum number to a lower one. These are the familiar transitions that have already been discussed in various contexts. For example, a rich variety of spectral lines from many different species was seen in the optical spectrum of an H II region in Figure 10.10.

If the hydrogen is ionized, a recombination line spectrum is observed as electrons continuously recombine with protons and then rapidly cascade down various bound–bound levels. As pointed out in Section 5.4.7, even if an HII region is highly ionized, it will experience a steady state of recombinations and ionizations so that a small fraction of particles are neutral at any given time (e.g. Example 7.6). It is the bound–bound transitions from this downwards cascading that give us some spectacular images of HII regions, especially from the Balmer lines, since these are in the optical part of the spectrum. The strongest of these, the Hα line, has already shown up in Figures 5.13, 7.10, and 8.6.

The highest frequency line that is possible from the hydrogen atom results from the largest energy difference between bound–bound states. This would be close to the ionization energy of 13.6 eV for which the corresponding wavelength, λ 91.18 nm, is in the ultraviolet part of the spectrum. At the other extreme, transitions between very high quantum levels result in radio wave emission since the energy levels of hydrogen become progressively closer together with increasing quantum number, n (Figure C.1 of the *online material*).

Thus, line emission from hydrogen progressively shifts in frequency across the electromagnetic spectrum from the UV to the radio band, depending on which principal quantum numbers are involved. At higher quantum numbers, however, the transition probabilities are lower, so it is common to refer to electronic transitions in atoms to be 'typically' in the optical or UV parts of the spectrum.

Electronic transitions occur in other atoms and in molecules as well. When all species are taken into account, spectral lines are seen across the entire electromagnetic spectrum from the X-ray to radio bands. Larger, heavier atoms can have higher energy transitions and also more of them, because there are more electrons associated with such elements. For example, several X-ray lines of iron were shown in the spectrum of a solar flare in Figure 8.7, and numerous X-ray lines from a variety of elements can be seen in Figure 10.6. Put together, a wide array of possible electronic transitions is available over a broad waveband, though only a subset of these may be excited at any time, depending on the physical conditions in the gas. Spectral lines are therefore probes of the physical conditions in the gas, as we will soon see. At the very least, the identification of spectral lines as belonging to certain species provides information as to which elements are present in the gas and what the abundance ratios might be.

11.1.2 Rotational and Vibrational Transitions – Molecules, IR and mm-Wave Spectra

In simple molecules, there are two additional types of transitions that can occur, besides electronic ones. These are *rotational transitions* and *vibrational transitions*. Rotation and vibration are illustrated for a diatomic molecule in Figure 11.1, but it should be noted that there are no exact quantum mechanical analogues to these motions. Just as we saw for an electron in 'orbit' about the nucleus and for the 'spin' of an electron (Appendix C of the *online material*), the rotation and vibration of a molecule are convenient models that allow us to picture these new energy levels and to derive their energies from a classical starting point.

If a molecule *rotates*, it has some energy associated with that rotation. A good *classical* approximation is that of a rigid rotator which, for a diatomic molecule, would be as if the molecule were a dumbbell with a nucleus at either end. As illustrated in Figure 11.1a, the rotation is about either of two principal axes[3]. If the two nuclei have different atomic weights, then the molecule will have a permanent electric dipole moment (Table I.2) due to the way in which charge is distributed in the molecule. For example, in the CO molecule, the oxygen atom has a slight negative charge and the carbon atom has a slight positive charge. In the classical picture, as the molecule rotates, so does the dipole moment and this would produce electromagnetic radiation continuously. However, just as we saw for 'orbiting' electrons, quantum mechanically, this is not the case. The rotation is quantized and radiation is emitted only when a transition occurs from one rotational state to another.

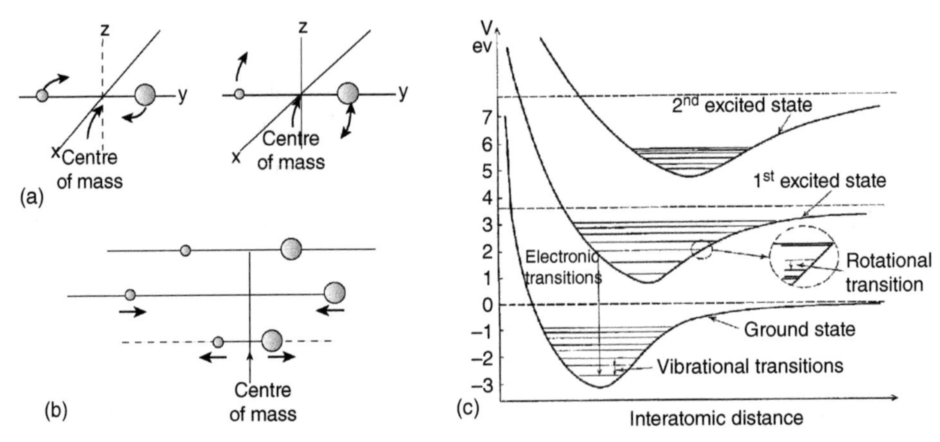

Figure 11.1 Molecular transitions in a diatomic molecule. (a) Illustration of rotation. Rotation is in the *x*–*y* plane at left (rotation about *z*) and in the *y*–*z* plane at right (rotation about *x*). The rotation is quantized and a spectral line is emitted only when there is a transition between two different rotational states. (b) Illustration of vibration. The motion resembles that of a simple harmonic oscillator. Large arrows show the direction of motion. (c) A sample energy level diagram showing the molecule's potential energy (including all nuclei and electrons) as a function of separation between nuclei. Electronic, vibrational, and rotational transitions are shown.

[3] Rotation about the third axis is not energetically important since the moment of inertia, I, is very small. For any object consisting of individual particles, $I = \sum m_i r_i^2$, where m_i is the mass of particle, i, and r_i is its distance from the axis of rotation.

It is nevertheless necessary for an electric dipole to be present in order for a rotational change to produce a rotational spectral line. Molecules like H_2, N_2, and O_2, for example, do not have electric dipole moments since their charge is equally distributed, and therefore, these molecules also display no rotationally induced electric dipole radiation. These molecules may have electric quadrupoles, however[4], and if so, much weaker (by many orders of magnitude) rotational lines can result from changes in the quadrupole moment. We will return to this point in Section 11.5.

As can be seen in the potential energy diagram of Figure 11.1c, rotational transitions correspond to the smallest energy changes in molecules and therefore these lines have the lowest frequencies, typically in the far-IR, sub-mm, or mm parts of the spectrum.

Aside from rotation, a molecule can also *vibrate*, as illustrated in Figure 11.1b and a classical analogue is that of simple harmonic motion. As with rotation, the vibrational states are quantized and a spectral line is emitted only if there are changes between these states[5]. The close-to-parabolic shape of simple harmonic motion can be seen in the potential energy diagram of Figure 11.1c. If the positive nuclei are separated by a small distance, there is little room for electrons between them and the force between the two protons is repulsive. At a large distance, more electrons will occupy the internuclear space and the force will be attractive. Thus, the molecule can oscillate over different internuclear separations if perturbed. If the separation becomes too great (Figure 11.1c) the molecule will dissociate, but if the separation is close to the minimum of the curve for any given electronic state, then it is stable[6].

Transitions between vibrational states are of much higher energy (typically in the IR) than those between rotational states. This means that the molecule could vibrate, perhaps 1000 times during a single rotation.

In reality, a single transition may involve a change in *both* vibrational and rotational states and these are called *ro-vibrational transitions*. The changes in quantum numbers must obey the same quantum mechanical selection rules as if they occurred individually. The total change in energy is the sum of the changes in the vibrational and rotational energies, $\Delta E_{\text{ro-vib}} = \Delta E_{\text{vib}} + \Delta E_{\text{rot}}$, but this will be dominated by ΔE_{vib} with ΔE_{rot} small in comparison. Observationally, for a given vibrational transition, all possible allowed rotational transitions also occur and therefore the vibrational line is broken into discrete components with spacings equal to those of the rotational steps, i.e. the rotational spectrum is superimposed on the vibrational line. Since the rotational steps are very closely spaced, this spreads out the vibrational 'line' into a spectral *band*. Similarly, it is possible for a transition to occur that involves a change in all three electronic, vibrational,

[4] Taking a simplified case in which the electrons and protons are along a straight line, the electric dipole moment is $p_1 = \sum q_i x_i$ where q_i is the charge of particle, i, and x_i is its position. The electric quadrupole moment is then $p_2 = \sum q_i x_i^2$. Even though the electric dipole moment may be zero, the electric quadrupole moment may be nonzero, depending on the configuration.

[5] One way in which vibrational modes in molecules can be probed is by shining laser light on a sample and observing the shift in wavelength that results from interactions with the molecules, a process called *Raman spectroscopy*, named after the Indian physicist, C. V. Raman.

[6] The minimum energy state is not *exactly* at the minimum of the curve. Even the ground state will have a slight vibration associated with it.

and rotational quantum numbers in one transition. In such a case, the total energy change is the sum of all three, with the electronic energy change dominant.

From the above discussion, it is clear that the observed spectrum of a single molecule will, in general, be far more complex than the spectrum of a single atom. A molecular cloud, moreover, will contain a variety of different kinds of neutral molecules as well as charged molecules and molecules containing isotopes, each of which has its own spectrum. Put together, molecular spectra tend to be so rich that *spectral surveys* are undertaken to search for and identify the lines in molecular clouds.

Figure 11.2 shows a dramatic example. The Orion KL[7] nebula contains a massive, young stellar cluster as well as abundant dust and molecular gas. The molecular gas reaches densities of $n \approx 10^7$ cm^{-3} and temperatures up to $T \approx 200$ K in the core of the cluster [22]. These conditions differ from those of typical interstellar molecular clouds that are far from star-forming regions ($n \approx 10^3$ cm^{-3}, $T \approx 20$ K, cf. Table 5.1). Thus, the excitation

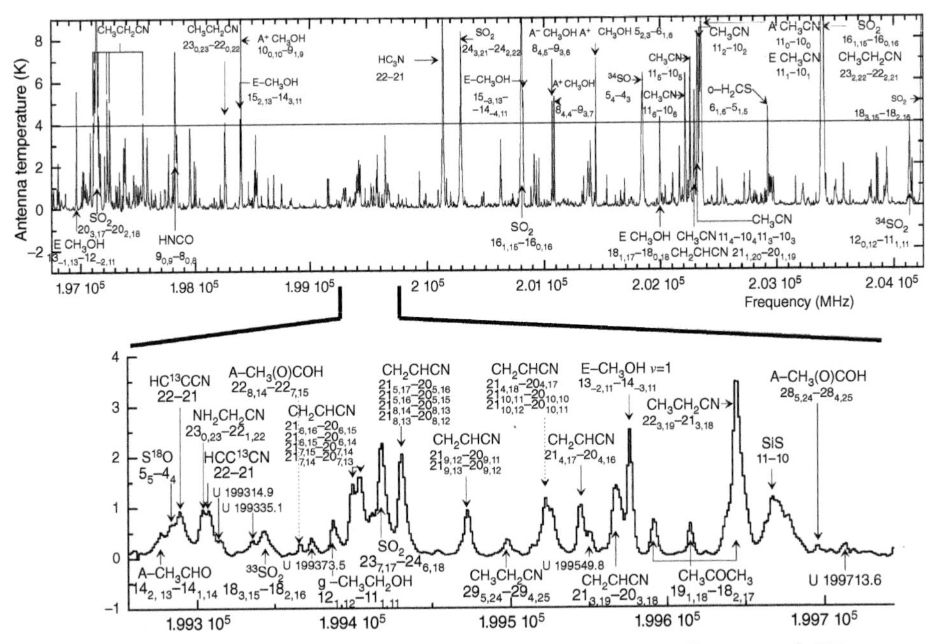

Figure 11.2 Spectrum of the molecular cloud, Orion KL, at a distance of 450 pc, over a small frequency range, 197 GHz (1.52 mm) to 204 GHz (1.47 mm) (top), obtained using the 30 m diameter Institut de Radio Astronomie Millimétrique (IRAM) telescope. The continuum has already been subtracted. Many more lines outside of this frequency range are also present in this source. A blow-up of a small frequency range (bottom) further reveals the richness of the spectrum. The lines have been identified by comparison with a catalogue of known spectral lines of over 1250 species. 'U' stands for 'unidentified' [41]. Source: Tercero et al. [41]. © 2005 Cambridge University Press.

[7] 'KL' stands for 'Kleinmann–Low'.

conditions are such that more lines are seen in the spectrum of Orion KL than would be the case in colder ISM clouds. Figure 11.2 shows only a tiny fraction of the lines that are actually present in this nebula. Another survey of Orion KL over the frequency range, 790 GHz (0.38 mm) to 900 GHz (0.033 mm), for example, reveals approximately 1000 spectral lines of which about 90% have been identified [6]. Still another survey between 41.5 GHz (0.072 mm) and 50 GHz (0.06 mm) has discovered over 200 lines representing 20 different molecules. Of these, 18 lines remain unidentified [32].

How many different molecules in space have actually been detected? As of 1 January 2019, around 200 molecules have been identified in the interstellar medium or circumstellar shells (see Table 11.1). The date is actually important because more molecules are actively being discovered as time goes by. Since the first edition of this textbook was published, the number of detected molecules has increased by 35%! From simple molecules like water (H_2O) and hydrogen peroxide (H_2O_2) to Buckminsterfullerenes ('Buckyballs', C_{60}), ionized [7] and neutral, it is clear that a rich variety of molecules are 'on offer' for scientific scrutiny.

Notice how many molecules include carbon, and therefore, an important part of the relatively new field of *astrochemistry* is actually *organic chemistry* in which carbon-bearing molecules are the focus of study. The polycyclic aromatic hydrocarbons (PAHs) introduced in Section 5.5.2 (see Figure 5.23) are part of this picture. Organic compounds such as *amino acids* (e.g. [14]) have also been detected in meteorites. These molecules do not imply the presence of life, but they do indicate that important molecular ingredients are present for life as we know it. Indeed, *astrobiology* is a growing subdiscipline in universities and institutes across North America[8] (see also Problem 11.15).

11.1.3 Nuclear Transitions – γ-Rays and High Energy Events

Probing still deeper into the heart of an atom's structure brings us to the nucleus within which further quantization occurs. If a nucleus finds itself in an excited state as a result of a *radioactive decay* or high-energy collision, then it may de-excite with the emission of a γ-ray photon at a specific energy. Various decay chains and daughter products are possible, depending on the specific nucleus and its stability. A mathematical description of nuclei and their energy levels is more difficult than has been discussed for the electronic, vibrational, and rotational transitions. Nevertheless, many nuclear transitions are well known. Table 11.2 lists some detected lines that are important in astrophysics, indicating the relevant decay chains and energies of the emitted γ-rays. Different isotopes of the same species produce lines at different energies, so it is possible to distinguish, not only between the element that is producing the line, but also which isotope is involved.

Gamma rays have the highest energies known for light (Table I.1), so the original energy required to excite the nucleus must also have been of the same order. Thus, γ-ray emission is associated with high-energy events. Examples are solar flares, novae, and supernovae

---

[8] For example, see https://create-astrobiology.mcgill.ca or https://astrobiology.nasa.gov.

Table 11.1 Molecules detected in astronomical sources[a]

2	3	4	5	6	7
H_2	C_3^*	$c\text{-}C_3H$	C_5^*	C_5H	C_6H
AlF	C_2H	$l\text{-}C_3H$	C_4H	$l\text{-}H_2C_4$	CH_2CHCN
AlCl	C_2O	C_3N	C_4Si	$C_2H_4^*$	CH_3C_2H
C_2^{**}	C_2S	C_3O	$l\text{-}C_3H_2$	CH_3CN	HC_5N
CH	CH_2	C_3S	$c\text{-}C_3H_2$	CH_3NC	CH_3CHO
CH^+	HCN	$C_2H_2^*$	H_2CCN	CH_3OH	CH_3NH_2
CN	HCO	NH_3	CH_4^*	CH_3SH	$c\text{-}C_2H_4O$
CO	HCO^+	HCCN	HC_3N	HC_3NH^+	H_2CCHOH
CO^+	HCS^+	$HCNH^+$	HC_2NC	HC_2CHO	C_6H^-
CP	HOC^+	HNCO	HCOOH	NH_2CHO	CH_3NCO
SiC	H_2O	HNCS	H_2CNH	C_5N	HC_5O
HCl	H_2S	$HOCO^+$	H_2C_2O	$l\text{-}HC_4H^*$	$HOCH_2CN$
KCl	HNC	H_2CO	H_2NCN	$l\text{-}HC_4N$	
NH	HNO	H_2CN	HNC_3	$c\text{-}H_2C_3O$	
NO	MgCN	H_2CS	SiH_4^*	H_2CCNH(?)	
NS	MgNC	H_3O^+	H_2COH^+	C_5N^-	
NaCl	N_2H^+	$c\text{-}SiC_3$	C_4H^-	HNCHCN	
OH	N_2O	CH_3^*	HC(O)CN	SiH_3CN	
PN	NaCN	C_3N^-	HNCNH	C_5S	
SO	OCS	PH_3	CH_3O		
SO^+	SO_2	HCNO	NH_4^+		
SiN	$c\text{-}SiC_2$	HOCN	H_2NCO^+		
SiO	CO_2^*	HSCN	$NCCNH^+$		
SiS	NH_2	H_2O_2	CH_3Cl		
CS	H_3^{+*}	C_3H^+			
HF	SiCN	HMgNC			
SH	AlNC	HCCO			
SH^+	SiNC	CNCN			
HD	HCP				
FeO(?)	CCP				
O_2	AlOH				
CF^+	H_2O^+				
SiH(?)	H_2Cl^+				
PO	KCN				
AlO	FeCN				
OH^+	HO_2				
CN^-	TiO_2				
HCl^+	C_2N				
TiO	Si_2C				
ArH^+	HS_2				
N_2	HCS				

(Continued)

Table 11.1 *(Continued)*

2	3	4	5	6	7
NO^+?	HSC				
NS^+	NCO				
8	**9**	**10**	**11**	**12**	**> 12 atoms**
CH_3C_3N	CH_3C_4H	CH_3C_5N	HC_9N	c-C_6H_6*	c-C_6H_5CN
$HC(O)OCH_3$	CH_3CH_2CN	$(CH_3)_2CO$	CH_3C_6H	$C_2H_5OCH_3$(?)	C_{60}*
CH_3COOH	$(CH_3)_2O$	$(CH_2OH)_2$	C_2H_5OCHO	n-C_3H_7CN	C_{70}*
C_7H	CH_3CH_2OH	CH_3CH_2CHO	$CH_3OC(O)CH_3$	i-C_3H_7CN	C_{60}^{+}*
H_2C_6	HC_7N	CH_3CHCH_2O			
CH_2OHCHO	C_8H	CH_3OCH_2OH			
l-HC_6H*	$CH_3C(O)NH_2$				
CH_2CHCHO(?)	C_8H^-				
CH_2CCHCN	C_3H_6				
H_2NCH_2CN	CH_3CH_2SH(?)				
CH_3CHNH	CH_3NHCHO(?)				
CH_3SiH_3	HC_7O				

[a]Numbers at the tops of the columns specify the number of atoms. A * specifies molecules that have been detected by their ro-vibrational spectrum. A ** specifies molecules that have been detected by electronic spectroscopy only. A (?) indicates a probable detection.
Source: Data from the Universität zu Köln, Physikalisches Institut, 1 January 2019. For a list of possible lines (though not necessarily detected yet), see https://cdms.astro.uni-koeln.de/cdms/portal.

(Section 5.3.3), and spallation of nuclei (Section 1.1.1) by the high-energy cosmic rays that permeate the Galaxy. Massive stellar winds from Wolf–Rayet stars (Figure 7.10) can also deposit nuclei that have been created via nucleosynthesis in their stellar interiors, into the ISM. Energies from nuclear transitions are typically in the MeV range which, although considered high energy, are at the *lower* end of the full γ-ray spectrum observed in nature.

Supernovae are a particularly important source of γ-ray lines since the nucleosynthesis that occurs during a supernova explosion leaves many nuclei in excited and unstable states. The γ-rays that result from de-exciting nuclei have been shown to drive supernova light curves which, like radioactivity, decay exponentially with time (Section 5.3.3). For example, the bolometric (recall, this means over all wavelengths, Section 3.1) light curve of Supernova 1987A[9], which occurred in the nearby galaxy, the Large Magellanic Cloud, decayed with a half-life of 77 days, nicely corresponding to the known 77 day half-life (111 day lifetime) of the decay of ^{56}Co to ^{56}Fe (part of the ^{56}Ni decay chain, Table 11.2). The γ-ray photons that are emitted during this decay interact with the surrounding gaseous material via Compton scattering (Section 7.1.2.2) and, with further interactions, the energy is converted to the observed bolometric luminosity[10]. SN1987A has now been

[9] This supernova, discovered in 1987 from Chile by Ian Shelton of the University of Toronto, Canada, has played a pivotal role in testing supernova models.
[10] This process is reminiscent of the diffusion and energy degradation of photons in the Sun as they travel from the core to the surface (Section 5.4.4).

Table 11.2 Some important detected nuclear γ-ray lines.

Isotope	Mean lifetime[a]	Half-life[a]	Decay chain[b]	Energy[c] (keV)	Sources[d]
^{7}Be	77 days	53 days	^{7}Be $\rightarrow$ ^{7}Li	478	Novae
^{56}Ni	111 days[e]	77 days	^{56}Ni $\rightarrow$ ^{56}Co $\rightarrow$ ^{56}Fe + e$^+$	847, 1238	SNe
				2598, 1771	SNe
^{57}Co	1.1 years	0.76 years	^{57}Co $\rightarrow$ ^{57}Fe	122, 136	SNe
^{22}Na	3.8 years	2.6 years	^{22}Na $\rightarrow$ ^{22}Ne + e$^+$	1275	Novae
^{44}Ti	85 years	59 years	^{44}Ti $\rightarrow$ ^{44}Sc $\rightarrow$ ^{44}Ca + e$^+$	1157, 78, 68	SNe
^{26}Al	1.04×10^6 years	0.72×10^6 years	^{26}Al $\rightarrow$ ^{26}Mg + e$^+$	1809	SNe, WR Novae, AGB stars
^{60}Fe	2.2×10^6 years	1.5×10^6 years	^{60}Fe $\rightarrow$ ^{60}Co	1173, 1332	SNe

[a]The 'lifetime', τ, is related to the *half-life*, $t_{1/2}$, by $t_{1/2} = \tau \ln 2$. The half-life is the amount of time required to reduce the quantity of matter to 1/2 of its original value, as first introduced in Section 1.1.1. In the event of two decays, the longer time is given.
[b]Not all possible decay routes are shown – just those that are relevant in the production of the γ-rays indicated. Decays that result in the emission of a positron (e^+) are also indicated.
[c]Energies of the lines.
[d]Sources in which the lines have been detected. WR means *Wolf–Rayet* (cf. Figure 7.10) and AGB means *asymptotic giant branch* (Section 5.3.2).
[e]The ^{56}Ni $\rightarrow$ ^{56}Co part of this chain has a lifetime of only 8.8 days.
Sources: From Diehl et al. [11] and Diehl [9].

observed for more than 30 years, and the rate of decline of its light curve has changed as the quantity of ^{56}Co has declined and other isotopes with longer half-lives have become more important sources of γ-ray photons. The current light curve is well-matched to the decay rate of ^{44}Ti, and the 68 and 78 keV γ-ray lines of this element have now been directly observed [8, 44].

Looking at the Milky Way, we can also see γ-ray line emission from Galactic supernovae. In the young ($\approx$ 312 year old) nearby (d = 3.4 kpc) supernova remnant (SNR), Cas A (Figure 3.2), all γ-ray lines listed in Table 11.2 from ^{44}Ti have been detected ([19, 44]). From the line flux, it has been possible to measure the total mass of ^{44}Ti ($\approx 2 \times 10^{-4}$ M$_\odot$) produced in the supernova explosion. Such studies not only contribute to our knowledge of elemental abundances (cf. Figure 5.9) but also provide a better understanding of the processes and conditions that occur during the SN explosion itself. Spontaneous nuclear de-excitations, themselves, are not sensitive to the temperature and density of the surrounding material (they depend on nuclear properties), but the rate at which nucleosynthesis can form these nuclei will be sensitive to physical conditions.

The ^{44}Ti γ-ray lifetime is not that different from the expected mean time between supernova explosions in our Milky Way (about one every 50 year, Section 5.3.4). If this line is detected and originates in supernovae, then the sources from which the line is observed should appear discrete. On the other hand, if a γ-ray line from a longer lived isotope with a supernova origin is observed, then its distribution should describe the distribution of Galactic supernovae *integrated* over a timescale corresponding to the lifetime of the line.

This is the case for the 1.809 MeV line from ^{26}Al which has a lifetime of a million years (Table 11.2). Over this period of time, with a mean supernova rate of one per 50 year, we might expect to see the *integrated* γ-ray luminosity from approximately 20 000 supernovae (SNe). These SNe should originate from different locations in the Galaxy, leading to distributed, rather than discrete, emission. Moreover, since the lifetimes of the massive progenitor stars are also of order, 10^6 year, the observed ^{26}Al emission will only be seen in locations where massive star formation has recently occurred. This line is therefore a tracer of massive star formation. We have encountered such tracers before (e.g. H II regions, Section 5.3.4), but compared to optical observations, γ-rays have an advantage in that they are highly penetrating and can travel through the ISM virtually unhindered.

A plot of the ^{26}Al distribution in the Milky Way is shown in Figure 11.3. Not all the emission in this figure results from supernovae alone. Between 20 and 50% may originate from Wolf–Rayet stars [25], but since Wolf–Rayet stars, themselves, are massive and short-lived, the bulk of the ^{26}Al emission still traces sites of massive star formation in the Galaxy. Doppler shifts of this line have also provided information on motions of the various sources (Section 9.1.1). Even the broad-scale rotation of the Galaxy has been measured from this line emission [10].

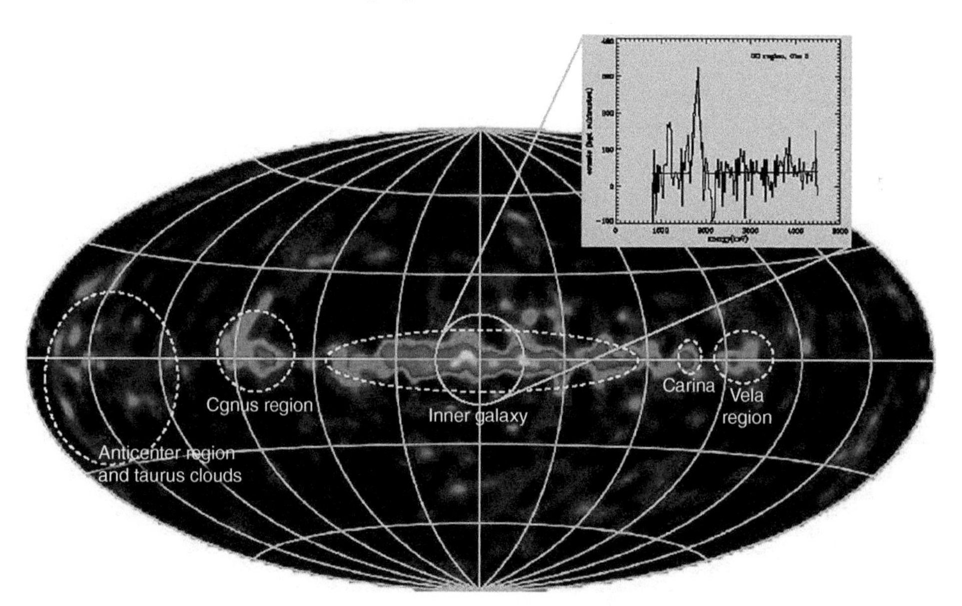

Figure 11.3 Map of the ^{26}Al 1.809 MeV γ-ray line emission in the Galaxy shown in a Hammer–Aitoff projection with the Galactic Centre at the map centre. This map [27] represents nine years of data collected by NASA's Compton Gamma Ray Observatory using the imaging Compton Telescope (COMPTEL). Specific regions in our Galaxy are circled and labelled. The inset illustrates the *spectral line* nature of the emission. Source: Al All-Sky Map, NASA, Retrieved from: https://heasarc.gsfc.nasa.gov/docs/cgro/epo/posters/Greatest_Hits/COMPTEL_26Al.html.

11.2 THE LINE STRENGTHS, THERMALIZATION, AND THE CRITICAL GAS DENSITY

Now that we have seen the variety of lines that could be present in astronomical sources, let us turn to some important concepts related to transitions that are seen in atoms and molecules.

In Chapter 10, we provided the emission coefficient, j_ν, for continuum radiation and then found the specific intensity, I_ν, by applying the relevant solution to the Equation of Transfer. Other quantities, such as flux density or luminosity, followed. The approach for spectral lines is essentially the same, but with the additional constraint that the line is present over a well-defined frequency interval. In order to characterize some *line strength*, then, we could either refer to the peak value of the line, or we could refer to the integral over the line. Both of these quantities are physically meaningful but, since the integral includes the total number of particles that are contributing to the line, we will use the integral to specify the 'line strength'. The emission line strength is determined by the *downwards transition rate* of whatever mechanism is producing the line, as well as the *number of particles that are in the upper level.*

Downwards transitions can occur spontaneously with a rate given by the Einstein *A* coefficient for a given transition (e.g. Table C.1 of the *online material*), or can be induced via a collision, whose rate depends on the collision cross-section for that particular transition as well as various physical parameters such as the density and temperature of the gas[11]. Example 7.1 provided a comparison of these two rates for the Ly α line of hydrogen under one set of conditions, showing the tendency of this line to undergo a spontaneous de-excitation, given the atom's short natural lifetime in the $n = 2$ state. On the other hand, the 21 cm line of H I is collisionally induced, since it has a longer spontaneous de-excitation timescale in comparison to a typical collision time. Since there is a different Einstein *A* coefficient and a different collision cross-section for each line, the probability of a downwards transition must be considered, spectral line by spectral line, for any given species under a set of physical conditions.

As discussed in Sections 5.4.4 and 5.4.5, if *all* transitions of an atom are collisionally induced, then the gas is in LTE. We now allow for the possibility that some transitions may be collisionally induced and some may not – a non-LTE situation. However, if a *specific* transition is collisionally induced, then it can be stated that the line is *thermalized*, as we first introduced in Section 5.4.5. In such a case, the Boltzmann Equation (Eq. 5.23) holds for the individual line being considered and the LTE solution of the Equation of Transfer (Eq. 8.19) applies for that line.

There is some importance in knowing whether a line results from a collisional or a spontaneous downwards transition because only if collisions are somehow involved can

[11] Downwards transitions can also be radiatively induced (called stimulated emission), but typically only if the radiation field is strong. More specifically, a radiatively induced downwards transition will occur if the timescale for this process is shorter than the timescales for both collisionally induced and spontaneous downwards transitions. Although this situation is not as common in astrophysics, stimulated emission must be included under some circumstances (cf. Footnote 4 in Chapter 8).

Table 11.3 Sample critical gas densities.[a]

Species	Wavelength or transition	Colliding species	Temperature (K)	n^*(cm^{-3})
HI[b]	Ly α (2p → 1 s)	electron	10^4	9.2×10^{16}
HI[c]	21 cm	HI	100	3×10^{-5}
OIII[d]	493.26 nm	electron	10^4	7.1×10^5
CO[e]	2.6 mm (J = 1 → 0)	H_2	30	2×10^3

[a]The critical gas density of the colliding species for the transition listed is determined by setting $\gamma_{j,i} n^* = A_{j,i}$, where $\gamma_{j,i}$ is the collisional rate coefficient (cm^3 s^{-1}) for a transition from upper state, j, to lower state, i, and $A_{j,i}$ (s^{-1}) is the Einstein A coefficient for that transition.
[b]Using γ from Table 5.2 and A_{ji} from Table C.1 of the *online material.*
[c]Using $\gamma = 9.5 \times 10^{-11}$ [39] and A_{ji} from Table C.1 of the *online material.*
[d]Using data from www.astronomy.ohio-state.edu/~pradhan/atomic.html with Eq. 4–11 from [39] and A_{ji} from the same reference.
[e]From data in [39]. J is the total angular momentum quantum number.

the observed line radiation tell us something about the gas kinetic temperature. The gas density at which the downwards collisional transition rate *equals* the downwards spontaneous rate for a given line is called the *critical gas density*[12], n^*. The transition rates also depend on temperature, but more weakly than density. Some sample values are given in Table 11.3.

From Table 11.3, the Ly α line, a recombination line which occurs in an ionized environment, will never be collisionally de-excited under typical interstellar conditions because the density of interstellar material (Table 5.1) does not achieve such high values. This does not mean that collisions are never important in an ionized hydrogen gas, however, since other recombination lines, especially at high quantum numbers, may indeed be collisionally induced at lower densities, as we shall see in Section 11.4.1. Other kinds of transitions in hydrogen can also be collisionally induced. For example, a transition between two n = 2 levels (2 s → 2p) becomes collisionally induced at a density of 1.5×10^4 cm^{-3} [24]. This is higher than the density of most interstellar ionized gas but can occur around active galactic nuclei. As for interstellar H I clouds, these have typical densities $n_H > 0.1$ cm^{-3}, so the H I λ21 cm line, with a critical gas density of 10^{-5} cm^{-3}, should always be thermalized (we have shown this already in Problem 8.5). This is also the case for the listed CO line, because most molecular clouds have densities, $n_{H_2} > 10^3$ cm^{-3}.

Given the large range of possible critical densities and the number of spectral lines that are observationally accessible, it is sometimes possible to *choose* the spectral line or lines that best probe the conditions in the source of interest. A well-designed observational programme will consider these points before time is spent at the telescope. We will explore how spectral lines can probe the physical conditions of the gas in Sections 11.4 and 11.5.

[12] The term, *critical density*, is more commonly used, but this term also has another meaning in cosmology, so we do not employ it here.

11.3 LINE BROADENING

An emitted spectral line cannot be arbitrarily narrow in frequency. It has a *natural line width* which is determined by the time the particle spends in the upper energy level before spontaneously de-exciting. The function that describes the shape of the line with frequency, regardless of the cause, is called the *line shape function* or the *line profile*. The line shape function that describes the natural line width is mathematically a *Lorentzian* profile, designated $\Phi_{\mathcal{L}}(\nu)$. Since all transitions have this quantum mechanical limitation, all spectral lines will have, at minimum, a natural line width. More information on the natural line profile is provided in Appendix D.1.3 of the *online material* and the mathematical function, $\Phi_{\mathcal{L}}(\nu)$, is given in Eq. D.14.

In *real* systems, however, the measured line width is always much greater than the natural line width. This means that one or more additional line-broadening mechanisms are present and are sufficiently important to dominate the quantum mechanical effects. Various line-broadening mechanisms are known. In astrophysics, however, the most important ones are *Doppler broadening and pressure broadening*. Of these two, Doppler broadening is most easily understood and most easily linked to the physical state of the gas. In the next two subsections, therefore, we describe Doppler broadening in some detail and then discuss pressure broadening and related effects more qualitatively.

11.3.1 Doppler Broadening and Temperature Diagnostics

Doppler broadening is due to the collective motions of the particles that are emitting the spectral line (recall that $\Delta\lambda/\lambda_0 = \upsilon/c$, where υ is a radial velocity relative to the observer, Eq. 9.3 or Eq. 9.4). Suppose a gas cloud (meaning any gaseous system) has no net motion towards or away from the observer, i.e. its systemic velocity (Section 9.1.1.1), $\upsilon_{sys} = 0$. Nevertheless, the individual particles within the cloud will still have some motion, be it thermal, turbulent, or perhaps from internal systematic motions such as rotation, expansion, or contraction or motions related to shock waves. If each particle within the cloud has a different motion, then the various radial velocity components of these motions will produce different Doppler shifts. The net result will be a superposition of all the individual Doppler-shifted lines, resulting in a *broadened* line (see Figure 11.4). If $\upsilon_{sys} = 0$, then the widened line will be centred on the line's rest frequency, and if $\upsilon_{sys} \neq 0$, then the line is centred at the Doppler-shifted frequency of the line for that υ_{sys}.

We have already seen a gross example of this for a rotating galaxy shown in Figure 9.4d. When emission from the entire galaxy is detected within a single beam, the galaxy's rotation causes the line profile to be spread out over a wide frequency range, or a wide velocity range if the x-axis is expressed in velocity units. Also, a rapidly expanding supernova remnant, if unresolved, should have a line profile whose width reflects the maximum velocities

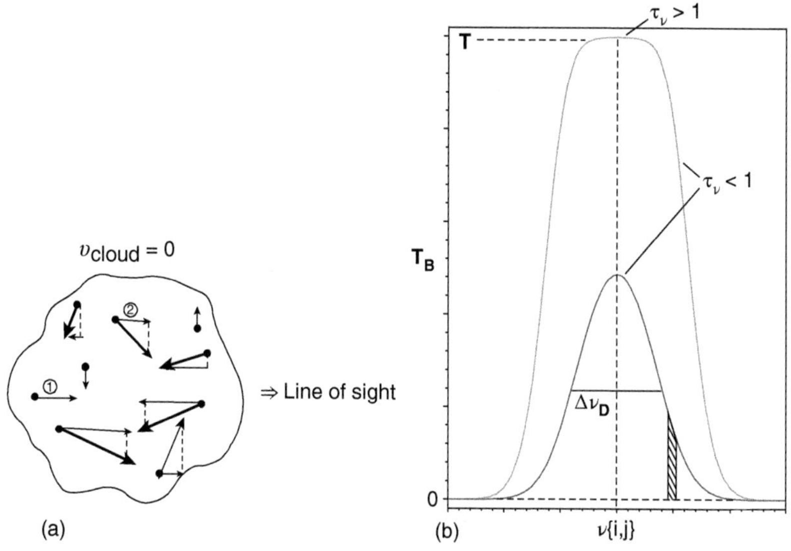

Figure 11.4 The thermal motions in a gas cloud that is globally at rest form a Gaussian-shaped spectral line, if the line is optically thin.
(a) Sketch of a gas cloud at rest, showing some of the internal particles (dots) and their velocity vectors (heavy arrows). The thin arrows show the components of velocity along the line of sight that broaden the spectral line. The two labelled particles have identical radial velocities and so will contribute to the line intensity at the same frequency.
(b) Two spectral lines plotted on a brightness temperature scale. The lower dark curve shows the Gaussian line shape given by Eq. (11.1) with the FWHM labelled. The hatched area (with width exaggerated) corresponds to all particles that have equal radial velocities, like those of particles 1 and 2 in (a). The grey higher curve shows the same line, but for a higher density cloud which has resulted in optically thick emission. In that case, $T_B = T$ near the line centre.

of both the advancing and receding sides of the shell. The *expansion velocity* is then half of the line width (Problem 11.4).

Not every cloud may be expanding or contracting or have other peculiar motions associated with it, but spectral lines that are formed in a gas at some temperature, T, will always experience at least *thermal line broadening* due to the Doppler shifts of particles that are in a Maxwell–Boltzmann velocity distribution (Figure 5.12). Taking a simple case in which the cloud has no systemic velocity, an optically thin spectral line from such a cloud has a *Gaussian* shape centred on the rest frequency, $\nu_{i,j}$, as illustrated in Figure 11.4b (lower curve)[13].

[13] As specified at the beginning of this chapter, we are only looking at emission lines here, but similar arguments could be applied to the shape of absorption lines. The subscripts, i and j, simply refer to a transition between state i and state j.

The Gaussian line shape function (line profile) is described by,

$$\Phi_{\mathcal{D}}(\nu) = \frac{1}{\sqrt{2\pi}\sigma_{\mathcal{D}}} \exp\left(-\frac{(\nu - \nu_{i,j})^2}{2{\sigma_{\mathcal{D}}}^2}\right) \qquad \mathrm{Hz}^{-1} \tag{11.1}$$

where

$$\sigma_{\mathcal{D}} = \frac{1}{\sqrt{2}}\frac{\nu_{i,j}}{c}\left(\frac{2\,kT}{m}\right)^{1/2} = \frac{1}{\sqrt{2}}\frac{\nu_{i,j}}{c}\,b = \frac{1}{\sqrt{2}\,\lambda_{i,j}}\,b \qquad \mathrm{Hz} \tag{11.2}$$

T is the gas kinetic temperature (K), m is the mass (g) of the particle emitting the spectral line, b (cm s^{-1}) is the *velocity parameter* which is defined by Eq. (11.2), i.e. $b = (2\,kT/m)^{1/2}$, and we have used the relation, $\lambda_{i,j}\,\nu_{i,j} = c$.

The full width at half-maximum (FWHM) intensity, $\Delta\nu_{\mathcal{D}}$, is related to $\sigma_{\mathcal{D}}$ and is given (in units of Hz) by,

$$\Delta\nu_{\mathcal{D}} = 2.3556\,\sigma_{\mathcal{D}} = 7.14\times10^{-7}\nu_{i,j}\left(\frac{T}{A}\right)^{1/2} = \frac{2.14\times10^{4}}{\lambda_{i,j}}\left(\frac{T}{A}\right)^{1/2} \tag{11.3}$$

where we have used Eq. (11.2), set $m = A\,m_H$, with m_H the mass (g) of the hydrogen atom, A (unitless) is the atomic weight, and the constants have been evaluated. The value of the dispersion, $\sigma_{\mathcal{D}}$, if it is desired, can also be obtained from Eq. (11.3). Note that higher frequency lines have broader widths in frequency, a simple consequence of the fact that $\Delta\nu = \nu_{i,j}\,\upsilon/c$ from the Doppler shift. The peak of the line shape function is

$$\Phi_{\mathcal{D}}(\nu_{i,j}) = \frac{0.9394}{\Delta\nu_{\mathcal{D}}} \qquad \mathrm{Hz}^{-1} \tag{11.4}$$

and the integral over all frequencies (a unitless quantity) is,

$$\int_{-\infty}^{\infty} \Phi_{\mathcal{D}}(\nu)d\nu = 1 \tag{11.5}$$

Notice that, if one measured the line peak and multiplied by the line width, the result would closely approximate the integral, to within 6% (e.g. Problem 11.10).

Equation (11.3) provides a relation between the line width in frequency, $\Delta\nu_{\mathcal{D}}$, which is a straightforward quantity to measure, and the temperature of the gas, assuming that thermal broadening is the dominant line-broadening mechanism. It is sometimes more useful to have a measure of the line width in velocity units, however, especially since many spectra are plotted as a function of velocity, rather than frequency or wavelength. Upon applying the Doppler formula to the right hand side of Eq. (11.3), the velocity line width is

$$\Delta\,\upsilon_{\mathcal{D}} = 2.14\times10^{4}\left(\frac{T}{A}\right)^{1/2} \qquad \mathrm{cm\ s}^{-1} \tag{11.6}$$

Equation (11.6) is now independent of the specific spectral line being observed. This must indeed be the case, because any given atom, which could have many transitions at different frequencies, will move at only one velocity.

The thermal line width is larger than the natural line width (e.g. Problem 11.3) and is always present in a thermal gas. However, there may be other Doppler motions (e.g. from rotation, turbulence, etc. as indicated above) which could widen the line even more, and these often dominate the thermal motions. If turbulence is present and the turbulent motions are also Gaussian, then it can be shown that the resulting profile, a convolution of the two functions, will still be Gaussian, but wider. If other motions are present, the profile may no longer be Gaussian, as we saw in Figure 9.4d for a rotating galaxy. Thus, for Doppler broadening, the shape and width of the line profile provide information as to the internal velocities or temperature of the cloud.

Allowing for the possibility of additional internal motions, then, the *observed* FWHM of the line, $\Delta\nu_{\rm FWHM}$ or, $\Delta\upsilon_{\rm FWHM}$, will be greater than or equal to the thermal line width,

$$\Delta\nu_{\rm FWHM} \geq \Delta\nu_{\mathcal{D}}, \qquad \Delta\upsilon_{\rm FWHM} \geq \Delta\upsilon_{\mathcal{D}} \tag{11.7}$$

Together with Eqs. (11.3) or (11.6), this shows that the measured line width places an *upper limit on the temperature of the gas.* This is a useful result because it means that we need to only plot the spectral line and measure its width, without doing any other calculations, in order to determine an upper limit to T.

In Section 8.4.1, moreover, we noted that the peak of a spectral line for which LTE conditions apply provides a *lower limit* to the gas temperature. Thus, provided this condition is met, the gas temperature *range* can be immediately constrained by simply plotting the spectrum and taking two relatively straightforward measurements, as Example 11.1 describes.

Example 11.1

The λ21 cm line of hydrogen emitted from an interstellar cloud ($\upsilon_{\rm sys} = 0$) has a peak specific intensity of $I_\nu = 4.03 \times 10^{-17}$ erg s^{-1} cm^{-2} Hz^{-1} sr^{-1} and a line width of $\Delta\nu_{\rm FWHM} = 15.0$ kHz. Find the range of possible temperatures of the cloud.

The λ21 cm line is thermalized, so the LTE equation applies to this line (Section 8.4.3). Therefore, by Eq. (8.24), $T_B \leq T$. Since $h\nu \ll kT$ for this line, we can use the Rayleigh–Jeans formula (Eq. 6.6) to convert the specific intensity, I_ν, at the frequency of the line centre ($\nu_{i,j} = 1420.4$ MHz) to a brightness temperature, finding $T_B = 65$ K for the line peak. We now put the given value of $\Delta\nu_{\rm FWHM}$ into Eq. (11.3) with $\lambda_{i,j} = 21.106$ cm (the wavelength corresponding to the central frequency of the line) and an atomic weight of A = 1 to find an upper limit to the temperature of T = 219 K. We conclude that the temperature of this cloud is in the range, $65 \leq T \leq 219$ K.

It is worth emphasizing that the thermal line width applies to any transition in the gas, no matter how it is produced (e.g. collisionally, radiatively, or spontaneously) because it is the velocity distribution of the particles that is determining this width (not the physics of the spectral line). All that is required is that the gas be thermal (i.e. have a temperature, T, Section 5.4.1). Thus, the upper limit to T applies to any optically thin line. On the other hand, the condition, $T_B \leq T$ will only be true for a line for which LTE conditions apply, as outlined in Section 8.4.

We have so far considered only optically thin lines. It is also of interest to consider what a thermalized line will look like if there is a large density of particles that contribute to the line. For a specific transition in a given cloud, the optical depth equation (Eq. 7.8) indicates that τ_ν increases linearly with increasing density. Therefore, at some high density, the line should become optically thick ($\tau_\nu > 1$). From the LTE solution to the Equation of Radiative Transfer (Eq. 8.19), assuming no background source, then $I_\nu = B_\nu(T)$ at frequencies for which $\tau_\nu > 1$. This also means that $T_B = T$ at these frequencies (Section 6.1.1). Since more particles contribute to the line centre than its wings, the result is a flat-topped profile, as illustrated by the higher curve in Figure 11.4b[14]. The presence of flat-topped profiles, therefore, is a diagnostic for optically thick thermalized lines. Provided we can be sure that no other effects are producing unusual profiles[15], the peak brightness temperature of such a line will give the gas temperature directly.

11.3.2 Pressure Broadening

The second important line-broadening mechanism in astrophysics is *pressure broadening* which we consider here only qualitatively. Pressure broadening actually includes a number of possible effects which become important when densities are higher, such as in denser stellar atmospheres, and involve interactions between a radiating atom and other particles around it. The interacting particles essentially perturb the radiating atom in such a way as to cause a small shift in the frequency of an emitted spectral line. For an ensemble of particles, each shift may be different, resulting in a widened line.

There are a number of approaches that have been taken to understand this process, an important one being to treat the atom as a radiating harmonic oscillator, as is done in Appendix D.1.2 of the *online material*. Collisions (impacts) with surrounding particles then effectively 'interrupt' the emitted radiation, introducing a perturbation on the phase and amplitude of the oscillation. The collision itself does not produce a transition, but perturbs the transition that is taking place. This interruption in time results in a wider frequency response (recall the reciprocal relation between time and frequency). The collision may be with protons, electrons, or neutral particles, and the result is different, depending on the type of impactor. This approach has been widely used, and the line broadening is therefore called *collision broadening*. It can be shown that collisional broadening, for each type of impactor, leads to a *Lorentzian* profile (Eq. D.14), but with the quantum mechanical radiation damping constant, $\Gamma_{i,j}$, replaced by a *collisional damping constant*, $\Gamma_{P i,j}$. The latter is a function

---

[14] *Flat-bottomed* profiles also result from optically thick *absorption* lines. In such a case, all of the background continuum emission is removed by the foreground line emission at and near the central line frequency. A quantity called the *equivalent width*, *W*, is used to measure the strength of an absorption line for any τ_ν, i.e. $W \equiv \int [I_{\nu 0} - I_\nu)/I_{\nu 0}]\, d_\nu$ where $I_{\nu 0}$ is the specific intensity of the background continuum and I_ν is the specific intensity measured in the line. The integral is taken over the measurable line.

[15] A spherically symmetric expanding wind which has a cut-off in radius, for example, can produce flat-topped profiles, even if optically thin.

of temperature, density, the specific line being considered, and the nature of the impacting particle, and therefore, the Lorentzian FWHM, $\Delta\nu_P$, also depends on these quantities.

Because we know that thermal line broadening, which has a Gaussian profile, is always present and its width is much greater than the natural line width, then the presence of an *observed* Lorentzian profile suggests that collision broadening is occurring and is dominating. This, then, provides information about the environment within which the line is forming. In reality, the various line widths are difficult to model and compare with observation, especially because collision coefficients are not known for every type of encounter and spectral line. However, some lines have been successfully reproduced. For example, the widths of certain strong sodium lines in the Solar atmosphere can be adequately explained by impacts with neutral hydrogen. If *both* collision broadening and thermal Doppler broadening are important (one does not strongly dominate the other), then both the Gaussian and Lorentzian profiles are incorporated into the line profile. The resulting shape is called a *Voigt profile*.

An interaction may be considered an 'impact' if the duration of the collision is much less than the time between collisions. However, if the impact duration approaches the time between collisions, then the effect may be considered continuous. For example, perturbations can occur from the collective quasi-static effects of surrounding ions. These ions produce a net electric field which causes the energy levels of a radiating atom to split and/or shift. Since the energy levels of many particles are no longer at one fixed value but are spread out, the resulting spectral line will also acquire a width. This effect is called *Stark broadening* because the shifts in energy levels result from the application of an external electric field, an effect known as *the Stark effect*. (The analogy for a magnetic field is the Zeeman effect, as discussed in Appendix C.3 of the *online material*.) The resulting line profile may no longer be Lorentzian[16].

For some of the above interactions, the different approaches could be considered different ways of describing similar effects. For example, an impact of a free electron with a bound electron can be thought of as an encounter in which the trajectory of the free electron changes significantly. As the free electron encounters the atom, it briefly applies Stark broadening to the atom. Thus, line broadening due to electron or proton collisions might also be referred to as Stark broadening. Similarly, the net electric field produced by surrounding ions can be thought of as the collective effects of many slow impacts. Because of this, as well as the fact that these mechanisms all tend to be important in high-pressure environments, the terms pressure broadening, collisional broadening, and Stark broadening are sometimes used without distinction in the literature, especially the first two. More information on these mechanisms can be found in [16].

[16] Stark broadening can be of two types, called *linear* and *quadratic*, depending on the strength of the frequency shift with particle separation. The line profile shape is different for the two types.

11.4 PROBING PHYSICAL CONDITIONS VIA ELECTRONIC TRANSITIONS

In principle, it is straightforward to write down the emission coefficient of a spectral line that results from a spontaneous downwards transition from upper level, j, to lower level, i,

$$j_\nu = \frac{n_j A_{j,i}\, h\,\nu}{4\,\pi}\, \Phi(\nu) \tag{11.8}$$

where n_j (cm^{-3}) is the number density of atoms having electrons in level, j, $A_{j,i}$ (s^{-1}) is the Einstein A coefficient, h is Planck's constant, ν (Hz) is the frequency of the line, and $\Phi(\nu)$ (Hz^{-1}) is the line shape function describing the relevant line-broadening mechanism (e.g. Eq. (11.1) if it is Gaussian). A dimensional analysis of this equation will verify that the units are erg s^{-1} cm^{-3} Hz^{-1} s^{-1}, as required for j_ν (Section 8.2).

In practice, however, it is often a challenge to arrive at a full description of the intensity of a spectral line. For example, the population of the upper state, n_j, must be found, and collisional and radiative de-excitations, as well as ionization equilibrium must also be included, if important. The full equations of statistical equilibrium (Section 5.4.4) may therefore be required in order to compute emission coefficients for the various lines in the atom. To compute the observed spectral line intensities, I_ν, the emission coefficient must be used in the appropriate solution to the Equation of Radiative Transfer (Section 8.3) which requires that accurate optical depths be determined. The optical depths will vary from line to line and will also vary, according to some line profile shape, within each line. The type and extent of the line-broadening mechanism are also important. For example, a line that would normally be optically thick could be optically thin if the cloud is undergoing large-scale turbulence or other motions that spread out the line and lower the line peak. This, in turn, affects the radiative transfer through the cloud. Finally, the spectral lines may sit atop an underlying continuum (e.g. Figure 10.10).

Fortunately, however, some conditions exist for which simplifications can be made without badly compromising the results. We will now look at some of these and consider what the observed spectral lines can tell us about the physical state of the gas. We will start with high quantum numbers and work 'down the quantum number ladder', focussing on three main transitions or types of transitions: *radio recombination lines, optical recombination lines, and the* λ *21 cm line of H I.* Recombination lines will be seen in any ionized gas, including H II regions, planetary nebulae, and diffuse ionized gas in the interstellar medium. The λ 21 cm line will be seen in neutral H I clouds (Section 5.4.7). As before, we will focus mainly on the hydrogen atom, but there are many other constituents besides hydrogen that emit their own spectral lines, each providing further diagnostics of the physical state of the gas.

11.4.1 Radio Recombination Lines

At high quantum numbers (e.g. n $\gtrsim$ 40), the energy levels of the hydrogen atom become very close together (Appendix C.1 and Figure C.1 of the *online material*), so transitions between these levels result in radiation at radio wavelengths. These transitions represent the upper quantum number limit to the recombination line spectrum (designated Hnα for a transition from principal quantum number n + 1 to n, Hnβ for n + 2 to n, etc.) and are called *radio recombination* lines (RRLs).

At high n, hydrogen transition probabilities are low, so RRLs tend to be weak in comparison to the lower quantum number optical recombination lines that we will discuss in the next section (Section 11.4.2). Nevertheless, many such observations have been made, not only of hydrogen RRLs, but of many other species as well (Figure 11.5). RRLs from quantum numbers as high as n ≈ 1009 have been measured from carbon, for example [40]! For hydrogen, RRLs have so far been observed at lower n but, in principle, it is believed that quantum numbers up to n ≈ 1600 are possible [15]. The diameter of such an atom would be an astonishing 0.3 mm! This is larger than the thickness of a human hair and is three million times larger than the size of the hydrogen atom in its ground state.

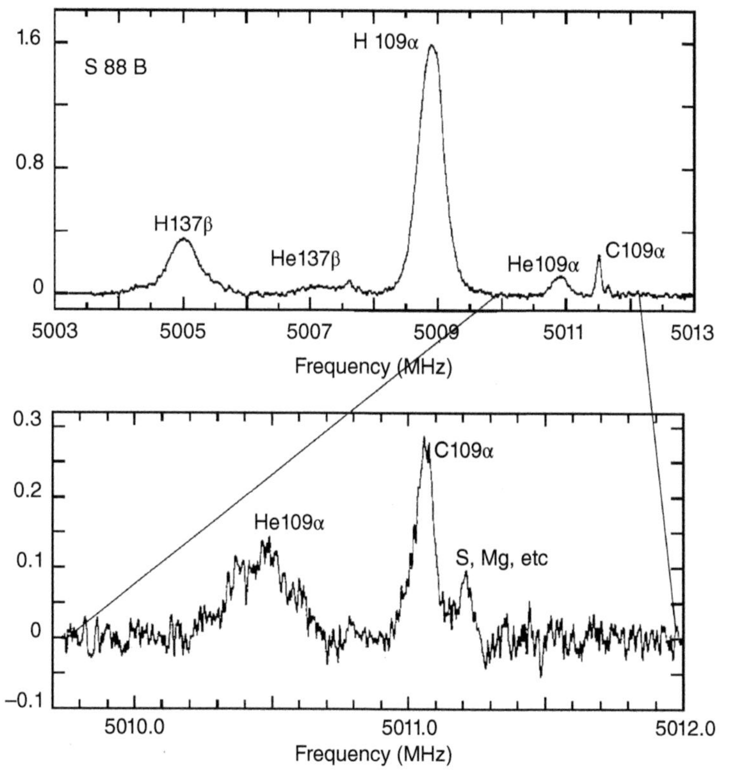

Figure 11.5 The radio recombination line spectrum of the H II region, Sharpless 88 B, near $\nu = 5$ GHz using the Arecibo radio telescope. The region that has been observed is circular on the sky of angular diameter $\theta_b = 59$ arcsec. The continuum has already been subtracted. Notice the presence of various weaker recombination lines, besides the H109α line of hydrogen. Source: Yervant Terzian.

As previously indicated (Section 10.2.2), radio waves are not hindered by dust and can therefore can be seen over great distances. If the optical emission of an H II region is obscured by extinction, radio (or long wavelength infrared) observations may be the only way of detecting that an HII region is present at all (see Figure 11.6).

Ultra-compact (UC) H II regions, for example, represent a class of H II regions which are almost always heavily obscured, optically. These are small, ($\lesssim$ 0.1 pc), dense ($\gtrsim 10^4 \mathrm{cm}^{-3}$) H II regions around O or B stars that are still embedded within their natal dusty molecular clouds [5]. They therefore represent a relatively early stage in the development of massive star-forming regions, and long wavelength data are required to probe their characteristics. RRLs also provide an opportunity to probe the diffuse ionized gas (DIG, Section 10.2.2) over large distances in the Galaxy. Together with a model of galaxy rotation, RRL Doppler shifts can be used to obtain distances to the line-emitting gas, something that cannot be achieved by continuum measurements alone (e.g. [2, 4]).

For RRLs, several assumptions can usually be made that simplify the analysis of these lines. Firstly, RRLs are weak lines and so are observed in the optically thin limit. Secondly, at high quantum numbers, the spontaneous downwards transition rate is a very strong function of quantum number, i.e. $A_{n+1\to n} \propto n^{-6}$ [12], so the timescale for spontaneous de-excitation becomes very large (e.g. the H109α line value in Table C.1 of the *online material* implies a lifetime of $t \approx 10^3$ s). This means that a collisional transition will likely occur before a spontaneous transition, and therefore, RRLs tend towards LTE conditions (Section 11.2, but see comments near the end of this subsection). And thirdly, at radio wavelengths, $h\nu \ll kT$ so the Rayleigh–Jeans approximation to the Planck function (Eq. 6.6) can be used. Thus, we can use Eq. (8.33) for any frequency in the line, $T_{B_\nu} = T_e\, \tau_\nu$. The optical depth, τ_ν, and the brightness temperature, T_{B_ν}, will vary with frequency according to the line shape function and the electron temperature, T_e, is usually taken to be constant.

We now assume that the line shape is Gaussian (applicable to thermal broadening alone or thermal broadening with Gaussian turbulence, Section 11.3.1) in order to write an expression for the optical depth at the *centre* of the line, τ_L. For transitions between quantum levels, n, and n + Δn, this is given by [13],

$$\tau_L = 1.01 \times 10^4 Z^2 \Delta n \left[\frac{f_{n,n+\Delta n}}{n}\right] T_e^{-2.5}\, e^{\left[\frac{\chi_n}{n^2 kT_e}\right]} \Delta\nu^{-1} \mathcal{EM}_L \tag{11.9}$$

where Z is the atomic number, $f_{n,\,n+\Delta n}$ is the *oscillator strength* for transition, $n \to n + \Delta n$ (Appendix D.1.3 of the *online material*), χ_n (erg) is the ionization potential (Section 7.2.2) of level n, $\Delta\nu$ (kHz) is the FWHM of the Gaussian line, and $\mathcal{EM}_L$ (cm^{-6} pc) is the emission measure of the line given by Eq. (10.6), where it is assumed that the ion and electron densities are roughly equal. The emission measure contains information on the electron density, n_e. Taking $\Delta n = 1$, $Z = 1$, $(f_{n,\,n+1})/n \approx 0.1908$ for large n when $\Delta n = 1$ [43], and noting that the value of the exponential approaches 1 for large n, then Eq. (11.9) becomes, for the centre of a Hnα RRL,

$$\tau_L = 1.92 \times 10^3 \left[\frac{T_e}{K}\right]^{-2.5} \left[\frac{\Delta\nu}{\mathrm{kHZ}}\right]^{-1} \left[\frac{\mathcal{EM}_L}{\mathrm{cm}^{-6}\mathrm{pc}}\right] \tag{11.10}$$

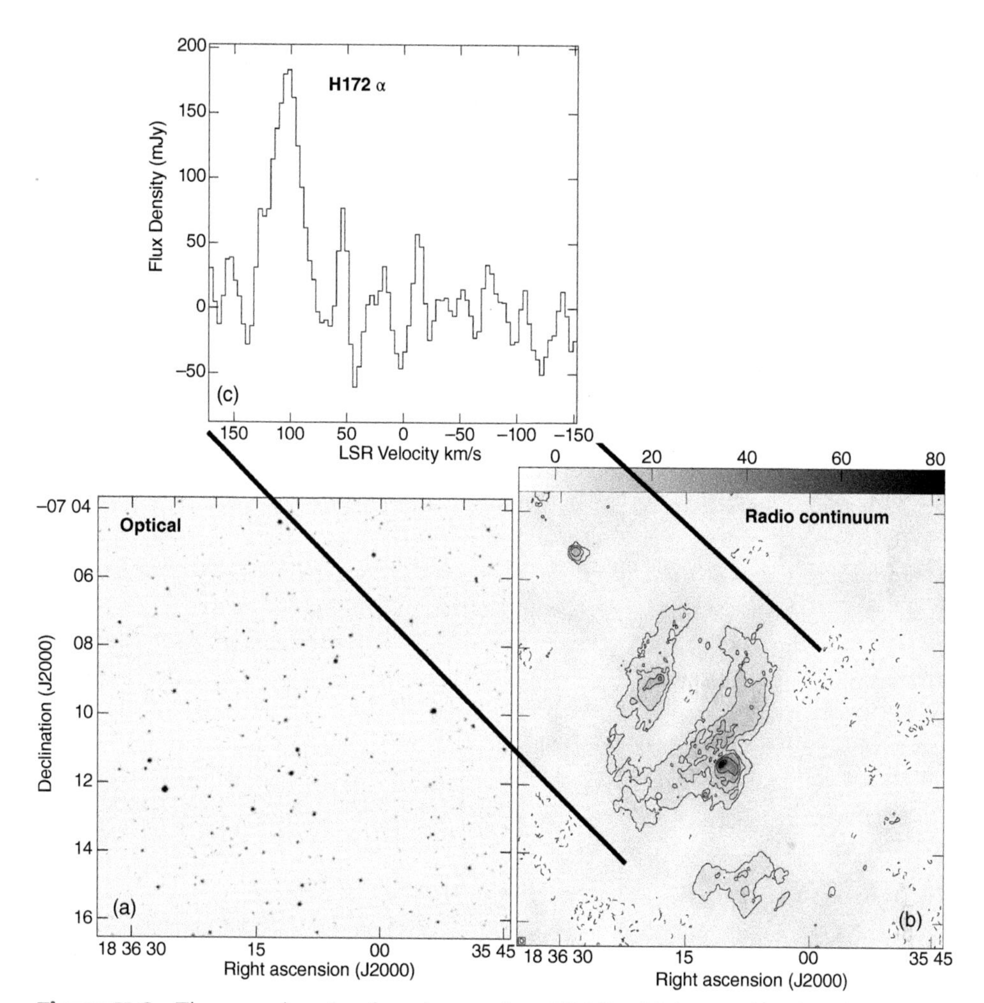

Figure 11.6 The massive star-forming region, G24.8 + 0.1, located in the plane of the Milky Way, is estimated to be at a distance of $D = 6.5$ kpc from us, and would be completely invisible were it not for observations at radio wavelengths. (a) This optical image shows only nearby stars since the optical emission from G24.8 + 0.1, itself, is heavily obscured by foreground dust. Source: http://skyview.gsfc.nasa.gov/cgi-bin/titlepage.pl. (b) This image, taken at $\nu = 1281.5$ MHz using the Giant Metrewave Radio Telescope (GMRT) in India, shows the same region as in (a). Here, we see complex radio continuum emission containing a number of HII regions. Most of the emission is shell-like, but a reasonable assumption is that it occupies a volume which is equivalent to a sphere of diameter, 4′. The radio continuum flux coming from this volume is $f_{1282\,\text{MHz}} = 7.7$ Jy. Source: GMRT images Reproduced by permission of N. Kantharia. (c) The H172 α recombination line, at the same frequency, from the region shown in (b) is displayed in this image. The underlying continuum has already been subtracted from this spectrum. LSR refers to Local standard of Rest (Section 9.1.1.1). Source: N. Kantharia.

The brightness temperature of the centre of an optically thin line can then be found by multiplying Eq. (11.10) by the electron temperature as noted above,

$$T_L = 1.92 \times 10^3 \left[\frac{T_e}{K}\right]^{-1.5} \left[\frac{\Delta\nu}{kHZ}\right]^{-1} \left[\frac{\mathcal{EM}_L}{cm^{-6}pc}\right] \quad (11.11)$$

As we have seen before, the emission from an optically thin thermal source is a function of *both* the temperature and density and this is no different for an RRL. As can be seen, Eq. (11.11) is a function of T_e and also of the density via the emission measure. Recall that for *continuum* emission, in principle and if technically feasible, it is sometimes possible to shift the frequency and make observations in both the optically thick and optically thin regimes. For thermal Bremsstrahlung radiation, this provides the extra information to solve for both temperature and density (Section 10.2.1). However, for a spectral line, the emission is at a specific frequency, so such shifts are not possible.

There is additional information associated with RRL observations, however, that aids our efforts in finding both the temperature and density of the gas. The same ionized gas that produces RRLs will also emit thermal Bremsstrahlung radiation. Therefore, the RRL sits on top of a thermal Bremsstrahlung continuum and T_L (Eq. (11.11) is a value that is measured from the level of the continuum 'upwards'. If the continuum emission is also optically thin[17], then the line-to-continuum brightness temperature ratio at the line centre, *r*, can be formed by dividing Eq. (11.11) by Eq. (10.14) or, equivalently, dividing Eq. (11.10) by Eq. (10.14). The result is [34],

$$r = \frac{T_L}{T_C} = \frac{\tau_L}{\tau_C} = 2.33 \times 10^4 \left[\frac{T_e}{K}\right]^{-1.15} \left[\frac{\Delta\nu}{kHz}\right]^{-1} \left[\frac{\nu}{GHz}\right]^{2.1} \quad (11.12)$$

where the subscripts, L and C, refer to the line and continuum, respectively.

In practice, we cannot measure the value of the continuum at exactly the line centre (e.g. Figure 8.3b), but we can measure it just adjacent to the line, as was done for the λ21 cm line in Section 8.4.3. In Eq. (11.12), we have assumed that $\mathcal{EM}_L/\mathcal{EM}_C = 1$ which would be the case for a pure hydrogen ionized gas. Thus, the line-to-continuum ratio can be used to find the electron temperature of the ionized gas explicitly. This is a very useful result because, if the assumptions are correct, a determination of T_e requires only a measurement of a single line together with its continuum, and the result is *independent of distance*. The electron density can then be found from either the line or continuum measurement, provided the distance is known, as Example 11.2 outlines.

Example 11.2

The $H93\alpha$ recombination line has been measured [1] within a $5'' \times 5''$ square region centred on the ultra-compact *H* II region, *G*45.12 (D = 6.4 kpc). The observing frequency is $\nu = 8.044$ GHz, the line FWHM is $\Delta\nu = 1497$ kHz, and the flux density in the continuum and line centre are $f_C = 1.72$ Jy and

[17] At very low frequencies, corresponding to very high quantum numbers, the line will be optically thin but the continuum could be optically thick.

$f_L = 0.059$ Jy, respectively. Assume that the line and continuum, both optically thin, are formed in the same physical region and find T_e, τ_L, τ_C, $\mathcal{EM}$, and n_e of this region.

From the given values, $f_L/f_C = 0.059/1.72 = 0.0343$, but using Eqs. (3.13) and (6.6), $f_L/f_C = T_L/T_C$, where T_L and T_C are the brightness temperatures of the line and continuum, respectively. We called this ratio, r, in Eq. (11.12), so $r = 0.0343$. Using the given values of $\Delta\nu$ and ν, Eq. (11.12) allows us to find the electron temperature of $T_e = 9198$ K.

The angular size of the observed 5″ × 5″ square region is $\Omega = 5.876 \times 10^{-10}$ sr, upon converting units. At a distance of $D = 6.4$ kpc, the angular diameter corresponds to (Eq. I.9) $l = 0.155$ pc which we also take to be the line of sight distance through the source. Thus, the geometry is that of a 'box' for this example.

The optical depths, τ_L and τ_C, can be found from $T_{B_\nu} = T_e\,\tau_\nu$ (Eq. 8.33) which requires that individual brightness temperatures be computed (only the ratio is known, from above). Converting the flux densities to specific intensities (Eqs. 3.13 and 3.8), yields $I_L = 1.00 \times 10^{-15}$ erg s^{-1} cm^{-2} Hz^{-1} sr^{-1} and $I_C = 2.93 \times 10^{-14}$ erg s^{-1} cm^{-2} Hz^{-1} sr^{-1} for the line and continuum, respectively. Now we need to convert from specific intensities to brightness temperatures (Eqs. 6.4 and 6.6), the result being $T_L = 50.3$ K and $T_C = 1475$ K for the line and continuum, respectively. With the known value of T_e, this gives $\tau_L = 0.0055$ and $\tau_C = 0.16$.

It remains to find $\mathcal{EM}$ and n_e. Either the line (Eqs. (11.10 or 11.11)) or continuum emission (Eqs. 10.13 or 10.14) can be used[18], both of which lead to $\mathcal{EM} = 34.7 \times 10^6$ cm^{-6} pc. We now use Eq. (10.6) with the value of l from above to obtain $n_e = 1.5 \times 10^4$ cm^{-3}.

Note that the calculated values are averages that are representative of the region as a whole. The density, for example, likely varies considerably within the region. A more accurate calculation for this source, which includes non-LTE effects, leads to $T_e = 8300$ K and $n_e = 1.8 \times 10^4$ cm^{-3} [1], i.e. errors of order 10% in temperature and a 15% in density have resulted from an LTE assumption.

As Example 11.2 has illustrated, some error is introduced into the calculations by making the assumptions outlined in this section. The magnitude of this error will vary, depending on the RRL adopted and the physical conditions in the gas (e.g.[38]). Thus, Eq. (11.12) can be used to obtain only an initial estimate of the electron temperature. For example, *stimulated emission,* which has been neglected here, could also be important (cf. footnote 4 in Chapter 8 and [33]). In addition, singly ionized helium will provide extra free electrons which contribute to the thermal Bremsstrahlung continuum but not to the line emission; to account for ionized helium, the right hand side of Eq. (11.12) should by multiplied by the factor, $1/[1 + N(\mathrm{HeII})/N(\mathrm{HII})]$.

The most important correction, however, involves departures from LTE. Typically, if $n > 168\, n_e$, where n_e is the electron density, LTE has been assumed [12]. However, departures cannot always be ignored and considerable effort has been expended to understand and take these effects into account (e.g. [26, 48]). For example, *maser* action (Section 5.4.5, cf. Footnote 27 in Chapter 5) has been observed in some RRLs (e.g. [35]) which is far from an LTE situation.

[18] Equation (10.18) is not applicable for this example because it uses a different geometry from what has been assumed here.

It is interesting that, at low radio frequencies, some RRLs can be seen in *absorption*. The background, in such a case, need not be a thermal continuum. For example, the absorption line of singly ionized carbon, C575 α, has been observed at $\nu = 32.4$ MHz [20] against the brighter Galactic background, the latter due primarily to synchrotron radiation [20]. Another example is carbon recombination line absorption observed in front of the synchrotron continuum of Cygnus A [23], the latter shown in Figure 10.17 Top.

11.4.2 Optical Recombination Lines

As we step down the 'ladder' of quantum numbers, we eventually reach the more familiar Balmer lines of hydrogen which are observed in the optical part of the spectrum (Table C.1 of the *online material*). Given the lengthy history of optical observations in comparison to any other waveband (Section 4.1), it is not surprising that images in the Balmer lines, especially the red H α line, are plentiful. Just as we saw for the RRLs, optical recombination lines are formed in ionized gas, and many beautiful and striking images of H II regions, planetary nebulae, and other ionized regions have been obtained in the H α line as well as in emission lines from other species. Figures 5.7, 5.13, 7.10, and 8.6 all provide some striking examples. The spectrum of an H II region was also shown in Figure 10.10, illustrating the rich abundance of lines available for study. What can these lines tell us about the physical conditions in the ionized gas?

For hydrogen in the relatively low-density interstellar medium, optical recombination lines are not in LTE. The lifetimes of electrons in their various states are very short (see the Einstein A coefficients in Table C.1 of the *online material*), so downwards transitions will be spontaneous. This means that the line emission coefficient, j_ν, can be expressed by Eq. (11.8). In fact, everything in this equation is known or understood *except for* the number of particles in the upper state, n_j. The problem, then, reduces to one of finding n_j.

This task is less trivial and requires a consideration of the recombination of free electrons with protons, non-LTE conditions, and quantum mechanical selection rules as electrons cascade down various levels in the atom. The result, integrated over the line, is,

$$\int_\nu j_\nu \, d\nu = \frac{1}{4\pi} h\nu_{j,i}\alpha_{j,i}\, n_e\, n_i \approx \frac{1}{4\pi} h\nu_{j,i}\alpha_{j,i}\, n_e^{\,2} \qquad \text{erg s}^{-1}\,\text{cm}^{-3}\,\text{sr}^{-1} \tag{11.13}$$

where $\nu_{j,i}$ is the frequency of the line centre, $\alpha_{j,i}$ is the effective *recombination coefficient for transition*, $j \rightarrow i$ (cm^3 s^{-1}), n_i, n_e are the ion and electron densities, respectively, and the approximation in Eq. (11.13) applies if the ion and electron densities are equal, which we take to be the case. Notice that, since the line shape function, $\Phi(\nu)$, is defined such that the integral over the line equals one (an example is in Eq. (11.5)), this function has now disappeared from Eq. (11.13).

The recombination coefficient has appeared before when we considered the ionization equilibrium of an H II region (see Example 7.4) for which α_r included recombinations to *all* levels of the atom. In the above case, $\alpha_{j,i}$ is different for different transitions and is *defined* such that the quantity, $\alpha_{j,i}\, n_i\, n_e$, is the total number of photons per second per cubic centimetre in transitions from level j to i. Because recombination is a collisional process, the

Table 11.4 Effective recombination coefficients[a]

$\alpha_{j,i}$ (cm^3 s^{-1})	Temperature (K)		
	5000	**10 000**	**20 000**
$\alpha_{3,2}$ (H α)	2.21×10^{-13}	1.17×10^{-13}	5.97×10^{-14}
$\alpha_{4,2}$ (H β)	5.41×10^{-14}	3.03×10^{-14}	1.61×10^{-14}

[a]Case B recombination has been assumed. Hα values are from [39] and Hβ values are from [24]. For the latter reference, averages for the densities, 10^2 and 10^4 cm^{-3}, are given.
Sources: Spitzer [39] and Osterbrock [24].

effective recombination coefficients, as we saw for α_r (Table 5.2), are functions of temperature. Several values are provided in Table 11.4.

For optically thin lines, in the absence of a background source, the specific intensity of a line simply requires a multiplication of j_ν by the line of sight depth through the cloud, l, as we saw in Eq. (8.17), i.e. $I_\nu = j_\nu\, l$. Therefore, we can now multiply Eq. (11.13) by l to obtain the intensity (erg s^{-1} cm^{-2} sr^{-1}) over the line, assuming $n_i \approx n_e$,

$$\int_\nu I_\nu \, d\nu = \frac{1}{4\pi}\, h\,\nu_{j,i}\,\alpha_{j,i}\, n_e^{\,2}\, l = 2.46 \times 10^{17}\, h\,\nu_{j,i}\,\alpha_{j,i}\, \mathcal{EM} \tag{11.14}$$

where we have expressed the emission measure, $\mathcal{EM} \approx n_e^{\,2}\, l$ (Eq. 10.6) in the usual units of cm^{-6} pc. As usual for thermal gas in the optically thin limit, the emission is a function of both temperature (via α_{ji}) and density (via $\mathcal{EM}$).

We now come to an interesting point. The effective recombination coefficients are well known and have been tabulated for a wide range of conditions. Also, for any two lines in a given atom, the emission measure should be the same since the line of sight distance through the source and the electron density do not depend on choice of line. This means that the *ratio* of the intensities of any two lines should be a function of frequency and the effective recombination coefficient, only. Forming this ratio from Eq. (11.14),

$$\frac{\int_\nu I_\nu \, d\nu\, (j \rightarrow i)}{\int_\nu I_\nu \, d\nu\, (k \rightarrow m)} = \frac{\nu_{j,i}\alpha_{j,i}}{\nu_{k,m}\,\alpha_{k,m}} \tag{11.15}$$

where j $\rightarrow$ i and k $\rightarrow$ m represent the two different line transitions.

Thus, the various line ratios can be calculated and tabulated as a function of temperature, because $\alpha_{j,i}$ is a function of temperature. In principle, then, measuring the line ratio of any two such lines can provide us with the temperature by comparison with the tabulated values. Since the ratio is formed from the integrals over two lines, the result is not dependent on the type of line broadening that is present, as was the case for RRLs for which a line/continuum ratio was formed at the line peak. Thus, we have yet another temperature diagnostic for H II regions or other interstellar ionized gas clouds. The density could then be found via Eq. (11.14) with a knowledge of T_e, though the distance to the object would also have to be known in order to ascertain the line of sight distance, l.

This is a simple and straightforward route to obtaining important parameters of an ionized gas. Unfortunately, as is often the case in real situations, there are a few complications to using hydrogen Balmer lines that we must now outline.

If all hydrogen lines were optically thin (a situation called *Case* A), then there should be no density dependence to $\alpha_{j,i}$ or to the line ratios of Eq. (11.15). However, as pointed out in Section 7.1.1.2, the Lyman lines of hydrogen, which emit in the UV, tend to be trapped in an ionized nebulae. For example, after several scatterings, there is some probability that a Ly β line (a transition from n = 3 to n = 1) will be degraded into an Hα (n = 3 to n = 2) and then a Ly α (n = 2 to n = 1) line. Even though the Balmer lines themselves may be optically thin, the fact that the gas is optically thick in the Lyman lines means that the Balmer line ratios will be altered in comparison to what they would be if all lines were optically thin.

When this more realistic situation is taken into account, the situation is called *Case* B. Because the optical depth depends on density, so will the line ratios. Therefore, the values of $\alpha_{j,i}$ given in Table 11.4, expected line ratios of Table 11.5, and the sample line ratio calculation of Example 11.3 are all provided for Case B. The corrections are not large. Over the range of temperature and density listed in Table 11.5, the maximum difference between Case A and Case B is only 3% and, as can be seen in the table, the maximum error introduced by changing the density a factor of 100 is even less.

Of greater concern is that the dependence of the Balmer line ratios on temperature is actually rather weak. Looking again at Table 11.5, a change in temperature of a factor of two results in changes in line ratios of only a few per cent – about the same as changing the assumption from Case A to Case B. The line ratios as well as model calculations for $\alpha_{j,i}$ would have to be measured and calculated, respectively, to very high accuracy for these ratios to provide useful temperature diagnostics. In other words, these Balmer line ratios are not too helpful for finding temperatures and densities.

Example 11.3

Compute the expected line ratio, Hα/Hβ, for an HII region of temperature, 10^4 K.

From Eq. (11.15), we can write,

$$\frac{H\alpha}{H\beta} = \frac{\nu_{3,2}}{\nu_{4,2}}\frac{\alpha_{3,2}}{\alpha_{4,2}} = \frac{\lambda_{4,2}}{\lambda_{3,2}}\frac{\alpha_{3,2}}{\alpha_{4,2}} = \left[\frac{486.132}{656.280}\right]\left[\frac{1.17\times10^{-13}}{3.03\times10^{-14}}\right] = 2.86 \tag{11.16}$$

where we have used $c = \lambda\,\nu$ and data from Tables 11.4 and C.1 of the *online material*.

Table 11.5 Sample line ratios for the hydrogen Balmer series[a]

T_e (K)	5000		10 000		20 000	
n_e (cm^{-3})	10^2	10^4	10^2	10^4	10^2	10^4
Hα/Hβ	3.04	3.00	2.86	2.85	2.75	2.74
Hγ/Hβ	0.458	0.460	0.468	0.469	0.475	0.476
Hδ/Hβ	0.251	0.253	0.259	0.260	0.264	0.264
Hϵ/Hβ	0.154	0.155	0.159	0.159	0.162	0.163

[a]Case B recombination has been assumed.
Source: Dopita and Sutherland [12]. © 2003 Springer Nature.

Fortunately, there are many other lines in HII regions besides those from hydrogen. The solution, then, is to find other line ratios that show stronger temperature dependences and for which all lines are optically thin.

Such lines do exist, the most commonly used ones being associated with [OIII] and [NII]. The square brackets refer to the fact that the relevant lines are *forbidden* (Appendix C.4 of the *online material*). For forbidden lines, the lifetime in the upper state is relatively long yet, in interstellar space, the densities are still low enough that emission results from spontaneous transitions. These lines were first discovered in nebulae, and the substance responsible for them was originally named 'nebulium' because the same lines could not be seen in the higher density gases of Earth-based laboratories in which collisionally induced transitions dominated. The unknown lines were later identified as being due to forbidden [OIII] and [NII] transitions.

It is useful to form ratios using three lines for each species, designated R_O and R_N for [OIII] and [NII], respectively [21],

$$R_O = \frac{I(\lambda 4959) + I(\lambda 5007)}{I(\lambda 4363)} = \frac{7.73\ \exp[(3.29 \times 10^4)/T_e]}{1 + 4.5 \times 10^{-4}[n_e/(T_e)^{1/2}]} \quad ([\text{OIII}]) \tag{11.17}$$

$$R_N = \frac{I(\lambda 6548) + I(\lambda 6853)}{I(\lambda 5775)} = \frac{6.91\ \exp[(2.50 \times 10^4)/T_e]}{1 + 2.5 \times 10^{-4}[n_e/(T_e)^{1/2}]} \quad ([\text{NII}]) \tag{11.18}$$

where the ratio is most sensitive to temperature in comparison to density (Problem 11.7).

These are not the only optical diagnostics of ionized regions. The line-to-continuum ratio can also be used, as was done for RRLs (though this ratio involves more complexity given the different processes that contribute to the optical continuum), and other line ratios, for example, from [SII] and [OII], are particularly sensitive to electron density, n_e. Thus, by carefully choosing the appropriate lines, a variety of optical observations can successfully probe the physical conditions in HII regions, planetary nebulae, and the diffuse ionized interstellar gas.

There remains one more important complication that must be considered when dealing with optical observations–that of obscuration by dust. Even if the spectral lines are optically thin, there will still be dust throughout ionized regions, as illustrated in a variety of figures in this text (e.g. Figure 5.13). If the extinction (Section 5.5.1) were independent of wavelength, then in forming a line ratio, the effect would be the same for both lines and the ratio would not be affected. However, extinction is wavelength dependent. Optical transitions that fall into the blue part of the spectrum will be affected by dust more than transitions in the red part of the spectrum, hence the term 'reddening' to describe this effect. As Figure 5.20 showed, $E_{\lambda-V} \propto 1/\lambda$, roughly, in the optical part of the spectrum, but there can be variations depending on location, so a 'location-specific' correction is desirable.

One way to deal with this problem is to use lines, if possible, that are very close together in frequency (for example, the [OII] λ 3727/λ 3726 ratio) so that extinction is approximately the same for both lines. Another approach, however, is to *use* line ratios to *correct* for the reddening. We found that the Balmer lines, for example, are not strongly dependent on temperature and density, so if we actually observe significant departures from the expected ratios of Table 11.5, then it is likely that those departures are due to reddening rather than to intrinsic variations in nebular properties. Therefore, a measurement of Balmer line ratios can provide an estimate of the reddening which can then be used to correct the ratios of Eqs. (11.17, 11.18).

Thus, the rich spectra seen in the optical waveband yield the properties of the nebulae, from information on the elements that they contain, to their densities and temperatures, and even their dust content. Notice how we can 'pick and choose' from amongst the plentiful spectral lines that are present in ionized nebulae (e.g. Figure 10.10), observing just the right lines to find the physical properties of interest. These are all bits and pieces of a broader mosaic to which we can appeal in an attempt to form a coherent picture of star formation and evolution, such as described in Chapter 5.

11.4.3 The 21 Cm Line of Hydrogen

In 1944, the Dutch astronomer, Jan Oort, turned to a young H. C. van de Hulst and said, '. . . by the way, radio astronomy can really become very important if there were at least one line in the radio spectrum' [42]. Within the year, van de Hulst had predicted the presence of the λ21 cm line of HI and, by 1951, three independent teams (Dutch, Australian, and American) successfully detected this line, each publishing their results in the same issue of the scientific journal, *Nature*.

The λ21 cm line originates from the bottom rung of the hydrogen atom's ladder. Resulting from a tiny electron that flips its spin, the corresponding transition is between two hyperfine levels of the ground state (Appendix C.4 of the *online material*). Because it is a radio line, it is not hindered by dust, so this emission can be seen clear across the Galaxy or, for that matter, through any other galaxy or intergalactic space. Oort's prediction was prophetic. The importance of this apparently insignificant quantum event cannot be overstated. It is largely through observations in the λ21 cm line that we know about the dynamic 'frothy' nature of the neutral interstellar medium (Figures 5.17, 8.10), the large-scale structure and rotation of the Milky Way (Section 9.1.1.2), the structure of the ISM in other galaxies, the rotations of other galaxies (Figure 4.21) and implied presence of associated dark matter (Section 5.2), tidal interactions between galaxies (Figure 11.7), cosmology [28], and the expansion of the Universe itself (Section 9.1.2). In this section, we concentrate on relating the observed λ21 cm line to the *mass* of H I.

We have already seen that essentially all hydrogen atoms in a neutral cloud will have their electrons in the ground state (Section 5.4.5). This provides an important simplification because it means that all levels in the atom higher than principal quantum number $n = 1$ can be ignored. Either an electron is in the upper hyperfine level of the ground state, or it is in the lower hyperfine level of the ground state. We have also seen that, because the lifetime in the upper state is long under typical astrophysical conditions, the line is

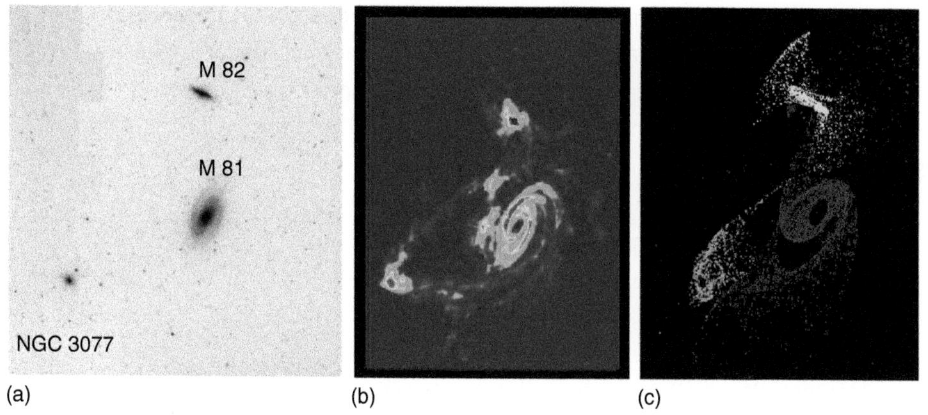

Figure 11.7 (a) Optical image of the M 81 Group (distance, D = 3.6 Mpc). (b) This HI image clearly shows that the three galaxies are interacting, with tidal bridges between the galaxies and other intergalactic neutral hydrogen gas. For M 81, the average column density is $N_{HI} \approx 4.0 \times 10^{20}$ cm^{-2} within an elliptical region of semi-major x semi-minor axes, 20 x 15 arcmin, respectively. (c) This numerical simulation has reproduced the interaction between the galaxies. The gas originally associated with each galaxy is represented by a different colour ([46, 47]). Source: Reproduced by permission of M. Yun.

collisionally dominated[19] and therefore thermalized (Section 8.4.3). This means that the Boltzmann Equation (Eq. 5.23) can be used to describe the relative populations of the two hyperfine levels.

The Boltzmann Equation is the product of two terms: the ratio of statistical weights of the two levels and the Boltzmann factor, $\exp(-\Delta E_n/kT) = \exp(-h\nu/kT)$, where ΔE_n is the energy difference between the two levels. We have previously found (e.g. Problem 8.5) that $h\nu \ll kT$ for this low-frequency line even if the temperature is very low, and therefore, the Boltzmann factor is ≈ 1. We are left with a very simple expression for the Boltzmann equation, that is, the population of states is simply the ratio of their statistical weights,

$$\frac{N_2}{N_1} = \frac{n_2}{n_1} = \frac{g_2}{g_1} = \frac{3}{1} \tag{11.19}$$

where the subscripts, 2, and 1, refer to the upper and lower hyperfine energy levels, respectively, and we have also expressed this ratio in terms of the ratio of number densities, n_2/n_1 (a division of both numerator and denominator by volume). Therefore, the number density fraction of all particles in the upper and lower states, respectively, is

$$\frac{n_2}{n_{HI}} = \frac{3}{4} \qquad \frac{n_1}{n_{HI}} = \frac{1}{4} \tag{11.20}$$

where n_{HI} is the total number density of neutral particles.

[19] More technically, a collision may induce an exchange of electrons between particles, after which the spin states may be different [29].

The optical depth (Eqs. 7.8, 7.9) refers to the number of particles in the *lower* state (potentially available for absorption) and is therefore

$$\tau_\nu = \sigma_\nu \int_l n_1 \, dl = \frac{1}{4} \sigma_\nu \int_l n_{\rm HI} \, dl = \frac{1}{4} \sigma_\nu \mathcal{N}_H \tag{11.21}$$

where σ_ν is the effective cross-section at the line frequency (cm^2) (presumed not to vary along the line of sight) dl is an element of line of sight distance through the cloud (cm), and the *column density*, $\mathcal{N}_{\rm HI}$ of hydrogen is *defined* by,

$$\mathcal{N}_{\rm HI} \equiv \int_l n_{\rm HI} dl \qquad ({\rm cm}^{-2}) \tag{11.22}$$

Thus, the column density of some species is the number density of that species integrated over the line of sight. Notice that we have now expressed the optical depth in terms of the more desirable quantity of the total number of neutral hydrogen atoms – not just the number in the upper or lower states.

For the λ21 cm line, it can be shown that the effective cross-section is given by [34],

$$\sigma_\nu = \frac{h\, c^2}{8\, \pi\, \nu_{1,2}\, k} \frac{g_2}{g_1} A_{2,1} \frac{\Phi_\nu}{{\rm T_S}} = 1.04 \times 10^{-14} \frac{\Phi_\nu}{{\rm T_S}} \tag{11.23}$$

where $\nu_{1,2}$ is the frequency of the transition, $A_{2,1}$ is the Einstein *A* coefficient for the transition (Table C.1 of the *online material*), Φ_ν is the line shape function (Eq. (11.1) being one example), and $\rm T_S$ is called the *spin temperature* of the gas, presumed to equal the kinetic temperature for a thermalized line (Section 5.4.5).

We can now put Eq. (11.23) into Eq. (11.21) and integrate over the line,

$$\int_\nu \tau_\nu \, d\nu = \frac{1}{4} (1.04 \times 10^{-14}) \frac{\mathcal{N}_{\rm HI}}{{\rm T_s}} \int_\nu \Phi_\nu \, d\nu \tag{11.24}$$

By definition, the integral over the line shape function is 1 (e.g. Eq. (11.5)), so Eq. (11.24) can be rearranged to solve for the HI column density,

$$\mathcal{N}_{\rm HI} = 3.85 \times 10^{14} {\rm T_s} \int_\nu \tau_\nu \, d\nu \tag{11.25}$$

$\rm T_S$ is presumed to be constant along a line of sight[20] but no such assumption is necessary for the density. When the λ21 cm line spectrum is plotted, the Doppler relation (Eq. 9.4) is often applied so that velocity, rather than frequency, is along the *x*-axis. In such a case, Eq. (11.25) becomes,

$$\mathcal{N}_{\rm HI} = (1.82 \times 10^{18})\, {\rm T_s} \int_\upsilon \tau_\upsilon \, d\upsilon \tag{11.26}$$

[20] The variation in temperature within some region of the ISM is usually less than the variation in density. However, should the spin temperature vary significantly along the line of sight, then the appropriate temperature to be used in these equations is a harmonic mean temperature, weighted by HI gas density, n_υ. More explicitly, $\left\langle \frac{1}{{\rm T_S}(\upsilon)} \right\rangle = \left[\int_l \frac{n_\upsilon(l)}{{\rm T_S}(l)} \, dl \right] / \left[\int_l n_\upsilon(l) \, dl \right]$ where l refers to position along the line of sight and υ is the velocity (linked to frequency via the Doppler formula, Eq. 9.4) of the gas. See [34] for further details.

where $\mathcal{N}_{HI}$ is still in cgs units of cm^{-2} but the velocity is in $km\ s^{-1}$. We now use the solution to the equation of transfer as applicable to this line (Eq. (8.31) to substitute for τ_v so that we can express the result in terms of the *observable* quantity, the brightness temperature, $T_B(v)$,

$$\mathcal{N}_{HI} = -(1.82 \times 10^{18})\, T_S \int_v \ln\left[1 - \frac{T_B(v)}{T_S}\right] dv \tag{11.27}$$

The column density can now be found, provided there is some knowledge of T_S. Note that, since $T_B \leq T_S$ (Eq. 8.24), the logarithmic term will be negative, so the column density is positive, as required.

As we have done before, let us consider the optically thick and optically thin limits of Eq. (11.27). If the line is optically thick at some velocity, v, then $T_B(v) = T_S$ (Eq. 8.32), and the logarithm evaluates is undefined. In such a case, the column density of the cloud cannot be obtained because we can only see into the front of the cloud (see Figure 6.2). On the other hand, if the line is optically thin, then $T_B(v) = T_S \tau_v$ (Eq. (8.33)). (As usual, in the optically thin limit, the emission is a function of both temperature, T_S, and density via τ_v.) We can now substitute $T_B(v)$ into Eq. (11.26) to find,

$$\left[\frac{\mathcal{N}_{HI}}{cm^{-2}}\right] = (1.82 \times 10^{18}) \int_v \left[\frac{T_B(v)}{K}\right]\left[\frac{dv}{km\ s^{-1}}\right] \qquad (\tau_v \ll 1) \tag{11.28}$$

where we explicitly indicate the units. Now the column density depends only on the observed brightness temperature integrated over the line, *without the need to know the gas temperature*. This arises because of the inverse dependence of the cross-section on temperature in Eq. (11.23).

This is an important result and is used widely because $\lambda 21$ cm lines are very often optically thin. Once the column density is known, the volume density and/or H I mass can be found, provided the distance is known, as Example 11.4 indicates.

Example 11.4

The HI line emission from the galaxy, NGC 2903, is optically thin and subtends a solid angle of $\Omega_S = 1.8 \times 10^{-5}$ sr. This emission, averaged over the source solid angle and integrated over all line velocities, is $\int I_v\, dv = 0.0458$ Jy beam^{-1} km s^{-1} (Figure 4.21 shows how this emission is distributed in the data cube). The beam solid angle, corrected for its Gaussian response (Example 4.4), is $\Omega_b = 5.21 \times 10^{-9}$ sr. Using information given in Figure 4.11, find the HI mass of NGC 2903. Compare this result to the total mass of the galaxy: $M_{tot} \approx 1.6 \times 10^{11}\ M_\odot$.

To use Eq. (11.28), we require the integral over the line to be in units of K km s^{-1}. From the definition of the Jy (Eq. 3.8) and the given beam solid angle, Ω_b, we find, $\int I_v\, dv = 8.79 \times 10^{-17}$ erg s^{-1} cm^{-2} Hz^{-1} sr^{-1} km s^{-1}. Then, using the Rayleigh–Jeans relation (Eq. 6.6) to convert to a brightness temperature at the line centre frequency, we have $\int_v T_B\, dv = 142$ K km s^{-1} as an average over Ω_S. Putting this into Eq. (11.28) yields an average column density of $\mathcal{N}_{HI} = 2.58 \times 10^{20}$ cm^{-2}.

Now if this column density is multiplied by the *area* (cm^2) of the source, the result will be the total number of neutral hydrogen atoms in the galaxy. The area is given by

$\sigma_s = r^2\,\Omega_s$ (Eq. I.10), where r is the distance to the galaxy. The caption to Figure 4.11 indicates that $r = 8.6$ Mpc, so $\sigma_s = 1.27 \times 10^{46}$ cm^2. Therefore, the total number of H I atoms is $N = \mathcal{N}_{HI}\,\sigma_s = 3.27 \times 10^{66}$. Multiplying this by the mass of an HI atom and dividing by the mass of the Sun (Tables T.1, T.2) yields a neutral hydrogen mass of $M_{HI} = 2.7 \times 10^9\ M_\odot$. This constitutes about 2% of the total mass given in the question.

A number of the steps in Example 11.4 can be folded into a single equation for the HI mass. The result is,

$$\left[\frac{M_{HI}}{M_\odot}\right] = 2.35 \times 10^5 \left[\frac{D}{\text{Mpc}}\right]^2 \int_\upsilon \left[\frac{f_\upsilon\, d\upsilon}{\text{Jy km s}^{-1}}\right] \tag{11.29}$$

where f_υ is the total flux density of the source at some velocity in the line and D is the distance to the source, in the specified units. Eq. (11.29) is commonly used to obtain the HI mass of an object once the flux density has been measured at each velocity in the line.

11.5 PROBING PHYSICAL CONDITIONS VIA MOLECULAR TRANSITIONS

Stars are formed from dense, dusty molecular clouds in the interstellar medium and, as Figure 11.2 showed, many molecular lines can be excited in the warm, dense molecular envelopes around newborn stars or in the nearby environment. Farther from local sites of star formation, however, are extended molecular clouds that permeate the disc of the Milky Way. These may also include embedded star formation, but the extended clouds (for example, *giant molecular clouds*, or GMCs, which may be 100 pc in size) have conditions that are less amenable to the excitation of so many lines. Typical temperatures and densities are T = 20 K and $n = 10^3$ cm^{-3}, both of which are considerably lower than the highest temperatures and densities found in the Orion KL cloud (Section 11.1.2). Nevertheless, many lines are still accessible and provide probes of the molecular conditions, especially rotational transitions which do not require as much energy to be excited as do vibrational ones.

Hydrogen is the most abundant element in the Universe, and it is not surprising the most abundant molecule, by far, is H_2[21]. Determining the mass of H_2 gas is equivalent to finding the mass of the molecular cloud, because the quantities of all other molecules and dust are much less. However, because H_2 has no permanent electric dipole moment, it cannot emit dipole radiation from rotational transitions.

As indicated in Section 11.1.2, however, a variety of quadrupolar transitions are possible and many have indeed been observed, although they tend to be weak. A number of mechanisms can excite H_2 including both radiative excitations as well as collisions. The low-frequency, pure rotational lines of H_2 in the ground vibrational state, for example,

[21] The process of formation of this molecule is still uncertain. It is currently thought that this molecule is unlikely to have formed by gas phase reactions. It may be formed in the presence of a catalyst, probably on the surfaces of dust grains, after which the molecule evaporates into the gas phase again [18]. Formation by gas-phase reactions, however, is possible for a variety of other molecules, e.g. [31].

have already been observed in the mid-IR range ($\approx$ 10 to 30 μm) [18]. And the new space-based instrument, the *James Webb Space Telescope,* has targeted these and other molecular hydrogen lines amongst its science goals, especially at high redshift where galaxies are forming [17]. These low-energy rotational lines, when collisionally induced, still require temperatures higher than about 100 K for excitation, however. This is certainly much lower than in circumstellar regions, but still do not reach the low temperature, molecular hydrogen that is in the deep ISM.

How then can we observe cold interstellar molecular clouds? The alternative is to consider the next most abundant molecule, which is carbon monoxide.

11.5.1 The Carbon Monoxide (CO) Molecule

The CO molecule has an electric dipole moment (Section 11.1.2) and, as indicated in Table 11.3, the critical density of the CO ($J = 1 \rightarrow 0$) rotational line at λ2.6 mm, at a molecular cloud temperature of 30 K is approximately the same as the typical density that is observed in such clouds. This implies that this particular rotational CO line is collisionally excited, and the species with which it collides will be the most abundant molecule, i.e. H_2. The λ2.6 mm line is a strong line, is easily observed, and in fact, is often optically thick. Other CO lines may also be present, depending on excitation conditions (for example, there are lines at λ1.3 mm, and λ0.87 mm), and lines of the $C^{18}O$ (e.g. Figure 5.19) and ^{13}CO isotopes, which are weaker and optically thin, may also be observable. Therefore, a variety of CO lines provide important diagnostics of interstellar molecular clouds. Moreover, the dynamics of the cloud can also be accessed via the Doppler shifts of these lines.

Because the density of the molecular cloud is essentially the density of its constituent H_2 molecules, much work has been expended in an attempt to determine the density of H_2 that is required to excite the various CO lines to their observed values, especially the CO ($J = 1 \rightarrow 0$) line, which is most readily observed[22]. This process is not straightforward because, unlike the λ21 cm line for which the quantity of H I can be determined directly from a line of the same atom, the observed CO lines are used to *infer* the quantity of the colliding species, H_2. For example, non-LTE conditions may apply and optical depth effects are important. The weaker, optically thin isotopic lines may be used, but they are not always observed and isotopic ratios may not be as well known as might be desired. Moreover, unknown amounts of carbon may be tied up in grains rather than existing only in the gas phase. Variations in the environmental conditions can also affect the CO-to-H_2 relation. If the metallicity is low, for example, the CO lines will be weaker for a given H_2 density. A high UV radiation field or cosmic ray flux will also be important because the two molecules can be dissociated, but CO tends to be dissociated more readily than H_2.

All things considered, there are many potential pitfalls in this process. Yet when the molecular mass is determined in other ways, such as via the assumption that the cloud is gravitationally bound and then compared to measures of its mass via a line ratio analysis,

[22] The wavelength of this line is long enough that it does not need to be observed from a high altitude, as can be seen in Figure 4.2.

the conclusion is that the relationship between H_2 density and CO ($J = 1 \rightarrow 0$) line strength is essentially linear. The relationship between the CO ($J = 1 \rightarrow 0$) brightness temperature and the H_2 column density is found to be

$$\left[\frac{\mathcal{N}_{H_2}}{\text{cm}^{-2}}\right] = X \int_{\upsilon} \left[\frac{T_B[\text{CO}(J = 1 \rightarrow 0)]}{\text{K}}\right] \left[\frac{d\upsilon}{\text{km s}^{-1}}\right] \tag{11.30}$$

Equation (11.30) applies only to the CO ($J = 1 \rightarrow 0$) λ 2.6 mm line and X is the sought-after conversion factor. For the Milky Way, the 'standard' conversion factor, X, is [3]

$$X = (2.0 \pm 0.6) \times 10^{20} \qquad \text{cm}^{-2}\ (\text{K km s}^{-1})^{-1} \tag{11.31}$$

It is well known that X increases if the metallicity decreases. This must be the case because a low metallicity, Z (Eq. 5.1), lowers the strength of the CO line but should not significantly affect the quantity of molecular hydrogen. Early work suggests that the variation in X is

Figure 11.8 The galaxy, NGC 5194 (alias M 51 or the Whirlpool Galaxy, see also Figure 5.8), is shown in the CO(J = 1–0) line in this image, taken from the Plateau de Bure Interferometer operated by the Institut de Radioastronomie Millimétrique (IRAM), from [36]. The spatial resolution is determined by the elliptically shaped Gaussian beam which has major × minor axes FWHM of 1.16 arcsec × 0.97 arcsec, respectively, and is seen as a small white dot below the map scale at lower left. The average integrated intensity within the 250 pc diameter white circle in the upper left corner is approximately 150 K km s^{-1}. A linear scale is shown at bottom left. Source: Image courtesy of Eva Schinnerer.

about a factor of 5 for a metallicity variation of a factor of 10 [45]. More recent observational and theoretical effort confirms the increase but shows a fairly wide scatter in the results, pointing to further limitations in using CO as a proxy for H_2 [3].

The CO molecule has now been observed in a wide variety of Galactic environments as well as in other galaxies. The same molecule that, in excess, is an unwelcome constituent of smog on the Earth 'lights up' the otherwise invisible molecular clouds in our own and other galaxies, providing us with a probe of the nature and properties of these cold clouds. Moreover, since stars are formed within molecular clouds (Section 5.3.2), CO and other molecules allow us to trace stellar birthplaces. The molecular gas distribution, in general, follows the light distribution of young stars in galaxies[23], as Figure 11.8 shows. The CO emission in the spiral galaxy, M 51, is particularly obvious along the spiral arms where shock waves have compressed the molecular gas, enhancing star formation in these regions. It is interesting to compare this figure to the optical image of the same galaxy shown in Figure 5.8.

PROBLEMS

11.1. [*Online*] Compute the frequencies of the following hydrogen transitions and indicate in which part of the electromagnetic spectrum the resulting spectral lines would be found, (a) Ly δ, (b) H_9, (c) Bε, (d) H110 α, (e) H239γ. Table C.1 and Eq. C.7 may be useful.

11.2. (a) Beginning with $\Phi_{\mathcal{D}}(\nu)$ (Eq. (11.1)), derive the line shape function, $\Phi_{\mathcal{D}}(\upsilon)$, where υ is the velocity of particles in a gas cloud. Express the result as a function of the velocity parameter, *b*, assume that the cloud, itself, is at rest and that all $\upsilon \ll c$.

(b) From the result of part (a), find an expression for the line width (FWHM) and confirm that it is the same as Eq. (11.6).

(c) Verify that $\int_{-\infty}^{\infty} \Phi_{\mathcal{D}}(\upsilon) d\upsilon = 1$

11.3. [*Online*]

(a) Find the ratio of natural to thermal line width, $\Delta\nu_{\mathcal{L}}/\Delta\nu_{\mathcal{D}}$, for the Balmer H α emission line from a pure hydrogen cloud and typical ISM conditions.

(b) A sodium line (λ589 nm) in the Solar atmosphere at a temperature of 7500 K is observed to have a line width of 10 nm. Find the thermal line width, $\Delta\lambda_{\mathcal{D}}$. What can you conclude about the dominant line-broadening mechanism in the Solar atmosphere?

11.4. The ^{44}Ti line measured from the Cas A supernova remnant (Figure 3.2) is centred at 1152 ± 15 keV and has a FWHM of 84.6 ± 10 keV [19]. What is the velocity of expansion of this supernova remnant, as measured by the ^{44}Ti ejecta? Does this result agree or disagree with the maximum known expansion velocity of $\approx$10 000 km s^{-1} for this SNR measured from higher resolution observations?

[23] There are differences on small scales when the distribution is scrutinized in detail, however.

11.5. [*Online*]

(a) With the help of a spreadsheet or other computer software, compute the frequencies (MHz) of all hydrogen radio recombination lines (RRLs), H$n\alpha$, where $50 \le n \le 300$. Compute the thermal line widths of each of these lines, $\Delta\nu_{\mathcal{D}}$ (kHz), assuming that they originate in a typical H II region.

(b) A radio telescope can observe in 'L band' over a frequency range, $1240 \le \nu \le 1700$ MHz. How many H$n\alpha$ RRLs fall within this band?

(c) Once the telescope's receiver has been set to a given frequency, ν_0, within L band, observations at any one time can only be carried out within a sub-band which has some bandwidth, ΔB, centred at ν_0, consisting of N_{ch} individual channels of width, $\Delta\nu_{ch}$. Bandwidth possibilities include – *Sub-band* 1: $\Delta B = 50$ MHz, $N_{ch} = 8$, $\Delta\nu_{ch} = 6250$ kHz; *Sub-band* 2: $\Delta B = 25$ MHz, $N_{ch} = 16$, $\Delta\nu_{ch} = 1562.5$ kHz; *Sub-band 3*: $\Delta B = 6.25$ MHz, $N_{ch} = 64$, $\Delta\nu_{ch} = 97.656$ kHz; *Sub-band* 4: $\Delta B = 1.5625$ MHz, $N_{ch} = 256$, $\Delta\nu_{ch} = 6.104$ kHz. Given these options, what is the highest number of H$n\alpha$ RRLs that can be observed at any one time and still be spectrally resolved, i.e. there needs to be three channels or more over the width of the line?

11.6. From the information in Figure 11.6 about the ionized gas in the G24.8 + 0.1 star-forming region, do the following:

(a) Estimate $\Delta\nu$ (kHz) from the line profile.

(b) Estimate the electron temperature, T_e, adopting the assumptions given in Section 11.4.1. Correct this value for non-LTE and other effects by increasing it by a factor of 1.5.

(c) Examine and compare Eq. (10.14) with Eq. (11.11) and suggest reasons as to why it might be better to use the continuum measurement to calculate the $\mathcal{EM}$, rather than the line measurement (assuming that the continuum is optically thin).

(d) From the continuum flux density, the corrected electron temperature from (b) above, the adopted spherical geometry, and assuming uniform density, determine $\mathcal{EM}$, n_e, and M_{HII}. Equations (10.16)–(10.18) will be useful.

(e) Assuming Solar abundance, determine the total mass of the ionized gas.

11.7. (a) Plot a graph of the optical line ratios, R_O, and R_N, as a function of electron temperature over the range, $5000 \le T_e \le 20\,000$, for two values of electron density, $n_e = 10^2$ and $n_e = 10^4$ cm^{-3}. Use a linear scale for T_e and a logarithmic scale for the ratio.

(b) Comment on the ability of these line ratios to distinguish between electron temperatures and electron densities over the ranges plotted.

(c) Indicate what the minimum dynamic range (Section 4.5) would have to be in order to measure the ratio, R_O, of an HII region with a temperature, $T_e = 5000$ K

11.8. Use the information in the question of Example 11.4 to find the HI mass of NGC 2903 using Eq. (11.29) instead of Eq. (11.28). [Hint: It is useful to find the number of beams over the source.]

11.9. From the information of Figure 11.7, find the HI mass ($M_\odot$) of M 81.

11.10. A spherical Galactic HI cloud at a distance of $D = 1$ kpc has an angular diameter of $\theta = 1.2$ arcmin. This cloud has a $\lambda 21$ cm line width that is due to thermal motions only, at a temperature of $T = 85$ K. The flux density of the cloud at the peak of the optically thin line is $f_{max} = 0.1$ Jy. For this cloud, find the following:

(a) The line FWHM, Δv (km s^{-1}).

(b) Its HI mass, M_{HI} ($M_\odot$).

(c) Its diameter, d (pc), and volume, V (cm^3).

(d) The number density of hydrogen atoms, n (cm^{-3}).

(e) Its brightness temperature, T_B (K), at the line peak.

(f) The optical depth, τ, at the line peak.

11.11. Estimate the molecular gas mas ($M_\odot$) contained in the molecular cloud complex that is circled in Figure 11.8.

11.12. A map of a galaxy giving Hα emission measures, $\mathcal{EM}$, in units of cm^{-6} pc, has been presented to you. An electron temperature of $T_e = 10^4$ K was assumed. At the location of an HII region in this galaxy, the value is $\mathcal{EM} = 6000$ cm^{-6} pc. You wish to compare this result to your map of the radio continuum emission. However, your radio map ($\nu = 1.575$ GHz) is in units of Jy beam^{-1}. At the location of the HII region, presumed to be entirely due to thermal Bremsstrahlung emission, you measure a specific intensity of 2 mJy beam^{-1}, where the beam has a Gaussian response and is elliptical on the sky with FWHM of 4.0×3.8 arcsec (major x minor axis, respectively).

(a) Calculate the 1.575 GHz beam solid angle in steradians.

(b) Find the brightness temperature of the HII region at 1.575 GHz (recall that the Rayleigh–Jeans formula should hold).

(c) Determine the emission measure of the HII region from the radio data.

(d) Compare the radio and Hα results. How can you account for any differences?

JUST FOR FUN

11.13. How dense is a 'dense' molecular cloud? Let us consider a molecular cloud core with a very high particle density of 10^6 cm^{-3}. Look up the best vacuum that can be obtained on Earth with pumps, e.g. an 'extreme ultrahigh vacuum', and compare the results.

11.14. Imagine that all of the spectral lines of hydrogen that have transitions from $n + 1$ to n were posts in a picket fence.

(a) On the far right of the fence, the last post corresponds to the longest radio wavelength that is detectable from the ground. Approximately what transition does this correspond to? (Eq. C.7 of the *online material* and Figure 4.2 will be of help.)

(b) What is the transition corresponding to the next post immediately to the left of the one in (a)? What is its frequency?

(c) Each post is 4 inches wide, and the next post is flush against the far right post. Work out the scale, i.e. 4 inches = x Hz.

(d) How long is your fence?

11.15. Many scientists have been thinking about what spectral signatures in an exoplanet atmosphere could indicate the presence of life as we know it. Figure 11.9 shows what the Earth's spectrum would look like from a distance and is a good starting template. Research what one might look for in the spectrum of an exoplanet and suggest some possibilities ([37] provides a nice overview).

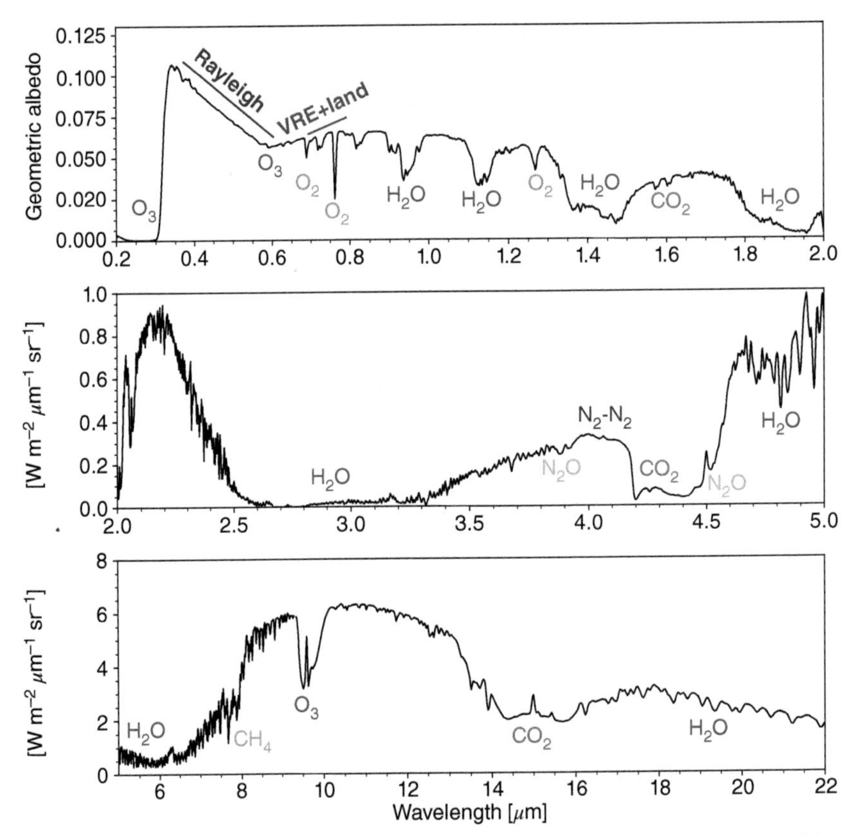

Figure 11.9 Synthetic spectrum of the Earth. The top panel is shown with respect to geometric albedo, while the bottom two panels show the specific intensity. Strong absorption features are labelled as well as Rayleigh scattering. Source: Schwieterman et al. [37]. © 2018 Mary Ann Liebert, Inc.

PART VI

The Signal Decoded

As the previous chapters have shown, considerable effort is required to extract the basic parameters of astronomical objects. The signal must be defined, measured, corrected for instrumental and atmospheric effects, possibly corrected for the effects of interstellar particles and fields, and understood at a physical level that is deep enough to relate the observed emission to the properties of the source. Now that the basics are all in place, how do we actually separate one type of process from the other, and how do we piece together reasonable scenarios so that some scientific insight may be gleaned, beyond a simple description of the astronomical source? The next chapter attempts to address these questions.

Chapter 12
Forensic Astronomy

> ... a list of observations, a catalogue of facts, however precise, does not constitute science. The reason is that they tell no story. Only facts arranged as narrative qualify as science.
>
> –Nobel Laureate, John Polanyi [16]

In this chapter, our goal is to put together the bits and pieces of information that we have gleaned from earlier chapters and, as the prologue suggests, discover a story. The signal may come to us 'coded' but, with knowledge of the physics of emission processes and interactions that occur en route to our telescopes, we may be able to decipher these codes and gain insight into the nature of the astronomical objects that both puzzle as well as fascinate us. Our tools are not just the sophisticated instruments that detect the signal, but also the equations that relate the light that we observe to parameters of the source. It is sometimes possible to go even further. By considering as much evidence as is available, the origin,

Astrophysics: Decoding the Cosmos, Second Edition. Judith A. Irwin.

Companion website: www.wiley.com/go/irwin/astrophysics2e

evolution, energy source, lifetime, relation to other sources, or other information that goes beyond basic properties, might be deduced. But first, we need to disentangle the mixed messages that are sometimes sent to us from space.

12.1 COMPLEX SPECTRA

12.1.1 Isolating the Signal

Sometimes we are lucky enough that a signal comes to us uncorrupted and also unmixed with other types of signals. An example is the λ 21 cm line from an isolated interstellar HI cloud. Not only does the radio emission travel through interstellar space unperturbed by dust, but also no other significant emission at or near that wavelength is emitted by the cloud. It is then straightforward to interpret the result. Often, however, this is not the case. Even if the signal has not been altered in its long journey towards Earth, or even if all corrections have been applied so that the signal is as close as possible to the state in which it was emitted, the result may still include a mixture of different kinds of radiation.

Consider, for example, the spectrum of NGC 2903 that was shown early on in Figure 4.3. The total flux density ($f_\nu = I_\nu\,\Omega$, Eq. 3.13) of the continuum emission from this spiral galaxy is plotted as a function of frequency in that figure. At the lowest frequencies up to $\nu = 10^{10}$ Hz, the radio continuum emission declines as a power law with increasing frequency. From our knowledge of emission processes at radio wavelengths, this suggests that the radio spectrum is dominated by synchrotron emission due to cosmic ray electrons interacting with the magnetic field in this galaxy. As the frequency increases, we see a broad peak resembling a (rough) Planck curve and, since this peak is in the infrared, this suggests that the emission is dominated by the collective contributions of dust grains (Problem 12.1). Above $\nu = 10^{14}$ Hz, another broad peak is seen, but is less well delineated, given the number and spread of data points in this part of the plot. This peak extends through the optical part of the spectrum, and we can infer that it is dominated by the collective Planck curves from different types of stars. Finally, several low-intensity points between 10^{17} and 10^{18} Hz (the X-ray band) indicate that high energy and/or temperature emission is also present, likely from a variety of sources.

In fact, the NGC 2903 spectrum is quite typical of what is seen for any normal spiral galaxy, including the Milky Way. From our knowledge of continuum processes, it is possible to provide a rather rough interpretation of the global spectrum, as described above. However, there are many objects in a galaxy, each emitting and thereby providing information that we would like to obtain. The emission from individual objects is hidden in plots like Figure 4.3, being swamped by the collective emission of many different objects[1].

[1] An exception might be if one type of object clearly dominates the spectrum, such as a very strong active galactic nucleus at the core of a distant galaxy, or a supernova that has recently gone off.

Spectral line emission has also been left out of the figure although the lines might have been plotted, had the data been available. If a dominant component can be identified, such as the infrared peak from dust, we are still left with many questions. What range of grain sizes and temperatures are represented? Is most of the dust associated with star-forming regions, with old stars, or with molecular clouds? Answering such questions helps us to grapple with deeper issues, such as the origin of the dust. We therefore need to *isolate* the dust (or other) emission from the more complex global signal.

An obvious approach is to isolate the data *spectrally*. Observations in different frequency bands require different observational techniques, so isolating the signal spectrally is in fact a by-product of any observation. This immediately limits the number of emission processes that are important in the band. Black body radiation from stars is negligible at centimetre wavelengths, for example. Further narrowing the band may also help. For example, observations of soft X-rays (lower frequency X-rays) are more likely to detect emission from thermal hot gas (Section 10.2.3) whereas hard X-rays (higher frequency X-rays) are more likely to pick up individual energetic sources, such as X-ray emitting pulsars. Each radiation process has its own characteristic spectrum, so it is possible to 'tune' the observations to focus on a specific type or types of emission. In the narrowband limit, one tunes to a specific spectral line.

The data can also be isolated *spatially*. High spatial resolution allows the targeting of specific regions or objects for careful study. In external galaxies that are not too distant, individual HII regions, globular clusters, bright stars, and interstellar clouds have all been spatially isolated and studied. The CO emission in M 51, as another example, shows spatially resolved interstellar molecular clouds quite well (Figure 11.8).

Even better is to isolate the emission *both* spectrally and spatially. An example has been shown in Figure 4.21, in which a data cube for the λ 21 cm emission from the galaxy, NGC 2903 is presented. This line has been isolated both spectrally and spatially, although *individual* HI clouds have not been resolved.

Even with much effort devoted to spectral and spatial resolution, however, it is still possible for a signal to contain a mixture of different types of emission. It is not always possible, for example, to arbitrarily improve the spatial resolution of a telescope (see Chapter 4) and, even if a specific object can be isolated, there may be foreground or background emission along the line of sight. Different types of radiation may also be emitted from the same object or spatial region. We must then consider whether there are other ways to isolate a given signal.

A possibility presents itself if the signals are so different that they can easily be recognized. An example is when spectral lines clearly sit on top of continuum emission, such as the optical lines in the HII region of Figure 10.10 or the X-ray emission lines seen in a Solar flare (Figure 8.7). The continuum and line can then be dealt with separately, both providing useful information about the source. Another is if the signal consists of spectral lines that are Doppler-shifted (Section 9.1.1.1). Various clouds along a given line of sight can then be separated because their relative velocities prevent them from occurring at the same frequency. HI and molecular interstellar clouds in our own Galaxy can often be isolated this way. A more dramatic example is the separation of intergalactic clouds in the Universe due to the expansion of space, as shown by the Lyman α forest in Figure 7.6.

In spite of every best effort, the fact remains that some signals simply cannot be isolated from others, either because the resolution is insufficient, because there are many signals from a line of sight at the same frequency, or because several types of signals are mixed at the source. For such cases – and there are many – we must *build a model.*

12.1.2 Modelling the Signal

The process of building a model spectrum involves guessing at a set of physical parameters for the source, including any constraints that might already be known, calculating a model spectrum using these parameters using equations such as those given in Chapters 10 and 11, and then comparing the model spectrum to the real one. Input parameters might include temperature, density, source size, fraction of emission due to various processes, or whatever else is required to calculate a spectrum. If the model and real spectra differ from each other, the inputs to the model must be adjusted and the process repeated until the model spectrum matches the real one as closely as possible. If a good match can be found, then the input parameters may provide a good description of the source. The problem is often considered to be solved at this point, although concerns about *uniqueness* should still be kept in mind. That is, it is possible that more than one set of input parameters could reproduce the observed spectrum equally well. For a thorough analysis, the set of input parameters will be carefully and sequentially adjusted (this is called 'searching through parameter-space') to see whether other combinations might also reproduce the spectrum.

An example of a model was shown in Figure 10.10 for the optical continuum emission from an HII region. This model included contributions from free-bound radiation and two-photon emission, as labelled in the figure. The difference between the calculated spectrum and the observed spectrum was then attributed to scattered light from dust.

A more complex example of modelling is illustrated by Figure 12.1 which shows the total γ-ray emission from the inner region of our Galaxy obtained from a variety of γ-ray instruments (upper data points in the plot). At γ-ray energies, a wider variety of astrophysical processes that have so far been described in this text, can occur. Let us consider each the important players in this impressive plot.

Neutral pion (π^0) decay (red curve) is clearly an important contributor between about 10 and 10^5 MeV. These pions result from high energy collisions between high energy cosmic rays (mostly CR protons) and atomic nuclei of the ISM (again mostly protons). We saw pions previously in Section 1.2.1 when cosmic rays (CRs) were first introduced (cf. Footnote 13 in that section). The neutral pion has a rest mass of 135 MeV and its most likely decay, after a mean lifetime of only 8×10^{-17} s, is into two γ-rays.

Inverse Compton (IC) emission (green curve) is familiar from Section 10.6. In this case, relativistic CR electrons (or positrons) scatter off of stellar photons of the *interstellar radiation field* which are then bumped up to γ-ray energies. The re-emission of absorbed

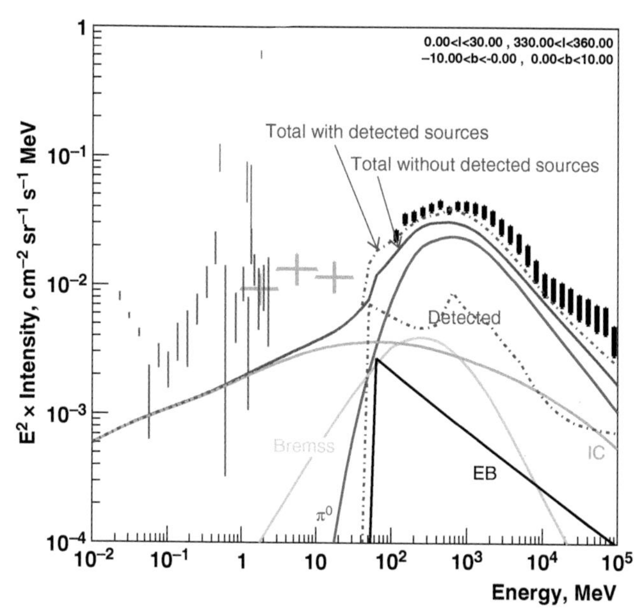

Figure 12.1 The γ-ray spectrum of the inner Galaxy showing data from a variety of γ-ray instruments. The data are shown along the top of the curves and are represented as blue, magenta and thick black vertical bars, as well as large green 'plus' signs. The 'intensity' (number of photons per unit time per unit area per unit solid angle per unit interval of energy) has been multiplied by E^2 on the ordinate axis. KEY: π^0 (red curve) is neutral pion decay, **IC** (green curve) is inverse Compton emission, **EB** (black curve) is the isotropic extragalactic background, **Bremss** (cyan curve) means relativistic Bremsstrahlung radiation, **detected** (lower dash-dotted magenta curve) refers to specific sources that have been detected, and **totals** with (upper dash-dotted magenta curve) and without (blue curve) the detected sources are also shown. A narrow peak at 0.511 MeV is from the γ-rays that are produced when electrons and positrons annihilate. Source: Strong [24]. © 2010 World Scientific Publishing Co Pte Ltd (see also [8]).

starlight by dust also contributes [18]. Almost all emission from the low energy end of the plot to about 30 MeV is dominated by γ-rays that have resulted from IC interactions.

An important component is Bremsstrahlung emission (cyan curve) from relativistic CR electrons (or positrons) scattering from interstellar gas (primarily hydrogen). This is called *relativistic Bremsstrahlung* and the scattering can be off of protons or other nuclei that still have bound electrons. The spectrum of relativistic Bremsstrahlung emission is not like the thermal Bremsstrahlung spectrum discussed in Section 10.2 due to the relativistic nature of the interaction and also because the CR electrons do *not* follow a Maxwell–Boltzmann velocity distribution which is a requirement for a gas to be labelled *thermal*.

There remains a contribution from detected sources (lower magenta dash-dotted curve), a narrow peak at 0.511 MeV due to electron–positron annihilation, and the isotropic extragalactic background (EB, solid black curve). The background is thought to be a superposition of many different unresolved sources such as active galactic nuclei, starburst galaxies, γ-ray bursts, or possibly other more esoteric sources [1][2]. Notice the close link between these high energy *photons* and the high energy CR *particles* that were first introduced in Chapter 1.

In a successful model, the sum of all emission from individual γ-ray mechanisms should add up to the data points that are shown as vertical bars or plus signs in Figure 12.1. The model representation is called 'total'. As can be seen, the model is quite good from about 30–10^5 MeV. However, at the lower energies, the total model falls short of the data, even considering the large error bars. Modifications to the displayed model can give some improvement at these lower energies, for example, if secondary electrons are included[3]. However, there is still a possibility that many compact sources are present that are currently unresolved and have not been included in the models. Examples are anomalous X-ray pulsars (pulsar with extremely powerful magnetic fields) and radio pulsars that have spectra that extend to a few hundred keV (e.g. [18]).

Gamma-ray telescopes must be located in space (Figure 4.7) and, of all wavebands, this is one of the newest on the astronomical scene. One of the problems plaguing γ-ray instruments has been their low spatial resolution which makes it difficult to pinpoint individual sources. This problem will become less important as instruments improve.

To illustrate how the modelling process works, we need to choose a spectral band in which there are fewer emission processes at work than we have just seen in the γ-ray band of Figure 12.1. A good choice is the radio spectrum, for which there are only two main contributors to the continuum in normal galaxies: thermal Bremsstrahlung emission and nonthermal synchrotron emission. Figure 12.2 presents a calculated radio spectrum, typical of a normal spiral galaxy, which shows the total emission as well as both the thermal and nonthermal components that contribute to the total. Throughout most of the plot, the nonthermal component dominates but, at high radio frequencies, the thermal component dominates. While this radio spectrum is typical for a normal galaxy (Figure 4.3), the specific fraction of emission contributed by thermal and nonthermal components can vary between galaxies and the value of α_{NT}, which is related to the cosmic ray energy spectrum (Eq. 10.44), can also differ between galaxies. How, then, can we determine the thermal/nonthermal contributions and extract α_{NT} from the observations? Example 12.1 suggests an approach.

---

[2] These could include hypothesized sources such as possible signatures of large scale structure formation, interactions of very high energy CRs with 'relic' photons, the decay of dark matter, etc.
[3] Secondary electrons are produced when primary CRs interact with ISM nuclei (cf. Section 1.2).

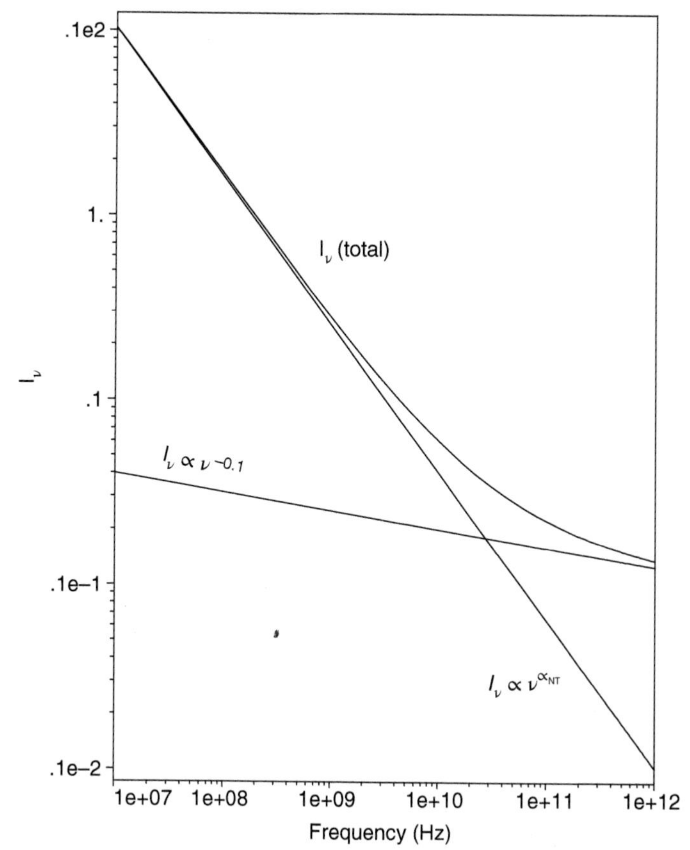

Figure 12.2 A model radio spectrum, showing the total observed emission (I_ν[total] $\propto \nu^\alpha$), and its two contributors, a thermal Bremsstrahlung component ($I_\nu \propto \nu^{-0.1}$) and a nonthermal synchrotron component ($I_\nu \propto \nu^{\alpha_{NT}}$). The observed spectral index, α, varies with frequency.

Example 12.1

The total continuum emission from a galaxy at radio wavelengths consists only of thermal Bremsstrahlung radiation whose spectrum can be described by $I_\nu(\mathrm{T}) = a\nu^{-0.1}$ (Eq. 10.15) and nonthermal synchrotron radiation whose spectrum is $I_\nu(\mathrm{NT}) = b\,\nu^{\alpha_{NT}}$ (Eq. 10.43, where we have now included the subscript, NT, on the spectral index, α). Both emission processes are optically thin. Describe a way to determine the constants, a, b, and α_{NT}, when only the total emission is measured.

The total emission can be represented by the sum of the thermal and nonthermal components (Figure 12.2),

$$I_\nu(\mathrm{total}) = I_\nu(\mathrm{T}) + I_\nu(\mathrm{NT}) = a\,\nu^{-0.1} + b\,\nu^{\alpha_{NT}} \quad (12.1)$$

There are three unknowns, *a*, *b*, and α_{NT}, and therefore, three equations are required. These equations can be set up by measuring I_{ν_i} (*total*) at three different frequencies, ν_i, and writing Eq. (12.1) for each frequency. The set of three equations can then be solved for *a*, *b*, and α_{NT}. Note that a similar equation could be written for the flux density (Problem 12.2) since $f_\nu = I_\nu \Omega$ (Eq. 3.13) and we take Ω to be constant with frequency. Best results can be obtained if the frequencies are spread out so that the contribution from the admixtures of thermal and nonthermal emission shows significant variation.

The constant, *a*, is related to the physical quantities associated with thermal Bremsstrahlung emission via Eqs. (10.14) and (10.15) so, depending on what other information is available, it may then be possible to extract information about the thermal part of the source. Similarly, *b* and α_{NT} are related to the nonthermal parameters by Eqs. (10.37) and (10.43).

A warning: it is possible for, α_{NT}, itself to vary with frequency! For example, higher energy relativistic electrons lose energy more quickly in a magnetic field than lower-energy electrons. This has the effect of *steepening* the nonthermal spectral index as ν increases. The only solution is to increase the number of frequencies at which the measurements are made.

With the help of computer software, this concept of modelling can be extended to 'build' spectra of nebulae, galaxies, or whatever other system is of interest. For example, there are many stars within any given resolution element in a typical optical observation of a galaxy. Extracting knowledge of the admixture of different kinds of stars requires modelling the total spectrum by including different numbers of stars of various spectral types (e.g. Problem 12.3) until a good match results. This approach is called *population synthesis* and provides important information about the galaxy. For example, if the spectrum can be modelled without including any hot massive stars, then the galaxy is not undergoing any significant star formation at the present time (Section 5.3.4). On the other hand, if many hot massive stars are required to describe the spectrum, then the galaxy is an active star-forming system. Extending this modelling concept to include HII regions, dust, or other contributors to the spectrum is also possible and is currently actively pursued by researchers interested in knowing the makeup of galaxies.

As a final comment, we note that the process of model-building is not restricted to spectra. The *spatial distribution* of light can also be modelled, something that is routinely carried out over a variety of scales, from planetary systems, to large scale structure in the Universe. Modelling spectra, however, is a particularly productive approach. This is because one can imagine an enormous number of possible spatial distributions for a variety of sources. On the other hand, *there are only a limited number of known emission mechanisms and corresponding spectral behaviour.* As we have shown in this text so far, once we have an understanding of these processes, it becomes possible to understand the vast majority of systems that present themselves in astronomy. And this is what we will do next – focus on some case studies.

12.2 CASE STUDIES – THE ACTIVE, THE YOUNG, AND THE OLD

How can we build links between objects and events in astrophysics and what scenarios can be pieced together? A good approach is to focus on specific objects or regions and apply the astrophysical principles that we have already seen. To this end, we now present *case studies* of three different sources.

12.2.1 Case Study 1: The Galactic Centre (the Active)

Eight kiloparsecs from the Sun in the direction of the constellation, Sagittarius (the Archer) is the location of the Galactic Centre (GC, see also Figure 5.3). Since the Sun is close to the midplane, our view of the GC is right through the Milky Way disc. This means that, at optical wavelengths, the GC is obscured by many magnitudes of extinction from dust. However, at infrared, radio, hard X-ray, and γ-ray wavelengths, which are not grossly affected by dust, it is possible to catch a glimpse of the heart of our Galaxy. 'Galactic Centre' is loosely used to refer to the very nucleus of the Milky Way as well as the vicinity around it, of order a few hundred parsec, within which the properties differ markedly from the rest of the Galaxy.

Numerous observations of the GC have revealed that this is a unique, active region. Here, we find X-ray and IR flaring, evidence for outflows, strong organized magnetic fields, high stellar densities, high gas pressures and temperatures, and high stellar and gaseous velocity dispersions. Given that the density of the Galaxy, as a whole, increases towards the centre, it is not surprising that the properties of the GC are more extreme than elsewhere in the Galaxy. There are unique features, however, that are seen nowhere else, as we will soon describe.

Figure 12.3 shows two views at radio wavelengths, 1.28 GHz at the top using the Meerkat radio interferometer in South Africa, and 8.6 GHz (λ 3.5 cm) at the bottom using the Green Bank single-dish radio telescope in the USA. In general, the top image appears like a sharpened version of the bottom, mostly because of the improvement in spatial resolution in going from a single dish to an interferometer (Section 4.2.5). The total radio emission is shown, including both nonthermal (synchrotron) and thermal (Bremsstrahlung) components but, because the wavelengths are different, the image samples different admixtures of those two components, as Figure 12.2 indicates. In addition, different sources in the field may have different intrinsic fractions of thermal and nonthermal flux. It is possible, for example, to distinguish between an obvious HII region (thermal emission) and the many supernova remnants (nonthermal) in these fields by looking at the relative brightness of the various sources as the frequency changes (Problem 12.5).

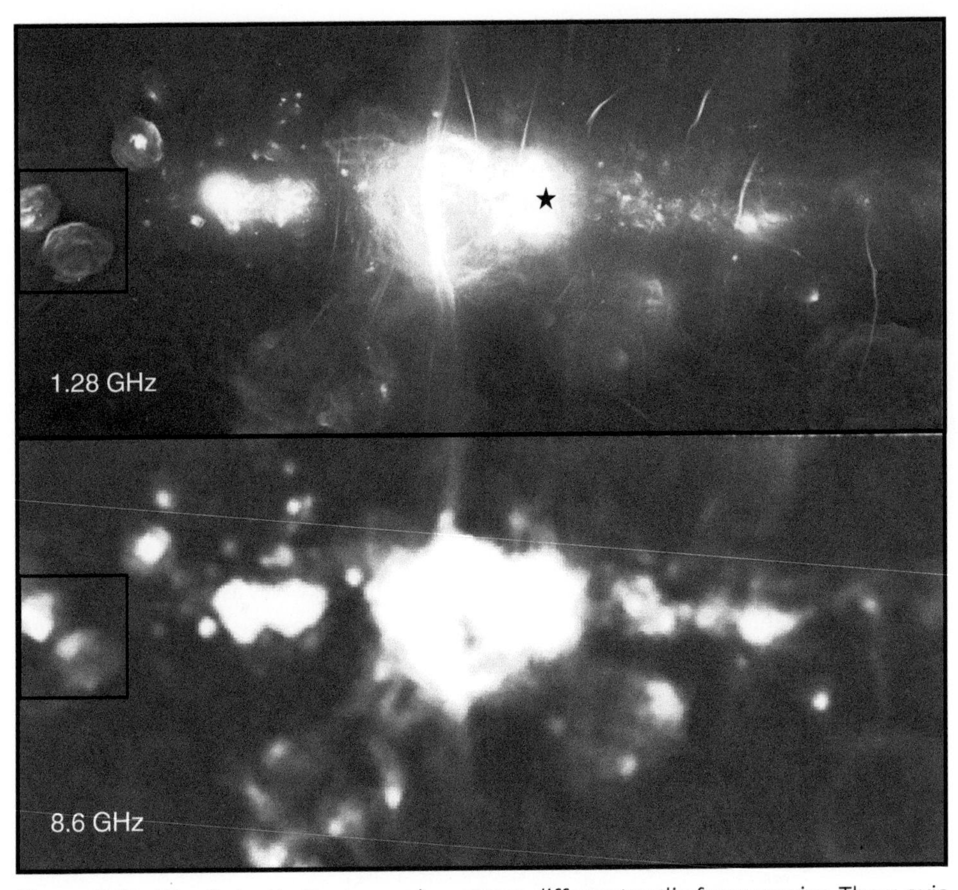

Figure 12.3 The Galactic Centre region at two different radio frequencies. The x-axis spans approximately two degrees (~280 pc) with the approximate location of the nucleus, containing a supermassive black hole, marked with a black star in the top picture. Many supernova remnants are visible as well as linear 'threads' which are aligned by magnetic fields. Two distinct objects are boxed. *Top:* Exquisite inaugural 1.28 GHz image from the 64-antenna MeerKAT radio telescope in South Africa. MeerKAT is a precursor to the international SKA radio telescope (Section 10.5.1). The spatial resolution is 6 arcsec [10]. Source: South African Radio Astronomy Observatory (SARAO). *Bottom:* Green Bank Telescope image of the same region at 8.6 GHz. Source: Reproduced using data from [14] available at the Harvard Dataverse (https://dataverse.harvard.edu), courtesy of C. J. Law. The spatial resolution is 1.5 arcmin.

The GC is a rich region containing many sources. Some of these sources are unique to the GC, such as magnetically aligned *threads* visible as thin vertical lines in Figure 12.3. In addition, however, there is now abundant evidence for nuclear *outflows*.

Figure 12.4 shows a remarkable example. Here, we see the *Fermi Bubbles* discovered using the Fermi γ-ray space telescope that was shown in Figure 4.7. These are

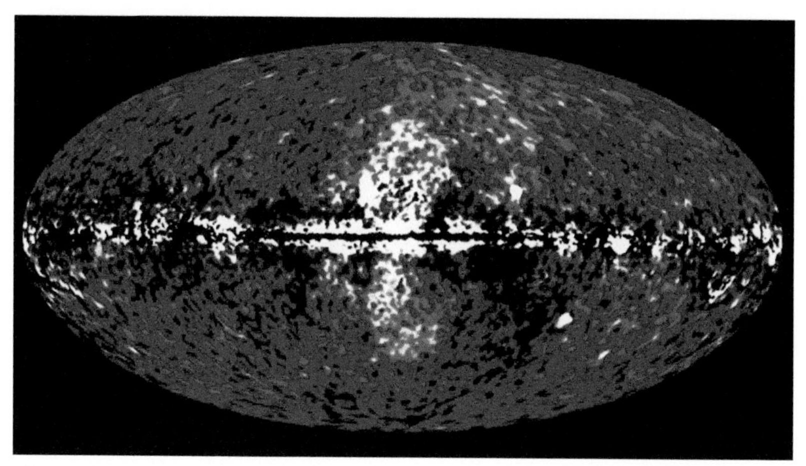

Figure 12.4 *Fermi Bubbles* are seen as a bipolar structure in this processed, false-colour γ-ray image which includes data from 1 to 10 GeV. The plane of the Galaxy runs horizontally in this Aitoff projection, and the Galactic Centre (GC) is right at the centre of the image. The bubbles are seen as white and red lobes extending 8–10 kpc above and below the GC. Source: NASA/DOE/Fermi LAT/D. Finkbeiner et al. See also [25].

bipolar-shaped bubbles that extend 8–10 kpc above and below the plane centred at the GC (e.g. [2]). The Fermi Bubbles trace outflows at speeds of 1000–1300 km s^{-1} and have an estimated age of $\sim$6 to 9 Myr from velocity measurements [4], though an age of 5–6 Myr has been estimated from numerical modelling [27][4]. They have been observed over γ-ray energies from 1 to 1000 GeV [9] but have also been observed in Hα emission [13]. Such distinct γ-ray features have been seen nowhere else.

Evidence for nuclear outflow extends to many other wavebands. Recent smaller radio lobes have been observed to extend to a height of 430 pc and are thought to be a few million years in age [10]. X-ray lobes and loops of molecular gas [5] centred on the GC have also been observed. It is likely that there has been episodic outflow activity at our Galactic Centre (e.g. [10]).

What is the source of this activity? Put together, the collective observations imply that the GC harbours an active galactic nucleus (AGN), similar to (but weaker than) those observed in other galaxies and discussed at various times throughout this text (e.g. Figure 7.5 or Figure 10.17). It is not yet clear to what extent the collective effects of star-forming activity (e.g. supernovae and stellar winds) could also contribute, but the AGN is likely a requirement.

The AGN paradigm requires the presence of a 'driver' for the high energies that are associated with it. In our Milky Way – that driver is located at a position called Sgr A*[5]

[4] A wider range of ages have also been suggested.
[5] Surprisingly Sgr A* itself is currently actually underluminous for an AGN of its mass.

marked with a black star in Figure 12.3 (Top). This is the closest AGN to us and its proximity allows us to study it with a linear resolution that surpasses anything possible for external galaxies.

At the heart of Sgr, A* is the true central engine: a *supermassive black hole*. The black hole itself does not emit light, but the strong gravitational field around it provides the energies required to drive the other observed phenomena. For the Milky Way AGN, a black hole mass of about 4 million solar masses is implied. Example 12.2 outlines the case.

Example 12.2

Figure 12.5 shows, on the left-hand side, a λ 2 μm (near-IR) view of a star field in the GC that is only 2 arcsec on a side. For a distance to the GC of $D = 8$ kpc, the corresponding linear scale is $d = 0.078$ pc (Eq. I.9). Almost all of the stars in this image belong to a dense stellar cluster that is near the Galactic nucleus, because the probability of finding a disc star (one that is closer to us along the line of sight) in such a small field of view is small (Problem 12.4). A star, S2, is labelled as well as the position of Sgr A*, the two blended together.

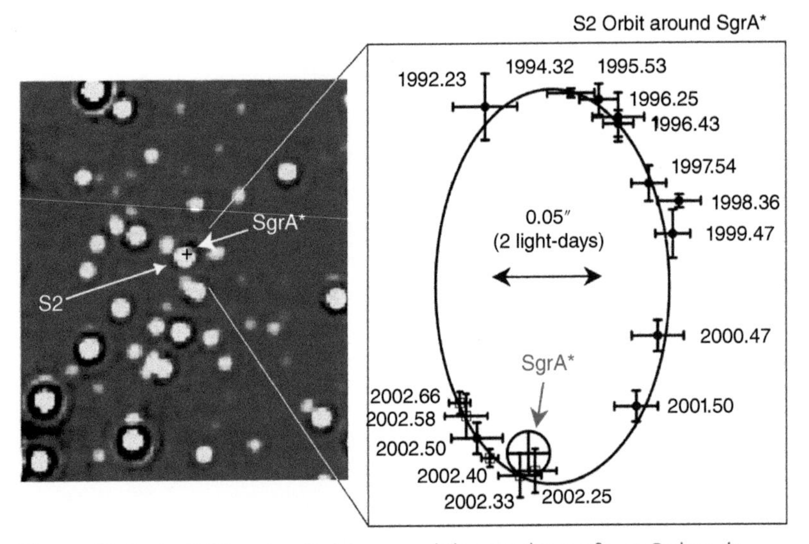

Figure 12.5 Left: The star field around the nucleus of our Galaxy is shown in this 2.1 μm (near-IR) image over a field of view, 2 arcsec on a side with an angular resolution of 0.06 arcsec. The nucleus, located at the position of Sgr A* and a star called S2, whose light is blended with Sgr A*, are both marked. Right: The elliptical orbit of S2 about Sgr A* (as seen on the sky) is shown, after having been observed over the course of 10 years. Notice that the speed of the orbit is greatest at closest approach. Source: From Schödel et al. [23]. European Southern Observatory.

The proper motion (Eq. 9.5) of S2 has been measured and its 10-year orbit about Sgr A* is shown in the right-hand image. Notice that the velocity of S2 is highest at closest approach to Sgr A*, consistent with Kepler's 2nd Law[6] and indicating that the interior mass is concentrated at the position of Sgr A*. The true orbit, which has some inclination with respect to the plane of the sky, is an ellipse with Sgr A* at one focus, in accordance with Kepler's 1st Law (see Appendix A.5 of the *online material* for the properties of an ellipse). The apparent (observed) orbit is also an ellipse, but Sgr A* is no longer at the focus of the apparent orbit because of the projection onto the plane of the sky. This fact can be used, with some geometry, to determine the true orbit. For S2, it is found that the inclination is $i = 46°$, the eccentricity is $e = 0.87$, the semi-major axis is $a = 5.5$ light days (1.42×10^{16} cm), and the orbital period is $\mathcal{T} = 15.2$ year.

The values of a and $\mathcal{T}$ determined from the orbit, together with Kepler's 3rd Law (Eq. 9.6), yield a mass interior to the orbit of, $M = (3.7 \pm 1.5) \times 10^6\ M_\odot$. The separation of S2 from the nucleus at closest approach is only 17 light-hours (124 AU). The only known object that could contain this much mass in such a small volume is a supermassive black hole. A stellar cluster or any other conceivable type of object would require a much larger volume. Moreover, any other type of object should be shining, and we should have seen it.

Since the time of this definitive observation, additional constraints have been obtained. From Doppler shifts of spectral absorption lines in the orbiting star, the star's radial velocity, v_r, has been determined and, together with the measured transverse motion on the sky, v_t, this has allowed a determination of its true space velocity (Figure 9.1). In addition, orbits for many more stars and 25 years of monitoring have now been determined. The result has been a more precise black hole mass of $M_{bh} = (4.28 \pm 0.10) \times 10^6\ M_\odot$ [6][7]. In addition, an X-ray flare has been observed from Sgr A* which showed some variability on a timescale of only 200 seconds [17]. This places a limit on the size of the black hole of $d < 0.4$ AU, via Eq. (9.25). Thus, 4 million Solar masses are within a region that is smaller than the orbit of Mercury! The conclusion that the Milky Way harbours a supermassive black hole at its centre seems inescapable.

12.2.2 Case Study 2. The Cygnus Star-Forming Complex (the Young)

In the constellation of Cygnus (the Swan), within the disc of the Milky Way and about 1.8 kpc distant, is the *Cygnus star-forming complex*. A picture of this region, from combined radio and far-infrared images, is shown in Figure 12.6. The enormous angular size of this region is truly impressive. The diameter of the field of view covers fully 10° of sky, about 20 times the diameter of the full moon. If our eyes were sensitive to the wavebands

[6] Kepler's laws were stated in Section 9.1.1.2.

[7] An adjusted distance to the Galactic Centre of 8.3 ± 0.1 kpc was also determined.

Figure 12.6 A complex star-forming region in our Milky Way in the direction of the constellation, Cygnus. The image diameter spans 10° of the sky. In this image, the colour, rose, represents emission at λ 74 cm, green is λ 21 cm emission, turquoise is λ 60 μm emission, and blue is λ 25 μm emission, all continuum emission. Two supernova remnants are labelled. Source: Composed for the Canadian Galactic Plane Survey by Jayanne English (CGPS/U. Manitoba) with the support of A. R. Taylor (CGPS/U. Calgary)

shown, it is clear that the sky would look nothing like the star-filled firmament that we see each night.

The Cygnus star-forming complex is a vast stellar nursery in which stars are born from dense molecular gas, live their lives, and die, recycling their chemically enriched material back into the interstellar medium. In the process, winds from hot massive stars, radiation, and explosive energy from supernovae compress, heat, ionize, and disrupt the surrounding gas clouds. For a star-forming region in the disc of the Milky Way, this is one of the most active. How can such a picture about this region emerge? For this, we need to draw on a number of resources, including the information contained in Figure 12.6 itself, data from other wavebands, and a knowledge of stellar evolution that has been painstakingly pieced together from both observation and theory (Section 5.3). We begin with a discussion of the figure itself, given in Example 12.3.

Example 12.3

In Figure 12.6, four different wavelengths of *continuum* emission are represented in *false colour*, the radio wavelengths, λ74 cm ($\nu = 408$ MHz) in *rose/red* and λ 21 cm ($\nu = 1420$ MHz) in *green*, and the far-infrared wavelengths, λ 60 μm in *turquoise* and λ 25 μm in *blue*.

Careful quantitative analysis of the individual images that contribute to Figure 12.6 is needed to properly extract the information that it contains. Nevertheless, it is possible to piece together some information, with just a careful visual examination. The mix of colours of the various sources, for example, provides us with clues about the emission processes, and therefore about the most likely sources.

Any object that is reddish or yellow-red, depending on other emission in the vicinity, must be *strongest* at the longest radio wavelength, λ 74 cm. Otherwise, it would appear to be a different colour. Since optically thin synchrotron emission becomes stronger at longer wavelengths, this suggests that the reddish sources are steep-spectrum, optically thin synchrotron-emitting sources. Two such circular features are labelled in the figure. The source, SNR γ Cygni, so named because it lies near the star, γ Cygni, is 1° in diameter. As the name indicates, this is a supernova remnant, which we could infer from its nonthermal radio spectrum and bubble-like shape. A second labelled SNR, G84.2–0.8, is also quite distinct and shows the *edge-brightened* morphology expected from a shell of material. At the very top of the picture, we see part of another edge-brightened shell which appears redder than the other two because it is more isolated from the other emission.

Throughout the image are also many reddish point sources. These may look like stars, but they are clearly not, since their reddish colour indicates that they emit most strongly at the longest radio wavelength, again consistent with a synchrotron spectrum. Stellar continuum spectra, by contrast, are Planck curves that peak from near-IR, optical, and UV wavelengths (e.g. Table T.5 and Eq. 6.8) and stellar emission would be negligible at all wavelengths shown in the figure. The red point sources are more broadly distributed in the image (more obvious in a wider field of view), rather than clustered in the Cygnus star-forming complex, and they are unresolved. This information suggests that these are background sources, likely quasars, in the distant Universe.

As the emission shifts to green (λ 21 cm *continuum*, not the line) through blue (λ 25 μm), more blending is seen between colours and sources. The green emission is still in the radio part of the spectrum and because green, rather than rose/red is seen, this colour emphasizes the contribution from flat-spectrum thermal Bremsstrahlung emission (Figure 12.2). This means that green is selecting ionized gas (HII). In fact, ionized gas is abundant in this complex, both as diffuse emission throughout the region as well as discrete cocoon-like 'knots' in the map. The existence of HII regions requires the presence of hot O and B stars (Section 7.2.2) and also means that there is sufficient interstellar gas available to be ionized.

Finally, the emission from the far-infrared wavelengths, shown in turquoise and blue, permeates the region. At these wavelengths, we are seeing collective Planck curves from dust. At λ 60 and λ 25 μm, the characteristic temperatures (Eq. 6.8) are

T = 48 K and T = 116 K, respectively, both far too cool to represent photospheric emission from stars. Some very blue 'knots' of emission are also seen, possibly very small dusty regions around forming stars.

Thus, in this image alone, we see supernova remnants, abundant interstellar gas and dust, and HII regions. We can also infer the presence of hot, O and B stars, although they are not seen directly. From our knowledge of stellar evolution, we know that HII regions, O and B stars and supernovae are all tracers of recent star formation (Section 5.3.4) because the lifetimes of O and B stars are so short (of order 10^6 years). They should also, therefore, still be associated with the molecular material from which they formed. Moreover, since dust and molecular gas are correlated spatially in the Milky Way (e.g. Figure 5.21), this also suggests that molecular gas should be present throughout this region.

The rudimentary treatment of Example 12.3 should never take the place of a proper quantitative analysis. Yet, with some careful scrutiny and an understanding as to which emission processes dominate in various wavebands, it is sometimes possible to extract useful information, as outlined in the example. Are there any *other* data, then, that support the conclusions of Example 12.3?

The γ Cygni SNR has been detected at X-ray and γ-ray wavelengths [26], confirming the high energy nature of this source. Its expansion velocity has also been measured to be $v_{\text{exp}} = 900 \text{ km s}^{-1}$ and the result of the expansion analysis indicates that the supernova exploded some 5000 years ago. This is a very recent event, by astronomical standards.

The molecular gas, inferred to be present in the example, has been directly observed and is extensive throughout the region in a *giant molecular cloud (GMC) complex*. From CO observations (Section 11.5.1), the estimated total molecular gas mass is $M_{\text{GMC}} = 5 \times 10^6 \text{ M}_{\odot}$ [22], making it one of the most massive molecular cloud complexes known in our Galaxy.

Finally, the hot O and B stars themselves, assumed to be responsible for the thermal radio emission, have also been directly observed. Several star clusters containing O and B stars, called *OB associations* are embedded in the cloud. One of them, the Cygnus OB2 association, is one of the most massive and richest clusters in the Galaxy, containing 2600 cluster member stars of which about 100 are O or B stars [11]. The total mass of this cluster alone is $4\text{–}10 \times 10^4 \text{ M}_{\odot}$.

There is little doubt that this complex plays host to a rich array of processes, of which star formation and the evolution of massive stars are key elements. While the details of star formation are not completely understood, complexes like the Cygnus cloud provide astrophysical 'laboratories' that may reveal the secrets of star formation in the future.

12.2.3 Case Study 3: The Globular Cluster, NGC 6397 (the Old)

In the southern hemisphere constellation of Ara (the Altar), located out of the disc of our Galaxy is NGC 6397, one of the closest globular clusters to us ($D = 2.6$ kpc). The globular clusters are spherical in appearance, typically contain a few hundred thousand stars each, and have a typical absolute magnitude of $M_V = -7.3$. There are about 150 known that dot the halo region and bulge of our Milky Way (Figure 5.3). The stellar densities are very high. For example, half of the mass of NGC 6397 (of order, 10^5 stars) is within a diameter of only about 3.5 pc.

These clusters are believed to be the oldest stellar systems in the Universe, and NGC 6397 is one of the oldest, at an age of 13.4 Gyr [7]. This is only marginally less than the age of the Universe (13.8 Gyr, Section 5.1), and the two values are in agreement if typical error bars of 0.6 Gyr for stellar ages are taken into account (e.g. [20])! Thus, these systems must have formed in the very early Universe and would have been among the first types of objects that took part in the assembly of massive galaxies like the Milky Way. They provide the means of dating the Milky Way itself, whose age (~13.6 Gyr old) is taken to be the age of the oldest clusters plus some allowance of time for the clusters to form. What are the characteristics of this system that lead to these conclusions? We begin, in Example 12.4, with a discussion of the colour-magnitude (CM) diagram of NGC 6397, shown in Figure 12.7.

Example 12.4

It is instructive to compare the CM diagram of the globular cluster, NGC 6397 (Figure 12.7 Left), with the CM diagram that was obtained by the Gaia satellite (Figure 3.15). The Gaia data are dominated by stars that are in the gaseous dusty disc of the Milky Way. If the mix of stars in the globular cluster were the same as in the disc, there should a close resemblance between the figures[8]. What we find, however, is that the two CM diagrams differ significantly from each other.

First of all, the globular cluster CM diagram lacks any trace of hot, luminous massive stars which we know, from their short lifetimes, to be young. All stars on the main sequence occur below a *turn-off point* which is the location that the main sequence turns off into the red giant branch (Section 5.3.2). If any recent star formation had been occurring in NGC 6397, then there should be evidence of hot, massive stars in this diagram, and they should have been be easily detected, being more luminous than the other detected stars on the lower main sequence. From a knowledge of stellar evolution, we know how long a star can remain on the main sequence before it exhausts its fuel and turns off to the red giant branch. This fact can be used to determine how long it has been since stars formed in the cluster. Model colour-magnitude diagrams for a range of ages have been constructed from theory, and they are then compared to the data, the best-fit result giving the cluster age[9]. This is called *cluster fitting*.

Secondly, the turn-off point is quite well defined and narrow, unlike the broad turn-off region that Gaia mapped in Figure 3.15. For NGC 6397, the turn-off corresponds to a stellar temperature of $T_{eff} = 6254$ K [12]. If the stars that formed long ago had formed continuously over some period of time, then different stars of different temperatures would now be turning off the main sequence, thickening the turn-off

[8] Note that the filter bands of the two figures differ, but this does not change the argument.
[9] Metallicity also affects the turnoff point so knowledge of the metallicity, if available, is important for this process to give the most accurate results.

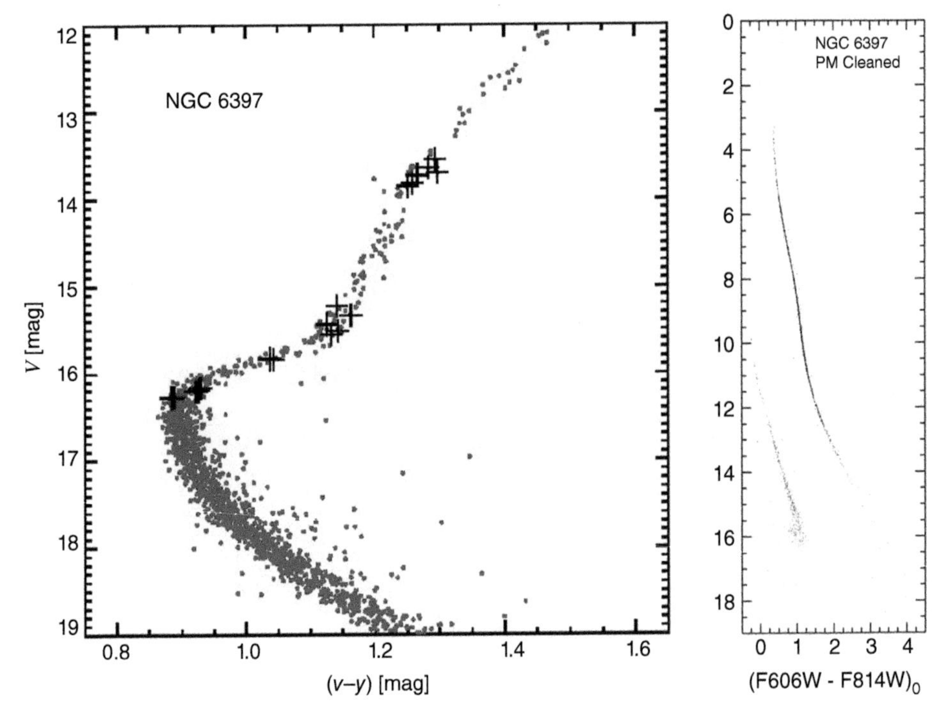

Figure 12.7 Colour-magnitude diagrams of the globular cluster, NGC 6397. Left: CM diagram obtained via with the Danish 1.54 m telescope in La Silla, Chile. Crosses mark specific stars that were targeted for spectroscopy in the study of [12]. Apparent V magnitudes are shown and typical temperatures along the red giant branch are ~5130 K. Source: Korn et al. [12]. © 2006 The Messenger. Right: CM diagram from [21] showing the white dwarf sequence below the main sequence. The ordinate axis plots absolute magnitudes from the F606W filter which is similar to V, although accurate work would require careful transformations (e.g. [3]). The faintest apparent magnitude shown on the Left plot (V = 19) corresponds to an absolute magnitude of 6.4. Therefore, the right plot is shown to a level that is almost 10 magnitudes fainter than the left plot! Source: Richer et al. [21]. © 2013 American Astronomical Society.

point. A sharp turn-off point, then, indicates that all the stars in the cluster were formed in a relatively short period of time. In other words, the cluster itself has a unique age.

The one close similarity, however, is the white dwarf curve. Both the globular cluster (Figure 12.7 Right) as well as the Gaia (Figure 3.15) CM diagrams clearly show this white dwarf sequence. Apparently, the end points of stellar evolution for low-mass stars are similar both in and out of the Milky Way disc.

Other evidence, besides the cluster fitting technique described in Example 12.4, also points to the conclusion that globular clusters are old systems in which star formation ceased long ago. There is currently no atomic or molecular gas in globular clusters and,

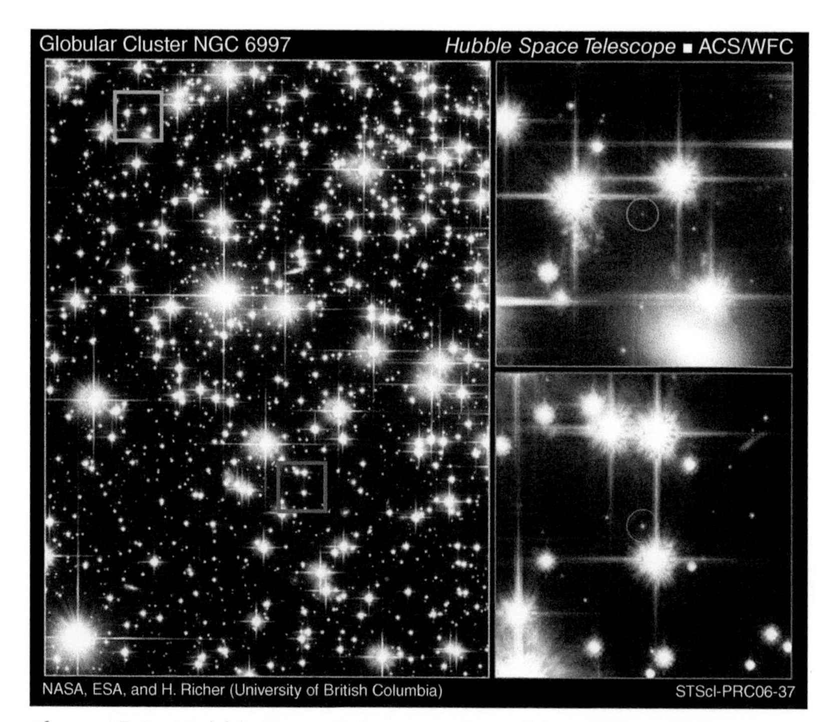

Figure 12.8 Hubble Space Telescope view of the centre of the globular cluster, NGC 6397, shown with exquisite detail. This image reaches the faintest magnitudes ever observed for low mass stars and white dwarfs. Two small squares in the left picture mark fields that are blown up at the right. In the image at the top right, a faint white dwarf star is circled. This white dwarf is at 28th magnitude and is blueish in colour, indicating its hot surface temperature. In the lower right image, a faint red dwarf star, at 26th magnitude, is circled. Close scrutiny reveals faint background galaxies that are seen through the cluster. Source: Reproduced by permission of NASA, ESA, and H. Richer (University of British Columbia)

therefore, no supply of material from which stars could form. Images of NGC 6397 (e.g. Figure 12.8) reveal only a population of old stars and stellar remnants. White dwarfs, for example, are the relics of stars that have long since left the main sequence and come to the end of their nuclear burning lifetimes (Section 5.3.2). A significant population of white dwarfs has been detected in NGC 6397 to an extremely faint magnitude limit, as the lower curve in Figure 12.8 (Right) shows. Because the luminosity of a white dwarf declines with time as it cools, the location of the white dwarf *cooling curve* on the CM diagram also allows the cluster to be dated, with consistent results [21].

Other creative dating techniques have also been applied. For example, observations of the element, beryllium, in cluster members are useful. Beryllium is neither produced in the Big Bang, nor by nucleosynthesis in stars in any significant quantity. The source of its production is by spallation from cosmic rays (Section 1.1.1). Over a period of time, the quantity of beryllium is expected to increase in any gas from which the cluster originally formed. Therefore, a measure of the beryllium abundance in the cluster provides a way of timing when the cluster formed, with some assumptions related to calibration. Beryllium absorption lines in the atmospheres of stars in NGC 6397 have now been measured, resulting in an age estimate that is in agreement with the other values [15].

Another important clue as to the age of a globular cluster is its very low metallicity (Section 5.3.1). For NGC 6397, for example, the metallicity is only 1/100th of the value observed in the Sun. We know that successive generations of star formation will increase the metallicity over time, so this implies that globular clusters formed early and that many successive generations of stars have not occurred within them. As pointed out in Section 5.3.4, however, it is important to note that the metallicity is not zero. This means that some metal enrichment took place even before the formation of the globular clusters. A population of massive stars that rapidly exploded as supernovae could have accomplished this, but the exact nature of this population is unknown and has not yet been observed[10].

All indicators are that there was an early rapid burst of star formation at which time the globular clusters formed, and star formation then ceased within these systems. Current data suggest that the globular cluster formation epoch began within about 1.7 Gyr of the Big Bang, lasted for 2.6 Gyr, and ended 10.8 ± 1.4 Gyr ago [7]. The spatial distributions and velocities of globular clusters in the spherical halo of the Milky Way agree with this scenario, pointing to an early formation period in low metallicity gas, before material settled into the spiral disc in which we now *do* see active star formation.

It is clear that globular clusters contain information whose implications go far beyond the boundaries of the cluster, itself. These systems provide clues as to the conditions that were present in the very early Universe at the time that our Milky Way was just forming. NGC 6397 has, in addition, proven to be a laboratory for the study of the lowest mass stars. The faint end of the stellar main sequence in NGC 6397 has been completely detected, showing that the main sequence has a low luminosity cut-off (Figure 12.7 Right). From these observations, a low mass cut-off of $0.083\ M_\odot$ has been determined [19]. This is in beautiful agreement with the theoretical low mass limit of $M_{*min} = 0.08\ M_\odot$ required for the nuclear burning discussed in Section 5.3.2. It also hints that there may be a population of substellar brown dwarfs in which nuclear burning has not, and never will, occur.

12.3 THE MESSENGER AND THE MESSAGE

The 'signals' that we have tracked over the scope of this text have not been imbued with meaning, though the meaning may indeed be there for those who seek its poetry. What *has* been assumed, implicitly, is that it is actually possible to measure a signal – a message – and discern something concrete about the objects that have emitted it or altered it along the way. For this, we require our detection instruments and a knowledge

[10] These are referred to as 'Population III' stars.

of the underlying foundations of the physical processes that are involved. More than that, our theoretical models form a framework for 'telling a story', as the prologue to this chapter implies. If not perfect, at least these stories provide as close a description as possible of our physical world.

The achievements and power of this approach are impressive. The spins of some electrons in a distant galaxy change their orientation, and we learn that the galaxy's mass is one hundred billion times that of the Sun. Electrons alter their course slightly while passing by nuclei, and we find that the hot gas in a cluster of galaxies far outweighs the mass of the visible stars. Atoms in the atmospheres of stars show absorption lines that are shifted slightly in frequency, and we conclude that a supermassive black hole lurks at the core of the Milky Way. Slight warps in space–time perturb photons travelling through a cluster of galaxies implying that dark matter dominates all other material. And glimmers of light from the dawn of time hint of the origin and evolution of the Universe on vast cosmological scales.

Perhaps in the future, such a world view may be superseded by something even more perfect and profound. For now, the signals that link our detectors to the cosmos continue to whisper their secrets, as the prologue to Chapter 1 suggests, 'both subtle and gross'.

PROBLEMS

12.1. The infrared peak in the spectrum of NGC 2903 is due to the collective contributions of many different types of grains over a variety of temperatures. Figure 4.3, however, suggests that the bulk of the grains might be described by one 'characteristic' temperature. Estimate this temperature from the figure.

12.2. At the three frequencies, 4850, 2700, and 1465 MHz, the radio continuum flux densities of a galaxy are, respectively, $f_{4850} = 3.94$ Jy, $f_{2700} = 5.37$ Jy, and $f_{1465} = 7.69$ Jy.

(a) Write three equations in three unknowns that show the explicit contributions of thermal Bremsstrahlung and nonthermal synchrotron radiation to the total emission (see Example 12.1, and note that cgs units are not necessary).

(b) With the help of computer algebra software, solve for the three unknown constants. (Hint: Include the restriction, $\alpha_{NT} < -0.15$.) Specify the units of each of the constants.

(c) What fraction of the total emission is due to non-thermal emission at $\nu = 4850$ MHz and at $\nu = 1465$ MHz?

12.3. With the help of a spreadsheet or other computer software, calculate the shape of the total spectrum of a small galaxy that consists only of three different types of stars: 10^6 M5V stars, 5000 A0V stars, and 20 K0Ib stars (see Tables T.5, T.7). We can only measure the flux density of stars, not their specific intensity (e.g. Figure 3.9). Therefore, for all stars at the same distance, the black body curves of the stars need to be multiplied by R^2, where R is the stellar radius ($f_\nu \propto R^2 B_\nu(T)$, Eqs. 3.10, 3.14) in order see their relative contributions. Plot $R^2 B_\nu(T)$ for each of the three stellar components, as well as the total, as a function of frequency. You may wish to experiment with different admixtures of stars to see how the result varies in this simple example of population synthesis.

12.4. As we look towards the Galactic Centre, we are looking through 8 kpc of disk before seeing the region of the GC itself. How many stars in the left image of Figure 12.5. might be due to foreground disk stars instead of stars at the GC? To answer this, assume that the number density of *disk* stars in the Milky Way is approximately $n = 0.1\ \mathrm{pc}^{-3}$.

12.5. In Figure 12.3, there are two distinct, roughly circular features that are boxed. Identify which one is the HII region and which one is the supernova remnant. Give your reasons.

12.6. For the following *radio* sources, indicate whether the emission is due to synchrotron radiation, thermal Bremsstrahlung radiation, or whether it is possible to tell. Provide an explanation.
(a) A source that has a steep, negative power law spectrum.
(b) A source that has a flat spectrum and is polarized.
(c) A source that has a flat spectrum and is not polarized.

12.7. Indicate whether the following lines are more likely to have originated in an HI cloud or an HII region and provide a reason. (When 'absorption' is indicated, assume that the required background source is available.)
(a) The λ 21 cm line in absorption.
(b) The λ 21 cm line in emission.
(c) The Lyα line in absorption.
(d) The Lyα line in emission.
(e) The Hα line in absorption.
(f) The Hα line in emission.

12.8. Suggest a possible object or objects that could produce spectra with the following characteristics (each part represents a different object or objects):
(a) There are many spectral lines in the mm-wave part of the spectrum. The underlying continuum shows a slope that suggests a Planck curve which peaks in the infrared.
(b) The spectrum contains a broad Planck-like peak in the near-infrared and optical only. No other significant continuum is observed.
(c) The optical emission resembles that of a star and is polarized.
(d) The X-ray continuum shows a declining spectrum with increasing frequency. Some X-ray emission lines are seen.
(e) The radio spectrum is flat and has emission lines.
(f) An emission line is observed at a wavelength of λ 21.232 cm.
(g) Numerous optical absorption lines are observed against a Planck continuum.

12.9. Using the concept of minimum energy (see Section 10.5.3), 'hot spots' in the radio lobes of Cygnus A have been found to have lifetimes (at $\nu = 89$ GHz) of

$t = 3 \times 10^4$ yr. From the information given in Figure 10.17 (top), find the distance to Cygnus A and the projected distance to the hot spots. Determine whether relativistic particles in the hot spots could have been accelerated at the nucleus of the galaxy and then transported out, or whether they were accelerated in the lobe itself.

12.10. Visit the *Astronomy Picture of the Day* (APOD) website at https://apod.nasa.gov/apod/astropix.html and, as we did in Example 12.3, attempt to identify the type of emission that is displayed in an image that interests you. Some questions to ask might be:
  (a) Does this picture represent a single waveband or is it a composite?
  (b) Does the image show continuum or line emission?
  (c) Is the emission thermal or nonthermal in nature?
  (d) If a spectrum in the displayed waveband were available to you, what information could you obtain about the source?

12.11. Visit the *Multi-wavelength Milky Way* website at https://asd.gsfc.nasa.gov/archive/mwmw. For each of the wavebands shown (e.g. in the poster), (a) identify the important emission mechanism(s), and (b) provide examples of the types of sources that contribute to the observed emission.

12.12. A point source of magnitude, $V = 18$, is observed in a galaxy at a distance of $D = 30$ Mpc. Is this source an individual star, a globular cluster, or a supernova?

JUST FOR FUN

12.13. In our daily life, almost every source of optical light entering our eyes comes originally from a black-body emitting source, either directly or via reflection or scattering. Try to find sources of illumination that are *not* from a black body. What emission processes do they represent? Internet searches may help.

12.14. Suppose you were a human-like creature that has evolved on an Earth-like planet in the globular cluster, NGC 6397. What would the night sky look like? What might *you* look like?

Appendix T

Table T.1 Fundamental physical data[a]

Symbol	Meaning	Value
Physical constants		
c	Speed of light in vacuum	$2.997\,924\,58 \times 10^{10}$ cm s^{-1}
G	Universal gravitational constant	$6.6742\,(10) \times 10^{-8}$ cm^3 g^{-1} s^{-2}
g	Standard gravitational acceleration (Earth)	$9.806\,65 \times 10^2$ cm s^{-2}
k	Boltzmann constant	$1.380\,6505\,(24) \times 10^{-16}$ erg K^{-1}
		$8.617\,343\,(15) \times 10^{-5}$ eV K^{-1}
h	Planck's constant	$6.626\,0693\,(11) \times 10^{-27}$ erg s
N_A[b]	Avogadro's constant	$6.022\,1415\,(10) \times 10^{23}$ mol^{-1}
$\mathcal{R}$	Molar gas constant	$8.314\,472\,(15) \times 10^7$ erg mol^{-1}
R_∞	Rydberg constant	$109\,737.315\,68\,525\,(73)$ cm^{-1}
Atomic and nuclear data		
e[c]	Atomic unit of charge	$4.803\,250\,(21) \times 10^{-10}$ esu
eV	Electron volt	$1.602\,176\,53\,(14) \times 10^{-12}$ erg
m_e	Electron mass	$9.109\,3826\,(16) \times 10^{-28}$ g
m_n	Neutron mass	$1.674\,927\,28\,(29) \times 10^{-24}$ g
m_p	Proton mass	$1.672\,621\,71\,(29) \times 10^{-24}$ g
u_{amu}	Unified atomic mass unit	$1.660\,538\,86\,(28) \times 10^{-24}$ g
r_e[d]	Classical electron radius	$2.817\,940\,325\,(28) \times 10^{-13}$ cm
λ_C	Compton wavelength	$2.426\,310\,238\,(16) \times 10^{-10}$ cm
λ_{Cn}	Neutron Compton wavelength	$1.319\,590\,9067\,(88) \times 10^{-13}$ cm
λ_{Cp}	Proton Compton wavelength	$1.321\,409\,8555\,(88) \times 10^{-13}$ cm
a_0	Bohr radius	$0.529\,177\,2108\,(18) \times 10^{-8}$ cm
σ_T	Thomson cross section	$0.665\,245\,873\,(13) \times 10^{-24}$ cm^2

(Continued)

Astrophysics: Decoding the Cosmos, Second Edition. Judith A. Irwin.

Companion website: www.wiley.com/go/irwin/astrophysics2e

Table T.1 *(Continued)*

Symbol	Meaning	Value
Radiation constants		
σ	Stefan–Boltzmann constant	$5.670\,400\,(40) \times 10^{-5}$ erg s^{-1} cm^{-2} K^{-4}
b^e	Wien's Displacement law constant	$2.897\,7685\,(51) \times 10^{-1}$ cm K
a^f	Radiation constant	$7.565\,91\,(25) \times 10^{-15}$ erg cm^{-3} K^{-4}

[a]Data have been taken from [10], converted to cgs units, unless otherwise indicated. The values in parentheses, where provided, represent the standard error of the last digits which are not in parentheses. Updates to these values can be found at [11].
[b]This is the number of particles in one *mole* (mol) of material.
[c]From [7]. An esu is an 'electrostatic unit'. The corresponding SI unit of charge is 1.602×10^{-19} C. Note that the force equation in cgs units, $F = (q_1\, q_2)/r^2$, does not have a constant of proportionality for F in dynes, q_1, q_2 in esu, and r in cm.
[d]$r_e = e^2/(m_e\, c^2)$.
[e]From Eq. (6.8).
[f]Ref. [2].

Table T.2 Astronomical data[a]

Symbol	Meaning	Value
Distance		
AU	Astronomical unit	$1.495\,978\,706(2) \times 10^{13}$ cm
pc	Parsec	$3.085\,677\,582 \times 10^{18}$ cm
ly	Light year	$9.460\,730\,47 \times 10^{17}$ cm
Time		
yr	Tropical year[b]	$3.155\,692\,58 \times 10^{7}$ s
yr	Sidereal year[c]	$3.155\,814\,98 \times 10^{7}$ s
day	Mean sidereal day[c]	23 h 56 m 04.090524 s
Earth		
$M_\oplus$	Earth mass	$5.973\,70(76) \times 10^{27}$ g
$R_\oplus$	Mean Earth radius[d]	$6.371\,000 \times 10^{8}$ cm
R_{equ}	Earth equatorial radius[d]	$6.378\,136 \times 10^{8}$ cm
R_{pol}	Earth polar radius[d]	$6.356\,753 \times 10^{8}$ cm
$\rho_\oplus$	Mean Earth density[d]	5.515 g cm^{-3}

(Continued)

Table T.2 *(Continued)*

Symbol	Meaning	Value
Moon		
M_m	Moon mass[d]	7.353×10^{25} g
R_m	Mean Moon radius[d]	$1.738\,2 \times 10^{8}$ cm
ρ_m	Mean Moon density[d]	3.341 g cm^{-3}
r_m	Mean Earth – Moon distance[d]	$3.844\,01(1) \times 10^{10}$ cm
Sun		
$M_\odot$	Solar mass	$1.989\,1(4) \times 10^{33}$ g
$R_\odot$	Solar radius	6.955×10^{10} cm
$\rho_\odot$	Mean solar density	1.41 g cm^{-3}
$L_\odot$	Solar luminosity[d]	$3.845\,(8) \times 10^{33}$ erg s^{-1}
$T_{\mathrm{eff}\odot}$	Solar effective temperature[e]	5781 K
$S_\odot$	Solar constant[f]	1.367×10^{6} erg s^{-1} cm^{-2}
$m_{V\odot}$	Solar apparent visual magnitude[e]	−26.76
$M_{V\odot}$	Solar absolute visual magnitude[e]	4.81
$M_{b\odot}$	Solar absolute bolometric magnitude[e]	4.74
$\theta_\odot$	Angular diameter of Sun from Earth	32.0′
Galaxy		
$R_\odot$	Sun Galactocentric distance[g]	8.5 (5) kpc
υ_{LSR}	Velocity of the Local Standard of Rest[h]	220 (20) km s^{-1}
Extragalactic		
H_0	Hubble constant[i]	71^{+4}_{-3} km s^{-1} Mpc^{-1}
T_{CMB}	Cosmic Microwave Background temperature[j]	2.7260 ± 0.0013 K

[a]Ref. [7] unless otherwise indicated. Values in parentheses indicate the one standard deviation uncertainties in the last digit or digits.
[b]Equinox to equinox.
[c]Fixed star to fixed star.
[d]Ref. [2].
[e]Ref. [1].
[f]Flux of the Sun at 1 AU.
[g]Distance of Sun from the centre of the Galaxy. A value of 8 kpc is often used (e.g. Section 12.2.1).
[h]Velocity of an object at a Galactocentric distance of $R_\odot$ in circular orbit about the centre of the Galaxy.
[i]Rate of expansion of the Universe. The value is from [9]; see Section 9.1.2 for a discussion of range of values.
[j]Ref [3].

Table T.3 Planetary data[a].

Planet	Equatorial Diameter (km)	Oblateness	Mass (Earth = 1)	Density (g cm^{-3})	Rotation period[b] (days)	Incl.[c]	Albedo[d] (Bond)	a[e] (AU)	e[f]
Mercury	4 879	0	0.055 274	5.43	58.646	0.0	0.119	0.387 1	0.205 6
Venus	12 104	0	0.815 005	5.24	243.019	2.6	0.750	0.723 3	0.006 8
Earth	12 756	1/298	1.000 000	5.52	0.9973	23.4	0.306	1.000 0	0.016 7
Mars	6 792	1/148	0.107 447	3.94	1.0260	25.2	0.250	1.523 7	0.093 5
Jupiter	142 980[g]	1/15.4	317.833	1.33	0.4101[h]	3.1	0.343	5.202 0	0.049 0
Saturn	120 540[g]	1/10.2	95.163	0.69	0.4440	26.7	0.342	9.575 2	0.056 8
Uranus	51 120[g]	1/43.6	14.536	1.27	0.7183	82.2	0.300	19.131 5	0.050 1
Neptune	49 530[g]	1/58.5	17.149	1.64	0.6712	28.3	0.290	29.968 1	0.008 6
Pluto[i]	2 390	<0.6%[j]	0.002 2	1.8	6.3872	57.4	0.4–0.6	39.546 3	0.250 9

[a] Ref. [6] unless otherwise indicated.
[b] Sidereal.
[c] Inclination of equator to orbital plane.
[d] Albedo is the fraction of incident light that is reflected. This table gives the *Bond albedo* (from http://nssdc.gsfc.nasa.gov), defined as the total fraction of reflected Sunlight over all wavebands.
[e] Mean distance to the Sun (equal to the semi-major axis of the ellipse).
[f] Mean orbital eccentricity. The perihelion and aphelion distances are given by $r_{min} = a(1 - e)$ and $r_{max} = a(1 + e)$, respectively.
[g] At 1 atm (101.325 kPa).
[h] For the most rapidly rotating equatorial region.
[i] At the 26th General Assembly of the International Astronomical Union held in Prague in August, 2006, 'Resolution 6' was adopted, indicating that 'Pluto is a "dwarf planet" . . . and is recognized as the prototype of a new category of Trans-Neptunian Objects.'
[j] Ref. [8].

Table T.4 Data for human vision[a]

Quantity	Value
Pupil diameter (dark-adapted)	0.8 cm
(daylight)	0.2 cm
Approx. number of photoreceptor cells[b]	
Rods	10^8
Cones	6.5×10^6
Retina effective area	10 cm^2
Focal length	2.5 cm
Resolution of human eye	1′[c]
Wavelength response of human eye:	
Bright-adapted (photopic) visible range	400–700 nm
Dark-adapted (scotopic) visible range	400–620 nm
Wavelength of peak sensitivity of eye:	
Bright-adapted	555 nm
Dark-adapted	507 nm
Representative wavelengths of colours:	
Violet	420 nm
Blue	470 nm
Green	530 nm
Yellow	580 nm
Orange	610 nm
Red	660 nm

[a]Wavelengths are from [6]. Note that values for the eye are typical, but vary with individual.
[b]In order to register as a separate source, each photoreceptor cell must be separated by at least one other unexcited photoreceptor cell. Dark-adapted vision response is primarily due to rods.
[c]This value varies depending on the individual and lighting conditions.

Table T.5 Stellar data[a] Luminosity Class V (Dwarfs).

Spectral type	B–V	V–R	M_V	BC	T_{eff} (K)	$R/R_{\odot}$	$M/M_{\odot}$
O3				−4.3?	48 000		50?
O5			−5?	−4.3?	44 000		30?
O6	−0.32	−0.15	−4.8	−4.25	43 000	12?	25?
O8	−0.31	−0.15	−4.1	−3.93	37 000	10	20?
B0	−0.29	−0.13	−3.3	−3.34	31 000	7.2	17
B1	−0.26	−0.11	−2.9	−2.6	24 100	5.3	10.7
B2	−0.24	−0.1	−2.5	−2.2	21 080	4.7	8.3
B3	−0.21	−0.08	−2	−1.69	18 000	3.5	6.3
B4	−0.18	−0.07	−1.5	−1.29	15 870	3	5
B5	−0.16	−0.06	−1.1	−1.08	14 720	2.9	4.3
B8	−0.1	−0.02	0	−0.6	11 950	2.3	3
A0	0	0.02	0.7	−0.14	9572	1.8	2.34
A2	0.06	0.08	1.3	0	8985	1.75	2.21
A5	0.14	0.16	1.9	0.02	8306	1.69	2.04
A7	0.19	0.19	2.3	0.02	7935	1.68	1.93
F0	0.31	0.3	2.7	0.04	7178	1.62	1.66
F2	0.36	0.35	3	0.04	6909	1.48	1.56
F5	0.44	0.4	3.5	0.02	6528	1.4	1.41
F8	0.53	0.47	4	−0.01	6160	1.2	1.25
G0	0.59	0.5	4.4	−0.05	5943	1.12	1.16
G2	0.63	0.53	4.7	−0.08	5811	1.08	1.11
G5	0.68	0.54	5.1	−0.11	5657	0.95	1.05
G8	0.74	0.58	5.6	−0.16	5486	0.91	0.97
K0	0.82	0.64	6	−0.23	5282	0.83	0.9
K2	0.92	0.74	6.5	−0.3	5055	0.75	0.81
K3	0.96	0.81	6.8	−0.33	4973	0.73	0.79
K5	1.15	0.99	7.5	−0.43	4623	0.64	0.65
K7	1.3	1.15	8	−0.54	4380	0.54	0.54
M0	1.41	1.28	8.8	−0.72	4212	0.48	0.46
M2	1.5	1.5	9.8	−0.99	4076	0.43	0.4
M5	1.6	1.8	12	−1.52	3923	0.38	0.34

[a] Ref. [4]. These values should be considered typical, but individual stars may differ from the exact values quoted here. Blanks imply that the data are not sufficiently well known to be quoted here.

Table T.6 Stellar data[a] Luminosity Class III (Giants).

Spectral type	B–V	V–R	M_V	BC	T_{eff}(K)	$R/R_\odot$
F0	0.31		1	0.04	7178	7
F2	0.36		0.9	0.04	6909	
F5	0.44		0.8	0.02	6528	
F8	0.54		0.7	−0.02	6160	8
G0	0.64		0.6	−0.09	5943	9
G2	0.76		0.5	−0.17	5811	10
G5	0.9	0.69	0.4	−0.29	5657	11
G8	0.96	0.7	0.3	−0.33	5486	12
K0	1.03	0.77	0.2	−0.37	5282	14
K2	1.18	0.84	0.1	−0.45	5055	17
K3	1.29	0.96	0.1	−0.53	4973	21
K5	1.44	1.2	0	−0.81	4623	40
K7	1.53		−0.1	−1.15	4380	60
M0	1.57	1.23	−0.2	−1.36	4212	100
M2	1.6	1.34	−0.2	−1.52	4076	130
M5	1.58	2.18	−0.2		3923	

[a]Ref. [4]. See notes to Table A.5. Masses of Type III giants are not well known but are likely $\approx$ 2–2.5 $M_\odot$.

Table T.7 Stellar data[a] Luminosity Class Ib (Supergiants).

Spectral type	B–V	V–R	M_V	BC	T_{eff}(K)	$R R_\odot$
A0	0.01	0.03	−5	−0.12	9550	70
A2	0.05	0.07	−5	−0.02	9000	
A5	0.1	0.12	−5	0.01	8500	
A7	0.13		−4.9	0.02	8300	
F0	0.16	0.21	−4.8	0.02	8030	80
F2	0.21	0.26	−4.8	0.02	7780	
F5	0.33	0.35	−4.7	0.04	7020	
F8	0.55	0.45	−4.6	−0.02	6080	
G0	0.76	0.51	−4.6	−0.18	5450	100
G2	0.87	0.58	−4.6	−0.26	5080	
G5	1	0.67	−4.5	−0.35	4850	
G8	1.13	0.69	−4.5	−0.41	4700	
K0	1.2	0.76	−4.5	−0.47	4500	110
K2	1.29	0.85	−4.5	−0.53	4400	
K3	1.38	0.94	−4.5	−0.68	4230	130
K5	1.6	1.2	−4.5	−1.52	3900	200
K7	1.62		−4.5		3870	
M0	1.65	1.23			3850	230
M2	1.65	1.34			3800	

[a]Ref. [4]. See notes to Table A.5. Masses of Type Ib supergiants are not well known but are likely $\approx$10 $M_\odot$.

Table T.8 Solar interior model[a]

$R/R_{\odot}$ [b]	$M/M_{\odot}$ [c]	L^{d} (erg s^{-1})	T^{e}(K)	ρ^{f}(g cm^{-3})
1.14E−03	1.64E−07	5.66E+27	1.58E+07	1.56E+02
2.48E−02	1.64E−03	5.50E+31	1.56E+07	1.49E+02
1.03E−01	8.29E−02	1.84E+33	1.30E+07	8.57E+01
1.87E−01	3.00E−01	3.52E+33	9.82E+06	3.95E+01
2.73E−01	5.44E−01	3.83E+33	7.40E+06	1.62E+01
3.55E−01	7.21E−01	3.85E+33	5.80E+06	6.50E+00
4.39E−01	8.36E−01	3.85E+33	4.64E+06	2.56E+00
5.29E−01	9.09E−01	3.85E+33	3.72E+06	1.01E+00
6.25E−01	9.53E−01	3.85E+33	2.94E+06	4.02E−01
7.22E−01	9.77E−01	3.85E+33	2.09E+06	1.76E−01
8.04E−01	9.90E−01	3.85E+33	1.33E+06	8.80E−02
8.66E−01	9.96E−01	3.85E+33	8.41E+05	4.41E−02
9.09E−01	9.99E−01	3.85E+33	5.34E+05	2.21E−02
9.39E−01	1.00E+00	3.85E+33	3.41E+05	1.12E−02
9.59E−01	1.00E+00	3.85E+33	2.19E+05	5.58E−03
9.72E−01	1.00E+00	3.85E+33	1.43E+05	2.76E−03
9.81E−01	1.00E+00	3.85E+33	9.57E+04	1.34E−03
9.86E−01	1.00E+00	3.85E+33	6.44E+04	6.52E−04
9.90E−01	1.00E+00	3.85E+33	4.48E+04	3.10E−04
9.93E−01	1.00E+00	3.85E+33	3.35E+04	1.39E−04
9.95E−01	1.00E+00	3.85E+33	2.65E+04	5.93E−05
9.96E−01	1.00E+00	3.85E+33	2.18E+04	2.43E−05
9.97E−01	1.00E+00	3.85E+33	1.84E+04	9.69E−06
9.98E−01	1.00E+00	3.85E+33	1.58E+04	3.81E−06
9.988E−01	1.00E+00	3.85E+33	1.36E+04	1.50E−06
9.993E−01	1.00E+00	3.85E+33	1.16E+04	6.04E−07
9.998E−01	1.00E+00	3.85E+33	8.72E+03	2.76E−07
1.0000E+00	1.00E+00	3.85E+33	5.67E+03	1.55E−07
1.0001E+00	1.00E+00	3.85E+33	5.11E+03	9.61E−08
1.0002E+00	1.00E+00	3.85E+33	4.82E+03	5.62E−08
1.0003E+00	1.00E+00	3.85E+33	4.70E+03	3.10E−08
1.0004E+00	1.00E+00	3.85E+33	4.65E+03	1.68E−08
1.0005E+00	1.00E+00	3.85E+33	4.64E+03	9.07E−09
1.0006E+00	1.00E+00	3.85E+33	4.63E+03	4.87E−09
1.0007E+00	1.00E+00	3.85E+33	4.63E+03	2.58E−09
1.0008E+00	1.00E+00	3.85E+33	4.63E+03	1.34E−09
1.0009E+00	1.00E+00	3.85E+33	4.63E+03	6.67E−10
1.0011E+00	1.00E+00	3.85E+33	4.63E+03	3.03E−10

[a] Modelled conditions for the interior of the Sun adopting $T_{eff} = 5779.6$ K, $L_{\odot} = 3.851 \times 10^{33}$ erg s^{-1}, $X = 0.708$, $Z = 0.02$, $R_{\odot} = 6.9598 \times 10^{10}$ cm and $M_{\odot} = 1.98910 \times 10^{33}$ g [5].
[b] Radius as a fraction of the Solar radius.
[c] Mass as a fraction of the Solar mass.
[d] Luminosity.
[e] Temperature.
[f] Mass density.
Source: Reproduced by permission of D. Guenther. Updated information on the *Standard Solar Model* can be found at [12].

Table T.9 Solar photosphere model[a].

$\log(\tau_0)$[b]	T[c] (K)	$\log(P_g)$[d] [log(dyn cm^{-2})]	$\log(P_e)$[e] [log(dyn cm^{-2})]	$\log(\kappa_0/P_e)$[f] [log(cm^3 s^2 g^{-2})]	x[g] (km)
−4.0	4310	3.13	−1.42	−1.22	−476
−3.8	4325	3.29	−1.28	−1.23	−443
−3.6	4345	3.42	−1.15	−1.24	−415
−3.4	4370	3.54	−1.03	−1.25	−389
−3.2	4405	3.66	−0.92	−1.27	−365
−3.0	4445	3.78	−0.80	−1.28	−340
−2.8	4488	3.89	−0.69	−1.30	−317
−2.6	4524	4.00	−0.58	−1.32	−293
−2.4	4561	4.11	−0.48	−1.33	−269
−2.2	4608	4.22	−0.37	−1.35	−245
−2.0	4660	4.33	−0.26	−1.37	−221
−1.8	4720	4.44	−0.14	−1.40	−196
−1.6	4800	4.54	−0.01	−1.42	−172
−1.4	4878	4.65	0.11	−1.45	−147
−1.2	4995	4.76	0.26	−1.49	−122
−1.0	5132	4.86	0.42	−1.54	−96
−0.8	5294	4.96	0.60	−1.58	−73
−0.6	5490	5.04	0.83	−1.64	−51
−0.4	5733	5.12	1.10	−1.71	−31
−0.2	6043	5.18	1.43	−1.79	−14
0.0	6429	5.22	1.80	−1.88	0
0.2	6904	5.26	2.21	−1.99	11
0.4	7467	5.28	2.64	−2.10	19
0.6	7962	5.30	2.96	−2.17	26
0.8	8358	5.32	3.20	−2.22	33
1.0	8630	5.43	3.34	−2.25	41
1.2	8811	5.36	3.46	−2.26	50

[a] Modelled conditions for the surface region of the Sun [4]. The model assumes that the solar effective temperature is $T_{eff} = 5770$ K and that $\log(g) = 4.44$, where g (cm s^{-2}) is the surface gravity.
[b] Log of the optical depth at the reference wavelength of 500 nm. $\log(\tau_0) = 0$ represents the 'surface' of the Sun's photosphere.
[c] Temperature at the corresponding optical depth.
[d] Log of the gas pressure.
[e] Log of the electron pressure.
[f] Log of the ratio of the mass absorption coefficient (Section 7.4.1), κ_0 (cm^2 g^{-1}) at the reference wavelength of 500 nm, to the electron pressure, P_e.
[g] Geometrical depth, where 0 corresponds to the photosphere surface and positive values are below the surface.

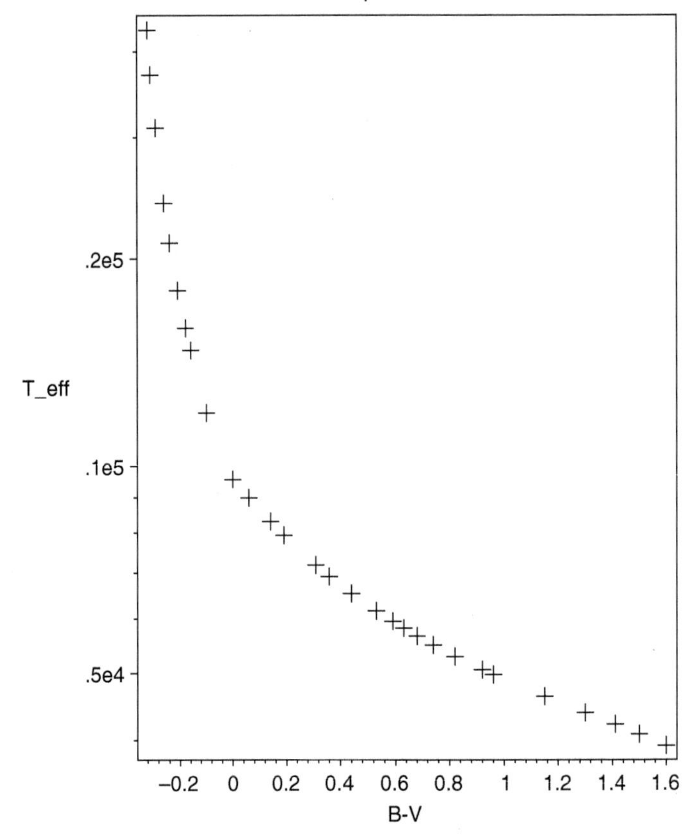

Figure T.1 B–V colour – temperature calibration for the Main Sequence, from the data of Table A.5.

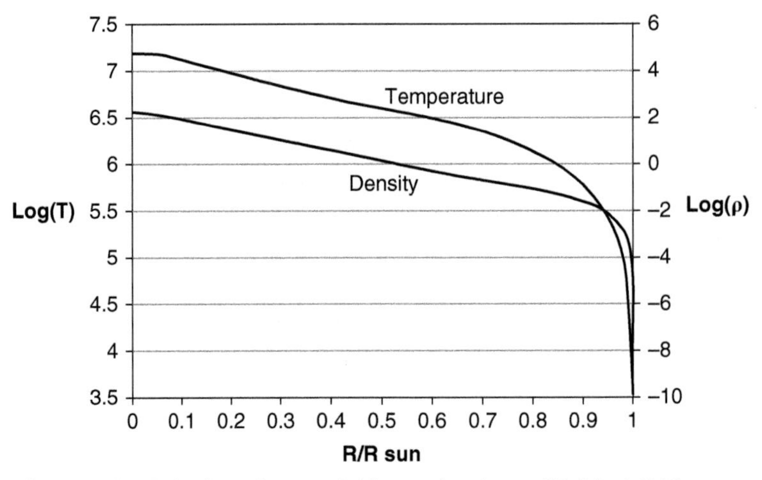

Figure T.2 Solar interior model from the data of Table A.8. The temperature is in K and density in g cm^{-3}.

Acronym Key to Bibliography

A&A	Astronomy & Astrophysics
A&ARv	Astronomy & Astrophysics Review
A&AS	Astronomy and Astrophysics Supplement Series
Adv. Space Res.	Advances in Space Research
AIP Conf. Proc	American Institute of Physics Conference Proceedings
AJ	The Astronomical Journal
Ann. Phys.	Annals of Physics
Annu. Rev. Nucl. Part. Sci.	Annual Review of Nuclear and Particle Science
Int. J. Mod. Phys. D	International Journal of Modern Physics D
AO	Applied Optics
ApJ	The Astrophysical Journal
ApJL	The Astrophysical Journal Letters
ApJS	Astrophysical Journal Supplement Series
ARAA	Annual Review of Astronomy and Astrophysics
AREPS	Annual Review of Earth and Planetary Sciences
ASP Conf. Ser.	Astronomical Society of the Pacific Conference Series
ASSB	Annales de la Société Scientifique de Bruxelles
Astron. Nachr.	Astronomische Nachrichten
Chem. Soc. Rev.	Chemical Society Reviews
Class. Quantum Grav.	Classical and Quantum Gravity
GeoRL	Geophysical Research Letters
Int. J. Mod. Phys.	International Journal of Modern Physics
JApA	Journal of Astrophysics and Astronomy
J. Astron. Inst.	Journal of Astronomical Instrumentation
JASTP	Journal of Atmospheric and Solar-Terrestrial Physics
JCAP	Journal of Cosmology and Astroparticle Physics

Astrophysics: Decoding the Cosmos, Second Edition. Judith A. Irwin.

Companion website: www.wiley.com/go/irwin/astrophysics2e

JGR	Journal of Geophysical Research
J. Geophys. Res. Atmos.	Journal of Geophysical Research Atmospheres
J. Geophys. Res. Planets	Journal of Geophysical Research Planets
JHEAp	Journal of High Energy Astrophysics
JKAS	Journal of the Korean Astronomical Society
JOSA	Journal of the Optical Society of America
JQSRT	Journal of Quantitative Spectroscopy & Radiative Transfer
J. Phys. Meteorit. Planet. Sci.	Journal of Physics Meteoritics and Planetary Science
MNRAS	Monthly Notices of the Royal Astronomical Society
Nat. Astron.	Nature Astronomy
Nat. Commun.	Nature Communications
Nat. Geosci.	Nature Geoscience
NewA	New Astronomy
NewAR	New Astronomy Reviews
NuPhA	Nuclear Physics A
OE	Optical Engineering
PASJ	Publications of the Astronomical Society of Japan
PASP	Publications of the Astronomical Society of the Pacific
Phys. Rev.	Physical Review
Phys. Rev. D	Physical Review D
Phys. Rev. Lett.	Physical Review Letters
PRL	Physical Review Letters
Phys. Today	Physics Today
P&SS	Planetary and Space Science
PNAS	Proceedings of the National Academy of Sciences of the United States of America
Proc. RoyalSoc. A	Proceedings of the Royal Society A
PoS	Proceedings of Science
Rev. Geophys.	Reviews of Geophysics
RPPh	Report on Progress in Physics
Rev. Mod. Phys.	Reviews of Moden Physics
Rev. Part. Phys.	Review of Particle Physics
Sci. Am.	Scientific American
SIAMR	Society of Industrial and Applied Mathematics Review
SSRv	Space Science Reviews

REFERENCES AND IN-DEPTH READING

CHAPTER 1

1. Abbasi, R. U., Abe, M., Abu-Zayyad, T., et al. 2014, *Indications of Intermediate-scale Anisotropy of Cosmic Rays with Energy Greater than 57 EeV in the Northern Sky Measured with the Surface Detector of the Telescope Array Experiment*, *ApJL*, 790, L21, DOI: https://doi.org/10.1088/2041-8205/790/2/L21

2. Alexandrov, A., Alexeev, V., Bagulya, A., et al. 2019, *Natural superheavy nuclei in astrophysical data*, arXiv:1908.02931v1
3. Amelin, Y., Krot, A. N., Hutcheon, I. D. et al. 2002, *Lead Isotopic Ages of Chondrules and Calcium-Aluminum-Rich Inclusions, Science*, 297, 1678
4. Anders, E. 1962, *Meteorite Ages, Rev. Mod. Phys.*, 34, 287
5. Arnett, W. D., Bahcall, J. N., Kirshner, R. P., et al. 1989, *Supernova 1987A, ARAA*, 27, 629, DOI: https://doi.org/10.1146/annurev.aa.27.090189.003213
6. Beck, R., & Krause, M. 2005, *Revised Equipartition and Minimum Energy Formula for Magnetic Field Strength Estimates from Radio Synchrotron Observations, Astron. Nachr.*, 326, 414, DOI: https://doi.org/10.1002/asna.200510366
7. Banfield, D., Spiga, A., Newman, C., et al. 2020, *The Amosphere of Mars as Observed by InSight, Nat. Geosci.*, 13, 190, DOI: https://doi.org/10.1038/s41561-020-0534-0
8. Bedard Jr., A. J., & Georges, T. M. 2000, *Atmospheric Infrasound, Phys. Today*, 53, 32
9. Beer, T. 1974, *Atmospheric Waves*, New York: Halsted Press, a Division of John Wiley & Sons, Inc.
10. Blasi, P. 2013, *The Origin of Galactic Cosmic Rays, A&ARv*, 21, 70, DOI: https://doi.org/10.1007/s00159-013-0070-7
11. Brown, P., Spalding, R. E., ReVelle, D. O., et al. 2002, *The Flux of Small Near-Earth Objects Colliding with the Earth, Nature*, 420, 294
12. Brown, P. G., Assink, J. D., Astiz, L., et al. 2013, *A 500-kiloton Airburst Over Chelyabinsk and an Enhanced Hazard from Small Impactors, Nature*, 503, 238, DOI: https://doi.org/10.1038/nature12741
13. Cairns, I., McCusker, C. B. A., Peak, L. S., et al. 1969, *Lightly Ionizing Particles in Air-Shower Cores, Phys. Rev.*, 186, 1394, DOI: https://doi.org/10.1103/PhysRev.186.1394
14. Casadei, D., & Bindi, V. 2004, *The Origin of Cosmic Ray Electrons and Positrons, ApJ*, 612, 262, DOI: https://doi.org/10.1086/422514
15. Cleveland, B. T., Daily, T., Davis, R., Jr., et al. 1998, *Measurement of the Solar Electron Neutrino Flux with the Homestake Chlorine Detector, ApJ*, 496, 505, DOI: https://doi.org/10.1086/305343
16. Edwards, W. N., Brown, P. G., & ReVelle, D. O. 2006, *Estimates of Meteoroid Kinetic Energies from Observations of Infrasonic Airwaves, JASTP*, 68, 1136, DOI: https://doi.org/10.1016/j.jastp.2006.02.010
17. Ellison, D. C., Drury, L. O'C., Meyer, J.-P., 1997, *Galactic Cosmic Rays from Supernova Remnants. II. Shock Acceleration of Gas and Dust, ApJ*, 487, 197, DOI: https://doi.org/10.1086/304580
18. Ens, T. A., Brown, P. G., Edwards, W. N., et al. 2012, *Infrasound Production by Bolides: A Global Statistical Study, JASTP*, 80, 208, DOI: https://doi.org/10.1016/j.jastp.2012.01.018
19. Eugster, O., Herzog, G. F., Marti, K., et al. 2006, *Irradiation Records, Cosmic-Ray Exposure Ages, and Transfer Times of Meteorites*, in *Meteorites and the Early Solar System II* (eds. D. S. Lauretta and H. Y. McSween Jr.), 829–851, Tucson: University of Arizona Press.

20. Flynn, G. J. 1994, *Interplanetary Dust Particles Collected from the Stratosphere: Physical, Chemical, and Mineralogical Properties and Implications for Their Sources, P&SS*, 42, 1151, DOI: https://doi.org/10.1016/0032-0633(94)90014-0
21. Formaggio, J. A., & Zeller, G. P. 2012, *From eV to EeV: Neutrino Cross Sections Across Energy Scales, Rev. Mod. Phys.* 84, 1307, DOI: https://doi.org/10.1103/RevModPhys.84.1307
22. Fransson, C., Larsson, J., Migotto, K., et al. 2015, *The Destruction of the Circumstellar Ring of SN 1987A, ApJL*, 806, L19, DOI: https://doi.org/10.1088/2041-8205/806/1/L19
23. George, J. S., Lave, K. A., Wiedenbeck, M. E., et al. 2009, *Elemental Composition and Energy Spectra of Galactic Cosmic Rays During Solar Cycle 23, ApJ*, 698, 1666, DOI: https://doi.org/10.1088/0004-637X/698/2/1666
24. Garriazzo, S., Archidiacono, M., de Salas, P. F., et al. 2018, *Neutrino Masses and Their Ordering: Global Data, Priors and Models, JCAP*, 2018, DOI:https://doi.org/10.1088/1475-7516/2018/03/011
25. Gi, N. 2017, *Using Bolide Airwaves To Estimate Meteoroid Source Characteristics and Window Damage Potential,* MSc Thesis, University of Western Ontario, https://ir.lib.uwo.ca/etd/4688/
26. Gi, N., & Brown, P. 2017, *Refinement of Bolide Characteristics from Infrasound Measurements, P&SS*, 143, 169, DOI: https://doi.org/10.1016/J.pss.2017.04.021
27. Grady, M. M., & Wright, I. 2006, *Types of Extraterrestrial Material Available for Study*, in *Meteorites and the Early Solar System II* (eds. D. S. Lauretta and H. Y. McSween Jr.), 3–18, Tucson: University of Arizona Press
28. Grenier, I. A., Black, J. H., & Strong, A. W. 2015, *The Nine Lives of Cosmic Rays in Galaxies, ARAA*, 53, 199, DOI: https://doi.org/10.1146/annurev-astro-082214-122457
29. Hackett, J. 2016, *How to Find Tiny Meteorites at Home, Sci. Am.*, 314, 4, 15, DOI: https://doi.org/10.1038/scientificamerican0416-15
30. Halzen, F. 2007, *Neutrino Astrophysics Experiments Beneath the Sea and Ice, Science*, 315, 66, DOI: https://doi.org/10.1126/science.1136504
31. Hanlon, W. F. 2008, *Energy Spectrum of Ultra High Energy Cosmic Rays Measured by the High Resolution Fly's Eye Observatory in Stereoscopic Mode*, PhD Thesis, Utah University, http://inspirehep.net/record/843833?ln=en
32. Herzog, G. F., & Caffee, M. W. 2014, *Cosmic-Ray Exposure Ages of Meteorites*, in *Treatise on Geochemistry* (Second Edition) (eds. H.D. Holland and K.K. Turekian), 419–453, Elsevier Ltd., DOI: https://doi.org/10.1016/B978-0-08-095975-7.00110-8
33. Höorandel, J. R. 2006, *A Review of Experimental Results at the Knee, J. Phys. Conf. Ser.* 47, 41, DOI: https://doi.org/10.1088/1742-6596/47/1/005
34. Hörandel, J. R. 2008, *Cosmic-ray Composition and Its Relation to Shock Acceleration by Supernova Remnants, Adv. Space Res.*, 41, 442, DOI: https://doi.org/10.1016/j.asr.2007.06.008
35. Hughes, D. H. 2003, *The Approximate Ratios Between the Diameters of Terrestrial Impact Craters and the Causative Incident Asteroids, MNRAS* 338, 999, DOI:https://doi.org/10.1046/j.1365-8711.2003.06157.x

36. IceCube Collaboration 2018a, *Neutrino Emission from the Direction of the Blazar TXS 0506+056 Prior to the IceCube-170922A Alert, Science*, 361, 147, DOI: https://doi.org/10.1126/science.aat2890
37. IceCube Collaboration 2018b, *Multimessenger Observations of a Flaring Blazar Coincident with High-energy Neutrino IceCube-170922A, Science*, 361, 1378, DOI: https://doi.org/10.1126/science.aat1378
38. Jelley, N., McDonald, A. B., & Robertson, R. G. H. 2009, *The Sudbury Neutrino Observatory, Annu. Rev. Nucl. Part. Sci.*, 59, 431, DOI: https://doi.org/10.1146/annurev.nucl.55.090704.151550
39. Kalyashova, M. E., Bykov, A. M., Osipov, S. M., et al. 2019, *Wolf-Rayet Stars in Young Massive Star Clusters as Potential Sources of Galactic Cosmic Rays, J. Phys. Conf. Ser.*, 1400, 022011, DOI: https://doi.org/10.1088/1742-6596/1400/2/022011
40. Larsson, J., Fransson, C., & Wheeler, J. C. 2011, *X-ray Illumination of the Ejecta of Supernova 1987A, Nature*, 474, 484, DOI: https://doi.org/10.1038/nature10090
41. Leya, I., Wieler, R., Aggrey, K., et al. 2001, *Exposure History of the St-Robert (H5) Fall*, Meteorit. *Planet. Sci.*, 36, 1479, DOI: https://doi.org/10.1111/j.1945-5100.2001.tb01840.x
42. Longair, M. S. 1992, *High Energy Astrophysics, Vol.* I, (Second Editon), Cambridge: Cambridge University Press
43. Mathews, J. D., Janches, D., Meisel, D. D., et al. 2001, *The Micrometeoroid Mass Flux into the Upper Atmosphere: Arecibo Results and a Comparison with Prior Estimates, GeoRL*, 28, 1929–1932, DOI: https://doi.org/10.1029/2000GL012621
44. MacPherson, G. J. 2007, *Calcium-Aluminum-Rich Inclusions in Chondritic Meteorites*, in *Treatise on Geochemistry* (eds. H.D. Holland and K.K. Turekian), 1–47, Elsevier Ltd., DOI: https://doi.org/10.1016/B0-08-043751-6/01065-3
45. Pacini, A. A. 2017, *Cosmic Rays: Bringing Messages from the Sky to the Earth's Surface, Revista Brasileira de Ensino de Fisica*, 39, e1306, DOI: https://doi.org/10.1590/1806-9126-RBEF-2016-0168
46. The Pierre Auger Collaboration, 2017, *Observation of a Large-scale Anisotropy in the Arrival Directions of Cosmic Rays Above* 8×10^{18} *eV, Science*, 357, 1266, DOI: https://doi.org/10.1126/science.aan4338
47. Plane, J. M. C. 2012, *Cosmic Dust in the Earth's Atmosphere, Chem. Soc. Rev.*, 41, 6507, DOI: https://doi.org/10.1039/C2CS35132C
48. Prasad, M. S., Rudraswami, N. G., Alexandre de Araujo, A., et al. 2018, *Characterisation, Sources and Flux of Unmelted Micrometeorites on Earth During the Last* $\sim$*50,000 Years, Nature, Sci. Rep.*, 8, 8887, DOI: https://doi.org/10.1038/s41598-018-27158-x
49. Rubincam, D. P. 2015, *Space Erosion and Cosmic Ray Exposure Ages of Stony Meteorites, Icarus*, 245, 112, DOI: https://doi.org/10.1016/j.icarus.2014.09.005
50. Schlickeiser, R. 2002, *Cosmic Ray Astrophysics*, Berlin: Springer-Verlag
51. Silber, E. A., & Brown, P. G. 2019, *Infrasound Monitoring as a Tool to Characterize Impacting Near-Earth Objects (NEOs)*, in *Infrasound Monitoring for Atmospheric Studies, Challenges in Middle Atmosphere Dynamics and Societal Benefits* (Second Edition)

(eds. A. Le Pichon, E. Blanc, and A. Hauchecorne), 939–986, Cham, Switzerland: Springer Nature
52. Silber, E. A., Brown, P. G., & Krzeminski, Z. 2015, *Optical Observations of Meteors Generating Infrasound: Weak Shock Theory and Validation, J. Geophys. Res. Planets*, 120, 413–428, DOI: https://doi.org/10.1002/2014JE004680
53. Vitagliano, E., Tamborra, I., & Raffelt, G. 2019, *Grand Unified Neutrino Spectrum at Earth: Sources and Spectral Components.* arXiv:1910.11878v1
54. Westphal, A. J., Bridges, J. C., Brownlee, D. E., et al. 2017, *The Future of Stardust Science, Meteorit. Planet. Sci.*, 52, 1859, DOI: https://doi.org/10.1111/maps.12893
55. Zolensky, M., Bland, P., Brown, P., et al. 2006, *Flux of Extraterrestrial Material*, in *Meteorites and the Early Solar System II* (eds. D. S. Lauretta and H. Y. McSween Jr.), 869–888, Tucson: University of Arizona Press

CHAPTER 2

1. Abbott, B. P., et al., (LIGO Scientific Collaboration and Virgo Collaboration) 2016, *Observation of Gravitational Waves from a Binary Black Hole, Phys. Rev. Lett.* 116, 061102, DOI: https://doi.org/10.1103/PhysRevLett.116.061102
2. Abbott, B. P., et al., (The LIGO Scientific Collaboration and Virgo Collaboration) 2016, *Properties of the Binary Black Hole Merger GW150914, Phys. Rev. Lett.* 116, 241102, DOI: https://doi.org/10.1103/PhysRevLett.116.241102
3. Abbott, B. P., et al. (The LIGO Scientific Collaboration and Virgo Collaboration) 2019, *GWTC-1: A Gravitational-Wave Transient Catalog of Compact Binary Mergers Observed by Ligo and Virgo During the First and Second Observing Runs, Phys. Rev. X*, 9, 031040. arXiv:1811.12907
4. Belczynski, K., Bulik, T., Fryer, C. L., et al. 2010, *On the Maximum Mass of Stellar Black Holes, ApJ*, 714, 1217, DOI: doi:org/10.1088/0004-637X/714/2/1217
5. Beniamini, P., & Piran, T. 2019, *The Gravitational Waves Merger Time Distribution of Binary Neutron Star Systems, MNRAS*, 487, 4847, DOI: https://doi.org/10.1093/mnras/stz1589
6. Bethe, H. A., & Brown, G. E. 1998, *Evolution of Binary Compact Objects That Merge, ApJ*, 506, 780, DOI: https://doi.org/10.1086/306265
7. Camp, J. B., & Cornish, N. J. 2004, *Gravitational Wave Astronomy, Annu. Rev. Nucl. Part. Sci.*, 54, 525, DOI: https://doi.org/10.1146/annurev.nucl.54.070103.181251
8. Cameron, A. D., Champion, D. J., Kramer, M., et al. 2018, *The High Time Resolution Universe Pulsar Survey – XIII. PSR J1757-1854, the Most Accelerated Binary Pulsar, MNRAS*, 475, L57, DOI: https://doi.org/10.1093/mnrasl/sly003
9. Carroll, B. W., & Ostlie, D. A. 1996, *An Introduction to Modern Astrophysics*, Reading, Massachusetts: Addison-Wesley Publishing Co. Inc.
10. Cutler, C., & Flanagan, E. E. 1994, *Gravitational Waves from Merging Compact Binaries: How Accurately Can One Extract the Binary's Parameters from the Inspiral Waveform?, Phys. Rev. D*, 49, 2658, DOI: https://doi.org/10.1103/PhysRevD.49.2658

11. Einstein, A. 1916, Doc. 182, as quoted in *The Collected Papers of Albert Einstein*, Princeton: Princeton University Press, Translation Volume 8, English Translation Supplement, 1998
12. Finn, L. S., & Thorne, K. S. 2000, *Gravitational Waves from a Compact Star in a Circular, Inspiral Orbit, in the Equatorial Plane of a Massive, Spinning Black Hole, as Observed by LISA, Phys. Rev. D*, 62, 124021, DOI: https://doi.org/10.1103/PhysRevD.62.124021
13. Hartle, J. B. 2003, *Gravity, An Introduction to Einstein's General Relativity*, San Francisco: Addison-Wesley, ISBN 0-8053-8662-9
14. Henriksen, R. N. 2011, *Practical Relativity*, Chichester, UK: Wiley, A John Wiley & Sons Ltd Publication, ISBN 978-0-470-74142-2
15. Henriksen, R. N. 2019, *Private Communication*
16. Hulse, R. A., & Taylor, J. H. 1975, *Discovery of a Pulsar in a binary System, ApJ*, 195, L51, DOI: https://doi.org/10.1086/181708
17. Isi, M., Giesler, M., Farr, W. M., et al. 2019, *Testing the No-hair Theorem with GW150914, Phys. Rev. Lett.* 123, 111102, DOI: https://doi.org/10.1103/PhysRevLett.123.111102
18. Kennefick, D. 2019, *No Shadow of a Doubt: The 1919 Eclipse That Confirmed Einstein's Theory of Relativity*, Princeton, New Jersey: Princeton University Press, ISBN: 978-0-691-18386-2
19. Bisnovatyi-Kogan, G. S., & Moiseenko, S. G. 2017, *Gravitational Waves and Core-collapse Supernovae, Physics-Uspekhi*, 60, 843, DOI: https://doi.org/10.3367/UFNe.2016.11.038112
20. LIGO Scientific and VIRGO Collaborations 2017, *The Basic Physics of the Binary Black Hole Merger GW150914, Ann. Phys.*, 529, 1600209, DOI: https://doi.org/10.1002/andp.201600209
21. LIGO Scientific Collaboration, Virgo Collaboration, Fermi GBM, et al. 2017, *Multi-messenger Observations of a Binary Neutron Star Merger, ApJ*, 848, L12, DOI: https://doi.org/10.3847/2041-8213/aa91c9
22. Lynch, R. S., Freire, P. C. C., Ransom, S. M., et al. 2012, *The Timing of Nine Globular Cluster Pulsars, ApJ*, 745, 109, DOI: https://doi.org/10.1088/0004-637X/745/2/109
23. Manchester, R. N., Hobbs, G. B., Teoh, A., et al. 2005, *The Australia Telescope National Facility Pulsar Catalogue, AJ*, 129, 1993, DOI: https://doi.org/10.1086/428488
24. Mapelli, M., Zampieri, L, Ripamonti, E., et al. 2013, *Dynamics of Stellar Black Holes in Young Star Clusters with Different Metallicities – I. Implications for X-ray Binaries, MNRAS*, 429, 2298, DOI: https://doi.org/10.1093/mnras/sts500
25. Martinez, J. G., Stovall, K., Freire, P. C. C., et al. 2015, *Pulsar J0453+1559: A Double Neutron Star System with a Large Mass Asymmetry, ApJ*, 812, 143, DOI: https://doi.org/10.1088/0004-637X/812/2/143
26. Moore, C. J., Cole, R. H., & Berry, C. P. L. 2015, *Gravitational-Wave Sensitivity Curves, Class. Quantum Grav.*, 32, 015014, DOI: https://doi.org/10.1088/0264-9381/32/1/015014

27. Nissanke, S., Holz, D. E., Hughes, S. A., et al. 2010, *Exploring Short Gamma-Ray Bursts as Gravitational-Wave Standard Sirens, ApJ*, 725, 496, DOI: https://doi.org/10.1088/0004-637X/725/1/496
28. Peters, P. C., & Mathews, J. 1963, *Gravitational Radiation from Point Masses in a Keplerian Orbit, Phys. Rev.*, 131, 435, DOI: https://doi.org/10.1103/PhysRev.131.435
29. Peters, P. C. 1964, *Gravitational Radiation and the Motion of Two Point Masses, Phys. Rev.*, 136, B1224, DOI: https://doi.org/10.1103/PhysRev.136.B1224
30. Phinney, E. S. 1991, *The Rate of Neutron Star Binary Mergers in the Universe: Minimal Predictions for Gravity Wave Detectors, ApJ*, 380, L17, DOI: https://doi.org/10.1086/186163
31. Planck collaboration 2020, *Planck 2018 Results. VI. Cosmological parameters, A&A*, 641, A6, DOI: https://doi.org/10.1051/0004-6361/201833910
32. Pössel, M., 2005, *"Chirping neutron stars"* in Einstein Online, Vol. 01, 1013, http://www.einstein-online.info/spotlights/chirping_neutron_stars.html
33. Pound, R. V., & Rebka, G. A. 1959, *Gravitational Red-Shift in Nuclear Resonance, Phys. Rev. Lett.*, 3, 439, DOI: https://doi.org/10.1103/PhysRevLett.3.439
34. Ruhl, J. E. et al., 2003, *Improved Measurement of the Angular Power Spectrum of Temperature Anisotropy in the Cosmic Microwave Background from Two New Analyses of Boomerang Observations, ApJ*, 599, 786, DOI: https://doi.org/10.1086/379345
35. Schutz, B. F. 1996, *Gravitational-wave Sources, Class. Quantum Grav.*, 13, A219, DOI: https://doi.org/10.1088/0264-9381/13/11A/031
36. Schutz, B. 2007, *Gravity from the Ground Up*, Cambridge: Cambridge University Press, ISBN 978-0-521-45506-0, Chapter 22
37. Spera, M., Mapelli, M., & Bressan, A. 2015, *The Mass Spectrum of Compact Remnants from the PARSEC Stellar Evolution Tracks, MNRAS*, 451, 4086, DOI: https://doi.org/10.1093/mnras/stv1161
38. Spergel, D. N., Verde, L., Peiris, H. V., et al., 2003, *First Year* Wilkinson Microwave Anisotropy Probe (WMAP)* *Observations: Determination of Cosmological Parameters, ApJS*, 148, 175, DOI: https://doi.org/10.1086/377226
39. Taylor, J. H., & Weisberg, J. M. 1982, *A New Test of General Relativity: Gravitational Radiation and the Binary Pulsar PSR 1913+16, ApJ*, 253, 908, DOI: https://doi.org/10.1086/159690
40. Van Leeuwen, J., Kasian, L, Stairs, I. H., et al. 2015, *The Binary Companion of Young, Relativistic Pulsar J1906+0746, ApJ*, 798, 118, DOI: https://doi.org/10.1088/0004-637X/798/2/118
41. Weisberg, J. M., Nice, D. J., & Taylor, J. H. 2010, *Timing Measurements of the Relativistic Binary Pulsar PSR B1913+16, ApJ*, 722, 1030, DOI: https://doi.org/10.1088/0004-637X/722/2/1030
42. Westfall, R. S. 1987, Never at Rest – A Biography of Isaac Newton, Cambridge: Cambridge University Press
43. Yang, Y.-Y., Zhang, C.-M., Li, D., et al. 2019, *The Classifications of Double Neutron Stars and their Correlations with the Binary Orbital Parameters, PASP*, 131:064201, DOI: https://doi.org/10.1088/1538-3873/ab00ca

CHAPTER 3

1. Allen, G. E., Keohane, J. W., Gotthelf, E. V., et al. 1997, *Evidence of X-Ray Synchrotron Emission from Electrons Accelerated to 40 TeV in the Supernova Remnant Cassiopeia A, ApJ*, 487, L97, DOI: https://doi.org/10.1086/310878
2. Anderson, M., Rudnick, L., Leppik, P., et al. 1991, *Relativistic Electron Populations in Cassiopeia A, ApJ*, 373, 146, DOI: https://doi.org/10.1086/170033
3. Bessell, M. S. 1990, *UBVRI Passbands, PASP*, 102, 1181, DOI: https://doi.org/10.1086/132749
4. Bessell, M. S., Castelli, F., & Plez, B. 1998, *Model Atmospheres Broad-band Colors, Bolometric Corrections and Temperature Calibrations for O – M Stars, A&A*, 333, 231
5. Bessell, M. S., & Brett, J. M. 1988, *JHKLM Photometry: Standard Systems, Passbands, and Intrinsic Colors, PASP*, 100, 1134, DOI: https://doi.org/10.1086/132281
6. *Chandra Supernova Remnant Catalog* at http://snrcat.cfa.harvard.edu
7. Evans, D. W., Riello, M., De Angeli, F., et al. 2018, *Gaia Data Release 2, A&A*, 616, A4, DOI: https://doi.org/10.1051/0004-6361/201832756
8. Eyer, L., Dubath, P., Saesen, S., et al. 2012, *From Hipparcos to Gaia,* in IAU Symp. **285**, 2011, 'New Horizons in Time-Domain Astronomy', DOI: doi.org/10.1017/S1743921312000506
9. Gaia Collaboration 2018, *Gaia Data Release 2 – Observational Hertzsprung-Russell Diagrams, A&A*, 616, A10, DOI: https://doi.org/10.1051/0004-6361/201832843
10. Guinan, E., & Wasatonic, R. J. 2020, *Betelgeuse Updates,* The Astronomer's Telegram, ATel #13439
11. Gupta, R. (Ed.) 2004, *Observer's Handbook 2005,* Royal Astronomical Society of Canada, Toronto: University of Toronto Press
12. Hardcastle, M. J., Worrall, D. M., Kraft, R. P., et al. 2003, *Radio and X-Ray Observations of the Jet in Centaurus A, ApJ*, 593, 169, DOI: https://doi.org/10.1086/376519
13. Harrison, E. R., 1981, *Cosmology, The Science of the Universe,* Cambridge: Cambridge University Press
14. Longair, M. S. 1992, *High Energy Astrophysics*, Vol. 1,(Second Edition), Cambridge: Cambridge University Press
15. Ord, S. M., van Straten, W., Hotan, A. W., et al. 2004, *Polarimetric Profiles of 27 Millisecond Pulsars, MNRAS*, 352, 804, DOI: https://doi.org/10.1111/j.1365-2966.2004.07963.x
16. Vallée, J. P. 2004, *Cosmic Magnetic Fields – as Observed in the Universe, in Galactic Dynamos, and in the Milky Way, NewAR*, 48, 763, DOI: https://doi.org/10.1016/j.newar.2004.03.017
17. Weiler, M. 2018, *Revised Gaia Data Release 2 Passbands, A&A*, 617, A138, DOI: https://doi.org/10.1051/0004-6361/201833462
18. Willingale, R., Bleeker, J. A. M., van der Heyden, K. J., et al. 2002, *X-ray Spectral Imaging and Doppler Mapping of Cassiopeia A, A&A*, 381, 1039, DOI: https://doi.org/10.1051/0004-6361:20011614

CHAPTER 4

1. Atwood, W. B., Bagagli, R., Bellazzini, R., et al. 2007, *Design and Initial Tests of the Tracker-converter of the Gamma-ray Large Area Space Telescope, Astropart. Phys.*, 28, 422, DOI: https://doi.org/10.1016/j.astropartphys.2007.08.010
2. Baars, J. W. M., Genzel, R., Pauliny-Toth, I. I. K., et al. 1977, *The Absolute Spectrum of Cas A; An Accurate Flux Density Scale and a Set of Secondary Calibrators, A&A*, 61, 99
3. Broten, N. W., Legg, T. H., Locke, J. L., et al. 1967, *Long Base Line Interferometry: A New Technique, Science*, 23, 1592, DOI: https://doi.org/10.1126/science.156.3782.1592
4. Broten, N. W., Legg, T. H., Locke J. L. et al. 1967, *Radio Interferometry with a Baseline of 3074 km, AJ*, 72, 787, DOI: https://doi.org/10.1086/110530
5. Casagrande, L., & VandenBerg D. A. 2014, *Synthetic Stellar Photometry – I. General Considerations and New Transformations for Broad-Band Systems, MNRAS*, 444, 392, DOI: https://doi.org/10.1093/mnras/stu1476
6. A. N. Cox (Ed.) 1999, *Allen's Astrophysical Quantities,*New York: Springer Verlag
7. Dravins, D., Lindegren, L., Mezey, E., et al. 1997, *Atmospheric Intensity Scintillation of Stars. II. Dependence on Optical Wavelength, PASP*, 109, 725, DOI: https://doi.org/10.1086/133937
8. English, J. 2017, *Canvas and Cosmos: Visual Art Techniques Applied to Astronomy Data, Int. J. Mod. Phys. D*, 26, 1730010, DOI: https://doi.org/10.1142/S0218271817300105
9. The Event Horizon Telescope Collaboration 2019, *First M87 Event Horizon Telescope Results. II. Array and Instrumentation, ApJL*, 875, L2, DOI: https://doi.org/10.3847/2041-8213/ab0c96
10. The Event Horizon Telescope Collaboration 2019, *First M87 Event Horizon Telescope Results. IV. Imaging the Central Supermassive Black Hole, ApJL*, 875, L4, DOI: https://doi.org/10.3847/2041-8213/ab0e85
11. Goncharov, A. V., Owner-Petersen, M., Andersen, T. et al. 2002, *Adaptive Optics Schemes for Future Extremely Large Telescopes, Opt. Eng.*, 41, 1065, DOI: https://doi.org/10.1117/1.1466461
12. Haniff, C. A., Mackay, C. D., Titterington, D. J., et al. 1997, *The First Images from Optical Aperture Synthesis, Nature*, 328, 694, DOI: https://doi.org/10.1038/328694a0
13. Hippler, S. 2019, *Adaptive Optics for Extremely Large Telescopes, J.Astron. Inst.*, 8, #2, 1950001, DOI: https://doi.org/10.1142/S2251171719500016
14. Longair, M. S. 1992, *High Energy Astrophysics*, Vol. 1 (Second Edition, Cambridge: Cambridge University Press
15. NASA/IPAC Extragalactic Database (NED) operated by the Jet Propulsion Laboratory, California Institute of Technology, under contract with the National Aeronautics and Space Administration (http://ned.ipac.caltech.edu)
16. Osborn, J., Föhring, D., Dhillon, V. S., et al. 2015, *Atmospheric Scintillation In Astronomical Photometry, MNRAS*, 452, 1707, DOI: https://doi.org/10.1093/mnras/stv1400
17. Rieke, G. H., Lebofsky, M. J., & Low, F. H. 1985, *An Absolute Photometric System at 10 and 20μm, AJ*, 90, 900, DOI: https://doi.org/10.1086/113800

18. Ryle, M., & Vonberg, D. D. 1946, *Solar Radiation on 175 Mc./s, Nature*, 158, 339, DOI: https://doi.org/10.1038/158339b0
19. Rodeghiero, G., Pott, J.-U., Arcidiacono, C., et al. 2018, *The Impact of ELT Distortions and Instabilities on Future Astrometric Observations, MNRAS*, 479, 1974, DOI: https://doi.org/10.1093/mnras/sty1426
20. Walker, G. 1987, *Astronomical Observations – An Optical Perspective,* Cambridge: Cambridge University Press

CHAPTER 5

1. Asplund, M., Grevesse, N., Sauval, A. J., et al. 2009, *The Chemical Composition of the Sun, ARAA*, 47, 481, DOI: https://doi.org/10.1146/annurev.astro.46.060407.145222
2. Bennett, C. L., Hill, R. S., Hinshaw, G., et al. 2003, *First-Year Wilkinson Microwave Anisotropy Probe (WMAP) Observations: Foreground Emission, ApJS*, 148, 97, DOI: https://doi.org/10.1086/377252
3. Catelan, M. 2018, *The Ages of (the Oldest) Stars,* in *Rediscovering Our Galaxy,* Proc. IAU Symp. No. 334, 2017, eds., C. Chiappini, I. Minchev, E. Starkenburg, & M. Velentini, DOI: 10.1017/S1743921318000868
4. Chabrier, G. 2003, *Galactic Stellar and Substellar Initial Mass Function, PASP*, 115, 763, DOI: https://doi.org/10.1086/376392
5. Church, M. J., Reed, D., Dotani, T., et al. 2005, *Discovery of Absorption Features of the Accretion Disc Corona and Systematic Acceleration of the X-ray Burst Rate in XB1323-619, MNRAS*, 359, 1336, DOI: https://doi.org/10.1111/j.1365-2966.2005.08728.x
6. Coc, A., & Vangioni, E. 2017, *Primordial Nucleosynthesis, Int. J. Mod. Phys. E*, 26, no. 8, 1741002, DOI: https://doi.org/10.1142/S0218301317410026
7. Cornet, T., Bourgeois, O., Le Mouélic, S., et al. 2012, *Geomorphological Significance of Ontario Lacus in Titan: Integrated Interpretation of Cassini VIMS, ISS and RADAR Data and Comparison with the Etosha Pan (Namibia), Icarus*, 218, 788, DOI: https://doi.org/10.1016/j.icarus.2012.01.013
8. Dunne, L., Eales, S., Ivison, R., et al. 2003, *Type II Supernovae as a Significant Source of Interstellar Dust, Nature*, 424, 285, DOI: https://doi.org/10.1038/nature01792
9. Freeman, K., & Bland-Hawthorn, J. 2002, *The New Galaxy: Signatures of its Formation, ARAA*, 40, 487, DOI: https://doi.org/10.1146/annurev.astro.40.060401.093840
10. Frei, Z., Guhathakurta, P., Gunn, J. E., et al. 1996, *A Catalog of Digital Images of 113 Nearby Galaxies, AJ*, 111, 174, DOI: https://doi.org/10.1086/117771
11. Gray, D. F. 2005, *The Observation and Analysis of Stellar Photospheres* (Third Edition), Cambridge: Cambridge University Press
12. Jones, D., & Boffin, H. M. J. 2017, *Binary Stars as the Key to Understanding Planetary Nebulae, Nat Astron,* 1, 0117, DOI: https://doi.org/10.1038/s41550-017-0117
13. Kiminki, M. M., Kim, J. S., Bagley, M. B., et al. 2015, *The O- and B-Type Stellar Population in W3: Beyond the High-Density Layer, ApJ*, 813, 42, DOI: https://doi.org/10.1088/0004-637X/813/1/42

14. Kwok, S. 2005, *Planetary Nebulae: New Challenges in the 21st Century, JKAS*, 38, 27, DOI: https://doi.org/10.5303/JKAS.2005.38.2.271
15. Lada, C. U., Bergin, E. A., Alves, J. F., et al. 2003, *The Dynamical State of Barnard 68: A Thermally Supported, Pulsating Dark Cloud, ApJ*, 586, 286, DOI: https://doi.org/10.1086/367610
16. Lahav, O., & Liddle, A. R. 2019, *Cosmological Parameters, Rev. Part. Phys.* 2020, arXiv:1912.03687
17. Lang, K. R. 1999, *Astrophysical Formulae,* New York: Springer-Verlag
18. Licquia, T. C., & Newman, J. A. 2015, *Improved Estimates of the Milky Way's Stellar Mass and Star Formation Rate from Hierarchical Bayesian Meta-analysis, ApJ*, 809, 96, DOI: https://doi.org/10.1088/0004-637X/806/1/96
19. Macquart, J.-P., Prochaska, J. X., McQuinn, M., et al. 2020, *A Census of Baryons in the Universe From Localized Fast Radio Bursts, Nature*, 581, 391–395, DOI: https://doi.org/10.1038/s41586-020-2300-2
20. Nicastro, F., Kaastra, J., Krongold, Y., et al. 2018, *Observations of the Missing Baryons in the Warm – Hot Intergalactic Medium, Nature*, 558, 406, DOI: https://doi.org/10.1038/s41586-018-0204-1
21. Planck Collaboration 2020, *Planck 2018 results. VI. Cosmological parameters, A&A*, 641, A6, DOI: https://doi.org/10.1051/0004-6361/201833910
22. Rauch, M. 1998, *The Lyman Alpha Forest in the Spectra of QSOs, ARAA*, 36, 267, DOI: https://doi.org/10.1146/annurev.astro.36.1.267
23. Shore, S.N. 2003, *The Tapestry of Modern Astrophysics*, New Jersey: Wiley-Interscience, a John Wiley & Sons, Inc. Publication
24. Spitzer, Jr., L. 1978, *Physical Processes in the Interstellar Medium,* New York: John Wiley & Sons
25. Stil, J., & Irwin, J. A., 2001, *GSH 138-01-94: An Old Supernova Remnant in the Far Outer Galaxy, ApJ*, 563, 816, DOI: https://doi.org/10.1086/324036
26. Tanimura, H., Hinshaw, G., McCarthy, I. G., et al. 2019, *A Search for Warm/Hot Gas Filaments Between Pairs of SDSS Luminous Red Galaxies, MNRAS*, 483, 223, DOI: https://doi.org/10.1093/mnras/sty3118
27. Trotter, A. S., Greenhill, L. J., Moran, J. M., et al. 1998, *Water Maser Emission and the Parsec-Scale Jet in NGC 3079, ApJ*, 495, 740, DOI: https://doi.org/10.1086/305335
28. Whittet, D. C. B. 2003, *Dust in the Galactic Environment* (Second Edition), Bristol: Institute of Physics Publishing

CHAPTER 6

1. Campbell, B., Walker, G. A. H., & Yang, S. 1988, *A Search for Substellar Companions to Solar-type Stars, ApJ*, 331, 902, DOI: https://doi.org/10.1086/166608
2. Charbonneau, D., Allen, L. E., Megeath, S. T., et al. 2005, *Detection of Thermal Emission from an Extrasolar Planet, ApJ*, 626, 523, DOI: https://doi.org/10.1086/429991

3. Chauvin, G., Lagrange, A.-M., Dumas, C., et al. 2004, *A Giant Planet Candidate Near a Young Brown Dwarf. Direct VLT/NACO Observations Using IR Wavefront Sensing, A&A*, 425, L29, DOI: https://doi.org/10.1051/0004-6361:200400056
4. Deming, D., Seager, S., Richardson, L. J., et al. 2005, *Infrared Radiation from an Extra-solar Planet, Nature* 434, 740, DOI: https://doi.org/10.1038/nature03507
5. Fixsen, D. J. 2009, *The Temperature of the Cosmic Microwave Background, ApJ*, 707, 916, DOI: https://doi.org/10.1088/0004-637X/707/2/916
6. Gilbert, E. A., Barclay, T., Schlieder, J. E., et al. 2020, *The First Habitable Zone Earth-sized Planet from TESS. I: Validation of the TOI-700 System, AJ*, 160, 116, DOI: https://doi.org/10.3847/1538-3881/aba4b2
7. Hagiwara, K., Hikasa, K, Nakamura, K., et al. 2002, *Review of Particle Properties, Phys. Rev. D*, 66, 010001, DOI: https://doi.org/10.1103/PhysRevD.66.010001
8. Hansen, J., Ruedy, R., Sato, M., et al. 2010, *Global Surface Temperature Change, Rev. Geophys.*, 48, RG4004, DOI: https://doi.org/10.1029/2010RG000345
9. Hartmann, W. K. 2005, *Moons and Planets,* Belmont California: Thomson Books
10. Hunziker, S., Schmid, H. M., Mouillet, D., et al. 2020, *RefPlanets: Search for Reflected Light from Extrasolar Planets with SPHERE/ZIMPOL, A&A*, 634, A69, DOI: https://doi.org/10.1051/0004-6361/201936641
11. Keppler, J., Benisty, M., Müller, A., et al. 2018, *Discovery of a Planetary-mass Companion within the Gap of the Transition Disk Around PDS 70, A&A*, 617, A44, DOI: https://doi.org/10.1051/0004-6361/201832957
12. Kopparapu, R. K., Ramirez, R., Kasting, J. F., et al. 2013, *Habitable Zones Around Main-Sequence Stars: New Estimates, ApJ*, 765, 131, DOI: https://doi.org/10.1088/0004-637X/765/2/131
13. Lenssen, N., Schmidt, G., Hansen, J., et al. 2019, *Improvements in the GISTEMP Uncertainty Model, J. Geophys. Res. Atmos.*, 124, 6307, DOI: https://doi.org/10.1029/2018JD029522
14. Leyton, M., Dye, S., & Monroe, J. 2017, *Exploring the Hidden Interior of the Earth with Directional Neutrino Measurements, Nat. Commun.*, 8, 15989, DOI: https://doi.org/10.1038/ncomms15989
15. Lisenfeld, U., Isaak, K.G., & Hills, R. 2000, *Dust and Gas in Luminous Infrared Galaxies – Results from SCUBA Observations, MNRAS*, 312, 433, DOI: https://doi.org/10.1046/j.1365-8711.2000.03150.x
16. Lynch, C. R., Murphy, T., Lenc, E., et al. 2018, *The Detectability of Radio Emission from Exoplanets, MNRAS*, 478, 1763, DOI: https://doi.org/10.1093/mnras/sty1138
17. Montañés-Rodriguez, P., Pallé, E., Goode, P. R., et al. 2005, *Globally Integrated Measurements of the Earth's Visible Spectral Albedo, ApJ*, 628, 1175, DOI: https://doi.org/10.1086/431420
18. Neuhäuser, R., Guenther, E. W., Wuchterl, G., et al. 2005, *Evidence for a Co-moving Sub-stellar companion of GQ Lup, A&A*, 435, L13, DOI: doi.org/https://doi.org/10.1051/0004-6361:200500104

19. Webb, N. A., Leahy, D., Guillot, S., et al. 2019, *Thermal X-ray Emission Identified from the Millisecond Pulsar PSR J1909-3744, A&A*, 627, A141, DOI: https://doi.org/10.1051/0004-6361/201732040
20. Wilson, M. 2010, *Explaining the Two-toned Nature of IAPETUS, Phys. Today*, 63, 2, 15, DOI: https://doi.org/10.1063/1.3326978
21. Wolszczan, A., & Frail, D. A. 1992, *A Planetary System Around the Millisecond Pulsar PSR1257+ 12, Nature*, 355, 145, DOI: https://doi.org/10.1038/355145a0

CHAPTER 7

1. Aird, J., Coil, A. L., & Georgakakis, A. 2018, *X-rays Across the Galaxy Population – II. The Distribution of AGN Accretion Rates as a Function of Stellar Mass and Redshift, MNRAS*, 474, 1225, DOI: https://doi.org/10.1093/mnras/stx2700
2. Castelló-Mor, N., Netzer, H., & Kaspi, S. 2016, *Super- and sub-Eddington Accreting Massive Black Holes: a Comparison of Slim and Thin Accretion Discs Through Study of the Spectral Energy Distribution, MNRAS*, 458, 1839, DOI: https://doi.org/10.1093/mnras/stw445
3. Crowther, P. A., Caballero-Nieves, S. M., Bostroem, K. A., et al. 2016, *The R136 Star Cluster Dissected with Hubble Space Telescope/STIS. I. Far-ultraviolet Spectroscopic Census and the Origin of He II λ 1640 in Young Star Clusters, MNRAS*, 458, 624, DOI: https://doi.org/10.1093/mnras/stw273
4. D'Avanzo, P. 2015, *Short Gamma-ray Bursts: A Review, JHEAp*, 7, 73, DOI: https://doi.org/10.1016/j.jheap.2015.07.002
5. Elvis, M. 2000, *A Structure for Quasars, ApJ*, 545, 63, DOI: https://doi.org/10.1086/317778
6. Figer, D. F. 2005, *An Upper Limit to the Masses of Stars, Nature*, 434, 192, DOI: https://doi.org/10.1038/nature03293
7. Grosdidier, Y., Moffat, A. F. J., Joncas, G. et al. 1998, *HST WFPC2/Hα Imagery of the Nebula M1-67: A Clumpy LBV Wind Imprinting Itself on the Nebular Structure?, ApJ*, 506, L127, DOI: https://doi.org/10.1086/311647
8. Guenther, D. B., Demarque, P., Kim, Y.-C., et al. 1992, *Standard Solar Model, ApJ*, 387, 372, DOI: https://doi.org/10.1086/171090
9. Jackson, J. D. 1998, *Classical Electrodynamics* (Third Editon), Hoboken, New Jersey: John Wiley & Sons, Inc.
10. Kishimoto, M. 1999, *The Location of the Nucleus of NGC 1068 and the Three-dimensional Structure of its Nuclear Region, ApJ*, 518, 676, DOI: https://doi.org/10.1086/307290
11. Kollmeier, J. A., Onken, C. A., Kochanek, C. S., et al. 2006, *Black Hole Masses and Eddington Ratios at $0.3 < z < 4$, ApJ*, 648, 128, DOI: https://doi.org/10.1086/505646
12. Lucretius 1969, *The Way Things Are*, translated by R. Humphries, Bloomington: Indiana University Press
13. MAGIC collaboration 2019, *Observations of inverse Compton emission from a long γ-ray burst, Nature*, 575, 459, DOI: https://doi.org/10.1038/s41586-019-1754-6

14. Massey, P., & Hunter, D. A. 1998, *Star Formation in R136: A Cluster Revealed by Hubble Space Telescope Spectroscopy, ApJ*, 493, 180, DOI: https://doi.org/10.1086/305126
15. Montero-Camacho, P., & Mao, Y. 2020, *Lyα Forest Power Spectrum as an Emerging Window into the Epoch of Reionization and Cosmic Dawn*, MNRAS, 499, 1640, DOI: https://doi.org/10.1093/mnras/staa2918
16. Osterbrock, D. E. 1989, *Astrophysics of Gaseous Nebulae and Active Galactic Nuclei*, Mill Valley, California: University Science Books
17. Panagia, N. 1973, *Some Physical Parameters of Early-type Stars, ApJ*, 78, 929, DOI: https://doi.org/10.1086/111498
18. Pedlar, A., Booler, R. V., Spencer, R. E., et al. 1983, *High-resolution Radio Observations of the Seyfert Galaxy NGC 1068, MNRAS*, 202, 647, DOI: https://doi.org/10.1093/mnras/202.3.647
19. Peimbert, M., Peimbert, A., & Delgado-Inglada, G. 2017, *Nebular Spectroscopy: A Guide on HII Regions and Planetary Nebulae, PASP*, 129:082001, DOI: https://doi.org/10.1088/1538-3873/aa72c3
20. Santos, M. R. 2004, *Probing Reionization with Lyman α Emission Lines, MNRAS*, 349, 1137, DOI: https://doi.org/10.1111/j.1365-2966.2004.07594.x
21. Scalo, J., & Elmegreen, B. G. 2004, *Interstellar Turbulence II: Implications and Effects, ARAA*, 42, 275, DOI: https://doi.org/10.1146/annurev.astro.42.120403.143327
22. Schwarzschild, M. 1958, *Structure and Evolution of Stars*, New York: Dover Publications, Inc.
23. Shore, S. N. 2003, *The Tapestry of Modern Astrophysics*, New Jersey: Wiley-Interscience, a John Wiley & Sons, In. Publication
24. Sneep, M., & Ubachs, U. 2005, *Direct Measurement of the Rayleigh Scattering Cross Section in Various Gases, JQSRT*, 92, 293, DOI: https://doi.org/10.1016/j.jqsrt.2004.07.025
25. Spitzer, Jr., L. 1978, *Physical Processes in the Interstellar Medium*, New York: John Wiley & Sons
26. Springel, V., White, S. D. M., Jenkins, A., et al. 2005, *Simulations of the Formation, Evolution and Clustering of Galaxies and Quasars, Nature*, 435, 629, DOI: https://doi.org/10.1038/nature03597
27. Shu, F. H., 1982, *The Physical Universe: An Introduction to Astronomy*, Sausalito, CA: University Science Books

CHAPTER 8

1. Carroll, B. W., & Ostlie, D. A. 1996, *An Introduction to Modern Astrophysics*, Reading, Massachusetts.: Addison-Wesley Publishing Co., Inc.
2. Dickey, J. M., McClure-Griffiths, N. M., Gaensler, B. M. et al. 2003, *Fitting Together the HI Absorption and Emission in the Southern Galactic Plane Survey, ApJ*, 585, 801, DOI: http://doi.org/10.1086/346081
3. Gray, D. F. 2005, *The Observation and Analysis of Stellar Photospheres* (Third Edition), Cambridge: Cambridge University Press

4. Jacoby, G. H., Hunter, D. A., & Christian, C. A. 1984, *A Library of Stellar Spectra, ApJS*, 56, 257, DOI: http://doi.org/10.1086/190983
5. Kanekar, N., & Briggs, F. H. 2004, *21-cm Absorption Studies with the Square Kilometer Array, NewAR*, 48, 1259, DOI: http://doi.org/10.1016/j.newar.2004.09.030
6. Moore, D. 2020, *What Stars Are Made of*, 169, Cambridge, Massachusetts: Harvard University Press
7. Rivinius, T., Carciofi, A. C., & Martayan, C. 2013, *Classical Be stars. Rapidly Rotating B stars with Viscous Keplerian Decretion Disks, A&ARv*, 21, 69, DOI: http://doi.org/10.1007/s00159-013-0069-0
8. Rybicki, G. B., & Lightman, A. P. 1979, *Radiative Processes in Astrophysics*, New York: John Wiley & Sons
9. Taylor, A. R., Gibson, S. J., Peracaula, M., et al. 2003, *The Canadian Galactic Plane Survey, AJ*, 125, 3145, DOI: http://doi.org/10.1086/375301

CHAPTER 9

1. AbdelSalam, H. M., Saha, P., & Williams, L. L. R. 1998, *Nonparametric Reconstruction of Abell 2218 from Combined Weak and Strong Lensing, AJ*, 116, 1541, DOI: https://doi.org/10.1086/300546
2. Adamek, J., Clarkson, C., Coates, L., et al. 2019, *Bias and Scatter in the Hubble Diagram from Cosmological Large-scale Structure, Phys. Rev. D*, 100, 021301. DOI: https://doi.org/10.1103/PhysRevD.100.021301
3. Afonso, C., Albert, J. N., Andersen, J., et al. 2003, *Limits on Galactic Dark Matter with 5 years of EROS SMC Data, A&A*, 400, 951, DOI: https://doi.org/10.1051/0004-6361:20030087
4. Alcock, C., Allsman, R. A., Alves, D. R., et al. 2001, *MACHO Project Limits on Black Hole Dark Matter in the 1-30 $M_\odot$ Range, ApJ*, 550, L169, DOI: https://doi.org/10.1086/319636
5. Alcock, C., Allsman, R. A., Alves, D. R., et al. 1998, *EROS and MACHO Combined Limits on Planetary-Mass Dark Matter in the Galactic Halo, ApJ*, 499, L9, DOI: https://doi.org/10.1086/311355
6. Beaulieu, J.-P., Bennett, D. P., Fouque, P., et al. 2006, *Discovery of a Cool Planet of 5.5 Earth Masses Through Gravitational Microlensing, Nature*, 439, 437, DOI: https://doi.org/10.1038/nature04441
7. Belokurov, V., Evans, N. W., Moiseev, A., et al. 2007, *The Cosmic Horseshoe: Discovery of an Einstein Ring Around a Giant Luminous Red Galaxy, ApJ*, 671, L9, DOI: https://doi.org/10.1086/524948
8. Blandford, R. D., & Narayan, R. 1992, *Cosmological Applications of Gravitational Lensing, ARAA*, 30, 311, DOI: https://doi.org/10.1146/annurev.astro.30.1.311
9. Chen, G. C.-F., Fassnacht, C. D., Suyu, S. H., et al. 2020, *A SHARP View of HOLiCOW: H_0 from Three Time-delay Gravitational Lens Systems with Adaptive Optics Imaging, MNRAS*, 490, 1743, DOI: https://doi.org/10.1093/mnras/stz2547

10. Chevallier, M., & Polarski, D. 2001, *Accelerating Universes with Scaling Dark Matter, Int. J. Mod. Phys. D*, 10, 213, DOI: https://doi.org/10.1142/S0218271801000822
11. Han, C., Kim, D., Jung, Y. K., et al. 2020, *One Planet or Two Planets? The ultra-sensitive Extreme-Magnification Microlensing Event KMT-2019-BLG-1953, AJ*, 160, 17, DOI: https://doi.org/10.3847/1538-3881/ab91ac
12. Hubble, E. 1929, *A Relation Between Distance and Radial Velocity Among Extra-Galactic Nebulae, PNAS*, 15, 168, DOI: https://doi.org/10.1073/pnas.15.3.168
13. Jee, I., Komatsu, E., & Suyu, S. H. 2015, *Measuring Angular Diameter Distances of Strong Gravitational Lenses, JCAP*, 2015, 033, DOI: https://doi.org/10.1088/1475-7516/2015/11/033
14. Kelly, P. L., Rodney, S. A., Treu, T., et al. 2015, *Multiple Images of a Highly Magnified Supernova Formed by an Early-type Cluster Galaxy Lens, Science*, 347, 1123, DOI: https://doi.org/10.1126/science.aaa3350
15. Leibundgut, C., Schommer, R., Phillips, M., et al. 1996, *Time Dilation in the Light Curve of the Distant Type IA Supernova SN 1995K, ApJ*, 466, L21, DOI: https://doi.org/10.1086/310164
16. Lemaître, G. 1927, *Un Univers homogène de mass constante et de rayon croissant rendant compte de la vitesse radiale des nébuleuses extra-galactiques, ASSB*, 47, 49
17. Lemaître, G. 1931, *Expansion of the Universe, a Homogeneous Universe of Constant Mass and Increasing Radius Accounting for the Radial Velocity of Extra-galactic Nebulae, MNRAS*, 91, 483, DOI: https://doi.org/10.1093/mnras/91.5.483
18. Lusso, E., Piedipalumbo, E., Risaliti, G., et al. 2019, *Tension with the Flat λCDM Model from a High-redshift Hubble Diagram of Supernovae, Quasars, and Gamma-ray Bursts, A&A*, 628, L4, DOI: https://doi.org/10.1051/0004-6361/201936223
19. Newbury, P. R., & Spiteri, R. J. 2002, *Inverting Gravitational Lenses, SIAMR*, 44, 111, DOI: https://doi.org/10.1137/S0036144500380934
20. Paraficz, D., & Hjorth, J. 2009, *Gravitational Lenses as Cosmic Rulers: Ω_m, Ω_Λ from Time Delays and Velocity Dispersions, A&A*, 507, 49, DOI: https://doi.org/10.1051/0004-6361/200913307
21. Planck Collaboration 2018, *Planck 2018 results. VI. Cosmological parameters, A&A*, 641, A6, DOI: https://doi.org/10.1051/0004-6361/201833910
22. Schneider, P., Ehlers, J., & Falco, E. E., 1992, *Gravitational Lenses*, New York: Springer-Verlag
23. Suyu, S. H., Chang, T.-C., Courbin, F., et al. 2018, *Cosmological Distance Indicators, SSRv*, 214, 91, DOI: https://doi.org/10.1007/s11214-018-0524-3
24. Suzuki, N., Rubin, D., Lidman, C., et al. 2012, *The Hubble Space Telescope Cluster Supernova Survey. V. Improving the Dark-energy Constraints above z > 1 and Building an Early-type-hosted Supernova Sample, ApJ*, 746, 85, DOI: https://doi.org/10.1088/0004-637X/746/1/85
25. Treu, T. 2010, *Strong Lensing by Galaxies, ARAA*, 48, 87, DOI: https://doi.org/10.1146/annurev-astro-081309-130924

26. van der Marel, R. P., Fardal, M. A., Sohn, S. T., et al. 2019, *First Gaia Dynamics of the Andromeda System: DR2 Proper Motions, Orbits, and Rotation of M31 and M33, ApJ*, 872:24, DOI: https://doi.org/10.3847/1538-4357/ab001b
27. Zwaan, M. A., van Dokkum, P. G., & Verheijen, M. A. W. 2001, *Hydrogen 21-Centimeter Emission from a Galaxy at Cosmological Distance, Science*, 293, 1800, DOI: https://doi .org/10.1126/science.1063034

CHAPTER 10

1. Anderson, M., & Rudnick, L. 1996, Sites of Relativistic Particle Acceleration in Supernova Remnant Cassiopeia A, *ApJ*, 456, 234, DOI: https://doi.org/10.1086/176644
2. Baars, J. W. M., Genzel, R., Pauliny-Toth, I. I. K., et al. 1977, *The Absolute Spectrum of Cas A; an Accurate Flux Density Scale and a Set of Secondary Calibrators, A&A*, 61, 99
3. Bagenal, F. 1992, *Giant Planet Magnetospheres, AREPS*, 20, 289, DOI: https://doi.org/ 10.1146/annurev.ea.20.050192.001445
4. Beck, R. 2016, *Magnetic Fields in Spiral Galaxies, A&ARv*, 24, 4, DOI: https://doi.org/10 .1007/s00159-015-0084-4
5. Bekefi, B. 1966, *Radiation Processes in Plasmas,* New York: John Wiley & Sons, Inc., DOI: https://doi.org/10.1126/science.157.3796.1544
6. Bignami, G. F., Careveo, P. A., de Luca, A., et al. 2003, *The Magnetic Field of an Isolated Neutron Star from X-ray Cyclotron Absorption Lines, Nature*, 423, 725, DOI: https://doi .org/10.1038/nature01703
7. Bonamente, M., Lieu, R., Joy, M. K., et al. 2002, *The Soft X-ray Emission in a Large Sample of Galaxy Clusters with the ROSAT Position Sensitive Proportional Counter, Nature*, 423, 725, DOI: https://doi.org/10.1086/341806
8. Brown, R. L, & Matthews, W. G. 1970, *Theoretical Continuous Spectra of Gaseous Nebulae, ApJ*, 160, 939, DOI: https://doi.org/10.1086/150483
9. Brussaard, P. J., & van de Hulst, H. C. 1962, *Approximation Formulas for Nonrelativistic Bremsstrahlung and Average Gaunt Factors for a Maxwellian Electron Gas, Rev. Mod. Phys.*, 34, 507, DOI: https://doi.org/10.1103/RevModPhys.34.507
10. Carlstrom, J. E., Holder, G. P., & Reese, E. D. 2002, *Cosmology with the Sunyaev-Zel'dovich Effect, ARAA*, 40, 643, DOI: https://doi.org/10.1146/annurev .astro.40.060401.093803
11. Chluba, J., & Sunyaev, R. A. 2008, *Two-photon Transitions in Hydrogen and Cosmological Recombination, A&A*, 480, 628, DOI: https://doi.org/10.1051/0004-6361: 20077921
12. Darbon, S., Perrin, J.-M., & Sivan, J.-P. 1998, *Extended Red Emission (ERE) Detected in the 30 Doradus Nebula, A&A*, 333, 264
13. Dickey, J. M., & Lockman, F. J. 1990, *HI in the Galaxy, ARA&A*, 28, 215, DOI: https://doi .org/10.1146/annurev.aa.28.090190.001243
14. Ebeling, H., Edge, A. C., Bohringer, H., et al. 1998, *The ROSAT Brightest Cluster Sample – I. The Compilation of the Sample and the Cluster logN-logS Distribution, MNRAS*, 301, 881, DOI: https://doi.org/10.1046/j.1365-8711.1998.01949.x

15. Jackson, J. D. 1998, *Classical Electrodynamics* (Third Edition), Hoboken, New Jersey: John Wiley & Sons, Inc.
16. Kassim, N. E., Perley, R. A., Dwarakanath, K. S., et al. 1995, *Evidence for Thermal Absorption Inside Cassiopeia A, ApJ*, 455, L59, DOI: https://doi.org/10.1086/309802
17. Kozmanyan, A., Bourdin, H., Mazzotta, P., et al. 2019, *Deriving the Hubble Constant using Planck and XMM-Newton Observations of Galaxy Clusters, A&A*, 621, A34, DOI: https://doi.org/10.1051/0004-6361/201833879
18. Lallement, R., Welsh, B. Y., Vergely, J. L., et al. 2003, *3D mapping of the Dense Interstellar Gas Around the Local Bubble, A&A*, 411, 447, DOI: https://doi.org/10.1051/0004-6361:20031214
19. Lang, K. R. 1999, *Astrophysical Formulae*, New York: Springer-Verlag
20. Li, J.-T., & Wang, Q. D. 2013, *Chandra Survey of Nearby Highly Inclined Disc Galaxies – I. X-ray Measurements of Galactic Coronae, MNRAS*, 428, 2085, DOI: https://doi.org/10.1093/mnras/sts183
21. Liu, W., Chiao, M., Collier, M. R., et al. 2017, *The Structure of the Local Hot Bubble, ApJ*, 834, 33, DOI: https://doi.org/10.3847/1538-4357/834/1/33
22. Longair, M. S. 1992, *High Energy Astrophysics*, Vol. 1 (Second Edition), Cambridge: Cambridge University Press
23. Melia, F. 2016, *Constancy of the cluster Gas Mass Fraction in the $R_h = ct$ Universe, Proc. Royal Soc. A*, 472: 20150765, DOI: https://doi.org/10.1098/rspa.2015.0765
24. Mewe, R., Lemen, J. R., & van den Oord, G. H. J. 1986, *Calculated X-radiation from Optically Thin Plasmas. VI. Improved Calculations for Continuum Emission and Approximation Formulae for Nonrelativistic Average Gaunt Factors, A&AS*, 65, 511
25. Pacholczyk, A. G. 1970, *Radio Astrophysics*, San Francisco: W. H. Freeman and Co.
26. Pérez-Torres, M. A., Alberdi, A., & Marcaide, J. M 2001, *The Role of Synchrotron Self-absorption in the Late Radio Emission from SN 1993J, A&A*, 374, 997, DOI: https://doi.org/10.1051/0004-6361:20010774
27. Perley, R. A., Dreher, J. W., & Cowan, J. J. 1984, *The Jet and Filaments of Cygnus A, ApJ*, 285, L35, DOI: https://doi.org/10.1086/184360
28. Raymond, J. C., & Smith, B. W. 1977, *Soft X-ray Spectrum of a Hot Plasma, ApJS*, 35, 419, DOI: https://doi.org/10.1086/190486
29. Readhead, A. C. S. 1994, *Equipartition Brightness Temperature and the Inverse Compton Catastrophe, ApJ*, 426, 51, DOI: https://doi.org/10.1086/174038
30. Reiprich, T. H., & Böhringer, H. 2002, *The Mass Function of an X-Ray Flux-limited Sample of Galaxy Clusters, ApJ*, 567, 716, DOI: https://doi.org/10.1086/338753
31. Rengelink, R. B., Tang, Y., de Bruyn, A. G., et al. 1997, *The Westerbork Northern Sky Survey (WENSS), I. A 570 Square Degree Mini-Survey Around the North Ecliptic Pole, A&AS*, 124, 259, DOI: https://doi.org/10.1051/aas:1997358
32. Rohlfs, K. 1986, Tools of Radio *Astronomy*, Berlin: Springer-Verlag
33. Roussel, H., Sadat, R., & Blanchard, A. 2000, *The Baryon Content of Groups and Clusters of Galaxies, A&A*, 361, 429
34. Rybicki, G. B., & Lightman, A. P. 1979, *Radiative Processes in Astrophysics*, New York: Wiley-Interscience, John Wiley & Sons

35. Schraml, J., & Mezger, P. G. 1969, *Galactic HII Regions. IV. 1.95-CM Observations with High Angular Resolution and High Positional Accuracy, ApJ*, 156, 269, DOI: https://doi.org/10.1086/149964
36. Shu, F. H. 1991, *The Physics of Astrophysics, Vol. I. Radiation,* Mill Valley California: University Science Books
37. Snowden, S. L., Freyberg, M. J., Plucinsky, P. P., et al. 1995, *First Maps of the Soft X-Ray Diffuse Background from the ROSAT XRT/PSPC All-Sky Survey, ApJ*, 454, 643, DOI: https://doi.org/10.1086/176517
38. Sutherland, R. S. 1998, *Accurate Free-free Gaunt Factors for Astrophysical Plasmas, MNRAS*, 300, 321, DOI: https://doi.org/10.1046/J.1365-8711.1998.01687.x
39. van Weeren, R. J., Andrade-Santos, F., Dawson, W. A., et al. 2017, *The Case for Electron Re-acceleration at Galaxy Cluster Shocks, Nat. Astron.*, 1, 0005, DOI: https://doi.org/10.1038/s41550-016-0005
40. Verschuur, G. L., & Kellermann, K. I. 1988, *Galactic and Extragalactic Radio Astronomy* (Second Edition), Berlin: Springer-Verlag
41. Wang, Q. D., Immler, S., Walterbos, R., et al. 2001, *CHANDRA Detection of a Hot Gaseous Corona around the Edge-on Galaxy NGC4631, ApJ*, 555, L99, DOI: https://doi.org/10.1086/323179
42. Willingale, R., Hands, A. D. P., Warwick, R. S., et al. 2003, *The X-ray Spectrum of the North Polar Spur, MNRAS*, 343, 995, DOI: https://doi.org/10.1046/j.1365-8711.2003.06741.x
43. Woodfinden, A., Henriksen, R. N., & Irwin, J. A. 2019, *Evolving Galactic Dynamos and Fits to the Reversing Rotation Measures in the Halo of NGC 4631, MNRAS*, 487, 1498, DOI: https://doi.org/10.1093/mnras/stz1366
44. Zarka, P. 1998, *Auroral Radio Emission at the Outer Planets: Observations and Theories, JGR*, 103, 20159, DOI: https://doi.org/10.1029/98JE01323
45. Zarka, P., Lazio, T. J. W., Hallinan, G., et al. 2014, *Magnetospheric Radio Emissions from Exoplanets with the SKA,* in *Advancing Astrophysics with the Square Kilometre Array,* PoS, https://pos.sissa.it/215/120/pdf

CHAPTER 11

1. Afflerbach, A., Churchwell, E., Acord, J. M., et al. 1996, *Galactic Temperature and Metallicity Gradients from Ultracompact HII Regions, ApJS*, 106, 423, DOI: https://doi.org/10.1086/192343
2. Alves, M. I. R., Calabretta, M., Davies, R. D., et al. 2015, *The HIPASS Survey of the Galactic Plane in Radio Recombination Lines, MNRAS*, 450, 2025, DOI: https://doi.org/10.1093/mnras/stv751
3. Bolatto, A. D., Wolfire, M., & Leroy, A. K. 2013, *The CO-to-H_2 Conversion Factor, ARAA*, 51, 207, DOI: https://doi.org/10.1146/annurev-astro-082812-140944
4. Chen, H.-Y., Chen, X., Wang, J.-Z., et al. 2020, *A 4-6 GHz Radio Recombination Line Survey in the Milky Way, ApJS*, 248, 3, DOI: https://doi.org/10.3847/1538-4365/ab818e

5. Churchwell, E. 2002, *Ultra-Compact HII Regions and Massive Star Formation, ARAA*, 40, 27, DOI: https://doi.org/10.1146/annurev.astro.40.060401.093845
6. Comito, C., Schilke, P., Phillips, T. G., et al. 2005, *A Molecular Line Survey of Orion KL in the 350 Micron Band, ApJS*, 156, 127, DOI: https://doi.org/10.1086/425996
7. Cordiner, M. A., Linnartz, H., Cox, N. L. J., et al. 2019, *Confirming Interstellar C_{60}^+ Using the Hubble Space Telescope, ApJL*, 875, L28, DOI: https://doi.org/10.3847/2041-8213/ab14e5
8. Diehl, R. 2017, *Gamma-ray line measurements from supernova explosions, in Supernova 1987A: 30 years later – Cosmic Rays and Nuclei from Supernovae and their aftermaths*, IAU Symp., 331, 157, DOI: https://doi.org/10.1017/S1743921317004343
9. Diehl, R. 2018, *Chapter 1, Astrophysics with Radioactive Isotopes*, in *Astrophysics with Radioactive Isotopes* (Second Edition), (eds. R. Diehl, D. H. Hartmann, & N. Prantzos), 453–453, Cham, Switzerland: Springer Nature, Astrophysics and Space Science Library
10. Diehl, R., Halloin, H., Kretschmer, K., et al. 2006, *Radioactive ^{26}Al from Massive Stars in the Galaxy, Nature* 439, 45, DOI: https://doi.org/10.1038/nature04364
11. Diehl, R., Prantzos, N., & von Ballmoos, P. 2006, *Astrophysical Constraints from Gamma-ray Spectroscopy, NuPhA*, 777, 70, DOI: https://doi.org/10.1016/j.nuclphysa.2005.02.155
12. Dopita, M. A., & Sutherland, R. S. 2003, *Astrophysics of the Diffuse Universe,* New York: Springer
13. Dupree, A. K., & Goldberg, L. 1970, *Radiofrequency Recombination Lines, ARAA*, 8, 231, DOI: https://doi.org/10.1146/annurev.aa.08.090170.001311
14. Elsila, J. E., Burton, A. S., Aponte, J. C., et al. 2016, *The Diversity of Meteoritic Amino Acids: Variations in Abundance and Enantiomeric Composition and Implications for Exobiology*, in 47th Lunar and Planetary Science Conference, held in Woodlands, Texas, LPI Contribution No. 1903, 1533
15. Gordon, M. A., & Sorochenko, R. L. 2002, *Radio Recombination Lines,* Dordrecht: Kluwer Academic Publishers
16. Gray, D. F. 2005, *The Observation and Analysis of Stellar Photospheres* (Third Eition) Cambridge: Cambridge University Press
17. Guillard, P., Boulanger, F., Lehnert, M. D., et al. 2015, *Warm molecular Hydrogen at high redshift with the James Webb Space Telescope,* in SF2A, Proc. of the Annual meeting of the French Society of Astronomy & Astrophysics, Eds., F. Martins, S. Boissier, V. Buat, et al., p. 81
18. Habart, E., Walmsley, M., Verstraete, L., et al. 2005, *Molecular Hydrogen, SSRv*, 119, 71, DOI: https://doi.org/10.1007/s11214-005-8062-1
19. Iyudin, A. F., Diehl, R., Bloemen, H., et al. 1994, *COMPTEL Observations of ^{44}Ti gamma-ray Line Emission from CAS A, A&A*, 284, L1
20. Kantharia, N. G., & Anantharamaiah, K. R. 2001, *Carbon Recombination Lines from the Galactic Plane at 34.5 & 328 MHz, JApA*, 22, 51, DOI: https://doi.org/10.1007/BF02933590
21. Lang, K. R. 1999, *Astrophysical Formulae*, New York: Springer-Verlag

22. Lerate, M. R., Barlow, M. J., Swinyard, B. M., et al. 2006, *A Far-infrared Molecular and Atomic Line Survey of the Orion KL Region, MNRAS*, 370, 597, DOI: https://doi.org/10.1111/j.1365-2966.2006.10518.x
23. Oonk, J. B. R., van Weeren, R. J., Salgado, F., et al. 2014, *Discovery of Carbon Radio Recombination Lines in Absorption Towards Cygnus A, MNRAS*, 437, 3506, DOI: https://doi.org/10.1093/mnras/stt2158
24. Osterbrock, D. E. 1989, *Astrophysics of Gaseous Nebulae and Active Galactic Nuclei*, Mill Valley, California: University Science Books
25. Palacios, A., Meynet, G., Vuissoz, C., et al. 2005, *New Estimates of the Contribution of Wolf-Rayet Stellar Winds to the Galactic ^{26}Al, A&A*, 429, 613, DOI: https://doi.org/10.1051/0004-6361:20041757
26. Peters, T., Longmore, S. N., & Dullemond, C. P. 2012, *Understanding Hydrogen Recombination Line Observations with ALMA and EVLA, MNRAS*, 425, 2352, DOI: https://doi.org/10.1111/j.1365-2966.2012.21676.x
27. Plüschke, S., Diehl, R., Schönfelder, V., et al. 2001, *The COMPTEL 1.809 MeV Survey*, in *Exploring the Gamma-ray Universe*, Proceedings of the Fourth INTEGRAL Workshop (eds. Eds. B. Battrick, A. Gimenez, V. Reglero, & C. Winkler), 55, Noordwijk: ESA Publications
28. Pritchard, J. R., & Loeb, A. 2012, *21 cm Cosmology in the 21st Century, RPPh*, 75, 086901, DOI: https://doi.org/10.1088/0034-4885/75/8/086901
29. Purcell, E. M., & Field, G. B. 1956, *Influence of Collisions Upon Population of Hyperfine States in Hydrogen, ApJ*, 124, 542, DOI: https://doi.org/10.1086/146259
30. Ramaseshan, S., & Rao, C. R. (compilers) 1988, *C. V. Raman, A Pictorial Biography*, Bangalore: Indian Academy of Sciences
31. Redondo, P., Barrientos, C., & Largo, A. 2017, *Complex Organic Molecules Formation in Space Through Gas Phase Reactions: A Theoretical Approach, ApJ*, 836, 240, DOI: https://doi.org/10.3847/1538-4357/aa5ca4
32. Rizzo, J. R., Tercero, B., & Cernicharo, J. 2017, *A Spectroscopic Survey of Orion KL between 41.5 and 50 GHz, A&A*, 605, A76, DOI: https://doi.org/10.1051/0004-6361/201629936
33. Roelfsema, P. R., & Goss, W. M. 1992, *High Resolution Radio Recombination Line Observations, A&ARv*, 4, 161, DOI: https://doi.org/10.1007/BF00874056
34. Rohlfs, K. 1986, *Tools of Radio Astronomy*, Berlin: Springer-Verlag
35. Sánchez Contreras, C., Báez-Rubio, A., Alcolea, J., et al. 2019, *A Rotating Fast Bipolar Wind and Disk System Around the B[3]-Type Star MWC 922, A&A*, 629, A136, DOI: https://doi.org/10.1051/0004-6361/201936057
36. Schinnerer, E., Meidt, S. E., Pety, J., et al. 2013, *The PdBI Arcsecond Whirlpool Survey (PAWS). I. A Cloud-scale/Multi-wavelength View of the Interstellar Medium in a Grand-design Spiral Galaxy, ApJ*, 779, 42, DOI: https://doi.org/10.1088/0004-637X/779/1/42
37. Schwieterman, E. W., Kiang, N. Y., Parenteau, M. N., et al. 2018, *Exoplanet Biosignatures: A Review of Remotely Detectable Signs of Life, Astrobiology*, 18, 663, DOI: https://doi.org/10.1089/ast.2017.1729

38. Shaver, P. A. 1980, *Accurate Electron Temperatures from Radio Recombination Lines, A&A*, 91, 279
39. Spitzer Jr., L. 1978, *Physical Processes in the Interstellar Medium*, New York: John Wiley & Sons
40. Stepkin, S. V., Konovalenko, A. A., Kantharia, N. G., et al. 2007, *Radio Recombination Lines from the Largest Bound Atoms in Space, MNRAS*, 374, 852, DOI: https://doi.org/10.1111/j.1365-2966.2006.11190.x
41. Tercero, B., Cernicharo, J., & Pardo, J. R. 2005, in, *Astrochemistry Throughout the Universe: Recent Successes and Current Challenges, IAU Symp. 231*, p. 203
42. van de Hulst, H. C. 1957, in *Radio Astronomy*, IAU Symp. 4, (ed. H. C. van de Hulst), 3, Hew York: Cambridge University Press
43. Verschuur, G. L. & Kellermann, K. I., eds. 1988, *Galactic and Extragalactic Radio Astronomy* (Second Edition), Berlin: Springer-Verlag
44. Weinberger, C., Diehl, R., Pleintinger, M. M. M., et al. 2020, *^{44}Ti ejecta in Young Supernova Remnants, A&A*, 638, A83, DOI: https://doi.org/10.1051/0004-6361/202037536
45. Wilson, C. D. 1995, *The Metallicity Dependence of the CO-to-H_2 Conversion Factor from Observations of Local Group Galaxies, ApJ*, 448, L97, DOI: https://doi.org/10.1086/309615
46. Yun, M., Ho, P. T. P., & Lo, K. Y. 1994, *A High-resolution Image of Atomic Hydrogen in the M81 Group of Galaxies, Nature*, 372, 530, DOI: https://doi.org/10.1038/372530a0
47. Yun, M. 1999, in *Galaxy Interactions at Low and High Redshift*, IAU Symp. 186, (Eds. J. E. Barnes and D. Sanders), 81
48. Zhu, F.-Y., Zhu, Q.-F., Wang, J.-Z., et al. 2019, *Determining Electron Temperature and Density in a HII Region Using the Relative Strengths of Hydrogen Radio Recombination Lines, ApJ*, 881, 14, DOI: https://doi.org/10.3847/1538-4357/ab2a75

CHAPTER 12

1. Abdo, A. A., Ackermann, M., Ajello, M., et al. 2010, *Spectrum of the Isotropic Diffuse Gamma-Ray Emission Derived from First-Year Fermi Large Area Telescope Data, PRL*, 104, 101101, DOI: https://doi.org/10.1103/PhysRevLett.104.101101
2. Ackermann, M., Albert, A., Atwood, W. B., et al. 2014, *The Spectrum and Morphology of the Fermi Bubbles, ApJ*, 793, 64, DOI: https://doi.org/10.1088/0004-637X/793/1/64
3. Bedin, L. R., Cassisi, S., Castelli, F., et al. 2005, *Transforming Observational Data and Theoretical Isochrones into the ACS/WFC Vega-mag System, MNRAS*, 357, 1038, DOI: https://doi.org/10.1111/j.1365-2966.2005.08735.x
4. Bordoloi, R., Fox, A. J., Lockman, F. J., et al. 2017, *Mapping the Nuclear Outflow of the Milky Way: Studying the Kinematics and Spatial Extent of the Northern Fermi Bubble, ApJ*, 834, 191, DOI: https://doi.org/10.3847/1538-4357/834/2/191
5. Fukui, Y., Yamamoto, H., Fujishita, M., et al. 2006, *Molecular Loops in the Galactic Center: Evidence for Magnetic Flotation, Science* 314, 106, DOI: https://doi.org/10.1126/science.1130425

6. Gillessen, S., Plewa, P. M., Eisenhauer, F., et al. 2017, *An Update on Monitoring Stellar Orbits in the Galactic Center, ApJ*, 837, 30, DOI: https://doi.org/10.3847/1538-4357/aa5c41
7. Gratton, R. G., Bragaglia, A., Carretta, E., et al. 2003, *Distances and Ages of NGC 6397, NGC 6752 and 47 Tuc, A&A*, 408, 529, DOI: https://doi.org/10.1051/0004-6361:20031003
8. Grenier, I. A., Black, J. H., & Strong, A. W. 2015, *The Nine Lives of Cosmic Rays in Galaxies, ARAA*, 53, 199, DOI: https://doi.org/10.1146/annurev-astro-082214-122457
9. Herold, L, & Malyshev, D. 2019, *Hard and Bright Gamma-ray Emission at the Base of the Fermi Bubbles, A&A*, 625, 110, DOI: https://doi.org/10.1051/0004-6361/201834670
10. Heywood, I., Camilo, F., Cotton, W. D., et al. 2019, *Inflation of 430-parsec Bipolar Radio Bubbles in the Galactic Centre by an Energetic Event, Nature* 573, 235, DOI: https://doi.org/10.1038/s41586-019-1532-5
11. Kiminki, D. C., Kobulnicky, H. A., Kinemuchi, K., et al. 2007, *A Radial Velocity Survey of the Cyg OB2 Association, ApJ*, 664, 1102, DOI: https://doi.org/10.1086/513709
12. Korn, A. Grundahl, F., Richard, O., et al. 2006, *New Abundances for Old Stars – Atomic Diffusion at Work in NGC 6397, Messenger*, 125, 6
13. Krishnarao, D., Benjamin, R. A., & Haffner, L. M. 2020, *Discovery of High-Velocity H-Alpha Above Galactic Center: Testing Models of the Fermi Bubble, APJ*, 899, L11, https://doi.org/10.3847/2041-8213/aba8f0
14. Law, C. J., Yusef-Zadeh, F., Cotton, W. D., et al. 2008, *Green Bank Telescope Multiwavelength Survey of the Galactic Center Region, ApJS*, 177, 255, DOI: https://doi.org/10.1086/533587
15. Pasquini, L., Bonifacio, P., Randich, S., et al. 2004, *Beryllium in Turnoff Stars of NGC 6397: Early Galaxy Spallation, Cosmochronology and Cluster Formation, A&A*, 426, 651, DOI: https://doi.org/10.1051/0004-6361:20041254
16. Polanyi, J. 2006 in a talk delivered to a History class at the University of Toronto, on Oct. 2, 2006, reprinted in the *Toronto Star* newspaper, Sunday, October 22, 2006
17. Porquet, D., Predehl, P., Aschenbach, B., et al. 2003, *XMM-Newton Observation of the Brightest X-ray Flare Detected so far from Sgr A**, *A&A*, 407, L17, DOI: https://doi.org/10.1051/0004-6361:20030983
18. Porter, T., Moskalenko, I. V., Strong, A. W., et al. 2008, *Inverse Compton Origin of the Hard X-ray and Soft Gamma-ray Emission from the Galactic Ridge, ApJ*, 682, 400, DOI: https://doi.org/10.1086/589615
19. Richer, H. B., Anderson, J., Brewer, J., et al. 2006, *Probing the Faintest Stars in a Globular Star Cluster, Science*, 313, 936, DOI: https://doi.org/10.1126/science.1130691
20. Richer, H. B., Dotter, A., Hurley, J., et al. 2008, *Deep Advanced Camera for Surveys Imaging in the Globular Cluster NGC 6397: the Cluster Color-Magnitude Diagram and Luminosity Function, ApJ*, 135, 2141, DOI: https://doi.org/10.1088/0004-6256/135/6/2141

21. Richer, H. B., Goldsbury, R., Heyl, J., et al. 2013, *Comparing the White Dwarf Cooling Sequences in 47 Tuc and NGC 6397*, *ApJ*, 778, 104, DOI: https://doi.org/10.1088/0004-637X/778/2/104
22. Schneider, N., Bontemps, S., Simon, R., et al. 2006, *A New View of the Cygnus X region. KOSMA ^{13}CO 2 to 1, 3 to 2, and ^{12}CO 3 to 2 Imaging*, *A&A*, 458, 855, DOI: https://doi.org/10.1051/0004-6361:20065088
23. Schödel, R., Ott, T., Genzel, R., et al. 2002, *A Star in a 15.2-year Orbit Around the supermassive Black Hole at the Centre of the Milky Way*, *Nature*, 419, 694, DOI: https://doi.org/10.1038/nature01121
24. Strong, A. W. 2010, *Interstellar Gamma Rays and Cosmic Rays: New Insights from Fermi-LAT and Integral*, in *Proceedings of the ICATPP Conference on Cosmic Rays for Particle and Astroparticle Physics* (eds. S. Giani, C. Leroy, & P. G. Rancolta), Singapore: World Scientific, DOI: https://doi.org/10.1142/9789814329033_0059
25. Su, M., Slatyer, T. R., & Finkbeiner, D. P. 2010, *Giant Gamma-Ray Bubbles from FERMI-LAT: Active Galactic Nucleus Activity or Bipolar Galactic Wind?*, *ApJ*, 724, 1044, DOI: https://doi.org/10.1088/0004-637X/724/2/1044
26. Uchiyama, Y., Takahashi, T., Aharonian, F. A., et al. 2002, *ASCA View of the Supernova Remnant γ Cygni (G78.2+2.1): Bremsstrahlung X-Ray Spectrum from Loss-flattened Electron Distribution*, *ApJ*, 571, 866, DOI: https://doi.org/10.1086/340121
27. Zhang, R., & Guo, F. 2020, *Simulating the Fermi bubbles as Forward Shocks Driven by AGN Jets*, *ApJ*, 894, 117, DOI: https://doi.org/10.3847/1538-4357/ab8bd0

APPENDIX T

1. Bessell, M. S., Castelli, F., & Plez, B. 1998, *Model Atmospheres Broad-band Colors, Bolometric Corrections and Temperature Calibrations for O - M stars*, *A&A*, 333, 231
2. Cox, A. N. (Ed.) 1999, *Allen's Astrophysical Quantities*, New York: Springer-Verlag
3. Fixsen, D. J. 2009, *The Temperature of the Cosmic Microwave Background*, *ApJ*, 707, 916, DOI: https://doi.org/10.1088/0004-637X/707/2/916
4. Gray, D. F. 2005, *The Observation and Analysis of Stellar Photospheres* (Third Edition), Cambridge: Cambridge University Press
5. Guenther, D. B., Demarque, P., Kim, Y.-C., et al. 1992, *Standar Solar Model*, *ApJ*, 387, 372, DOI: https://doi.org/10.1086/171090
6. Gupta, R. (Ed.) 2005, *Observer's Handbook 2005*, Royal Astronomical Society of Canada, Toronto: University of Toronto Press
7. Lang, K. R., 1999, *Astrophysical Formulae*, New York: Springer-Verlag
8. Nimmo, F., Umurhan, O., Lisse, C. M., et al. 2017, *Mean Radius and Shape of Pluto and Charon from New Horizons Images*, *Icarus*, 287, 12, DOI: https://doi.org/10.1016/j.icarus.2016.06.027

9. Spergel, D. N., Verde, L., Peiris, H. V., et al. 2003, *First-year Wilkinson Microwave Anisotropy Probe (WMAP) Observations: Determination of Cosmological Parameters, ApJS*, 148, 175, DOI: https://doi.org/10.1086/377226
10. Taylor, B. N., & Mohr, P. J., compilers, 2002, *The NIST Reference on Constants, Units and Uncertainty*, (2002 version), http://physics.nist.gov/cuu
11. Tiesinga, E., Mohr, P. J., Newell, D. B., et al., compilers, 2019, *The NIST Reference on Constants, Units and Uncertainty* (2020 version), http://physics.nist.gov/cuu
12. Turck-Chièze, S. 2016, *The Standard Solar Model and Beyond, J. Phys. Conf. Ser.*, 665, Nuclear Physics in Astrophysics VI (NPA6) 19 - 24 May 2013, DOI: https://doi.org/10.1088/1742-6596/665/1/012078

Index

Items in *italic* refer to figures, while items appearing in **bold** refer to tables.

a

b

Astrophysics: Decoding the Cosmos, Second Edition. Judith A. Irwin.

Companion website: www.wiley.com/go/irwin/astrophysics2e

C

d

e

f

i

j

k

l

m

n

o

p

q

r

s

V

W

X

Z